Genetics and Regulation of Nitrogen Fixation in Free-Living Bacteria

Nitrogen Fixation: Origins, Applications, and Research Progress

VOLUME 2

Genetics and Regulation of Nitrogen Fixation in Free-Living Bacteria

Edited by

Werner Klipp

Lehrstuhl für Biologie der Mikroorganismen,
Fakultät für Biologie, Ruhr-Universität Bochum, Germany

Bernd Masepohl

Lehrstuhl für Biologie der Mikroorganismen,
Fakultät für Biologie, Ruhr-Universität Bochum, Germany

John R. Gallon

Biochemistry Research Group,
School of Biological Sciences, University of Wales, Swansea, U.K.

and

William E. Newton

Department of Biochemistry,
Virginia Polytechnic Institute & State University, Blacksburg, U.S.A.

KLUWER ACADEMIC PUBLISHERS
DORDRECHT / BOSTON / LONDON

A C.I.P. Catalogue record for this book is available from the Library of Congress.

ISBN 1-4020-2178-X (HB)
ISBN 1-4020-2179-1 (e-book)

Published by Kluwer Academic Publishers,
P.O. Box 17, 3300 AA Dordrecht, The Netherlands.

Sold and distributed in North, Central and South America
by Kluwer Academic Publishers,
101 Philip Drive, Norwell, MA 02061, U.S.A.

In all other countries, sold and distributed
by Kluwer Academic Publishers,
P.O. Box 322, 3300 AH Dordrecht, The Netherlands.

background figure caption:
"A seed crop of clover (*Trifolium hirtum*) in flower near Moora, Western Australia. Photograph courtesy of Mike Davies, Senior Technical Officer, Pasture Research Group of Agriculture WA and reproduced with permission."
vol. II-specific figure caption:
"The regulatory complexity of nitrogen fixation as currently understood for *Rhodobacter capsulatus*. This phototrophic purple bacterium may synthesize and utilize either of two nitrogenases, a Mo-based or an Fe-only nitrogenase, in response to ammonium availability through a highly controlled regulatory cascade encompassing a two-component system (NtrB/NtrC), two specific transcriptional activator proteins (NifA and AnfA), a system that reversibly modifies a nitrogenase component protein (DraT/DraG), and two PII-like signal-transduction proteins (GlnB, GlnK). For details, see Chapter 7. Cartoon courtesy of Bernd Masepohl, Ruhr University, Bochum, Germany."

Printed on acid-free paper

Printed in the Netherlands

TABLE OF CONTENTS

SERIES PREFACE

Nitrogen Fixation: Origins, Applications, and Research Progress

Nitrogen fixation, along with photosynthesis as the energy supplier, is the basis of all life on Earth (and maybe elsewhere too!). Nitrogen fixation provides the basic component, fixed nitrogen as ammonia, of two major groups of macromolecules, namely nucleic acids and proteins. Fixed nitrogen is required for the N-containing heterocycles (or bases) that constitute the essential coding entities of deoxyribonucleic acids (DNA) and ribonucleic acids (RNA), which are responsible for the high-fidelity storage and transfer of genetic information, respectively. It is also required for the amino-acid residues of the proteins, which are encoded by the DNA and that actually do the work in living cells. At the turn of the millennium, it seemed to me that now was as good a time as any (and maybe better than most) to look back, particularly over the last 100 years or so, and ponder just what had been achieved. What is the state of our knowledge of nitrogen fixation, both biological and abiological? How has this knowledge been used and what are its impacts on humanity?

In an attempt to answer these questions and to capture the essence of our current knowledge, I devised a seven-volume series, which was designed to cover all aspects of nitrogen-fixation research. I then approached my long-time contact at Kluwer Academic Publishers, Ad Plaizier, with the idea. I had worked with Ad for many years on the publication of the Proceedings of most of the International Congresses on Nitrogen Fixation. My personal belief is that congresses, symposia, and workshops must not be closed shops and that those of us unable to attend should have access to the material presented. My solution is to capture the material in print in the form of proceedings. So it was quite natural for me to turn to the printed word for this detailed review of nitrogen fixation. Ad's immediate affirmation of the project encouraged me to share my initial design with many of my current co-editors and, with their assistance, to develop the detailed contents of each of the seven volumes and to enlist prospective authors for each chapter.

There are many ways in which the subject matter could be divided. Our decision was to break it down as follows: nitrogenases, commercial processes, and relevant chemical models; genetics and regulation; genomes and genomics; associative, endophytic, and cyanobacterial systems; actinorhizal associations; leguminous symbioses; and agriculture, forestry, ecology, and the environment. I feel very fortunate to have been able to recruit some outstanding researchers as co-editors for this project. My co-editors were Mike Dilworth, Claudine Elmerich, John Gallon, Euan James, Werner Klipp, Bernd Masepohl, Rafael Palacios, Katharina Pawlowski, Ray Richards, Barry Smith, Janet Sprent, and Dietrich Werner. They worked very hard and ably and were most willing to keep the volumes moving along reasonably close to our initial timetable. All have been a pleasure to work with and I thank them all for their support and unflagging interest.

Nitrogen-fixation research and its application to agriculture have been ongoing for many centuries – from even before it was recognized as nitrogen fixation. The Romans developed the crop-rotation system over 2000 years ago for maintaining and improving soil fertility with nitrogen-fixing legumes as an integral component. Even though crop rotation and the use of legumes was practiced widely but intermittently since then, it wasn't until 1800 years later that insight came as to how legumes produced their beneficial effect. Now, we know that bacteria are harbored within nodules on the legumes' roots and that they are responsible for fixing N_2 and providing these plants with much of the fixed nitrogen required for healthy growth. Because some of the fixed nitrogen remains in the unharvested parts of the crop, its release to the soil by mineralization of the residue explains the follow-up beneficial impact of legumes. With this realization, and over the next 100 years or so, commercial inoculants, which ensured successful bacterial nodulation of legume crops, became available. Then, in the early 1900's, abiological sources of fixed nitrogen were developed, most notable of these was the Haber-Bosch process. Because fixed nitrogen is almost always the limiting nutrient in agriculture, the resulting massive increase in synthetic fixed-nitrogen available for fertilizer has enabled the enormous increase in food production over the second half of the 20th century, particularly when coupled with the new "green revolution" crop varieties. Never before in human history has the global population enjoyed such a substantial supply of food.

Unfortunately, this bright shiny coin has a slightly tarnished side! The abundance of nitrogen fertilizer has removed the necessity to plant forage legumes and to return animal manures to fields to replenish their fertility. The result is a continuing loss of soil organic matter, which decreases the soil's tilth, its water-holding capacity, and its ability to support microbial populations. Nowadays, farms do not operate as self-contained recycling units for crop nutrients; fertilizers are trucked in and meat and food crops are trucked out. And if it's not recycled, how do we dispose of all of the animal waste, which is rich in fixed nitrogen, coming from feedlots, broiler houses, and pig farms? And what is the environmental impact of its disposal? This problem is compounded by inappropriate agricultural practice in many countries, where the plentiful supply of cheap commercial nitrogen fertilizer, plus farm subsidies, has encouraged high (and increasing) application rates. In these circumstances, only about half (at best) of the applied nitrogen reaches the crop plant for which it was intended; the rest leaches and "runs off" into streams, rivers, lakes, and finally into coastal waters. The resulting eutrophication can be detrimental to marine life. If it encroaches on drinking-water supplies, a human health hazard is possible. Furthermore, oxidation of urea and ammonium fertilizers to nitrate progressively acidifies the soil – a major problem in many agricultural areas of the world. A related problem is the emission of nitrogen oxides (NO_x) from the soil by the action of microorganisms on the applied fertilizer and, if fertilizer is surface broadcast, a large proportion may be volatilized and lost as ammonia. For urea in rice paddies, an extreme example, as much as 50% is volatilized and lost to the atmosphere. And what goes up must come down; in the case of fertilizer nitrogen, it returns to Earth in the rain, often acidic in nature. This uncontrolled deposition has unpredictable environmental effects, especially in pristine environments like forests, and may also affect biodiversity.

Some of these problems may be overcome by more efficient use of the applied fertilizer nitrogen. A tried and tested approach (that should be used more often) is to ensure that a balanced supply of nutrients (and not simply applying more and more) is applied at the right time (maybe in several separate applications) and in the correct place (under the soil surface and not broadcast). An entirely different approach that could slow the loss of fertilizer nitrogen is through the use of nitrification inhibitors, which would slow the rate of conversion of the applied ammonia into nitrate, and so decrease its loss through leaching. A third approach to ameliorating the problems outlined above is through the expanded use of biological nitrogen fixation. It's not likely that we shall soon have plants, which are capable of fixing N_2 without associated microbes, available for agricultural use. But the discovery of N_2-fixing endophytes within the tissues of our major crops, like rice, maize, and sugarcane, and their obvious benefit to the crop, shows that real progress is being made. Moreover, with new techniques and experimental approaches, such as those provided by the advent of genomics, we have reasons to renew our belief that both bacteria and plants may be engineered to improve biological nitrogen fixation, possibly through developing new symbiotic systems involving the major cereal and tuber crops.

In the meantime, the major impact might be through agricultural sustainability involving the wider use of legumes, reintroduction of crop-rotation cycles, and incorporation of crop residues into the soil. But even these practices will have to be performed judiciously because, if legumes are used only as cover crops and are not used for grazing, their growth could impact the amount of cultivatable land available for food crops. Even so, the dietary preferences of developed countries (who eats beans when steak is available?) and current agricultural practices make it unlikely that the fixed-nitrogen input by rhizobia in agricultural soils will change much in the near-term future. A significant positive input could accrue, however, from matching rhizobial strains more judiciously with their host legumes and from introducing "new" legume species, particularly into currently marginal land. In the longer term, it may be possible to engineer crops in general, but cereals in particular, to use the applied fertilizer more efficiently. That would be a giant step the right direction. We shall have to wait and see what the ingenuity of mankind can do when "the chips are down" as they will be sometime in the future as food security becomes a priority for many nations. At the moment, there is no doubt that commercially synthesized fertilizer nitrogen will continue to provide the key component for the protein required by the next generation or two.

So, even as we continue the discussion about the benefits, drawbacks, and likely outcomes of each of these approaches, including our hopes and fears for the future, the time has arrived to close this effort to delineate what we know about nitrogen fixation and what we have achieved with that knowledge. It now remains for me to thank personally all the authors for their interest and commitment to this project. Their efforts, massaged gently by the editorial team, have produced an indispensable reference work. The content is my responsibility and I apologize upfront for any omissions and oversights. Even so, I remain confident that these volumes will serve well the many scientists researching nitrogen fixation and related fields, students considering the nitrogen-fixation challenge, and administrators wanting to either become acquainted with or remain current in this field. I also acknowledge the many scientists who were not direct contributors to this series of books, but whose contributions to the field are documented in their pages. It would be remiss of me not to acknowledge also the patience and assistance of the several members of the Kluwer staff who have assisted me along the way. Since my initial dealings with Ad Plaizier, I have had the pleasure of working with Arno Flier, Jacco Flipsen, Frans van Dunne, and Claire van Heukelom; all of whom provided encouragement and good advice – and there were times when I needed both!

It took more years than I care to remember from the first planning discussions with Ad Plaizier to the completion of the first volumes in this series. Although the editorial team shared some fun times and a sense of achievement as volumes were completed, we also had our darker moments. Two members of our editorial team died during this period. Both Werner Klipp (1953-2002) and John Gallon (1944-2003) had been working on Volume II of the series, *Genetics and Regulation of Nitrogen-Fixing Bacteria*, and that volume is dedicated to their memory. Other major contributors to the field were also lost in this time period: Barbara Burgess, whose influence reached beyond the nitrogenase arena into the field of iron-sulfur cluster biochemistry; Johanna Döbereiner, who was the discoverer and acknowledged leader in nitrogen-fixing associations with grasses; Lu Jiaxi, whose "string bag" model of the FeMo-cofactor prosthetic group of Mo-nitrogenase might well describe its mode of action; Nikolai L'vov, who was involved with the early studies of molybdenum-containing cofactors; Dick Miller, whose work produced new insights into MgATP binding to nitrogenase; Richard Pau, who influenced our understanding of alternative nitrogenases and how molybdenum is taken up and transported; and Dieter Sellmann, who was a synthetic inorganic chemistry with a deep interest in how N_2 is activated on metal sites. I hope these volumes will in some way help both preserve their scientific contributions and reflect their enthusiasm for science. I remember them all fondly.

Only the reactions and interest of you, the reader, will determine if we have been successful in capturing the essence and excitement of the many sterling achievements and exciting discoveries in the research and application efforts of our predecessors and current colleagues over the past 150 years or so. I sincerely hope you enjoy reading these volumes as much as I've enjoyed producing them.

William E. Newton
Blacksburg, February 2004

PREFACE

Genetics and Regulation of Nitrogen-Fixing Bacteria

This book is the second volume of a seven-volume series, which covers all fields of research related to nitrogen fixation - from basic studies through applied aspects to environmental impacts. Volume II provides a comprehensive and detailed source of information concerning the genetics and regulation of biological nitrogen fixation in free-living prokaryotes. This preface attempts to provide the reader with some insight into how this volume originated, how it was planned, and then how it developed over the several years of its production.

Once the editorial team was established, the first job was to decide which of the many free-living diazotrophs that have been subjected to genetic analysis should be included in this volume. Would we need to develop specific criteria for selection or would the organisms, in effect, select themselves? Of course, *Klebsiella pneumoniae* and *Azotobacter vinelandii*, which have served (and still serve) as the main model organisms for the genetic analysis of diazotrophy, plus some of the other bacteria described in this volume, did indeed select themselves. However, there was considerable discussion surrounding well-characterized fixing species, like *Azorhizobium caulinodans* and *Herbaspirillum seropedicae*, both of which are able to fix atmospheric N_2 under free-living conditions. Was this volume the right place for them? If they were omitted here, would it compromise the volume as a major reference work? After discussions both among ourselves and with the editorial teams of other volumes, it was finally agreed that bacteria such as these belong elsewhere. We decided, for better or worse, that their ability to fix N_2 in either a symbiotic or close associative interaction with a plant host was the significant differentiating factor and so, they are described in other volumes of this series that deal specifically with these interactions and the involved partners.

Similar concerns arose with, for example, coverage of the superoxide-dependent nitrogenase system of *Streptomyces thermoautotrophicus*. How should this system be treated? The decision again was that this unique nitrogenase should be described in volume I of this series, *Catalysts for Nitrogen Fixation: Nitrogenases, Relevant Chemical Models and Commercial Processes*, which deals with the structural and functional aspects of nitrogenases and related non-biological systems, because there is only very limited knowledge of the genetics and regulation in this system.

After the questions concerning which organisms should be included were resolved, there was still the question of what more general topics must be covered. As you will see, we decided to include, in addition to the individual chapters representing an organism-based view of selected well-characterized diazotrophic proteobacteria, cyanobacteria, Gram-positive clostridia, and archaea, the more general cross-organismic themes dealing with different regulatory aspects, electron transport to nitrogenase, and molybdenum metabolism. In all the chapters, whereever appropriate, historical aspects have been included to give the reader a sense of where things started and how much has been achieved, especially in the last 25 years or so.

The chapters are ordered with respect to the two themes mentioned above; a longer organism-based section and a shorter more general section. We are well aware that there are many possibilities for ordering the themes and the individual chapters of this volume and, no matter which is chosen, there are always many arguments either for or against any particular organization.

Our collective wisdom was to start with the organism-based theme and with chapters on the two model organisms. So first, the historical aspects of the genetics and regulation of nitrogen fixation are covered in Chapter 1, which is authored by Ray Dixon. Because of its close relationship to *Escherichia coli, K. pneumoniae* was the organism of choice for the initial genetics and regulation studies of nitrogen fixation. It turned out to be a very good choice for another, completely unexpected, reason; *K. pneumoniae* harbors only one nitrogenase and this is the classical molybdenum-containing nitrogenase. In contrast, some *Azotobacter* species, the subject of Chapter 2, which is co-authored by Christina Kennedy and Paul Bishop, including *A. vinelandii*, harbor as many as two alternative nitrogenases in addition to the classical molybdenum-nitrogenase. These two alternative nitrogenases, which contain either vanadium or iron in place of molybdenum, are regulated differently to the Mo-nitrogenase and so, of course, complicate the genetics and regulation within these organisms. The volume continues with two shorter papers on distantly related groups of prokaryotes outside the proteobacteria. These are the clostridia (Chapter 3, authored by J.-S. Chen) and archaea (Chapter 4, authored by John Leigh). Many of the organisms in these groups have been recalcitrant to genetic studies and much less is known about how nitrogen fixation is regulated. These chapters are followed by chapters about three different groups of photosynthetic bacteria, namely the heterocyst-forming cyanobacteria (Chapter 5, authored by Terry Thiel), the non-heterocystous cyanobacteria (Chapter 6, authored by John Gallon), and the purple bacteria (Chapter 7, co-authored by Bernd Masepohl, Thomas Drepper, and Werner Klipp).

The move to the more general cross-organismic theme starts with post-translational regulation of nitrogenase in photosynthetic bacteria (Chapter 8, co-authored by Stefan Nordlund and Paul Ludden), which allows a seamless transition from the chapter on purple bacteria to the review on regulation of nitrogen fixation in free-living diazotrophs (Chapter 9, authored by Mike Merrick). Two more chapters, one on molybdenum metabolism (Chapter 10, authored by Richard Pau) and the other on electron transport to nitrogenase (Chapter 11, authored by Kaz Saeki) complete the general overview section. The volume ends with a chapter (Chapter 12, co-authored by John Gallon and Bernd Masepohl) that considers the prospects for, and provides an outlook on, the future of this area of scientific endeavor.

We wish to thank all the authors of this volume for their enormous efforts to make it an indispensable reference work for all scientists working in the field, for those administrators with authority and responsibility in this arena, and for students who are brave enough to want to enter this challenging area of research. There are many other experts in the field who could have made contributions to this volume, but who were not asked to do so. We wish to make it crystal clear that our choices in no way reflect on them and their abilities. In fact, this volume could not have been produced without their contributions to the field, many of which are incorporated into these pages. In addition, we proffer our apologies in advance if any topics of interest are omitted. At this point, we would like to draw special attention to another volume in this series, which is entitled *Genomes and Genomics of Nitrogen-Fixing Organisms*. It deals with the whole genome of the many nitrogen-fixing bacteria (including species described in this volume), whose genomes have been completely sequenced. There, a considerable amount of information is gathered, much of it closely related to the topics of this volume, and it is also the volume in which the evolutionary aspects of nitrogen fixation are considered.

It took more than two years from the first phase of planning until the completion of this volume. Both the fun and the sense of achievement that should have accrued during this time were overshadowed by the deaths of three contributors to this volume. Werner Klipp, who helped start this project, died in May 2002 towards the end of the major planning phase. And, as if this wasn't enough, John Gallon, who had been recruited and was actively working on this volume after Werner´s death, died in August 2003 during the final preparation of the individual chapters. Then, in February 2004 after completion of the volume, Richard Pau passed away.

With great sadness, some months ago, we decided to dedicate this volume to our two co-editors and good friends, who were lost on the journey to complete this project. And now, as we look back, we fondly remember all three of them, not just for their keen scientific insight and research prowess, but for their enthusiasm for science, for the joy they took from scientific discovery, and for sharing the fun with us along the way.

Bernd Masepohl
Bochum, February 2004

William E. Newton
Blacksburg, February 2004

LIST OF CONTRIBUTORS

Paul E. BISHOP
Department of Microbiology, North Carolina State University and USDA, Raleigh, NC 27695, U.S.A.
Email: paul_bishop@ncsu.edu

Jiann-Shin CHEN
Department of Biochemistry, Virginia Polytechnic Institute and State University, Blacksburg, VA 24061, U.S.A.
Email: chenjs@vt.edu

Ray DIXON
Department of Molecular Microbiology, John Innes Centre, Norwich NR2 3DR, U.K.
Email: ray.dixon@bbsrc.ac.uk

Thomas DREPPER
Ruhr-Universität Bochum, Fakultät für Biologie, Lehrstuhl für Biologie der Mikroorganismen, D-44780 Bochum, Germany.
Email: thomas.drepper@ruhr-uni-bochum.de

John R. GALLON
Biochemistry Research Group, School of Biological Sciences, University of Wales Swansea, Singleton Park, Swansea, SA2 8PP, U.K.

Christina KENNEDY
Department of Plant Pathology, University of Arizona, Tucson, AZ 85721, U.S.A.
Email: ckennedy@u.arizona.edu

Werner KLIPP
Ruhr-Universität Bochum, Fakultät für Biologie, Lehrstuhl für Biologie der Mikroorganismen, D-44780 Bochum, Germany

John A. LEIGH
Department of Microbiology, University of Washington, Seattle, WA 98195, U.S.A.
Email: leighj@u.washington.edu

Paul W. LUDDEN
Department of Plant and Microbial Biology, 111 Koshland Hall, University of California, Berkeley, CA 94720-3102, U.S.A.
Email: pludden@nature.berkeley.edu

Bernd MASEPOHL
Ruhr-Universität Bochum, Fakultät für Biologie, Lehrstuhl für Biologie der Mikroorganismen, D-44780 Bochum, Germany.
Email: bernd.masepohl@ruhr-uni-bochum.de

Mike MERRICK
Department of Molecular Microbiology, John Innes Centre, Norwich, NR4 7UH, U.K.
Email: mike.merrick@bbsrc.ac.uk

William E. NEWTON
Department of Biochemistry, Virginia Polytechnic Institute and State University, Blacksburg, VA 24061, U.S.A.
Email: wenewton@vt.edu

Stefan NORDLUND
Department of Biochemistry and Biophysics, Stockholm University, SE-106 91 Stockholm, Sweden.
Email: stefan@dbb.su.se

Richard N. PAU
Department of Biochemistry and Molecular Biology, The University of Melbourne, Parkville 3010, Australia.
Email: r.pau@unimelb.edu.au

Kazuhiko SAEKI
Department of Biology, Graduate School of Science, Osaka University, 1-1 Machikaneyama, Toyonaka, Osaka 560-0043, Japan.
Email:ksaeki@bio.sci.osaka-u.ac.jp

Teresa THIEL
Department of Biology, University of Missouri-St. Louis, 8801 Natural Bridge Rd., St. Louis, MO 63121, U.S.A.
Email: thiel@umsl.edu

John Gallon
(1944-2003)

Werner Klipp
(1953-2002)

We dedicate this volume to two of our colleagues, Werner Klipp and John Gallon, both of whom died during the period when we were preparing this work for publication. Both were well-known and well-respected members of the international research community with interests that ranged widely across many aspects of science related to nitrogen-fixation research. Werner's interests covered many of the molecular aspects related to both free-living and symbiotic bacteria, including molybdenum metabolism, but he was especially interested in genetics and regulation. His enthusiasm for his work was obvious to those of us who were fortunate enough to know him well and was matched only by his brilliant rapid-fire seminars. Without his input, the field would be poorer by far with respect to our detailed knowledge of nitrogen fixation, especially in photosynthetic bacteria. John's interests encompassed the whole organism, particularly the non-heterocystous cyanobacteria, rather than individual cell components. His "bigger picture" interests led him to research areas where complex interactions are the norm. His work at the interface of the organism with its environment has helped us in our understanding of how the environment impacts the organism's ability to survive and grow and, just as importantly, how organisms impact their environment. Both Werner and John enjoyed their science immensely; discussions with either of them were always great fun. Many of us will remember them both as first-class researchers, constructive collaborators, fine teachers, and good friends.

Chapter 1

HISTORICAL PERSPECTIVE – DEVELOPMENT OF *NIF* GENETICS AND REGULATION IN *KLEBSIELLA PNEUMONIAE*

R. DIXON
Department of Molecular Microbiology, John Innes Centre, Norwich NR2 3DR, UK

"History never looks like history when you are living through it."
John W. Gardner

1. INTRODUCTION

By the end of the 1960's, research on biological nitrogen fixation had progressed to the extent that genetic analysis of this process and its regulation had clearly become feasible. Biochemical studies during this decade had established methods for the purification of nitrogenase component proteins and the advent of the acetylene reduction test had revolutionised measurement of nitrogen fixation both *in vitro* and *in vivo*. Pathways for ammonia assimilation under nitrogen-limited conditions had been characterised and physiological studies had suggested a role for fixed nitrogen in "repression" of nitrogenase synthesis.

Mutants of *Azotobacter vinelandii* incapable of fixing N_2 were first isolated as early as 1950, but such mutants were not particularly useful at that time, because there was no genetic method available to map the mutations, nor was it possible to examine the properties of mutant proteins at the biochemical level. Clearly, genetic analysis would require the development of gene transfer systems in nitrogen-fixing bacteria. The rapid advances in microbial genetics during the 1960's provided the impetus to investigate the genetics of nitrogen fixation and, in the early 1970's, *Klebsiella pneumoniae* became the model organism for genetic analysis of diazotrophy. In considering the development of *nif* genetics from a historical standpoint, the 1970's therefore provide an obvious starting-point for this review.

W. Klipp, B. Masepohl, J. R. Gallon and W. E. Newton (eds.), Genetics and Regulation of Nitrogen Fixation in Free-Living Bacteria, 1-25.

However, in writing this Chapter, it has been more difficult to decide where the story should end, as past developments clearly form a continuum with current research. I have arbitrarily chosen the mid-1990's as a closing point, primarily because the most recent advances in the field will be covered by other authors in this volume.

2. THE EARLY YEARS

2.1. nif Gene Transfer

In 1970, two laboratories independently initiated genetic studies on nitrogen fixation in the gram-negative bacterium *Klebsiella pneumoniae* strain M5a1. This organism was chosen primarily because it was an enteric diazotroph related to *Escherichia coli* and *Salmonella typhimurium*, both of which had been extensively studied at the genetic level. This diazotrophic non-mucoid *K. pneumoniae* strain had also been used for biochemical and physiological studies of nitrogen fixation. Genetic recombination either by transduction or conjugation had been reported in some *Klebsiella* strains but not in those which were competent to fix N_2. Stanley Streicher, a graduate student working in Ray Valentine's laboratory at La Jolla isolated *nif* mutants of *K. pneumoniae* M5a1 unable to grow on N_2 as sole nitrogen source and showed that these mutants were defective in catalysing acetylene reduction (Streicher *et al.*, 1971). Although strain M5al was not sensitive to bacteriophage active on other *Klebsiella* strains, Streicher *et al.* found that this diazotrophic strain was sensitive to phage P1, which had been used previously as a tool for generalised transduction in *E. coli.* P1 lysates prepared on wild-type strain M5a1, transduced the *nif*$^-$ mutants to a Nif$^+$ phenotype, allowing the transductants to regain the ability to grow on N_2 as sole nitrogen source. A series of two-point transductional crosses with 30 mutants revealed that *nif* genes were located in one region of the chromosome, close to the histidine (*his*) biosynthetic operon (Streicher *et al.*, 1971).

In contrast to the transduction approach, I used conjugation to establish *nif* gene transfer in *K. pneumoniae* M5a1 while working as a graduate student in John Postgate's laboratory at Sussex (Dixon and Postgate, 1971). The self-transmissible R plasmid, R144drd3, was observed to promote chromosome mobilisation in *K. pneumoniae* and, when transferred from a wild-type donor strain to a *nif*$^-$ recipient strain, gave rise to Nif$^+$ recombinants at frequencies around 10^{-5}. These studies also demonstrated that *nif* mutations were located close to the *his* operon in *K. pneumoniae.* In order to increase the frequency of conjugal gene transfer, we irradiated the M5a1 (R144drd3) strain with UV light to obtain an Hfr-like strain, which was subsequently shown to mobilise the *his* region of the chromosome in a polarised fashion (Dixon *et al.*, 1975; Dixon and Postgate, 1972). This high frequency donor strain was used to attempt conjugal transfer of the *K. pneumoniae his* region to *E. coli.* Remarkably, some of the His$^+$ *E. coli* transconjugants gained the ability to fix N_2 and thus the first genetically engineered diazotroph had been created (Dixon and Postgate, 1972).

At the time, the genetic transfer of nitrogen fixation to this extensively studied model prokaryote was regarded as an exceptional breakthrough. It indicated that all the genes required to enable *E. coli* to fix N_2 were located close to the histidine biosynthetic operon in *K. pneumoniae* and that rapid genetic analysis of nitrogen fixation would be facilitated by the powerful genetic tools available in *E.coli*. It also led to speculation concerning future prospects for genetic engineering of nitrogen fixation and the ultimate goal of generating autonomous nitrogen-fixing plants (Shanmugam and Valentine, 1975b; Streicher *et al.*, 1972). During this period, genetic transfer of nitrogen fixation from *Rhizobium trifolii* to a non-nitrogen fixing strain of *K. aerogenes* was also reported (Dunican and Tierney, 1974), although this finding has not been confirmed by other laboratories.

Analysis of the nitrogen-fixing *E. coli* transconjugants revealed that *K. pneumoniae nif* genes were either integrated into the *E. coli* chromosome or were located on autonomously replicating plasmids (Cannon *et al.*, 1974a, 1974b). Selection for recombinant plasmids *in vivo* led to the construction of a wide host range plasmid of the P incompatibility group carrying the *his-nif* region of *K. pneumoniae*, which was transmissible to *Agrobacterium tumefaciens* and *Rhizobium meliloti* (Dixon *et al.*, 1976). The most useful property of this plasmid, however, was not its wide host range, but its stability in *K. pneumoniae* and it was used extensively for subsequent complementation analysis and cloning of the *nif* gene cluster (see below).

2.2. Regulation by Ammonium

Tempest's discovery of a new route for ammonia assimilation, involving the enzymes glutamate synthase (GOGAT) and glutamine synthetase (GS) in *Klebsiella aerogenes* (Tempest *et al.*, 1970), was soon extended to nitrogen-fixing bacteria. Mutants of *K. pneumoniae* M5a1 defective in glutamate synthase (*asm* mutants) were unable to assimilate nitrogen under nitrogen-fixing conditions but were able to grow in media containing ammonia as the nitrogen source (Nagatani *et al.*, 1971). These observations strongly suggested that the GS-GOGAT pathway was essential for ammonia assimilation under diazotrophic conditions and also focussed attention on the potential role of GS in regulating nitrogenase synthesis in response to ammonium. Based upon studies on a variety of mutants with lesions in *glnA*, the structural gene for GS, Magasanik and his colleagues had proposed that GS was not only a key enzyme in ammonia assimilation, but also played a key role in regulating transcription of nitrogen-regulated operons (Magasanik *et al.*, 1974). This proposal was backed up by the observation that purified GS stimulated transcription of the histidine utilisation operon *in vitro* (Tyler *et al.*, 1974). The proposed regulatory role for GS in nitrogen regulation was supported by studies on ammonium regulation of nitrogen fixation. Independent studies by a graduate student (Roy Tubb) working in Postgate's laboratory and Stanley Streicher, who had moved from Valentine's group to MIT, led to similar conclusions. Mutants defective in glutamine synthetase showed no detectable nitrogenase activity, but mutants with constitutive levels of GS activity synthesised nitrogenase in the presence of ammonium (Shanmugam *et al.*, 1975; Streicher *et al.*, 1974; Tubb, 1974). These findings led to the hypothesis that GS positively controlled *nif* transcription as proposed for other nitrogen-regulated genes (Magasanik *et al.*, 1974; Prival *et al.*, 1973). This hypothesis survived in the literature for ten years until it was eventually realised that nitrogen regulation was mediated not by GS but by the products of regulatory genes located in the same operon as *glnA* itself (see below).

The early observations on nitrogen assimilation in diazotrophic bacteria were, however, further exploited to isolate mutant *K. pneumoniae* strains that excreted ammonia when grown under nitrogen-fixing conditions (Shanmugam and Valentine, 1975a). Although this was heralded as a potential route for microbial production of ammonium-based fertiliser, the energetic cost of biological nitrogen fixation precluded any exploitation of this finding on a large scale.

3. DEFINING THE *K. PNEUMONIAE NIF* GENES

Biochemical studies on nitrogenase had shown that the enzyme was comprised of two component proteins, suggesting that only a few genes would be required for nitrogenase biosynthesis. In the latter half of the 1970s, several laboratories embarked on identifying *nif* genes and determining their operon structure. The major competition was between Winston Brill's group in Madison and the Sussex group. Brill's group had initially mapped *K. pneumoniae nif* mutants by three-factor transduction crosses with phage P1 using *his* as an outside marker (St John *et al.*, 1975). This approach was followed by complementation analysis of *nif* mutants with plasmid pRD1 and further P1 transduction experiments, which suggested that the *nif* region was comprised of at least seven cistrons (Dixon *et al.*, 1977), divided into a proximal *his* cluster, containing *nifB*, *nifA* (*nifL*), *nifF*, and a more distal cluster, comprising *nifE*, *nifK*, *nifD* and *nifH* (Kennedy, 1977). Brill's group exploited phage Mu to isolate a series of *nif* deletion mutants both in the chromosome and on a derivative of pRD1 (Bachhuber *et al.*, 1976; Macneil *et al.*, 1978a). These deletions were used to map an extensive collection of point mutants that closed the gap between the proximal and distal *nif* clusters identified by (Kennedy, 1977) and identified a total of 14 *nif* genes designated as *nifQ, nifB, nifA, nifL, nifF, nifM, nifV, nifS, nifN, nifE, nifK, nifD, nifH and nifJ* (Macneil *et al.*, 1978b). A group in China led by Shen also mapped a series of *nif* mutations to close this gap (Hsueh *et al.*, 1977).

Meanwhile, the Sussex group had identified four new *nif* genes, which they designated as *nifM, nifN, nifI* and *nifJ* (Merrick *et al.*, 1978). This nomenclature caused some confusion at the time, because *nifN* was equivalent to the gene designated *nifS* by Brill's group and *nifI* was equivalent to *nifN*. This conflict was resolved after a heated discussion at the 3rd International Symposium on Nitrogen Fixation in Madison, Wisconsin, in June 1978, whereby the Brill nomenclature was accepted. Around this time, Claudine Elmerich's laboratory at the Institut Pasteur in Paris isolated Mu insertions in *nif* genes. By examining the transcriptional polarity of Mu or other transposon-induced mutations in complementation tests, all three laboratories suggested a similar operon structure for the *nif* gene cluster

(Elmerich *et al.*, 1978; Macneil *et al.*, 1978b; Merrick *et al.*, 1978). Further collaboration between the Sussex and Paris groups led to a more detailed fine-structure map which included a newly identified gene designated *nifU,* located between *nifS* and *nifN* (Merrick *et al.*, 1980).

Although *nifH*, *nifD* and *nifK* were soon identified as the structural genes for nitrogenase, the complexity of the *nif* gene cluster came as a surprise to the nitrogen fixation community and the immediate challenge was to determine the functions of each of the gene products. Gary Roberts working in Brill's laboratory identified *nif* polypeptides on two-dimensional gels and examined nitrogenase activity in mutant crude extracts, following *in vitro* complementation with purified Fe protein, MoFe protein or iron-molybdenum cofactor (FeMoco). Fifteen *nif*-encoded polypetides were identified on gels. The *nifB, nifE* and *nifN* polypeptides were shown to be required for FeMoco synthesis and the *nifM* and *nifS* genes for the synthesis of active Fe protein (Roberts and Brill, 1980; Roberts *et al.*, 1978). Although *nifF* and *nifJ* were both required for nitrogenase activity *in vivo*, crude extracts of these mutants were active *in vitro,* implying a role for these proteins in electron transport to nitrogenase (Hill and Kavanagh, 1980; Nieva-Gomez *et al.*, 1980; Roberts *et al.*, 1978). Sequencing of *nif* genes provided useful clues to the possible functions of other *nif*-encoded polypeptides (see below).

4. THE RECOMBINANT DNA ERA

Several nitrogen-fixation scientists were invited to the Asilomar Conference in 1975, which led to the establishment of guidelines for working with the newly discovered recombinant DNA technology. Once the guidelines were established, it was not long before the cloning of *nif* DNA began. Frank Cannon at Sussex, working in collaboration with Fred Ausubel's laboratory at Harvard, isolated DNA from the *nif* plasmid pRD1 and after partial digestion with EcoRI, inserted fragments into the amplifiable plasmid pMB9. Ligated DNA was transformed into a restriction-deficient *K. pneumoniae* strain containing mutations in *hisD* and *nifB*. His$^+$ transformants were screened for those which complemented the *nifB* mutation. One plasmid, designated as pCRA37, complemented *nifB* and *nifF* mutations but not *nifD* or *nifH* mutants; the physical map of this recombinant plasmid correlated well with genetic map derived by P1 co-transductional analysis (Cannon *et al.*, 1977). The remaining *nif* genes were cloned in subsequent stages, culminating in the construction of multicopy plasmids carrying the entire *nif* region (Cannon *et al.*, 1979; Pühler *et al.*, 1979; Reidel *et al.*, 1979). The availability of *K. pneumoniae nif* DNA fragments immediately facilitated the cloning of *nif* genes from other diazotrophs using Southern hybridisation (Ruvkun and Ausubel, 1980). The following year, the entire *K. pneumoniae nif* gene cluster was stably integrated into the genome of yeast following transformation with an *E. coli*-yeast shuttle vector (Zamir *et al.*, 1981). Although this experiment was an interesting feat of genetic engineering, it had little practical relevance because no attempt was made to express the prokaryotic nif genes from yeast promoters. Expression of *nifH*, *nifD*, and *nifK* was subsequently achieved in yeast but the nitrogenase component proteins were apparently inactive in this host (Berman *et al.*, 1985; Holland *et al.*, 1987).

The advent of the recombinant DNA era and the possibilities for exploiting the technology to improve biological nitrogen fixation generated considerable interest during the late 1970's and a conference on "Genetic Engineering for Nitrogen Fixation" sponsored by the US National Science Foundation was held at the Brookhaven National Laboratory in 1977 to discuss these issues. Although the potential for transferring nitrogen-fixation genes from prokaryotes directly into plants was considered at this meeting, in retrospect this conference was far too premature because plant transformation had not been developed. It is interesting, however, to consider the motivation for engineering biological nitrogen fixation at that time, because there was rising concern about the energy costs of fertiliser production. One projection suggested: "assuming a steady supply of natural gas at current levels, in the year 2000, 10% of the entire U.S. supply (2 trillion cubic feet) will be consumed by fertiliser factories" (Ausubel, 1977). This was, of course, a gross over-estimation because current industrial fertiliser production accounts for only around 1-2% of worldwide energy consumption.

The new chemical and dideoxy methods for sequencing DNA (Maxam and Gilbert, 1977; Sanger *et al.*, 1977) were soon applied to determine the nucleotide sequences of *nif* genes. The first sequence to be published was that of *nifH,* from the cyanobacterium *Anabaena* 7120 by Bob Haselkorn's laboratory in Chicago (Mevarech *et al.*, 1980). This sequence was closely followed by that of *K. pneumoniae nifH* and *nifD* (Scott *et al.*, 1981; Sundaresan and Ausubel, 1981). Further sequence analysis revealed five new *nif* genes *nifT*, *nifW, nifX, nifY*, *nifZ* and the complete DNA sequence of the entire 24.2 Kb *K. pneumoniae nif* gene cluster was completed in 1988 (Arnold *et al.*, 1988). The availability of the sequence data provided an extremely useful platform to establish the functions of *nif* gene products, for example, the role of *nifM* in Fe protein biosynthesis (Paul and Merrick, 1987, 1989) and the *nifU* and *nifS* genes in iron-sulphur cluster biosynthesis (Beynon *et al.*, 1987; Dean and Jacobson, 1992). A description of the role of *nif* genes in nitrogenase biosynthesis and function is given in Table 1.

5. *NIF* GENE REGULATION

5.1. Identification of the Major Regulatory Genes

The observation that nitrogenase component proteins were absent in *nifA* mutants led to the suggestion that the *nifA* gene product was a positive activator of *nif* gene expression (Dixon *et al.*, 1977). A new gene, *nifL*, upstream of *nifA*, was defined on the basis of transductional mapping of a mutant allele, *nifL2265*, which was partially transdominant to the wild-type allele (Kennedy, 1977). Mu-induced insertion mutations in *nifL* were polar on *nifA*, but this polarity could be relieved by suppressor mutations, suggesting that *nifL* is not essential for nitrogen fixation O_2 (Macneil *et al.*, 1978b). Although repression of nitrogenase biosynthesis by ammonia was a recognised phenomenon at that time (section 2.2.), repression in

Table 1. Proposed functions of K. pneumoniae *nif genes*

Gene	*Product and/or known function*	*References*
nifJ	Pyruvate oxido-reductase; couples pyruvate oxidation to reduction of the *nifF* product	(Hill and Kavanagh, 1980; Nieva-Gomez *et al.*, 1980; Shah *et al.*, 1983)
nifH	Fe protein subunit	(Dean and Jacobson, 1992)
nifDK	MoFe protein α and β subunits	(Dean and Jacobson, 1992)
nifT	Function unknown; not essential for nitrogen fixation	(Simon *et al.*, 1996)
nifY	Associates with MoFe protein and dissociates upon FeMo cofactor insertion	(Homer *et al.*, 1993)
nifEN	Required for FeMo cofactor biosynthesis	(Allen *et al.*, 1994; Dean and Jacobson, 1992)
nifX	Not essential for nitrogen fixation; required for FeMo cofactor biosynthesis	(Shah *et al.*, 1999)
nifU	Fe-S cluster biosynthesis	(Yuvaniyama *et al.*, 2000)
nifS	Fe-S cluster biosynthesis	(Zheng *et al.*, 1993)
nifV	Encodes a homocitrate synthase. Homocitrate is an organic component of FeMo cofactor	(Hawkes *et al.*, 1984; Hoover *et al.*, 1987; McLean and Dixon, 1981)
nifW	Function unknown; interacts with the MoFe protein	(Kim and Burgess, 1996; Paul and Merrick, 1989)
nifZ	Function unknown; required for full activity of the MoFe protein	(Paul and Merrick, 1989)
nifM	Required for Fe protein maturation. Putative peptidyl-prolyl *cis/trans* isomerase	(Dean and Jacobson, 1992; Gavini and Pulukat, 2002)
nifF	Flavodoxin required for electron transfer to the Fe protein	(Deistung *et al.*, 1985; Thorneley *et al.*, 1992)
nifL	Negative regulatory protein	(Dixon, 1998)
nifA	Positive regulator of *nif* transcription	(Dixon, 1998)
nifB	Required for FeMo cofactor biosythesis	(Allen *et al.*, 1995; Shah *et al.*, 1994)
nifQ	Incorporation of Mo into FeMo cofactor	(Imperial *et al.*, 1984)

response to O_2 was not firmly established until 1978, when it was demonstrated that repression could be distinguished mechanistically from regulation by ammonium (Eady *et al.*, 1978). In 1979, an elegant technique had been developed to fuse any promoter to the *E. coli* lactose operon (Casadaban and Cohen, 1979) and this was soon exploited to isolate *nif-lac* fusions (Dixon *et al.*, 1980; Macneil *et al.*, 1981). These studies revealed that all *nif* promoters respond to repression by ammonium and all except the *nifLA* promoter were strongly repressed by O_2. Similar findings were observed subsequently when rates of *nif* transcription were measured (Cannon *et al.*, 1985; Collins and Brill, 1985). The *nifA* gene product was shown to act *in trans* as a positive activator of *nif* gene expression, but was not required to activate the *nifLA* promoter (Dixon *et al.*, 1980). A strange, and as yet unexplained phenomenon, is that Nif^+ strains give rise to a purple colour on nitrogen-free medium containing 6-cyanopurine and this phenotype requires the *nifA* gene product (Macneil and Brill, 1978). A few of the suppressors of polar insertion mutations in *nifL* were able to synthesise nitrogenase in the presence of ammonium and were purple on 6-cyanopurine plates (Macneil *et al.*, 1981). These mutations, which were closely linked to *nifL*, also suppressed the Nif^- phenotype of *gln* mutants, leading to the hypothesis that *gln*-encoded factors might activate the *nifLA* promoter and that the *nifA* product specifically activates transcription of the other *nif* operons (Dixon *et al.*, 1980; Macneil *et al.*, 1981).

In the late 1970's, the hypothesis that GS, in its non-adenlylylated form, was a positive regulator of nitrogen-regulated genes, including the *nif* operons, was beginning to wane, although there was some evidence that purified GS bound specifically to *nif* promoters (Janssen *et al.*, 1980). However, the simple model for GS as a nitrogen-responsive transcriptional activator was clearly incomplete because three new genes influencing nitrogen regulation had been discovered; *ntrA* (formerly *glnF*) unlinked to *glnA* (Garcia *et al.*, 1977) and *ntrB* (*glnL*) and *ntrC (glnG)* (McFarland *et al.*, 1981; Pahel and Tyler, 1979), which where shown to be in the same operon as *glnA* (Macneil *et al.*, 1982a; Pahel *et al.*, 1982). The earlier evidence that *glnA*-linked mutations influenced regulation of nitrogen-regulated genes could be explained on the basis of their polar effects on *ntrB* and *ntrC*. The first evidence for the role of these new genes in *nif* gene regulation was obtained by Leonardo and Goldberg (1980), who demonstrated that *ntrA* and *ntrC* mutants of *K. pneumoniae* were unable to synthesise nitrogenase. Subsequent studies in other laboratories confirmed these conclusions (de Bruijn and Ausubel, 1981; Espin *et al.*, 1982; Espin *et al.*, 1981) and at this stage, it became obvious that both *ntrC* and *ntrA* were required for positive control of *nif* transcription. Constitutive expression of *nifA* led to activation of *nif* transcription in the absence of *ntrC*, suggesting that *ntrC* is required to activate the *nifLA* promoter (Buchanan-Wollaston *et al.*, 1981a). However, constitutive expression of *nifA* did not by-pass the requirement for *ntrA*, suggesting that the *ntrA* product had a more general role in *nif* gene activation (Merrick, 1983; Ow and Ausubel, 1983; Sibold and Elmerich, 1982). These findings were elaborated into a cascade model for *nif* gene activation in which NtrC, acting in concert with NtrA, activated transcription of the *nifLA* operon, giving rise to the expression of NifA, which activated transcription of the other six *nif* operons, also in concert with NtrA.

While attention was focussed on the *ntrC* and *nifA* gene products as positive activators of *nif* transcription, there was increasing evidence that *nifL* was involved in negative control. Several regulatory mutants mapping in *nifL* enabled *nif* transcription in the presence of O_2 but not in the presence of fixed nitrogen (Hill *et al.*, 1981). Subsequently, *nifL* was shown also to mediate repression in response to fixed nitrogen (Merrick *et al.*, 1982). When present on a multicopy plasmid, *nifL* inhibited *nif* transcription, even in the absence of O_2 and fixed nitrogen, suggesting that over-expression of this protein might titrate out factors required to maintain it in an inactive form (Buchanan-Wollaston *et al.*, 1981b; Reidel *et al.*, 1983). Based on stability analysis of *nif* mRNA, it was also proposed that the *nifL* product might have a role in post-transcriptional control of *nif* genes by destabilising *nif* mRNA in response to O_2 and fixed nitrogen (Collins *et al.*, 1986). However, this model has since been disproved and current analysis indicates that the stability of *nif* structural mRNA is associated with nitrogenase activity (Simon *et al.*, 1999).

The newly-discovered *ntrB* gene provided a focus for analysis as a potential negative regulator of *ntrC*. Several laboratories had isolated mutations in this gene that gave rise to constitutive expression of GS and other operons normally subject to repression by ammonium (Chen *et al.*, 1982; Macneil *et al.*, 1982b; McFarland *et al.*, 1981). Mutations in *ntrB* also gave rise to *nif* transcription in the presence of ammonium (Ow and Ausubel, 1983). These results suggested that *ntrB,* like *nifL*, had a negative function when ammonium was present. It was also thought that *nif* transcription could be subject to the stringent response under nitrogen-limiting conditions because *relA* mutants, which are unable to accumulate ppGpp, gave low rates of *nif* expression under nitrogen-starved conditions (Riesenberg *et al.*, 1982). However, it was subsequently found that there was no correlation between the levels of ppGpp and the extent of *nif* depression (Nair and Eady, 1984).

5.2. Sequencing of nif *Promoters and Regulatory Genes*

Determination of transcription start-sites and sequencing of the *nif* promoters revealed some surprises. Analysis of the *nifLA* promoter demonstrated that it had no typical –35 region and, in agreement with this finding, some positive control by NtrC was maintained even in deletions extending to –28 (Drummond *et al.*, 1983). Analysis of promoters activated either by NtrC or NifA revealed a heptameric consensus sequence, TTTGCA, in the –15 region, which was proposed to be a binding sequence for transcriptional activation (Ow *et al.*, 1983). In a seminal paper, sequencing of five *nif* promoters revealed a characteristic primary structure with the consensus CTGG at –24 and TTGCA at –12 (Beynon *et al.*, 1983). The significance of this unique consensus sequence was not then fully realised, although it was postulated that this might provide a recognition sequence for a modified form of RNA polymerase containing the *ntrA* product, which might act as a novel sigma factor (de Bruijn and Ausubel, 1983). Footprinting experiments, using crude extracts from cells grown under nitrogen-fixing conditions, indicated that an upstream AT-rich region present in some *nif* promoters was protected from DNase I digestion (Beynon *et al.*, 1983). At the time, it was thought that this protection represented an interaction with RNA polymerase but later it was shown to be due to the binding of Integration Host Factor (IHF) (Hoover *et al.*, 1990) (see below).

Mutations located in invariant nucleotides in the –24 and –12 regions of *nif* promoters prevented transcriptional activation (Buck *et al.*, 1985; Khan *et al.*, 1986; Ow *et al.*, 1985). The spacing between the conserved GG and GC motifs in the –24 and –12 elements was found to be critical for promoter activity, reflecting a stringent spacing requirement (Buck, 1986). Some of the *nif* promoters, when cloned on multicopy plasmids, inhibited chromosomal *nif* expression in *K. pneumoniae*, resulting in a Nif$^-$ phenotype (Buchanan-Wollaston *et al.*, 1981b; Reidel *et al.*, 1983). This "multicopy effect" was attributed to titration of NifA by the excess promoters, thus preventing transcriptional activation of genomic *nif* promoters. Mutations that suppressed the multicopy effect were located either in the consensus –24 –12 region or in sequences further upstream (Brown and Ausubel, 1984; Buck *et al.*, 1985). Deletion analysis of the *nifH*, *nifU* and *nifB* promoters revealed that sequences upstream of –100 were required both for NifA-mediated activation and transcriptional activation. Sequence analysis of *nif* promoters identified an invariant TGT–N_{10}–ACA motif, which was proposed as a NifA binding site, designated as an Upstream Activator Sequence (UAS). Remarkably, the position and orientation of the UAS was not critical for promoter activity and it functioned when positioned up to a distance of 2 kb from the downstream –24 –12 element (Buck *et al.*, 1986).

At about the same time, it was found that activation of the *glnA* promoter by NtrC also occurred at a distance and that NtrC-binding sites could function far upstream of the promoter (Reitzer and Magasanik, 1986). NifA and NtrC were therefore brought into the limelight as eukaryotic-like transcriptional activators that bound to regulatory sequences similar to enhancers. This realization ultimately led to the classification of these transcriptional activators as Enhancer Binding Proteins (EBPs). Further analysis of the spatial requirements for UAS function showed that activation was face-of-the-helix dependent, indicating a stereospecific requirement for positioning of the activator with respect to the –24 –12 region. It was, therefore, proposed that NifA and NtrC activate transcription via a DNA looping mechanism (Buck *et al.*, 1987; Minchin *et al.*, 1989). However, activation of transcription was not entirely dependent on the UAS sequences because, for example, weak activation of the *nifH* promoter by NtrC had been detected (Buck *et al.*, 1985). In these cases, it was argued that neither activator binding at the UAS nor DNA loop formation are absolute prerequisites for transcriptional activation and that the activator might contact the downstream bound RNA polymerase from solution. Comparison of promoter sequences that were not strictly dependent on the UAS suggested that a run of T residues between –17 to –14 might be critical in the response of the promoter in the absence of the UAS. Conversion of this sequence in the *nifH* promoter from CCCT to TTTT suppressed the requirement for the UAS to be located on the same face of the helix with respect to the –24 –12 sequence and the promoter was far more responsive to a truncated form of NifA lacking the DNA-binding domain (Buck and Cannon, 1989). Whereas the binding of RNA polymerase to the wild-type *nifH* promoter could not be detected by *in vivo* footprinting, protection was observed with the mutant promoter. These observations suggested that relatively weak binding of RNA polymerase to the downstream sequence coupled with stereospecific binding of the activator at the UAS, ensures the fidelity of activation, thus, ensuring that this *nif* promoter is specifically activated by NifA (Morett and Buck, 1989).

Nucleotide sequencing of the *K. pneumoniae* regulatory genes, *ntrA, ntrB, ntrC, nifL* and *nifA*, was completed between 1985 and 1987, but initially provided only limited clues to their function because these were the first genes of their class to be sequenced. Although NtrA was proposed to be a sigma factor and appeared to compete with σ^{70} when overexpressed *in vivo* (Merrick and Stewart, 1985), the sequence of *K. pneumoniae ntrA* showed that the encoded protein was not similar to other sigma factors. However, it did contain potential DNA-binding regions, which conceivably could be involved in promoter recognition (Merrick and Gibbins, 1985). Comparison of the NifA and NtrC sequences revealed a strongly conserved central domain and a C-terminal domain containing a helix-turn-helix motif proposed to be required for DNA binding (Buikema *et al.*, 1985; Drummond *et al.*, 1986). The role of the C-terminal domains of NtrC and NifA in DNA binding was established subsequently by mutagenesis and *in vivo* footprinting experiments (Contreras and Drummond, 1988; Morett and Buck, 1988). The homology between these two proteins suggested a common mechanism of transcriptional activation, commensurate with previous observations that, when overexpressed, NifA could substitute for NtrC at promoters normally activated by the latter (Drummond *et al.*, 1983; Merrick, 1983; Ow and Ausubel, 1983). However, the amino-terminal domain of NtrC was clearly different to that of NifA, but was homologous to diverse bacterial regulatory proteins, including OmpR from *E. coli* and Spo0A from *Bacillus subtilis* (Drummond *et al.*, 1986). The sequence of *ntrB* initially revealed no homologues (MacFarlane and Merrick, 1985), but the subsequent sequencing of *nifL* revealed a common C-terminal domain present in NtrB and other regulatory proteins, including EnvZ, PhoR, CpxA and CheA (Drummond and Wootton, 1987). These proteins all had a corresponding regulatory partner belonging to the NtrC family. The existence of these regulatory pairs of proteins had also been noted in Fred Ausubel's laboratory. It was proposed that they had evolved from a common ancestral system that transduced environmental signals from the C-terminal domain of one protein (*e.g.*, NtrB) to the N-terminal domain of its partner (*e.g.*, NtrC). These protein pairs were called two-component regulatory systems (Nixon *et al.*, 1986; Ronson *et al.*, 1987). In retrospect, it is fascinating that the discovery of two-component regulation arose from studies on nitrogen control of nitrogen fixation.

5.3. Biochemical Studies on Regulatory Proteins

5.3.1. Nitrogen Regulatory Proteins and the Discovery of σ^{54}

The E. coli NtrC protein was first purified in Boris Magasanik's laboratory in 1983 (Reitzer and Magasanik, 1983) and was subsequently isolated from Salmonella typhimurium and K. pneumoniae (Ames and Nikaido, 1985; Hawkes et al., 1985). In all cases, the protein was shown to be a dimeric DNA-binding protein, which recognised sequences with dyad symmetry and the consensus sequence, 5'-TGCACTA(N)$_3$TGGTGCAA-3' (Ames and Nikaido, 1985; Dixon, 1984; Hawkes et al., 1985). The purified protein repressed transcription from both the ntrBC promoter and the upstream glnA promoter, which are both transcribed by σ^{70}-RNA polymerase holoenzyme (Hawkes et al., 1985; Reitzer and Magasanik, 1983). These results confirmed genetic studies, which indicated that the ntrC product acts as a repressor of transcription at these promoters, and provided an obvious mechanism to explain how NtrC mediates "negative" control, but gave no clues as to how this protein might activate transcription in concert with NtrA.

In late 1985, two exciting breakthroughs enabled this question to be addressed. Using a coupled *in vitro* transcription-translation system as an assay for activity, *S. typhimurium* NtrA was partially purified (Hirschman *et al.*, 1985). The purified NtrA-containing fraction activated transcription from the *glnA* promoter dependent on the addition of NtrC and *E. coli* core RNA polymerase. When σ^{70} was substituted for NtrA, no activation of the *glnA* promoter was detected. Conversely, NtrA could not substitute for σ^{70} at the *lac*UV5 promoter. These properties, and the observation that NtrA co-purified with RNA polymerase during the early stages of purification, strongly suggested that NtrA is an alternative RNA polymerase sigma factor (Hirschman *et al.*, 1985). At about the same time, the *E. coli* NtrA was purified and was demonstrated to bind core RNA polymerase. *In vitro* transcription experiments demonstrated that activation of the *glnA* promoter required NtrC and a mutant form of NtrB (NtrB2302), in addition to NtrA and core RNA polymerase (Hunt and Magasanik, 1985). These results suggested that NtrB is required to activate NtrC and also that the product of *ntrA* is a sigma factor. It was proposed that the name of *ntrA* should be changed to *rpoN* and its product designated as σ^{60} (Hunt and Magasanik, 1985). The latter was subsequently changed to σ^{54}, once the molecular weight had been more accurately derived from sequence data. Thus, an alternative sigma factor, which recognised the unique –24 and –12 consensus sequences present in *rpoN*-dependent promoters, had been discovered. This finding led to major new drives to understand the structure and function of this novel sigma factor (Merrick, 1993).

The purification of NtrB made it possible to investigate its role in modulating the activity of NtrC because previous genetic experiments had indicated that NtrB converts inactive NtrC into a form capable of activating transcription under nitrogen-limiting conditions. This response to nitrogen status involves a complex metabolic cascade mediated by the products of *glnB* (P_{II}) and *glnD* (uridylyl-transferase) (see Merrick, this volume). It was observed that, in a mixture containing NtrB, NtrC and ATP, NtrB catalyses the phosphorylation of NtrC and only the covalently modified form of the activator was competent to activate transcription at the *glnA*p2 promoter. When purified P_{II} protein was added to the mixture of NtrB and phosphorylated NtrC, NtrC became dephosphorylated and inactive as a transcriptional activator. These experiments suggested that NtrB is a protein kinase that phosphorylates NtrC to activate transcription and that the kinase activity of NtrB is antagonised by P_{II} in response to nitrogen status (Ninfa and Magasanik, 1986). This experiment was the first to demonstrate phosphotransfer in a two-component regulatory system. Subsequently, it was shown that the amino-terminal domain of NtrC is phosphorylated by NtrB and that phosphorylated NtrC has an autophosphatase activity (Keener and Kustu, 1988).

Further mechanistic insights into the mechanism of action of the NtrC protein were achieved when it was observed that ATP hydrolysis was required for transcriptional activation. In the absence of NtrC, the σ^{54}-RNA polymerase holoenzyme was shown to form a stable complex at the *glnA* promoter, referred to as the closed complex (Popham *et al.*, 1989). NtrC was, therefore, not required to stabilise the binding of the σ^{54}-holoenzyme to the promoter but appeared to be required at a subsequent stage to catalyse the conversion of the closed complex to the open complex in which the DNA strands surrounding the transcription start site are melted prior to initiation. To simplify the analysis of transcriptional initiation, a mutant form of the NtrC protein, which was competent to activate transcription in the absence of NtrB, was used. This altered NtrC avoided the complication of the ATP requirement for phosphorylation of NtrC. However, when studied in the absence of NtrB, the mutant NtrC protein still required ATP to catalyse formation of open promoter complexes (Popham *et al.*, 1989). Because non-hydrolysable analogues did not substitute for ATP, it was concluded that ATP hydrolysis catalysed by NtrC is necessary for the isomerization of the closed complex between σ^{54}-holoenzyme and the *glnA* promoter to the open promoter complex. Further analysis of the mechanism of long distance activation involved whether the NtrC enhancer sites could function *in vitro* when located in *trans* on a separate plasmid to that carrying the σ^{54}-recognition sequence. Whereas the enhancer sites could function *in cis* when located on the same plasmid as the downstream promoter sequence, they only functioned *in trans* when present on different circles of a singly-linked catenane. This finding suggested that one function of the enhancer sequences is to "tether" the NtrC in the vicinity of the promoter and, therefore, increase the frequency with which it encounters σ^{54}-holoenzyme (Wedel and Kustu, 1991). DNA looped structures formed by interaction between enhancer-bound NtrC and σ^{54}-holoenzyme were observed directly in the electron microscope. The DNA loops were only observed when ATP was also present, implying that stable loops are only formed when polymerase has completed the transition to the open complex (Su *et al.*, 1990).

The rapid developments in understanding the function of the nitrogen regulatory proteins at the *glnA* promoter were soon utilised to probe the mechanism of activation of *nif* transcription by NtrC. *In vitro* experiments with the *K. pneumoniae nifLA* promoter demonstrated that NtrB, NtrC and σ^{54} are required for transcriptional activation and the upstream promoter region contains tandem binding sites for NtrC located at –142 and –163 (Austin *et al.*, 1987; Wong *et al.*, 1987). These sites do not show strong homology with the NtrC DNA-binding site consensus and were shown to have a low affinity for NtrC compared with sites in the *glnA* promoter (Minchin *et al.*, 1988). Consistent with this finding, *nifLA* transcription *in vitro* requires a 10-fold higher activator concentration than transcription from *glnA*p2 (Austin *et al.*, 1987).

Although nitrogen regulation of *nifLA* transcription by NtrC was well defined by these experiments, the apparent transcriptional regulation of this promoter in response to O_2, observed previously with *lac* fusions *in vivo* (Dixon *et al.*, 1980), was not understood. Two groups had found that *nif* gene expression in *K. pneumoniae* was prevented by DNA gyrase inhibitors and, on the basis of this result, proposed that aerobic regulation of *nif* gene expression may in part be regulated through changes in DNA supercoiling because superhelical density decreases upon the shift between anaerobic to aerobic growth conditions (Dimri and Das, 1988; Kranz and Haselkorn, 1986). In contrast to the *glnA*p2 promoter, which appeared to be relatively insensitive to changes in DNA topology, transcription from the *nifLA* promoter was observed to be extremely sensitive to the topological nature of the DNA template (Dixon *et al.*, 1988; Whitehall *et al.*, 1992). The intrinsic curvature of the region between the NtrC-binding sites and the downstream promoter elements is also important for promoter activation (Cheema *et al.*, 1999). Both the downstream –24 –12 regions of the promoter and the promoter upstream elements were shown to be involved in the supercoiling response (Dixon *et al.*, 1988; Whitehall *et al.*, 1992, 1993). However, a direct relationship between the O_2 response of the promoter and the level of negative DNA supercoiling has not so far been proven.

5.3.2. Nitrogen Fixation-Specific Regulatory Proteins

Biochemical experiments on the NifL and NifA proteins were far more difficult than studies on NtrB and NtrC because the proteins were extremely difficult to purify and, when over-expressed, appeared in inclusion bodies. In the absence of purified proteins, early studies on NifA-mediated transcription relied upon *in vivo* footprinting techniques. Dimethyl sulphate protection experiments indicated that σ^{54}-holoenzyme bound to the -24 –12 region of the promoter and sequences around the transcription site showed chemical reactivity to potassium permanganate when the promoter was activated by NifA *in vivo*. These experiments suggested that, like NtrC, NifA catalysed the isomerization of closed complexes formed with σ^{54}-RNA polymerase holoenzyme to open promoter complexes (Morett and Buck, 1989). Although *K. pneumoniae* NifL and NifA overexpressed *in vivo* were very insoluble (Austin *et al.*, 1990; Tuli and Merrick, 1988), NifA activity was detectable when the protein was expressed *in vitro* using a coupled transcription-translation system. The *in vitro* synthesised protein activated expression of a *nifH-lac* fusion and activity was stimulated by the addition of purified σ^{54} to the extracts (Santero *et al.*, 1989). Both NifL and NifA activity were detected in S30 extracts derived from strains overproducing these proteins. The *in vitro* activity of NifA was dramatically decreased when wild-type NifL and NifA were co-expressed, but not when a mutant form of NifL (NifL2404), which is defective in the inhibition of NifA *in vivo*, was present (Austin *et al.*, 1990). In the attempts to demonstrate DNA binding by NifA *in vitro*, a factor was detected in crude extracts which footprinted the *nifH* promoter in the region between the UAS and the –24 –12 promoter recognition element

(Santero *et al.*, 1989). The DNAse I footprint was found to be identical to that observed previously (Beynon *et al.*, 1983) and this factor was identified as Integration Host Factor (IHF), which was known to bind DNA. IHF was shown to bind to *nif* promoters from many representatives of the proteobacteria and bending of the DNA in the *nifH* promoter region was demonstrated by electron microscopy. IHF stimulated the ability of NifA to activate transcription in the coupled *in vitro* transcription-translation system and it was proposed that it facilitated the interaction between NifA bound at the UAS with downstream bound σ^{54}-holoenzyme (Hoover *et al.*, 1990).

The inability to purify the *K. pneumoniae* NifA protein for detailed biochemical analysis hampered further analysis of its role in DNA binding and transcriptional activation. In the early 1990's, studies of the role of both the NifA- and IHF-binding sites in the various *nif* promoters were carried out primarily by detecting occupancy of the binding sites by *in vivo* footprinting (Cannon *et al.*, 1990, 1991; Charlton *et al.*, 1993; Molina-Lopez *et al.*, 1994). However, fusion of the maltose binding protein (MBP) to NifA did enable its purification, albeit in a highly aggregated form. Like NtrC, the MBP-NifA fusion protein catalysed the formation of open complexes in a nucleotide-dependent reaction, but the NifA fusion protein had broader nucleotide specificity than NtrC (Lee *et al.*, 1993a). The MBP-NifA fusion also bound to the *nifH* UAS as did the isolated C-terminal domain of NifA. When NifA was released from the MBP by protease cleavage, it was insoluble but the isolated central catalytic domain of NifA could be cleaved from an MBP fusion in an active form and was shown to exhibit an ATPase activity, congruent with its role in the catalysis of open complex formation (Berger *et al.*, 1994). Partially purified NifA was observed to bind stably to GroEL *in vitro* and was released by the action of GroES and ATP providing a potential route for improving the refolding of the protein (Govezensky *et al.*, 1991, 1994). Circular dichroism spectra of the isolated C-terminal domain of NifA suggest that it is primarily α-helical in solution (Missaillidis *et al.*, 1999). NMR studies indicate that this domain contains three helices, two of which fold into a classical helix-turn-helix motif which makes direct contact with UAS DNA (Ray *et al.*, 2002).

The poor solubility of the *K. pneumoniae* NifL protein also hampered analysis of its role in regulating the activity of NifA. Immunoprecipitation experiments suggested that NifL and NifA form a complex *in vivo* (Henderson *et al.*, 1989) congruent with the finding that expression of the two proteins is translationally coupled (Govantes *et al.*, 1996, 1998). *In vivo* footprinting experiments suggested that NifL prevented NifA from interacting with the *nifH* UAS (Morett *et al.*, 1990). NifL isolated from inclusion bodies by denaturation and subsequent refolding was competent to inhibit transcriptional activation either by an MBP-NifA fusion or by the isolated catalytic domain of NifA. The renatured NifL did not inhibit transcription by NtrC, implying that the inhibition was specific to the interaction with NifA. However, NifL did not inhibit the ATPase activity of the central domain of NifA, implying that NifL may interfere with the interaction between NifA and the σ^{54}-RNA polymerase holoenzyme (Berger *et al.*, 1994; Lee *et al.*, 1993b). The isolated C-terminal domain of NifL, which was more soluble than the intact protein, was competent to inhibit NifA activity (Narberhaus *et al.*, 1995).

The intractable nature of the *K. pneumoniae* NifL and NifA proteins contrasts strongly with their *Azotobacter vinelandii* homologues, which can be purified in a soluble form (Austin *et al.*, 1994), facilitating biochemical experiments on the mechanism by which NifL senses both O_2 and fixed nitrogen. These experiments have revealed that *A. vinelandii* NifL is a flavoprotein with an amino-terminal PAS domain required for redox sensing (Hill *et al.*, 1996) and a C-terminal nucleotide-binding domain, which interacts directly with the GlnK protein in response to the nitrogen source (Little *et al.*, 2000, 2002; Söderbäck *et al.*, 1998; Rudnick *et al.*, 2002) (see Kennedy, this volume). The *K. pneumoniae* NifL has also been shown to bind FAD (Klopprogge and Schmitz, 1999; Schmitz, 1997) but the properties of the C-terminal domain appear to be different from those of *A. vinelandii* NifL and the mechanism whereby the *K. pneumoniae* NifL-NifA system responds to fixed nitrogen via GlnK is also different, although direct biochemical studies on the interaction have not been feasible (see Merrick, this volume). Nevertheless, studies on *K. pneumoniae nifL* and *nifA* are continuing to yield very interesting results, for example, on the role of the Fnr protein in NifL-dependent O_2 control (Grabbe *et al.*, 2001) and the observation that NifL is primarily associated with the membrane under nitrogen-fixing conditions (Klopprogge *et al.*, 2002).

6. CODA

Looking back over the last thirty years, it is fascinating to see how the field of *nif* genetics has developed and how rapid advances made during this period have impacted upon other scientific areas. Fundamental research on the genetics of nitrogen fixation has resulted in many "spin-offs" and surprising discoveries. Developments in this area undoubtedly facilitated biochemical studies on nitrogenase because understanding the genetic requirements for nitrogen fixation provided valuable insights into the mechanism of co-factor biosynthesis and enzyme maturation, electron transport to nitrogenase, and more recently, enabled the use of site-directed mutagenesis to probe substrate binding and catalysis. Our current knowledge of iron-sulphur cluster biosynthesis in bacteria, animals and plants, stems primarily from the identification of the *nifU* and *nifS* genes and biochemical characterisation of their gene products (Frazzon *et al.*, 2002).

Research on *nif* genetics throughout the last thirty years has also been closely allied to studies on nitrogen assimilation and nitrogen regulation. This alliance is to be expected of course because fixed nitrogen is the product of nitrogen fixation and it was recognised early on that the synthesis of nitrogenase was likely to be regulated by the supply of fixed nitrogen. Historically, there have been many interactions between laboratories interested in the regulation of nitrogen fixation and nitrogen metabolism and complementary research in these two areas has led to rapid developments. In particular, the laboratories of Boris Magasanik, Sydney Kustu and more recently Alex Ninfa, have had a major influence on our understanding of *nif* gene regulation as a consequence of their work on nitrogen regulation. Conversely, research on *nif* regulation has had a pervasive impact on bacterial gene regulation in general. For example, the sequencing of *nif* promoters undoubtedly contributed to the discovery of σ^{54} and transcriptional activation at a distance in bacteria. These findings established the role of UAS sequences as bacterial enhancers and the concept of σ^{54}-dependent transcriptional activators as Enhancer Binding Proteins. As mentioned in Section 5.2., research on *nif* regulation resulted in the discovery of the pervasive two-component regulatory systems, opening a new era in microbial and plant signal transduction.

In contrast to the predictions made in the 1970's, energy costs have not escalated to the extent anticipated and industrial fertilisers continue to make a substantial contribution to world agriculture. It is, therefore, not surprising that there has been relatively little interest recently in biotechnological exploitation of nitrogen fixation. The long-awaited "holy grail" of nitrogen fixation - to engineer *nif* genes and construct nitrogen-fixing plants - has not yet been realised, although preliminary studies indicate that the Fe protein of nitrogenase can be expressed in an active form in eukaryotic cells (Dixon *et al.*, 2000). However, as we enter the 21st century and recognise there is a need to improve agricultural sustainability and food security, research on *nif* genes and their regulation will have an important part to play in the realisation of these goals.

REFERENCES

Allen, R. M., Chatterjee, R., Ludden, P. W., and Shah, V. K. (1995). Incorporation of iron and sulfur from NifB cofactor into the iron-molybdenum cofactor of dinitrogenase. *J. Biol. Chem., 270*, 26890-26896.

Allen, R. M., Chatterjee, R., Madden, M. S., Ludden, P. W., and Shah, V. K. (1994). Biosynthesis of the iron-molybdenum cofactor of nitrogenase. *Crit. Rev. Biotechnol., 14*, 225-249.

Ames, G. F. L., and Nikaido, K. (1985). Nitrogen regulation in *Salmonella typhimurium*. Identification of an *ntrC* protein-binding site and definition of a consensus binding sequence. *EMBO J., 4*, 539-547.

Arnold, W., Rump, A., Klipp, W., Priefer, U. B., and Pühler, A. (1988). Nucleotide sequence of a 24,206 base-pair DNA fragment carrying the entire nitrogen fixation gene cluster of *Klebsiella pneumoniae*. *J. Mol. Biol., 203*, 715-738.

Austin, S., Buck, M., Cannon, W., Eydmann, T., and Dixon, R. (1994). Purification and *in vitro* activities of the native nitrogen fixation control proteins NIFA and NIFL. *J. Bacteriol., 176*, 3460-3465.

Austin, S., Henderson, N., and Dixon, R. (1987). Requirements for transcriptional activation *in vitro* of the nitrogen regulated *glnA* and *nifLA* promoters from *Klebsiella pneumoniae*: dependence on activator concentration. *Mol. Microbiol., 1*, 92-100.

Austin, S., Henderson, N., and Dixon, R. (1990). Characterisation of the *Klebsiella pneumoniae* nitrogen-fixation regulatory proteins NIFA and NIFL *in vitro*. *Eur. J. Biochem., 187*, 353-360.

Ausubel, F. (1977). Cloning nitrogen fixing genes from *Klebsiella pnumoniae in vitro* and the isolation of *nif* promoter mutants affecting glutamine sythetase regulation. In R. H. Burris, P. R. Day, R. W. F. Hardy, D. R. Helinski, M. Lamborg, L. Owens, *et al.* (Eds.), *Genetic engineering for nitrogen fixation* (Vol. 9, pp. 111-128). New York and London: Plenum Press.

Bachhuber, M., Brill, W. J., and Howe, M. M. (1976). Use of bacteriophage Mu to isolate deletions in the *his-nif* region of *Klebsiella pneumoniae*. *J. Bacteriol., 128*, 749-753.

Berger, D. K., Narberhaus, F., and Kustu, S. (1994). The isolated catalytic domain of NIFA, a bacterial enhancer-binding protein, activates transcription *in vitro*: activation is inhibited by NIFL. *Proc. Natl. Acad. Sci. U.S.A., 91*, 103-107.

Berman, J., Gershoni, J. M., and Zamir, A. (1985). Expression of nitrogen fixation genes in foreign hosts. assembly of nitrogenase Fe protein in *Escherichia coli* and in yeast. *J. Biol. Chem., 260*, 5240-5243.

Beynon, J., Ally, A., Cannon, M., Cannon, F. C., Jacobson, M., Cash, V., *et al.* (1987). Comparative organization of nitrogen fixation-specific genes from *Azotobacter vinelandii* and *Klebsiella pneumoniae*: DNA sequence of the *nifUSV* genes. *J. Bacteriol., 169*, 4024-4029.

Beynon, J., Cannon, M., Buchanan-Wollaston, V., and Cannon, F. (1983). The *nif* promoters of *Klebsiella pneumoniae* have a characteristic primary structure. *Cell, 34*, 665-671.

Brown, S. E., and Ausubel, F. M. (1984). Mutations affecting regulation of the *Klebsiella pneumoniae nifH* (nitrogenase reductase) promoter. *J. Bacteriol., 157*, 143-147.

Buchanan-Wollaston, V., Cannon, M. C., Beynon, J. L., and Cannon, F. C. (1981a). Role of the *nifA* gene product in the regulation of *nif* expression in *Klebsiella pneumoniae. Nature, 294*, 776-778.

Buchanan-Wollaston, V., Cannon, M. C., and Cannon, F. C. (1981b). The use of cloned *nif* (nitrogen fixation) DNA to investigate transcription regulation of *nif* expression in *Klebsiella pneumoniae. Mol. Gen. Genet., 184*, 102-106.

Buck, M. (1986). Deletion analysis of the *Klebseilla pneumoniae* nitrogenase promoter: importance of spacing between conserved sequences around positions -12 and -24 for activation by the *nifA* and *ntrC* (*glnG*) products. *J. Bacteriol., 166*, 545-551.

Buck, M., and Cannon, W. (1989). Mutations in the RNA polymerase recognition sequence of the *Klebsiella pneumoniae nifH* promoter permitting transcriptional activation in the absence of NifA binding to upstream sequences. *Nucleic Acids Res., 17*, 2597-2611.

Buck, M., Cannon, W., and Woodcock, J. (1987). Transcriptional activation of the *Klebsiella pneumoniae* nitrogenase promoter may involve DNA loop formation. *Mol. Microbiol., 1*, 243-249.

Buck, M., Khan, H., and Dixon, R. (1985). Site-directed mutagenesis of the *Klebsiella pneumoniae nifL* and *nifH* promoters and in vivo analysis of promoter activity. *Nucleic Acids Res., 13*, 7621-7638.

Buck, M., Miller, S., Drummond, M., and Dixon, R. (1986). Upstream activator sequences are present in the promoters of nitrogen fixation genes. *Nature, 320*, 374-378.

Buikema, W. J., Szeto, W. W., Lemley, P. V., Orme-Johnson, W. H., and Ausubel, F. M. (1985). Nitrogen fixation specific regulatory genes of *Klebsiella pneumoniae* and *Rhizobium meliloti* share homology with the general nitrogen regulatory gene *ntrC* of *K.pneumoniae. Nucleic Acids Res., 13*, 4539-4555.

Cannon, F. C., Dixon, R. A., Postgate, J. R., and Primrose, S. B. (1974a). Chromosomal integration of Klebsiella nitrogen fixation genes in *Escherichia coli. J. Gen. Microbiol., 80*, 227-239.

Cannon, F. C., Dixon, R. A., Postgate, J. R., and Primrose, S. B. (1974b). Plasmids formed in nitrogen-fixing *Escherichia coli-Klebsiella pneumoniae* hybrids. *J. Gen. Microbiol., 80*, 241-251.

Cannon, F. C., Reidel, G. E., and Ausubel, F. (1977). Recombinant plasmid that carries part of the nitrogen fixation (*nif*) gene cluster of *Klebsiella pneumoniae. Proc. Natl. Acad. Sci. U.S.A., 74*, 2963-2967.

Cannon, F. C., Reidel, G. E., and Ausubel, F. (1979). Overlapping sequences of *Klebsiella pneumoniae nif* DNA cloned and characterised. *Mol. Gen. Genet., 174*, 59-66.

Cannon, M., Hill, S., Kavanaugh, E., and Cannon, F. (1985). A molecular genetic study of *nif* expression in *Klebsiella pneumoniae* at the level of transcription, translation and nitrogenase activity. *Mol. Gen. Genet., 198*, 198.

Cannon, W., Charlton, W., and Buck, M. (1991). Organisation and function of binding sites for the transcriptional activator NifA in the *Klebsiella pneumoniae nifE* and *nifU* promoters. *J. Mol. Biol., 220*, 915-931.

Cannon, W. V., Kreutzer, R., Kent, H. M., Morett, E., and Buck, M. (1990). Activation of the *Klebsiella pneumoniae nifU* promoter: Multiple and overlapping upstream NifA binding sites and involvement of integration host factor. *Nucleic Acids Res., 18*, 1693-1701.

Casadaban, M. J., and Cohen, S. N. (1979). Lactose genes fused to exogenous promoters in one step using a Mu-lac bacteriophage: in vivo probe for transcriptional control sequences. *Proc. Natl. Acad. Sci. U.S.A., 76*, 4530-4533.

Charlton, W., Cannon, W., and Buck, M. (1993). The *Klebsiella pneumoniae nifJ* promoter: analysis of promoter elements regulating activation by the NifA promoter. *Mol. Microbiol., 7*, 1007-1021.

Cheema, A. K., Choudhury, N. R., and Das, H. K. (1999). A- and T-tract-mediated intrinsic curvature in native DNA between the binding site of the upstream activator NtrC and the *nifLA* promoter of *Klebsiella pneumoniae* facilitates transcription. *J. Bacteriol., 181*, 5296-5302.

Chen, Y.-M., Backman, K., and Magasanik, B. (1982). Characterisation of a gene, *glnL*, the product of which is involved in the regulation of nitrogen utilization in *Escherichia coli. J. Bacteriol., 150*, 214-220.

Collins, J. J., and Brill, W. J. (1985). Control of *Klebsiella pneumoniae nif* mRNA synthesis. *J. Bacteriol., 162*, 1186-1190.

Collins, J. J., Roberts, G. P., and Brill, W. J. (1986). Posttranscriptional control of *Klebsiella pneumoniae nif* mRNA stability by the *nifL* product. *J. Bacteriol., 168*, 173-178.

Contreras, A., and Drummond, M. (1988). The effect on the function of the transcriptional activator NtrC from *Klebsiella pneumoniae* of mutations in the DNA recognition helix. *Nucleic Acids Res., 16*, 4025-4039.

de Bruijn, F. J., and Ausubel, F. (1983). The cloning and characterization of the *glnF* (*ntrA*) gene of *Klebsiella pneumoniae*: role of *glnF* (*ntrA*) in the regulation of nitrogen fixation (*nif*) and other nitrogen assimilation genes. *Mol. Gen. Genet., 192*, 342-353.

de Bruijn, F. J., and Ausubel, F. M. (1981). The cloning and transposon Tn5 mutagenesis of the *glnA* region of *Klebsiella pneumoniae*: identification of *glnR*, a gene involved in the regulation of the *nif* and *hut* operons. *Mol. Gen. Genet., 183*, 289-297.

Dean, D. R., and Jacobson, M. R. (1992). Biochemical Genetics of Nitrogenase. In G. Stacey, R. H. Burris and H. J. Evans (Eds.), *Biological Nitrogen Fixation* (pp. 763-834). New York: Chapman and Hall.

Deistung, J., Cannon, F. C., Cannon, M. C., Hill, S., and Thorneley, R. N. (1985). Electron transfer to nitrogenase in *Klebsiella pneumoniae. nifF* gene cloned and the gene product, a flavodoxin, purified. *Biochem. J., 231*, 743-753.

Dimri, G. P., and Das, H. K. (1988). Transcriptional regulation of nitrogen fixation genes by DNA supercoiling. *Mol. Gen. Genet., 212*, 360-363.

Dixon, R. (1984). Tandem promoters determine regulation of the *Klebsiella pneumoniae* glutamine synthetase (*glnA*) gene. *Nucleic Acids Res., 12*, 7811-7830.

Dixon, R. (1998). The oxygen-responsive NIFL-NIFA complex: a novel two-component regulatory system contolling nitrogenase synthesis in γ-Proteobacteria. *Arch. Microbiol., 169*, 371-380.

Dixon, R., Cannon, F., and Kondorosi, A. (1976). Construction of a P plasmid carrying nitrogen fixation genes from *Klebsiella pneumoniae. Nature, 260*, 268-271.

Dixon, R., Cannon, F. C., and Postgate, J. R. (1975). Properties of the R factor R144drd3 in *Klebsiella pneumoniae* strain M5al. *Genet. Res., 25*, 327-338.

Dixon, R., Cheng, Q., Dowson-Day, M., and Day, A. (2000). Prospects for engineering nitrogen-fixing photosynthetic eukaryotes. In F. O. Pedrosa, M. Hungria, M. G. Yates and W. E. Newton (Eds.), *Nitrogen fixation:From molecules to crop productivity* (pp. 635-639). Dordrecht: Kluwer Academic Publishers.

Dixon, R., Eady, R. R., Espin, G., Hill, S., Iaccarino, M., Kahn, D., *et al.* (1980). Analysis of regulation of *Klebsiella pneumoniae* nitrogen fixation (*nif*) gene cluster with gene fusions. *Nature, 286*, 128-132.

Dixon, R., Henderson, N. C., and Austin, S. (1988). DNA supercoiling and aerobic regulation of transcription from the *Klebsiella pneumoniae nifLA* promoter. *Nucleic Acids Res., 16*, 9933-9946.

Dixon, R., Kennedy, C., Kondorosi, A., Krishnapillai, V., and Merrick, M. (1977). Complementation analysis of *Klebsiella pneumoniae* mutants defective in nitrogen fixation. *Mol. Gen. Genet., 157*, 189-198.

Dixon, R., and Postgate, J. R. (1972). Genetic transfer of nitrogen fixation from *Klebsiella pneumoniae* to *Escherichia coli. Nature, 237*, 102-103.

Dixon, R. A., and Postgate, J. R. (1971). Transfer of nitrogen-fixation genes by conjugation in *Klebsiella pneumoniae. Nature, 234*, 47-48.

Drummond, M., Clements, J., Merrick, M., and Dixon, R. (1983). Positive control and autogenous regulation of the *nifLA* promoter in *Klebsiella pneumoniae. Nature, 301*, 302-307.

Drummond, M., Whitty, P., and Wootton, J. (1986). Sequence and domain relationships of *ntr* and *nifA* from *Klebsiella pneumoniae*: homologies to other regulatory proteins. *EMBO J., 5*, 441-447.

Drummond, M. H., and Wootton, J. C. (1987). Sequence of *nifL* from *Klebsiella pneumoniae*: mode of action and relationship to two families of regulatory proteins. *Mol. Microbiol., 1*, 37-44.

Dunican, L. K., and Tierney, A. B. (1974). Genetic transfer of nitrogen fixation from *Rhizobium trifolii* to *Klebsiella aerogenes. Biochem. Biophys. Res. Commun., 57*, 62-72.

Eady, R. R., Issack, R., Kennedy, C., Postgate, J. R., and Ratcliffe, H. D. (1978). Nitrogenase synthesis in *Klebsiella pneumoniae*: Comparison of ammonium and oxygen regulation. *J. Gen. Microbiol., 104*, 277-285.

Elmerich, C., Houmard, J., Sibold, L., Manheimer, I., and Charpin, N. (1978). Genetic and biochemical analysis of mutants induced by bacteriophage Mu DNA integration into *Klebsiella pneumoniae* nitrogen fixation genes. *Mol. Gen. Genet., 165*, 181-189.

Espin, G., Alvarez-Morales, A., Cannon, F., Dixon, R., and Merrick, M. J. (1982). Cloning of the *glnA*, *ntrB* and *ntrC* genes of *Klebsiella pneumoniae* and studies of their role in regulation of the nitrogen fixation (*nif*) gene cluster. *Mol. Gen. Genet., 186*, 518-524.

Espin, G., Alvarez-Morales, A., and Merrick, M. (1981). Complementation analysis of *glnA*-linked mutations which affect nitrogen fixation in *Klebsiella pneumoniae. Mol. Gen. Genet., 184*, 213-217.

Frazzon, J., Fick, J. R., and Dean, D. R. (2002). Biosynthesis of iron-sulphur clusters is a complex and highly conserved process. *Biochem. Soc. Trans., 30*, 680-685.

Garcia, E., Bancroft, S., Rhee, S. G., and Kustu, S. (1977). The product of a newly identified gene, *glnF*, is required for synthesis of glutamine synthetase in Salmonella. *Proc. Natl. Acad. Sci. U.S.A., 74*, 1662-1666.

Gavini, N., and Pulukat, L. (2002). Role of NifM in maturation of the Fe protein of nitrogenase. In T. Finan, M. O'Brian, D. Layzell, K. Vessey and B. Newton (Eds.), *Nitrogen fixation: Global perspectives* (pp. 228-232). Wallingford, Oxford: CABI International.

Govantes, F., Andujar, E., and Santero, E. (1998) Mechanism of translational coupling in the *nifLA* operon of *Klebsiella pneumoniae. EMBO J., 17*, 2368-2377.

Govantes, F., Molina-Lopez, J. A., and Santero, E. (1996) Mechanism of coordinated synthesis of the antagonistic regulatory proteins NifL and NifA of *Klebsiella pneumoniae. J. Bacteriol., 178*, 6817-6823.

Govezensky, D., Bochkareva, E. S., Zamir, A., and Girshovich, A. S. (1994) Chaperonins as potential gene regulatory factors. *In vitro* interaction and solubilization of NifA, the *nif* transcriptional activator, with GroEL. *J. Biol. Chem. 269*, 14003-14006.

Govezensky, D., Greener, T., Segal, G., and Zamir, A. (1991) Involvement of GroEL in *nif* gene regulation and nitrogenase assembly. *J. Bacteriol. 173*, 6339-6346.

Grabbe, R., Klopprogge, K., and Schmitz, R. A. (2001). Fnr Is required for NifL-dependent oxygen control of *nif* gene expression in *Klebsiella pneumoniae. J. Bacteriol., 183*, 1385-1393.

Hawkes, T., McLean, P. A., and Smith, B. E. (1984). Nitrogenase from *nifV* mutants of *Klebsiella pneumoniae* contains an altered form of the iron-molybdenum cofactor. *Biochem. J., 217*, 317-321.

Hawkes, T., Merrick, M., and Dixon, R. (1985). Interaction of purified NtrC protein with nitrogen regulated promoters from *Klebsiella pneumoniae. Mol. Gen. Genet., 201*, 492-498.

Henderson, N., Austin, S. A., and Dixon, R. A. (1989). Role of metal ions in negative regulation of nitrogen fixation by the *nifL* gene product from *Klebsiella pneumoniae. Mol. Gen. Genet., 216*, 484-491.

Hill, S., Austin, S., Eydmann, T., Jones, T., and Dixon, R. (1996). *Azotobacter vinelandii* NIFL is a flavoprotein that modulates transcriptional activation of nitrogen-fixation genes via a redox-sensitive switch. *Proc. Natl. Acad. Sci. U.S.A., 93*, 2143-2148.

Hill, S., and Kavanagh, E. P. (1980). Roles of *nifF* and *nifJ* gene products in electron transport to nitrogenase in *Klebsiella pneumoniae. J. Bacteriol., 141*, 470-475.

Hill, S., Kennedy, C., Kavanagh, E., Goldberg, R., and Hanau, R. (1981). Nitrogen fixation gene (*nifL*) involved in oxygen regulation of nitrogenase synthesis in *K. pneumoniae. Nature, 290*, 424-426.

Hirschman, J., Wong, P. K., Sei, K., Keener, J., and Kustu, S. (1985). Products of nitrogen regulatory genes *ntrA* and *ntrC* of enteric bacteria activate *glnA* transcription *in vitro*: evidence that the *ntrA* product is a sigma factor. *Proc. Natl. Acad. Sci. U.S.A., 82*, 7525-7529.

Holland, D., Zilberstein, A., Govezensky, D., Salomon, D., and Zamir, A. (1987). Nitrogenase MoFe protein subunits from *Klebsiella pneumoniae* expressed in foreign hosts. *J. Biol. Chem., 262*, 8814-8820.

Homer, M. J., Paustian, T. D., Shah, V. K., and Roberts, G. P. (1993). The *nifY* product of *Klebsiella pneumoniae* is associated with apodinitrogenase and dissociates upon activation with the iron-molybdenum cofactor. *J. Bacteriol., 175*, 4907-4910.

Hoover, T. R., Robertson, A. D., Cerny, R. L., Hayes, R. N., Imperial, J., Shah, V. K., *et al.* (1987). Identification of the V factor needed for synthesis of the iron-molybdenum cofactor of nitrogenase as homocitrate. *Nature, 329*, 855-857.

Hoover, T. R., Santero, E., Porter, S., and Kustu, S. (1990). The integration host factor stimulates interaction of RNA polymerase with NIFA, the transcriptional activator for nitrogen fixation operons. *Cell, 63*, 11-22.

Hsueh, C. T., Chin, J. C., Yu, Y. Y., Chen, H. C., Li, W. C., Shen, M. C., *et al.* (1977). Genetic analysis of the nitrogen fixation system in *Klebsiella pneumoniae*. *Sci. Sin., 20*, 807-817.

Hunt, T. P., and Magasanik, B. (1985). Transcription of *glnA* by purified *Escherichia coli* components: core RNA polymerase and the products of *glnF*, *glnG*, and *glnL*. *Proc. Natl. Acad. Sci. U.S.A., 82*, 8453-8457.

Imperial, J., Ugalde, R. A., Shah, V. K., and Brill, W. J. (1984). Role of the *nifQ* gene product in the incorporation of molybdenum into nitrogenase in *Klebsiella pneumoniae*. *J. Bacteriol., 158*, 187-194.

Janssen, K., GE, R., Ausubel, F., and Cannon, F. (1980). Transcriptional studies with cloned nitrogen fixation genes. In W. E. Newton and W. H. Orme-Johnson (Eds.), *Nitrogen fixation* (Vol. 1, pp. 85-93). Baltimore: University Park Press.

Keener, J., and Kustu, S. (1988). Protein kinase and phosphoprotein phosphatase activities of nitrogen regulatory proteins NTRB and NTRC of enteric bacteria: Roles of the conserved amino-terminal domain of NTRC. *Proc. Natl. Acad. Sci. U.S.A., 85*, 4976-4980.

Kennedy, C. (1977). Linkage map of the nitrogen fixation (*nif*) genes in *Klebsiella pneumoniae*. *Mol. Gen. Genet., 157*, 199-204.

Khan, H., Buck, M., and Dixon, R. (1986). Deletion loop mutagenesis of the *nifL* promoter from *Klebsiella pneumoniae:* role of the -26 to -12 region in the promoter function. *Gene, 45*, 281-288.

Kim, S., and Burgess, B. K. (1996). Evidence for the direct interaction of the *nifW* gene product with the MoFe protein. *J. Biol. Chem., 271*, 9764-9770.

Klopprogge, K., Grabbe, R., Hoppert, M., and Schmitz, R. A. (2002). Membrane association of *Klebsiella pneumoniae* NifL is affected by molecular oxygen and combined nitrogen. *Arch. Microbiol., 177*, 223-234.

Klopprogge, K., and Schmitz, R. A. (1999). NifL of *Klebsiella pneumoniae*: redox characterization in relation to the nitrogen source. *Biochim. Biophys. Acta., 1431*, 462-470.

Kranz, R. G., and Haselkorn, R. (1986). Anaerobic regulation of nitrogen-fixation genes in *Rhodopseudomonas capsulata*. *Proc. Natl. Acad. Sci. U.S.A., 83*, 6805-6809.

Lee, H.-S., Berger, D. K., and Kustu, S. (1993a). Activity of purified NIFA, a transcriptional activator of nitrogen fixation genes. *Proc. Natl. Acad. Sci. U.S.A., 90*, 2266-2270.

Lee, H.-S., Narberhaus, F., and Kustu, S. (1993b). In vitro activity of NifL, a signal transduction protein for biological nitrogen fixation. *J. Bacteriol., 175*, 7683-7688.

Little, R., Colombo, V., Leech, A., and Dixon, R. (2002) Direct interaction of the NifL regulatory protein with the GlnK signal transducer enables the *Azotobacter vinelandii* NifL-NifA regulatory system to respond to conditions replete for nitrogen. *J. Biol. Chem. 277*, 15472-15481.

Little, R., Reyes-Ramirez, F., Zhang, Y., van Heeswijk, W. C., and Dixon, R. (2000) Signal transduction to the *Azotobacter vinelandii* NIFL-NIFA regulatory system is influenced directly by interaction with 2-oxoglutarate and the PII regulatory protein. *EMBO J., 19*, 6041-6050.

MacFarlane, S. A., and Merrick, M. (1985). The nucleotide sequence of the nitrogen regulation gene *ntrB* and the *glnA-ntrBC* intergenic region of *Klebsiella pnuemoniae*. *Nucleic Acids Res., 13*, 7591-7606.

Macneil, D., and Brill, W. J. (1978). 6-cyanopurine, a color indicator useful for isolating mutations in the *nif* (nitrogen fixation) genes of *Klebsiella pneumoniae*. *J. Bacteriol., 136*, 247-252.

Macneil, T., Brill, W. J., and Mowe, M. M. (1978a). Bacteriophage Mu-induced deletions in a plasmid containing the *nif* (N_2 fixation) genes of *Klebsiella pneumoniae*. *J. Bacteriol., 134*, 821-829.

Macneil, T., Macneil, D., Roberts, G. P., Supiano, M. A., and Brill, W. J. (1978b). Fine-structure mapping and complementation analysis of *nif* (nitrogen fixation) genes in *Klebsiella pneumoniae*. *J. Bacteriol., 136*, 253-266.

Macneil, T., Macneil, D., and Tyler, B. (1982a). Fine-structure deletion map and complementation analysis of the *glnA-glnL-glnG* region in *Escherichia coli*. *J. Bacteriol., 150*, 1302-1313.

Macneil, T., Roberts, G. P., Macneil, D., and Tyler, B. (1982b). The products of *glnL* and *glnG* are bifunctional regulatory proteins. *Mol. Gen. Genet., 188*, 325-333.

Macneil, D., Zhu, J., and Brill, W. J. (1981). Regulation of nitrogen fixation in *Klebsiella pneumoniae*: isolation and characterization of strains with *nif-lac* fusions. *J. Bacteriol., 145*, 348-357.

Magasanik, B., Prival, M. J., Brenchley, J. E., Tyler, B. M., DeLeo, A. B., Streicher, S. L., *et al.* (1974). Glutamine synthetase as a regulator of enzyme synthesis. In B. L. Horrecker and E. R. Stadtman (Eds.), *Current topics in cellular regulation* (Vol. 8, pp. 119-138). New York: Academic Press.

Maxam, A. M., and Gilbert, W. (1977). A new method for sequencing DNA. *Proc. Natl. Acad. Sci. U.S.A., 74*, 560-564.

McFarland, N., McCarter, L., Artz, S., and Kustu, S. (1981). Nitrogen regulatory locus "*glnR*" of enteric bacteria is composed of cistrons *ntrB* and *ntrC*: identification of their protein products. *Proc. Natl. Acad. Sci. U.S.A., 78*, 2135-2139.

McLean, P. A., and Dixon, R. A. (1981). Requirement of *nifV* gene for production of wild-type nitrogenase enzyme in *Klebsiella pneumoniae. Nature, 292*, 655-656.

Merrick, M., Filser, M., Dixon, R., Elmerich, C., Sibold, L., and Houmard, J. (1980). The use of translocatable genetic elements to construct a fine-structure map of the *Klebsiella pneumoniae* nitrogen fixation (*nif*) gene cluster. *J. Gen. Microbiol., 117*, 509-520.

Merrick, M., Filser, M., Kennedy, C., and Dixon, R. (1978). Polarity of mutations induced by insertion of transposons Tn5, Tn7, and Tn10 into the *nif* gene cluster of *Klebsiella pneumoniae. Mol. Gen. Genet., 165*, 103-111.

Merrick, M., Hill, S., Hennecke, H., Hahn, M., Dixon, R., and Kennedy, C. (1982). Repressor properties of the *nifL* gene product of *Klebsiella pneumoniae. Mol. Gen. Genet., 185*, 75-81.

Merrick, M. J. (1983). Nitrogen control of the *nif* regulon in *Klebsiella pneumoniae*: Involvement of the *ntrA* gene and analogies between *ntrC* and *nifA*. *EMBO J., 2*, 39-44.

Merrick, M. J. (1993). In a class of its own - the RNA polymerase sigma factor σ^{54} (σ^{N}). *Mol. Microbiol., 10*, 903-909.

Merrick, M. J., and Gibbins, J. R. (1985). The nucleotide sequence of the nitrogen-regulation gene *ntrA* of *Klebsiella pneumoniae* and comparison with conserved features in bacterial RNA polymerase sigma factors. *Nucleic Acids Res., 13*, 7607-7620.

Merrick, M. J., and Stewart, W. D. P. (1985). Studies on the regulation and function of the *Klebsiella pneumoniae ntrA* gene. *Gene, 35*, 297-303.

Mevarech, M., Rice, D., and Haselkorn, R. (1980). Nucleotide sequence of a cyanobacterial *nifH* gene coding for nitrogenase reductase. *Proc. Natl. Acad. Sci. U.S.A., 77*, 6476-6480.

Minchin, S. D., Austin, S., and Dixon, R. A. (1988). The role of activator binding sites in transcriptional control of the divergently transcribed *nifF* and *nifLA* promoters from *Klebsiella pneumoniae. Mol. Microbiol., 2*, 433-442.

Minchin, S. D., Austin, S., and Dixon, R. A. (1989). Transcriptional activation of the *Klebsiella pneumoniae nifLA* promoter by NTRC is face-of-the-helix dependent and the activator stabilizes the interaction of sigma 54-RNA polymerase with the promoter. *EMBO J., 8*, 3491-3499.

Missaillidis, S., Jaseja, M., Ray, P., Chittock, R., Wharton, C. W., Drake, A. F., *et al.* (1999). Secondary structure of the C-terminal DNA-binding domain of the transcriptional activator NifA from *Klebsiella pneumoniae*: Spectroscopic analyses. *Arch. Biochem. Biophys., 361*, 173-182.

Molina-Lopez, J. A., Govantes, F., and Santero, E. (1994). Geometry of the process of transcription activation at the sigma 54-dependent *nifH* promoter of *Klebsiella pneumoniae. J. Biol. Chem., 269*, 25419-25425.

Morett, E., and Buck, M. (1988). NifA-dependent *in vivo* protection demonstrates that the upstream activator sequence of *nif* promoters is a protein binding site. *Proc. Natl. Acad. Sci. U.S.A., 85*, 9401-9405.

Morett, E., and Buck, M. (1989). *In vivo* studies on the interaction of RNA polymerase-σ54 with the *Klebsiella pneumoniae* and *Rhizobium meliloti nifH* promoters: The role of NifA in the formation of an open promoter complex. *J. Mol. Biol., 210*, 65-77.

Morett, E., Kreutzer, R., Cannon, W., and Buck, M. (1990). The influence of the *Klebsiella pneumoniae* regulatory gene *nifL* upon the transcriptional activator protein NifA. *Mol. Microbiol., 4*, 1253-1258.

Nagatani, H., Shimizu, M., and Valentine, R. C. (1971). The mechanism of ammonia assimilation in nitrogen-fixing bacteria. *Arch. Microbiol., 79*, 164.

Nair, M. B., and Eady, R. R. (1984). Nitrogenase synthesis in *Klebsiella pneumoniae*: Enhanced *nif* expression without accumulation of guanosine 5'-diphosphate 3'-diphosphate. *J. Gen. Microbiol., 130*, 3063-3069.

Narberhaus, F., Lee, H.-S., Schmitz, R. A., He, L., and Kustu, S. (1995). The C-terminal domain of NIFL is sufficient to inhibit NIFA activity. *J. Bacteriol., 177*, 5078-5087.

Nieva-Gomez, D., Roberts, G. P., Klevickis, S., and Brill, W. J. (1980). Electron transport to nitrogenase in *Klebsiella pneumoniae. Proc. Natl. Acad. Sci. U.S.A., 77*, 2555-2558.

Ninfa, A. J., and Magasanik, B. (1986). Covalent modification of the glnG product, NR I, by the glnL product, NR II, regulates the transcription of the glnALG operon in Escherichia coli *Proc. Natl. Acad. Sci. U.S.A., 83*, 5909-5913.

Nixon, B. T., Ronson, C. W., and Ausubel, F. M. (1986). Two-component regulatory systems responsive to environmental stimuli share strongly conserved domains with the nitrogen assimilation regulatory genes *ntrB* and *ntrC*. *Proc. Natl. Acad. Sci. U.S.A., 83*, 7850-7854.
Ow, D. W., and Ausubel, F. (1983). Regulation of nitrogen metabolism genes by *nifA* gene product in *Klebsiella pneumoniae*. *Nature, 301*, 307-313.
Ow, D. W., Sundaresan, V., Rothstein, D. M., Brown, S. E., and Ausubel, F. (1983). Promoters regulated by the *glnG* (*ntrC*) and *nifA* gene products share a heptameric consensus sequence in the -15 region. *Proc. Natl. Acad. Sci. U.S.A., 80*, 2524-2528.
Ow, D. W., Xiong, Y., Gu, Q., and Shen, S. C. (1985). Mutational analysis of the *Klebsiella pneumoniae* nitrogenase promoter: sequences essential for positive control by *nifA* and *ntrC* (*glnG*) products. *J. Bacteriol., 161*, 868-874.
Pahel, G., Rothstein, D. M., and Magasanik, B. (1982). Complex *glnA-glnL-glnG* operon of *Escherichia coli*. *J. Bacteriol., 150*, 202-213.
Pahel, G., and Tyler, B. (1979). A new *glnA*-linked regulatory gene for glutamine synthetase in *Escherichia coli*. *Proc. Natl. Acad. Sci. U.S.A., 76*, 4544-4548.
Paul, W., and Merrick, M. (1987). The nucleotide sequence of the *nifM* gene of *Klebsiella pneumoniae* and identification of a new *nif* gene: *nifZ*. *Eur. J. Biochem., 170*, 259-265.
Paul, W., and Merrick, M. (1989). The roles of the *nifW*, *nifZ* and *nifM* genes in *Klebsiella pneumoniae* nitrogenase biosynthesis. *Eur. J. Biochem., 178*, 675-682.
Popham, D. L., Szeto, D., Keener, J., and Kustu, S. (1989). Function of a bacterial activator protein that binds to transcriptional enhancers. *Science, 243*, 629-635.
Prival, M. J., Brenchley, J. E., and Magasanik, B. (1973). Glutamine synthetase and the regulation of histidase formation in *Klebsiella aerogenes*. *J. Biol. Chem., 248*, 4334-4344.
Pühler, A., Burkardt, H. J., and Klipp, W. (1979). Cloning of the entire region for nitrogen fixation from *Klebsiella pneumoniae* on a multicopy plasmid vehicle in *Escherichia coli*. *Mol. Gen. Genet., 176*, 17-24.
Ray, P., Smith, K. J., Parslow, R. A., Dixon, R., and Hyde, E. I. (2002). Secondary structure and DNA binding by the C-terminal domain of the transcriptional activator NifA from *Klebsiella pneumoniae*. *Nucleic Acids Res., 30*, 3972-3980.
Reidel, G. E., Ausubel, F., and Cannon, F. C. (1979). Physical map of chromosomal nitrogen fixation (*nif*) genes of *Klebsiella pneumoniae*. *Proc. Natl. Acad. Sci. U.S.A., 76*, 2866-2870.
Reidel, G. E., Brown, S. E., and Ausubel, F. (1983). Nitrogen fixation by *Klebsiella pneumoniae* is inhibited by certain multicopy hybrid *nif* plasmids. *J. Bacteriol., 153*, 45-56.
Reitzer, L. J., and Magasanik, B. (1983). Isolation of the nitrogen assimilation regulator NRI, the product of the *glnG* gene of *Escherichia coli*. *Proc. Natl. Acad. Sci. U.S.A., 80*, 5554-5558.
Reitzer, L. J., and Magasanik, B. (1986). Transcription of *glnA* in *E.coli* is stimulated by activator bound to sites far from the promoter. *Cell, 45*, 785-792.
Riesenberg, D., Erdei, S., Kondorosi, E., and Kari, C. (1982). Positive involvement of ppGpp in derepression of the *nif* operon in *Klebsiella pneumoniae*. *Mol. Gen. Genet., 185*, 198-204.
Roberts, G. P., and Brill, W. J. (1980). Gene-product relationships of the *nif* regulon of *Klebsiella pneumoniae*. *J. Bacteriol., 144*, 210-216.
Roberts, G. P., Macneil, T., Macneil, D., and Brill, W. J. (1978). Regulation and characterization of protein products coded by the *nif* (nitrogen fixation) genes of *Klebsiella pneumoniae*. *J. Bacteriol., 136*, 267-279.
Ronson, C. W., Nixon, B. T., and Ausubel, F. M. (1987). Conserved domains in bacterial regulatory proteins that respond to environmental stimuli. *Cell, 49*, 579.
Rudnick, P., Kunz, C., Gunatilaka, M. K., Hines, E. R., and Kennedy, C. (2002). Role of GlnK in NifL-mediated regulation of NifA activity in *Azotobacter vinelandii*. *J. Bacteriol., 184*, 812-820.
Ruvkun, G. B., and Ausubel, F. (1980). Interspecies homology of nitrogenase genes. *Proc. Natl. Acad. Sci. U.S.A., 77*, 191-195.
Sanger, F., Nicklen, S., and Coulson, A. R. (1977). Sequencing with chain-terminating inhibitors. *Proc. Natl. Acad. Sci. U.S.A., 74*, 5463-5467.
Santero, E., Hoover, T., Keener, J., and Kustu, S. (1989). *In vitro* activity of the nitrogen fixation regulatory protein NIFA. *Proc. Natl. Acad. Sci. U.S.A., 86*, 7346-7350.
Schmitz, R. A. (1997). NifL of *Klebsiella pneumoniae* carries an N-terminally bound FAD cofactor, which is not directly required for the inhibitory function of NifL. *FEMS Microbiol. Lett., 157*, 313-318.

Scott, K. F., Rolfe, B. G., and Shine, J. (1981). Biological nitrogen fixation: Primary structure of the *Klebsiella pneumoniae nifH* and *nifD* genes. *J. Mol. Appl. Genet., 1*, 71-81.

Shah, V. K., Allen, J. R., Spangler, N. J., and Ludden, P. W. (1994). *In vitro* synthesis of the iron-molybdenum cofactor of nitrogenase. Purification and characterization of NifB cofactor, the product of NIFB protein. *J. Biol. Chem., 269*, 1154-1158.

Shah, V. K., Rangaraj, P., Chatterjee, R., Allen, R. M., Roll, J. T., Roberts, G. P., *et al.* (1999). Requirement of NifX and other *nif* proteins for *in vitro* biosynthesis of the iron-molybdenum cofactor of nitrogenase. *J. Bacteriol., 181*, 2797-2801.

Shah, V. K., Stacey, G., and Brill, W. J. (1983). Electron transport to nitrogenase. Purification and characterization of pyruvate:flavodoxin oxidoreductase. The *nifJ* gene product. *J. Biol. Chem., 258*, 12064-12068.

Shanmugam, K. T., Chan, I., and Morandi, C. (1975). Regulation of nitrogen fixation. Nitrogenase-dependent mutants of *Klebsiella pneumoniae. Biochim. Biophys. Acta, 408*, 101-111.

Shanmugam, K. T., and Valentine, R. C. (1975a). Microbial production of ammonium ion from nitrogen. *Proc. Natl. Acad. Sci. U.S.A., 72*, 136-139.

Shanmugam, K. T., and Valentine, R. C. (1975b). Molecular biology of nitrogen fixation. *Science, 187*, 919-924.

Sibold, L., and Elmerich, C. (1982). Constitutive expression of nitrogen fixation (*nif*) genes of *Klebsiella pneumoniae* due to a DNA duplication. *EMBO J., 1*, 1551-1558.

Simon, H. M., Gosink, M. M., and Roberts, G. P. (1999). Importance of *cis* determinants and nitrogenase activity in regulated stability of the *Klebsiella pneumoniae* nitrogenase structural gene mRNA. *J. Bacteriol., 181*, 3751-3760.

Simon, H. M., Homer, M. J., and Roberts, G. P. (1996). Perturbation of *nifT* expression in *Klebsiella pneumoniae* has limited effect on nitrogen fixation. *J. Bacteriol., 178*, 2975-2977.

Söderbäck, E., Reyes-Ramirez, F., Eydmann, T., Austin, S., Hill, S., and Dixon, R. (1998) The redox- and fixed nitrogen-responsive regulatory protein NIFL from *Azotobacter vinelandii* comprises discrete flavin and nucleotide-binding domains. *Mol. Microbiol., 28*, 179-192.

St. John, R. T., Johnston, H. M., Seidman, C., Garfinkel, D., Gordon, J. K., Shah, V. K., *et al.* (1975). Biochemistry and genetics of *Klebsiella pneumoniae* mutant strains unable to fix N_2. *J. Bacteriol., 121*, 759-765.

Streicher, S., Gurney, E., and Valentine, R. C. (1971). Transduction of the nitrogen fixation genes in *Klebsiella pneumoniae. Proc. Natl. Acad. Sci. U.S.A., 68*, 1174-1177.

Streicher, S., Gurney, E., and Valentine, R. C. (1972). The nitrogen fixation genes. *Nature, 239*, 495-499.

Streicher, S. L., Shanmugan, K. T., Ausubel, F., Morandi, C., and Goldberg, R. B. (1974). Regulation of nitrogen fixation in *Klebsiella pneumoniae*: evidence for a role of glutamine synthetase as regulator of nitrogenase synthesis. *J.Bacteriol., 120*, 815-821.

Su, W., Porter,S.,, Kustu, S., and Echols, H. (1990). DNA-looping and enhancer activity: Association between DNA-bound NtrC activator and RNA polymerase at the bacterial *glnA* promoter. *Proc. Natl. Acad. Sci. U.S.A., 87*, 5504-5508.

Sundaresan, V., and Ausubel, F. (1981). Nucleotide sequence of the gene coding for the nitrogenase iron protein from *Klebsiella pneumoniae. J. Biol. Chem., 256*, 2808-2812.

Tempest, D. W., Meers, J. L., and Brown, C. M. (1970). Synthesis of glutamate in *Aerobacter aerogenes* by a hitherto unknown route. *Biochem. J., 117*, 405-407.

Thorneley, R. N., Abell, C., Ashby, G. A., Drummond, M. H., Eady, R. R., Huff, S., *et al.* (1992). Posttranslational modification of *Klebsiella pneumoniae* flavodoxin by covalent attachment of coenzyme A, shown by 31P NMR and electrospray mass spectrometry, prevents electron transfer from the nifJ protein to nitrogenase. A possible new regulatory mechanism for biological nitrogen fixation. *Biochem., 31*, 1216-1224.

Tubb, R. S. (1974). Glutamine synthetase and ammonium regulation in *Klebsiella. Nature, 251*, 481.

Tuli, R., and Merrick, M. J. (1988). Over-production and characterisation of the *nifA* gene product of *Klebsiella pneumoniae* - the transcription activator of *nif* gene expression. *J. Gen. Microbiol., 134*, 425-432.

Tyler, B., DeLeo, A. B., and Magasanik, B. (1974). Activation of transcription of *hut* DNA by glutamine synthetase. *Proc. Natl. Acad. Sci. U.S.A., 71*, 225-229.

Wedel, A., Weiss, D. S., Popham, D., Droge, P., and Kustu, S. (1991). A bacterial enhancer functions to tether a transcriptional activator near a promoter. *Science, 248*, 486-490.

Whitehall, S., Austin, S., and Dixon, R. (1992). DNA supercoiling response of the σ^{54}-dependent *Klebsiella pneumoniae nifL* promoter *in vitro*. *J. Mol. Biol., 225*, 591-607.

Whitehall, S., Austin, S., and Dixon, R. (1993). The function of the upstream region of the σ^{54} -dependent *Klebsiella pneumoniae nifL* promoter is sensitive to DNA supercoiling. *Mol. Microbiol., 9*, 1107-1117.

Wong, P. K., Popham, D., Keener, J., and Kustu, S. (1987). *In vitro* transcription of the nitrogen fixation regulatory operon *nifLA* of *Klebsiella pneumoniae*. *J. Bacteriol., 169*, 2876-2880.

Yuvaniyama, P., Agar, J. N., Cash, V. L., Johnson, M. K., and Dean, D. R. (2000). NifS-directed assembly of a transient [2Fe-2S] cluster within the NifU protein. *Proc. Natl. Acad. Sci. U.S.A., 97*, 599-604.

Zamir, A., Maina, C. V., Fink, G. R., and Szalay, A. A. (1981). Stable chromosomal integration of the entire nitrogen fixation gene cluster from *Klebsiella pneumoniae* in yeast. *Proc. Natl. Acad. Sci. U.S.A., 78*, 3496-3500.

Zheng, L., White, R. H., Cash, V. L., Jack, R. F., and Dean, D. R. (1993). Cysteine desulfurase activity indicates a role for NIFS in metallocluster biosynthesis. *Proc. Natl. Acad. Sci. U.S.A., 90*, 2754-2758.

Chapter 2

GENETICS OF NITROGEN FIXATION AND RELATED ASPECTS OF METABOLISM IN SPECIES OF *AZOTOBACTER*: HISTORY AND CURRENT STATUS

C. KENNEDY[1] AND P. BISHOP[2]

[1] *Department of Plant Pathology, University of Arizona, Tucson, AZ 85721, USA*

[2] *Department of Microbiology, North Carolina State University and USDA, Raleigh, NC 27695, USA*

1. RESEARCH ON THE GENUS *AZOTOBACTER* (1901 – 2003)

The first paper describing two species of nitrogen-fixing soil bacteria representing the genus *Azotobacter, A. chroococcum* and *A. agilis,* was published in 1901 by the Dutch microbiologist, Martinus Beijerinck (1901). The species, which later became the genetic workhorse of the azotobacters, *A. vinelandii,* was discovered a century ago at Rutgers University by Jacob Lipman (1903), a Russian emigrant to the USA. Publications during the next several decades focussed on systematics and classification of these and several other soil isolates identified as species of *Azotobacter*, describing cultural and morphological characteristics (summarized in Thompson and Skerman, 1979).

In the 1960's and 1970's, two broad areas dominated research on *Azotobacter* species: (i) the physiology of O_2 tolerance in *A. vinelandii* and *A. chroococcum* (reviewed by Robson and Postgate, 1980), and (ii) the biochemistry of the nitrogenase enzyme purified from *A. vinelandii* (reviewed by Peters *et al.*, 1995). The past two decades have seen the tools of molecular biology and genetics applied to understanding nitrogen fixation and related aspects of metabolism and, in 2002, a draft of the sequence of the genome of *A. vinelandii* became available through the US Department of Energy, Joint Genome Institute at www.azotobacter.org and www.jgi.doe.gov/JGI_microbial/html/azotobacter/azoto/homepage/html). Automatic annotation of this genome and its assemblies has been carried out by Frank Larimer and associates at the DOE lab at Oak Ridge National Laboratories (ORNL)

W. Klipp, B. Masepohl, J. R. Gallon and W. E. Newton (eds.), Genetics and Regulation of Nitrogen Fixation in Free-Living Bacteria, 27-52.

and at The Institute for Genome Research (TIGR).

A discovery of major importance to studies of *Azotobacter* species and also to nitrogen fixation in general was the finding, originally published in 1980, that *A. vinelandii* contained an 'alternative' nitrogenase, which did not contain molybdenum (Bishop *et al.*, 1980). Further research, during the 1980's and 1990's, established that *A. vinelandii* and *A. paspali* (a species of *Azotobacter* associated with the rhizosphere of the tropical grass, *Paspalum notatum*) harbor three genetically distinct nitrogenase enzymes, and that *A. chroococcum* contains two distinct enzymes for nitrogen fixation. Other diazotrophs also carry genes for alternative nitrogenases. This work is reviewed later in this chapter.

Studies during the last three decades have focused primarily on *A. vinelandii* strain AvOP (ATCC13705; also named strain UW at the University of Wisconsin) and its derivatives because this organism is particularly amenable to genetic analysis, being an excellent recipient of DNA donated by either transformation or conjugation from donor strains. These characteristics are probably due to the inability of this strain to synthesize the exopolysaccharide, alginate; strain UW contains an insertion element in the *algU* gene, which encodes a sigma factor (AlgU) that associates with RNA polymerase to promote expression of other *alg* genes (Martinez-Salazar *et al.*, 1996). While there are many areas of interest in the physiological and metabolic aspects of *Azotobacter* biology, this chapter focuses on the genetics of nitrogenase biosynthesis and the regulation of nitrogen fixation and assimilation in response to environmental factors. It will also incorporate related aspects of physiology that influence the efficiency of nitrogen fixation in these organisms.

Advances and current topics include understanding the genetic basis of: (i) nitrogen fixation, the *nif* and related genes encoding nitrogenase (the classical Mo-containing enzyme as well as the 'alternative' non-Mo enzymes); (ii) the regulatory mechanisms that govern the ability of nitrogenase enzyme(s) to be synthesized, involving *nifL, nifA, glnK, glnD*, and other genes; (iii) the major enzyme of ammonium assimilation, glutamine synthetase, GS, encoded by *glnA*; (iv) O_2 protection by the respiratory cytochrome *d* pathway and the FeSII protein; and (v) the genetics of the alternative, non-Mo-containing enzymes for nitrogen fixation in *A. vinelandii*. The concomitant and preceding discoveries in gene-transfer technologies, including transformation in *A. vinelandii*, conjugation using wide host range plasmids, the use of transposons, and recombinant DNA techniques for insertional inactivation of genes, were all important developments during the 1980's and 1990's that led to advances in all aspects of understanding the genetic basis of nitrogen fixation and related aspects of metabolism in species of *Azotobacter*. These studies have now been augmented and extended into new realms of research by the recent sequencing of the genome of *A. vinelandii* strain AvOP, now at a rough draft stage, as mentioned above. Members of the *A. vinelandii* research community are actively involved in this project, which should result in a 'finished' and 'annotated' analysis of the genome for publication in 2004.

The genome of *A. vinelandii* strain AvOP consists of ~5.3 Mb of DNA; the GC content is ~65%. Evidence for a circular genome is based on genetic studies in which variously mutated and antibiotic-resistant strains served as donors and recipients during conjugational crosses (Blanco *et al*., 1990). Somewhat uniquely, the number of chromosomes in this organism can vary widely from one (or a few) per cell during early-to-mid-logarithmic growth phase to apparently more than 80 during late logarithmic/early stationary phase (Maldonado *et al.,* 1994). The mechanistic basis for and significance of this plasticity in chromosome content is not well-understood. *A. vinelandii* strain AvOP is not known to contain endogenous plasmids, whereas other species and strains of *Azotobacter* may contain up to five or six native plasmids, as occurs in *A. chroococcum* (Robson *et al*., 1984; also reviewed in Kennedy *et al*., 2003).

Another aspect of the uniqueness of *A. vinelandii* in the study of microbial genetics is its ability to reveal that certain mutations are lethal and that the genes in which they are located are essential. This situation has been demonstrated for several genes, including *glnA* (encoding glutamine synthetase), *glnK* and *glnD* (encoding regulatory proteins involved in ammonium-sensing regulatory pathways), and the *isc* genes (encoding proteins involved in iron-sulfur cluster formation), and others as discussed below.

In addition to being able to fix N_2, most *Azotobacter* species can utilize nitrate as a nitrogen source in an assimilatory fashion by reduction to nitrite and then to ammonium. Genes for both nitrate and nitrite reductases, *nasB* and *nasA,* have been isolated from *A. vinelandii* OP. The *nasAB* operon, regulated from a σ^{54}, NtrC-dependent promoter, is repressed by ammonium and requires nitrate for full expression (Ramos *et al*., 1993). Soluble nitrate reductases have been purified from *A. vinelandii* and *A. chroococcum* (Guerrero, 1975; Gangeswaren *et al*., 1993). Nitrate transport and reduction have been studied in *A. chroococcum*. Although the genes have not been characterized in this organism, the physiology of nitrate assimilation is similar to that in *A. vinelandii*. Ammonium prevents full synthesis/activity of the enzyme, and nitrate is required for full activity.

All *Azotobacter* species are urease-positive in physiological tests, a feature that distinguishes *Azotobacter* from *Azomonas* species (Thompson and Skerman, 1979; Kennedy *et al*., 2003). This feature is consistent with the fact that urea is a good N source in both *A. vinelandii* and *A. chroococcum*. It is also a useful trait, because urea added to media at 5-10 mM does not prevent nitrogenase synthesis, in contrast to the effects of ammonium. Consequently, urea allows studies of expression of these genes in liquid cultures (see, *e.g*., Walmsley and Kennedy, 1991).

2. APPLICATION OF THE TOOLS OF GENETICS AND MOLECULAR BIOLOGY IN SPECIES OF *AZOTOBACTER*

A. vinelandii can be readily studied genetically because of the ease of handling the widely-used, non-gummy strain AvOP (ATCC 13705), derived from AvO (ATCC 12518). AvOP was also named strain UW in studies at the University of Wisconsin (Bishop and Brill, 1977) and CA in studies at North Carolina State University (in the Bishop laboratory). This strain can receive plasmids donated from either *E. coli* or *A. vinelandii* donors at a high frequency during conjugational transfer, and can be transformed with either linear or plasmid DNA from donor strains. Competence for

DNA uptake is induced after growth on both Fe- and Mo-free medium (Page, 1982; Doran *et al.*, 1983; Page, 1985). Another widely-used derivative of UW is UW136, wild-type for nitrogen fixation but carrying a mutation for rifampicin resistance (Bishop and Brill, 1977). The features of UW and its derivatives make these strains genetically manipulable by either introduction of wide-host range plasmids for complementation analysis or by isolation of mutant strains. Mutant strains may be formed by spontaneous mutation, induced using mutagens, by transposition with Tn*5*, or by gene-replacement techniques using suicide plasmids (reviewed in Kennedy *et al.,* 2003).

Azotobacter chroococcum has also been studied genetically and, while successful transformation of this species has not been reported, the cells can receive plasmids by conjugation with *E. coli* donors. Strain MCD1, a non-gummy derivative of ATCC4412 (NCIB8033) cured of three of five indigenous plasmids, has been used for genetic studies (Robson *et al.*, 1984; Evans *et al.*, 1988).

The genome sizes of several *Azotobacter* species were determined by pulse-field gel electrophoresis of large chromosomal DNA fragments generated by rare-cutting restriction endonucleases. The sizes range from 3.10 Mb for *A. chroococcum* NCIB9043 to 4.57 Mb for *A. vinelandii* UW (reviewed in Kennedy *et al.,* 2003). In another study, Manna and Das (1994) found the M4 strain of *A. chroococcum* to have a genome size of 5.3 Mb. This size disparity supports previous findings of genomic variation within the *A. chroococcum* species (Becking, 1992). The presence of a single circular chromosome in *A. vinelandii* is indicated from genetic analysis (Blanco *et al.*, 1990). The size of the *A. paspali* genome is somewhat smaller than that of *A. vinelandii*, not unexpectedly because the former uses fewer compounds as C source and has a more restricted rhizosphere habitat than the free-living *A. vinelandii*. The sequence of the *A. vinelandii* AvOP genome indicates its size to be 5.3 Mb.

The origin of replication of *A. vinelandii* has been identified and sequenced, including a 200-bp segment capable of replicating plasmid pBR322 in an *E. coli polA* mutant strain. A larger fragment of 1652 bp, which includes the 200-bp Ori sequence, has 14 possible DnaA protein-binding sites, ten possible binding sites for the IHF protein, and 19 GATC boxes (putative sites for methylation by the Dam protein), which are involved in initiation of DNA replication in *E. coli* and other organisms (Singh *et al.*, 2000). Identified genes, which are thought to be involved in general recombination in *A. vinelandii,* include *recA* and *recF* (Venkatesh and Das, 1992; Badran *et al.*, 1999).

Multiple chromosomes are common in the two species examined, *A. vinelandii* and *A. chroococcum.* Although, in *A. vinelandii,* the chromosome number can reach about 80 copies per cell in late-exponential phase cells, in old stationary-phase cultures the amount of DNA is drastically reduced, possibly in preparation for cyst formation (Maldonaldo *et al.*, 1994). In exponential cultures of *A. chroococcum* MCD-1, cells contained 20 to 25 genome equivalents (Robson *et al.*, 1984).

No plasmids have been detected in *A. vinelandii* AvOP (UW) wild-type and derived mutant strains. Interestingly, a cryptic bacteriophage, which is similar to the *E. coli* phage P2, was found in the genome sequence of AvOP (B. Fane, personal communication). One strain of *A. paspali* did not harbor any plasmids (Do Nascimento and Tavares, 1987). Robson *et al.* (1984) reported finding between two and six plasmids in each of eight strains of *A. chroococcum* examined; these plasmids ranged in size from 10.5 kb to 300 kb. The *A. chroococcum* strains included NCIB8003 (ATCC4412); strain MCC-1, a Nal^r Sm^r and non-gummy derivative of NCIB8003; and six new isolates from soils in Sussex, UK. One plasmid in NCIB8003 is probably of the IncP class. Strains cured of plasmids, after growth in ethidium bromide, acriflavin, or mitomycin, showed no altered phenotypes that are usually associated with plasmids, except for possibly one trait, the production of agar-diffusible exopolysaccharides. Evidence for a plasmid encoding a gene or genes for degradation of 2,4-dichlorophenoxyacetic acid in strain MSB1 of *A. chroococcum* was reported (Balajee and Mahadevan, 1989).

Wide host-range plasmids of the IncQ and IncP classes can be transferred from *E. coli* by conjugation to *A. chroococcum* strain MCD1 and its derivatives (Kennedy and Robson, 1983; Ramos and Robson, 1987) and by either conjugation or transformation to *A. vinelandii* strain AvOP and mutant derivatives (Kennedy and Robson, 1983; Toukdarian and Kennedy, 1987; Doran *et al.*, 1987; Glick *et al.*, 1989). This observation provides the potential for isolation of genes from IncP cosmid libraries by mutant complementation and also analysis of gene function by specific genes cloned into either IncP or IncQ plasmids. Introducing wide host-range plasmids of the IncQ compatibility group resulted in decreased growth rates in *A. vinelandii* host strains (Glick *et al.*, 1986).

The state of competence for transformation of strains of *A. vinelandii* is associated with siderophore production and can be induced by growth on either Fe-deficient or Fe- and Mo-deficient media (Page, 1985; Page and Grant, 1987). Competent cells can be transformed with either plasmid or linear DNA (Doran *et al.*, 1987). Transformation frequency could be enhanced 10- to 50-fold when pKT210 (IncQ) carried an insert fragment of *A. vinelandii nif* DNA, suggesting that *A. vinelandii* might possess a homology-facilitated transformation system. Mutated genes can be introduced directly either on plasmids unable to replicate (suicide vectors) or on linear chromosomal DNA prepared from mutant strains; recombination with the chromosome at sites flanking the mutation leads to replacement of the wild-type genes (*e.g.*, as described by Toukdarian and Kennedy, 1986; Jacobson *et al.*, 1989; and in many other reports).

Transposon mutagenesis is easily achieved in *A. vinelandii* and *A. chroococcum* using Tn*5* carried on suicide vectors unable to replicate in these host organisms (for examples, see Kennedy *et al.*, 1986; Joerger *et al.*, 1989; Luque *et al.*, 1993; Tibelius *et al.*, 1993; Wu *et al.*, 1997). In addition, Tn*10* was successfully used to isolate mutants of *A. vinelandii*, including methionine auxotrophs (Contreras and Casadesus, 1987). When Tn*5* was introduced into *A. beijerinckii* (NCIB11292), the cells of most of the isolates had abnormal morphologies and rapidly lost viability when subcultured. In contrast, Tn*76* was successfully used to make mutants in this strain (Owen and Ward, 1985). Contreras *et al.* (1991) developed a method to 'freeze' Tn*5* mutations in *A. vinelandii* by replacing the original transposon with a defective Tn*5*, resulting in highly stable mutants.

Attempts to isolate auxotrophic mutants of *A. vinelandii* were unsuccessful in earlier decades. It was believed that the high chromosome copy number (first reported by Sadoff *et al.*, 1979) might lead to difficulty in segregation of mutant phenotypes. However, the successful isolation of many other types of mutations over the past 20 years by either transposon mutagenesis or cassette insertion in cloned genes has shown that isolation of mutant strains of *A. vinelandii* is not inherently difficult. The variable DNA content of this organism indicates that there are growth phases and conditions when the chromosome number is low, thereby allowing for 'normal' segregation of introduced, selectable mutations in non-essential genes. Although isolation of auxotrophs remains difficult, success has been achieved for the construction of Ade$^-$, Met$^-$, and Leu$^-$ mutants (Mishra and Wyss, 1968; Contreras and Casadesus, 1987; Manna and Das, 1997). *A. vinelandii* cannot, apparently, transport many amino acids (Toukdarian *et al.*, 1991; D. R. Dean, personal communication). Although Mishra *et al.* (1991) reported that threonine was transported at a rate similar to that found for *E. coli* and methionine at a rate at more than twice that in *E. coli*, five other amino acids tested were transported at a fraction of the rate. In *A. beijerinckii*, RP4::Tn*76* was successfully used to isolate Nif$^-$ mutants and also auxotrophs requiring either adenine or leucine (Owen and Ward, 1985). D. R. Dean and coworkers (personal communication) found that tryptophan auxotrophs could be isolated in *A. vinelandii* only if the strain harbored a wide-host-range plasmid carrying the tryptophan transport genes from *E. coli*. Also consistent with the idea that amino acids are not easily transported into species of *Azotobacter* and *Azomonas*, are data of Thompson and Skerman (1979), who found that none of the 17 amino acids tested could serve as C sources in any of the species of these genera. Thus, limited capacity for amino acid transport is probably a general characteristic of the genus *Azotobacter*.

Mutations have been introduced into dozens of genes of *A. vinelandii* and *A. chroococcum*, resulting in gene replacement and the construction of mutant strains. A Nif$^-$ mutant of *A. paspali* was constructed by the conjugation and recombination of a *kan*^{r} cassette with the *nifHDK* operon, leading to gene replacement (Fallik *et al.*, 1993). Some mutations are 'lethal' in *A. vinelandii*, due to the essential nature of the genes under study. As mentioned above, these genes include *glnA*, which encodes glutamine synthetase, the sole pathway for assimilation of ammonium (Kennedy and Toukdarian, 1987; Toukdarian *et al.*, 1990); the *isc* and *fdxD* genes, which encode enzymes and a ferredoxin for Fe-S cluster formation (Zheng *et al.*, 1998; Jung *et al.*, 1999); and the *glnK* and *glnD* genes, which encode proteins that sense and signal the status of the fixed-nitrogen supply to glutamine synthetase to either activate or inactivate the enzyme (Meletzus *et al.*, 1998; Colnaghi *et al.*, 2000). The ability of *A. vinelandii* to harbor high copy numbers of the chromosome is an advantage in that this feature allows identification of essential genes. Both wild-type and mutated copies marked by insertion of an antibiotic-resistance gene can be maintained and identified by either hybridization or PCR analysis. When selective pressure is removed by growth in medium without antibiotics, the chromosomes carrying the mutated essential genes are rapidly lost along with antibiotic resistance.

3. THE *nif* GENES ENCODING THE ENZYMES FOR STRUCTURE, FUNCTION, AND BIOSYNTHESIS OF Mo-CONTAINING NITROGENASE

Aerobic nitrogen fixation is one defining characteristic of *Azotobacter* and *Azomonas*, two genera of the family Pseudomonadaceae, a member of the phylum Gammaproteobacteria, in the domain Bacteria (Kennedy *et al.,* 2003). The following is a more or less chronological account of the history of the characterization of genes involved in nitrogenase biosynthesis and activity carried out during the past decades. The first reports of isolation of Nif⁻ mutants were published over 50 years ago (Wyss and Wyss, 1950; Green *et al.,* 1953). More useful Nif⁻ mutants of strain AvOP were isolated by Fisher and Brill (1969) and Shah *et al.* (1973). These strains of the UW series of mutants were analyzed biochemically in terms nitrogenase component activities (Shah *et al.,* 1973). Extracts of one mutant, UW45, could be reactivated for nitrogenase activity by addition of acid-treated MoFe protein (also known as Component 1 or dinitrogenase) (Nagatani *et al.,* 1974). This treatment is now known to release the FeMo cofactor, an essential component of the MoFe protein, which is the site of substrate binding and reduction. The UW45 strain was a necessary beginning for identifying the *nif* genes involved in and the complex pathway for FeMo cofactor biosynthesis (reviewed in Volume 1 of this series and by Rangaraj *et al.,* 2000).

Bishop and Brill (1977) determined, by transformation analysis, the genetic order of eight mutations, which suggested there might be two separate clusters of *nif* genes. This suggestion was verified later by DNA sequencing (see below). Genetic complementation and mapping studies carried out in *Klebsiella pneumoniae* had led to both designations for some of the genes and insight into some of the functions of the gene products in this organism (Dixon *et al.,* 1978; Kennedy, 1978; Roberts *et al.*, 1978; Roberts and Brill, 1980; see also Dixon, this Volume). Researchers in the *Azotobacter* community then agreed to assign the same designations to the corresponding genes in these organisms. The first such correspondence was reported by Bishop *et al.*, (1985), who used *nifHDK* genes cloned from *K. pneumoniae* to isolate by hybridization two plasmids from an *Eco*RI *A. vinelandii* genomic library. The *nifHDK* genes of *A. vinelandii* were physically mapped and also shown to rescue mutants UW6 and UW38 (both lacking Fe protein encoded by *nifH)* and UW10 (lacking MoFe protein encoded by *nifDK*). Further correspondence to *K. pneumoniae nif* genes was reported for *A. chroococcum* by hybridization and complementation of mutants; these included *nifF, nifMVS,* and *nifNE*, in addition to *nifHDK* (Jones *et al*., 1984; Evans *et al*., 1985). Functional equivalence of *nifH, nifN*, and *nifM* genes in *A. vinelandii* to those in *K. pneumoniae* was demonstrated by biochemical characterization of Tn*5* mutations in these three genes (Kennedy *et al*., 1986).

Another mutant among the UW series was UW1, which was unable to synthesize either of the component proteins of nitrogenase (Shah *et al.,* 1973). To explore the hypothesis that the mutation in UW1 was in a gene corresponding to the *nifA* gene in *K. pneuoniae*, the product of which was required for expression of all other *nif* genes in this organism, the *nifA* gene of *K. pneumoniae* was cloned under the control of a constitutive promoter in the IncQ plasmid pKT230, resulting in recombinant plasmid pCK1, and transferred by conjugation to mutant strain UW1 of *A. vinelandii* and also to mutant strain MCD1008 of *A. chroococcum*. Both transconjugant strains were able to fix N_2, and neither nitrogenase synthesis nor activity was repressed by either ammonium or O_2 (Kennedy and Robson, 1983). This finding was confirmed by Kennedy and Drummond (1985) using another construct, pCK3, a derivative of pRK290 in which the *K. pneumoniae nifA* gene was cloned and strongly expressed from the *lac* promoter of *E. coli*. This work was later confirmed by the finding of a *nifA*-like gene in *A. vinelandii* by sequence analysis (Bennett *et al.*, 1988). Bennett *et al.* (1988) also reported a partial sequence of the region upstream of *nifA* and found some similarities to the *nifL* gene of *K. pneumoniae*, which had been reported to encode a potential negative regulator of *nif* gene expression in that organism (Kennedy, 1978) and sequenced by Drummond and Wootton (1987). Mutations in *nifL* in *A. vinelandii* can result in derepressed nitrogenase activity in this organism, resulting in excretion of ammonium of up to 10 mM in liquid cultures (Bali *et al.*, 1992; Brewin *et al.*, 1999). The sequence of *nifL* in *A. vinelandii* was completed by Blanco *et al.*, (1993) and by Raina *et al.*, (1993). The function of NifA and NifL are discussed below.

The major cluster of genes encoding the nitrogenase enzyme synthesized under Mo-sufficient conditions in *A. vinelandii*, was sequenced and extensively characterized by mutational analysis by D. R. Dean's group (Brigle *et al.*, 1985; Dean and Brigle, 1985; Beynon *et al.*, 1987; Bennett *et al.*, 1988; Brigle and Dean, 1988; Jacobson *et al.*, 1989). In addition to the *nif* genes identified in *K. pneumoniae*, several other open reading frames (ORFs) were identified. This gene cluster is shown in Figure 1.

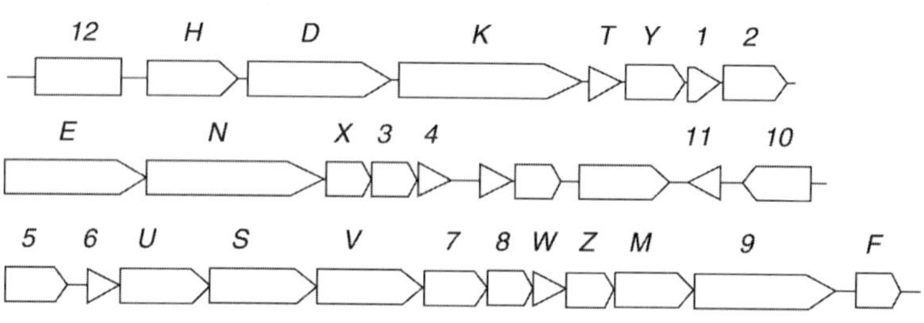

Figure 1. Major nif *cluster from* A. vinelandii. *Numbers indicate ORFs, whereas letters indicate* nif *genes. Genes are ordered left to right and top to bottom in a continuous sequence. Gene and ORF designations are as presented by Jacobson et al. (1989).*

A. chroococcum MCD-1 (derived from ATCC 4412) has a similar major cluster, but was less well-characterized in terms of mutational analysis (Jones *et al.*, 1984; Evans *et al.*, 1988). A second *nif* cluster, located elsewhere on the *A. vinelandii* chromosome and shown as Figure 2, carries two operons; *nifLA*, which encodes both the negative and positive regulatory proteins (described below), and *nifB-fdxN-ORF-nifQ*, which encodes the genes for molybdenum transport and processing (Joerger and Bishop, 1988; Rodriguez-Quinones *et al.*, 1993).

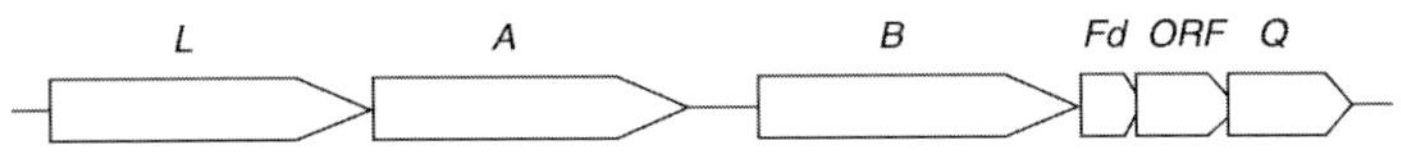

Figure 2. Minor nif *cluster from* A. vinelandii.

Specific functions of some of the *nif* gene products are outlined below. A more extensive review of them can be found in Rangaraj *et al.* (2000) and Dixon (this volume).

(i) The *nifHDK* genes encode the basic structural proteins of the nitrogenase enzyme; the *nifH* encodes the Fe protein (Component 2 or dinitrogenase reductase) and the *nifDK* genes encode the subunits of the MoFe protein (Component 1 or dinitrogenase).

(ii) The products of the *nifB, nifV, nifENX, nifQ,* and *nifH* genes are required for synthesis of the FeMo cofactor, an essential component of the enzyme for substrate (N_2) binding and reduction.

(iii) The *nifU* and *nifS* products are required for formation of the Fe-S clusters in the enzyme, some of which are involved in electron transfer to the FeMo cofactor.

(iv) The products of the *nifM* and *nifW* genes participate in other aspects of enzyme maturation.

(v). The products of the *nifLA* operon are involved in regulation of expression of all the other *nif* genes (see Merrick, this volume).

Unlike in most other diazotrophs of the Proteobacteria, ammonium inhibition of nitrogen fixation occurs at only one target in *A. vinelandii*, by inactivation of NifA by NifL. It does not involve either regulation of expression of the *nifLA* operon, as occurs in *K. pneumoniae,* or reversible inactivation of nitrogenase enzyme, as occurs in the photosynthetic proteobacterial diazotrophs.

In addition to using the classic Mo-containing enzyme for nitrogen fixation, all *Azotobacter* species examined have the potential to synthesize one or more alternative nitrogenases that do not contain Mo. Evidence for such alternative nitrogenases first came from studies with *A. vinelandii* mutants UW10 and UW91, derivatives of AvOP (UW) (ATCC 13705) carrying mutations in the *nifH* and *nifD* genes, which encode protein subunits of nitrogenase. The discovery and characterization of the alternative nitrogenases are discussed later in this chapter.

4. REGULATION OF EXPRESSION OF *nif* AND ASSOCIATED GENES BY AMMONIUM AND O_2: ROLES FOR PRODUCTS OF THE *nifA, nifL, glnD,* AND *glnK* GENES

NifL prevents expression of the other *nif* genes if either high levels of ammonium or O_2 is present, conditions that would make nitrogenase unnecessary for growth or become inactive, respectively. The NifL protein, which carries a flavin cofactor, inhibits the activity of NifA, which is a transcriptional activator of the other *nif* genes and operons under these conditions (for review, see Dixon, 1998). The presence of a *nifL* gene is characteristic of diazotrophs in the Gammaproteobacteria, which includes both *Klebsiella pneumoniae* and *Azotobacter vinelandii*, the two species most studied. The *nifL* gene has not been identified in other proteobacterial diazotrophs, with the exception of *Azoarcus* species in the class Betaproteobacteria (Egener *et al.*, 2002). As mentioned above, certain mutations in *nifL* can result in constitutive expression of *nif* genes, which leads to unregulated overexpression of the nitrogenase enzyme and excretion of ammonium in *A. vinelandii* (Bali *et al.*, 1992; Brewin *et al.*, 1999). Expression of the *nifLA* operon does not require NtrC in *A. vinelandii*, as it does in *K. pneumoniae* (Toukdarian and Kennedy, 1986). The mechanisms by which NifL inactivates NifA and how the GlnK and GlnD proteins interact to result in conversion of NifL from its inert to the inactivating conformation are discussed in detail by both Dixon, Chapter 1, and Merrick, Chapter 9, in this volume.

Unique aspects of the *glnK* and *glnD* genes encoding these proteins in *A. vinelandii* are that: (i) *glnK* is the only gene that encodes a PII-like protein that is reversibly uridylylated by GlnD (Meletzus *et al.*, 1998), unlike in most other Proteobacteria in which two or three genes encoding PII proteins are present; and (ii) *glnK* and *glnD* are essential genes and null mutations cannot be stably maintained (Colnaghi *et al.*, 2001; Rudnick *et al.*, 2002). In the case of *glnD*, null mutations can be introduced into a strain in which glutamine synthetase (GS) cannot be inactivated by adenylylation; this strain carries a mutation identified as *gln-71* which is probably an allele of *glnE*, the gene encoding the enzyme, GlnE, for adenylylation/deadenylylation of GS. GS is itself an essential enzyme in *A. vinelandii* and *glnA* null mutations could also not be isolated (Toukdarian *et al.*, 1990). That GlnK has two vital functions in *A. vinelandii* was indicated by the finding that, whereas *glnK* null mutations could not be stably maintained in a *gln-71* strain, a *glnKY51F* mutation could be introduced into this background. Thus, if GlnK cannot be uridylylated (by mutation of Y51 to F), then the influence of GlnK-UMP on GS deadenylylation is severe and GS remains inactive. However, another, yet unknown, vital function for GlnK is indicated because a full knock-out of function of *glnK* is lethal in either the wild-type or *gln-71* strains (Rudnick *et al.*, 2002).

5. ANCILLARY PROPERTIES OF *AZOTOBACTER* SPECIES THAT AID THE EFFICIENCY OF NITROGEN FIXATION: O_2 PROTECTION OF NITROGENASE, HYDROGENASE, AND FERREDOXINS

5.1. O_2 Protection

Nitrogen fixation is an O_2-sensitive process due to the O_2-lability of the nitrogenase component proteins. In *Azotobacter*, N_2 fixation occurs under fully aerobic growth conditions in all species examined except *A. salinestris*. Even the latter species can adapt to aerobic diazotrophy under conditions of Fe limitation, which increases production of a melanin-catechol that may trap Fe and be involved in O_2 tolerance (Page and Shivprasad, 1995). Aerotolerance in *A. vinelandii* is due mainly to a low O_2 affinity, high K_m cytochrome *d* terminal oxidase encoded by the *cydAB* genes. Mutation in either of these genes results in mutant strains unable to grow aerobically on N_2, which represents the loss of respiratory protection. Under low (2%) O_2, the mutants can fix N_2 (Kelly *et al.*, 1990). The CydR protein represses transcription of the *cydAB* operon under conditions of low O_2 (Wu *et al.*, 1997) during which the cytochrome *c* pathway provides respiration (Rey and Maier, 1997). A supply of ATP and reducing equivalents are additional factors that may be important to prevent O_2 damage to nitrogenase (Oelze, 2000).

Another factor of significance is the FeSII (or Shethna) protein. This protein binds to and protects Mo-nitrogenase from O_2 damage (by so-called 'conformational protection'). However, it apparently does not either bind to or protect either of the alternative Vnf and Anf nitrogenases during periods of O_2 stress (Dervartaniun *et al.*, 1968; Moshiri *et al.*, 1994). *A. vinelandii* strains with mutations in the gene, *fesII,* encoding the FeSII Shethna protein, show reduced viability under carbon deprivation if O_2 is present (Maier and Moshiri, 2000). This protein is found in *A. chroococcum*, where it also provides 'conformational protection' from O_2 damage to nitrogenase (Robson, 1979). An *fesII* DNA probe from *A. vinelandii* hybridized only to DNA prepared from *A. chroococcum* but not to that of several other diazotrophs tested (Moshiri *et al.*, 1994). The FeSII protein may, therefore, be found only in species of *Azotobacter* and, thus, may be a distinguishing characteristic. Production of alginates and an alginate capsule may provide a barrier to O_2 transfer (Sabra *et al.*, 2000), adding to the battery of mechanisms elaborated by *Azotobacter* species for protecting nitrogenase from damage by O_2 in these obligate aerobes.

5.2. Hydrogenase

H_2-uptake systems are similar in both *A. vinelandii* and *A. chroococcum*, having been best characterized physiologically in the latter species. H_2 recycling occurs in N_2-fixing cultures. H_2, a by-product of the reduction of N_2 to NH_4^+, is re-oxidized to H^+ by a Ni-dependent hydrogenase encoded by the 16-gene *hup* cluster in *A. chroococcum* (Lisheng and Tibelius, 1994; Lisheng *et al.*, 1994) and by the similar *hox/hyp* cluster in *A. vinelandii* (Enon *et al.*, 1992; Garg *et al.*, 1994). Hydrogenase activity is apparently beneficial in *A. chroococcum*, as shown by the ability of the Hup^+ strain MCD-1 to better survive than a Hup^- mutant under controlled carbon-limited chemostat conditions (Yates and Campbell, 1989). In *A. vinelandii,* both Hox^+ and Hox^- strains yielded very similar amounts of protein in chemostat cultures (Linkerhagner and Oelze, 1995). Both also had similar respiratory activities and, in the wild-type, only a small fraction of total respiratory activity was due to H_2-dependent O_2 consumption. This finding led to the conclusion that hydrogenase does not benefit *A. vinelandii* during carbon-limited growth under N_2-fixing conditions.

5.3. Ferredoxins

It is somewhat remarkable that *A. vinelandii* can synthesize at least 12 different ferredoxin-like proteins (Jung *et al.*, 1999), some of which (FeSI, FixX, FdxN, and FixFd) have been implicated in electron transfer to nitrogenase. However, no mutation in a single ferredoxin-encoding gene prevents electron transfer to nitrogenase, thus, there is probably redundancy of function among these proteins. The VnfFd ferredoxin is required for activity of the V-containing nitrogenase (Raina *et al.*, 1993). Another ferredoxin is the FeSII Shethna protein (described above), which protects nitrogenase from inactivation by O_2. Known ferredoxins in *A. chroococcum* include those encoded by a gene adjacent to *vnfH* and also FdI and FdII, corresponding to ferredoxins characterized in *A. vinelandii* (George *et al.*, 1984; Robson et al., 1989; Thomson, 1991).

6. DISCOVERY OF MOLYBDENUM-INDEPENDENT NITROGENASE SYSTEMS IN *A. VINELANDII*

The process of N_2 fixation had been studied for many years in *A. vinelandii* but it was not realized until the 1980s that this organism has three genetically distinct nitrogenase systems (Bishop and Premakumar, 1992). One of the enzyme systems is the well-characterized Mo-containing nitrogenase (nitrogenase-1) and the other two systems are Mo-independent nitrogenases; one is the vanadium(V)-containing nitrogenase (nitrogenase-2) and the other contains only iron (nitrogenase-3). Nitrogenase-1 is only expressed when *A. vinelandii* is grown in media containing Mo. Nitrogenase-2 is synthesized when this organism is grown in fixed-nitrogen-free media lacking Mo but with V (Hales *et al.*, 1986; Hales *et al.*, 1986; Robson *et al.*, 1986). This enzyme system consists of two components; dinitrogenase reductase-2 (also called Fe protein-2), a dimer of two identical subunits, and dinitrogenase-2 (or VFe protein), a hexamer (M_r of about 240,000) of two dissimilar pairs of large subunits (α and β) and a pair of small subunits (γ) (Robson *et al.*, 1989). Dinitrogenase reductase-2 has a M_r of about 62,000 and contains four Fe atoms and four acid-labile sulfides per dimer (Eady *et al.*, 1988; Hales *et al.*, 1986). Dinitrogenase-2 contains two V atoms, 23 Fe atoms, and 20 acid-labile sulfides per molecule (Eady *et al.*, 1987). A FeV cofactor, analogous to the FeMo cofactor, has been extracted from dinitrogenase-2 using N-methylformamide (Smith *et al.*, 1988).

Nitrogenase-3 does not appear to contain either Mo or V; it is made under Mo- and V-deficient diazotrophic conditions (Chisnell *et al.*, 1988). This enzyme is composed of two components, dinitrogenase reductase-3 (or Fe protein-3) and dinitrogenase-3 (or FeFe protein). Dinitrogenase reductase-3 is a dimer (M_r of approximately 65,000) of two identical subunits, whereas dinitrogenase-3 is a hexamer (M_r of about 216, 000) composed of two dissimilar pairs of subunits (α and β) and a pair of small subunits (γ) (Pau *et al.*, 1993). Dinitrogenase reductase-3 contains four Fe atoms and four acid-labile sulfides per dimer. Dinitrogenase-3 contains approximately 24 Fe atoms and 18 acid-labile sulfides per molecule (Chisnell *et al.*, 1988).

Nitrogenase-2 and nitrogenase-3 were unknown prior to 1980. Until that time, it was generally thought that Mo was absolutely required for N_2 fixation, even though scattered reports indicated that some diazotrophs could grow slowly in N-free medium lacking Mo. These low rates of N_2 fixation were usually attributed to the incorporation of trace amounts of contaminating Mo into nitrogenase. The early work of Bortels (1930) on *Azotobacter* species established the enhancement of N_2 fixation upon the addition of small amounts of Mo to growth media. Bortels (1936) also showed that low concentrations of V stimulated the growth of *Azotobacter* species under diazotrophic conditions. For many years, it seemed possible that V could substitute for Mo in dinitrogenase-1, because many of the chemical properties of these two metals are similar (Burns *et al.*, 1971; McKenna *et al.*, 1970). By the mid 1970s, it appeared that stimulation by V might be explained by other hypotheses such as the incorporation of V into dinitrogenase-1 with consequent stabilization of the enzyme and a more effective utilization of the small amount of Mo found in Mo-starved cells (Benemann *et al.*, 1972).

In the early 1980s, evidence was presented indicating that *A. vinelandii* contained at least two nitrogenase systems; the conventional Mo-containing nitrogenase system and an alternative nitrogenase system expressed in the absence of Mo. The evidence was based on the observation that Nif^- mutant strains underwent phenotypic reversal (*i.e.*, Nif^- to Nif^+) under conditions of Mo deprivation. Initially, these reports (Bishop *et al.*, 1980; Bishop *et al.*, 1982; Page and Collinson, 1982; Premakumar *et al.*, 1984) were received with skepticism because they challenged the long-held belief that Mo was absolutely required for nitrogen fixation and that nitrogenases were essentially the same regardless of their source. The latter notion was strengthened by Southern blot experiments (Ruvkun and Ausubel, 1980), which indicated that the structural genes encoding nitrogenases from diverse diazotrophic organisms were highly conserved at the nucleotide sequence level. In retrospect, it is interesting that in this same report multiple hybridizing bands were seen in *A. vinelandii* genomic DNA, which we now know are due to the presence of three nitrogenase systems in this organism.

Because the Nif^- strains of *A. vinelandii* first used to demonstrate phenotypic reversal contained point mutations, it was possible that phenotypic reversal resulted from increased leakiness of the mutant phenotypes when starved for Mo and not from derepression of an alternative N_2-fixation system. This hypothesis was ruled out when mutant strains carrying deletions in the structural genes for nitrogenase-1 were unequivocally shown to undergo phenotypic reversal under Mo-deficient conditions (Bishop *et al.*, 1986; Bishop *et al.*, 1986; Robson, 1986). Strains with deletions in the structural genes (*nifHDK*) have also facilitated the isolation of nitrogenase-2 from *A. chroococcum* (Robson *et al.*, 1986) and nitrogenases-2 and -3 from *A. vinelandii* (Chisnell *et al.*, 1988; Hales *et al.*, 1986; Hales *et al.*, 1986).

6.1. Characterization of Genes Encoding Mo-independent Nitrogenase Systems in A. vinelandii *and* A. chroococcum

Each of the two Mo-independent systems is described separately.

6.1.1. Genes Encoding Nitrogenase-2

The structural genes encoding dinitogenase-2 (the VFe protein) and dinitrogenase reductase-2 (Fe protein-2) have been described for both *A. chroococcum* (Robson *et al.*, 1986; Robson *et al.*, 1989) and *A. vinelandii* (Joerger *et al.*, 1990; Raina *et al.*, 1988). In contrast to the single operon (*nifHDK*) encoding the subunits for nitrogenase-1, the genes encoding nitrogenase-2 proteins are split between two operons (Figure 3). The *vnfH* gene encodes the dinitrogenase reductase-2 subunits and is part of a two-gene operon. The second gene (*vnfFd*) encodes a ferredoxin-like protein and is required for nitrogenase-2-dependent diazotrophic growth (Raina *et al.*, 1993). The *vnfH* gene is preceded by a potential promoter sequence that would be predicted to interact with core RNA polymerase containing the σ^N factor (*ntrA*, *rpoN*, or *glnF* gene product). The *vnfDGK* operon, located either 1.0 kb (*A. vinelandii*) or 2.5 kb (*A. chroococcum*) downstream from the *vnfHFd* operon, encodes the subunits for dinitrogenase-2. The v*nfD* gene encodes the α-subunit and *vnfK* encodes the β-subunit. In *A. vinelandii*, the 1.0-kb region between the *vnfHFd* and *vnfDGK* operons does not appear to contain any identifiable genes (Joerger *et al.*, 1990). The third subunit, γ, (M_r of 13,274) is encoded by the *vnfG* gene (Robson *et al.*, 1989). This gene does not have a counterpart in the Mo-containing nitrogenase system, however, it does share some sequence similarity to the predicted translation product of *anfG*, a gene located between *anfD* and *anfK*. This third subunit is required for diazotrophic growth in Mo-deficient medium containing V and is thought to be involved in the processing of apodinitrogenase-2 to functional dinitrogenase-2 (Chatterjee *et al.*, 1997; Waugh *et al.*, 1995).

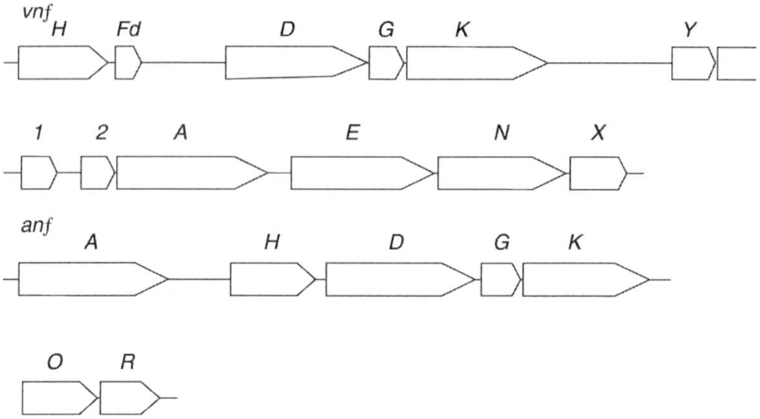

Figure 3. Organization of Mo-independent nitrogenase genes. Numbers indicate ORFs, whereas letters indicate either vnf *or* anf *genes. The* vnfAENX *region is located approximately 5.2 kb upstream of* vnfH. *The* anfOR *genes are located directly downstream of* anfK.

The sizes of transcripts from the *vnfHFd* operon are 1.4 and 1.0 kb and they are only partially NH_4^+-repressible. The 1.0-kb transcript contains the structural *vnfH* gene without the ferredoxin-like gene, whereas the 1.4-kb transcript represents the full-length *vnfHFd* transcript (Jacobitz and Bishop, 1992; Luque and Pau, 1991). The *vnfDGK* genes appear to be co-transcribed and transcription is only partially NH_4^+-repressible (Jacobitz and Bishop, 1992). Northern-blot experiments conducted with a *vnfDG* probe indicate that transcripts from the *vnfDGK* operon are 3.4 and 1.8 kb in size (Jacobitz and Bishop, 1992; Luque and Pau, 1991).

6.1.2. Genes Encoding Nitrogenase-3

The structural genes encoding nitrogenase-3 in *A. vinelandii* are organized as a single operon, *anfHDGKOR* (Joerger *et al.*, 1989b; Mylona *et al.*, 1996; see Figure 3). The subunits of dinitrogenase reductase-3 (Fe protein-3) are encoded by *anfH*, whereas the α– and β–subunits of dinitrogenase-3 (FeFe protein) are encoded by *anfD* and *anfK*, respectively. The *anfG* gene encodes a third subunit (γ) for dinitrogenase-3 (Pau *et al.*, 1993) and is required for growth under Mo- and V-deficient conditions (Waugh *et al.*, 1995). The *anfHDGKOR* operon is preceded by a potential promoter sequence that would interact with σ^N-containing RNA polymerase. The predicted protein products of the *anfO* and *anfR* genes do not show overall similarity to any known *nif* gene product. However, the predicted *anfO* product contains some sequence identity to the NH_2-terminal part of dinitrogenase reductase and another region exhibits identity to presumed heme-binding domains of P-450 cytochromes. The predicted product of *anfR* does not show any significant similarity with other amino acid sequences in the Genebank data base (Joerger *et al.*, 1989b; Mylona *et al.*, 1996). Mutants containing in-frame deletions in either *anfO* or *anfR* are Anf$^-$, thus, both genes are required for nitrogenase-3-dependent diazotrophic growth (Mylona *et al.*, 1996).

The sizes of transcripts from the *anfHDGKOR* operon are: (i) a 6.0-kb full-length transcript containing *anfHDGKOR*; (ii) a 4.3-kb transcript containing *anfHDGK*; (iii) a 3.5-kb transcript containing *anfDGK*; (iv) a 2.6-kb transcript containing *anfHD*; (v) a 1.0-kb transcript containing *anfH*; and (vi) a 0.6-kb transcript hybridizing to a probe containing *anfOR* (Luque and Pau, 1991; Premakumar *et al.*, 1992). At the present time, the origin of these individual mRNA species is unknown although several mechanisms could account for them. One possibility is that the 6.0-kb mRNA species (*anfHDGKOR*) is transcribed and subsequently cleaved at specific sites.

6.1.3. Genes Required for Active Mo-independent Nitrogenases

Other gene products that are required for active nitrogenases-2 and -3 are encoded by the *vnfENX* operon, which is located immediately downstream from *vnfA* (Figure 1). These gene products are homologous to the *nifENX* gene products that are required for FeMo cofactor synthesis (Wolfinger and Bishop, 1991). A V-Fe cluster has been shown to accumulate on VnfX during synthesis of the FeV cofactor, thus, VnfX is involved in VFe cofactor synthesis for dinitrogenase-2 (Rüttimann-Johnson *et al.*, 1999). *VnfY*, located downstream from *vnfK* is required for effective V-dependent diazotrophic growth and appears to play a role in the maturation of dinitrogenase-2 with a specific role in the formation of the FeV cofactor and/or its insertion into apodinitrogenase-2 (Rüttimann-Johnson *et al.*, 2003). Some genes are shared by all three nitrogenase systems; these are *nifB*, *nifU*, *nifS*, *nifV*, and *nifM* (Kennedy *et al.*, 1986; Joerger and Bishop, 1988; Kennedy and Dean, 1992). Their common requirement suggests that the three enzyme systems share several common features.

6.2. Regulation of the Molybdenum-independent Nitrogenase Systems in A. vinelandii

Transcription of genes involved in the nitrogenase-2 and -3 systems appears to initiate near potential promoter sites that conform to the RNA polymerase-σ^N recognition sequence, CTGG-N_8-TTGCA, where the underlined nucleotides are invariant (Beynon *et al.*, 1983). Such sites are found in the 5' non-coding region of all *vnf* and *anf* operons, except for *vnfA* where one potential promoter is situated within the 3' end of ORF2 (Figure 1) and another is located in the region between ORF1 and ORF2 (Joerger *et al.*, 1989a; Joerger *et al.*, 1989b; Joerger *et al.*, 1990). As expected from these observations, the *ntrA* gene, which encodes σ^N, is required for all three nitrogenase systems (Toukdarian and Kennedy, 1986).

The three nitrogenases found in *A. vinelandii* are expressed in response to the presence or absence of Mo or V. In the presence of Mo, nitrogenase-1 is expressed and nitrogenases-2 and -3 are repressed. Nitrogenase-2 is expressed under Mo-deficient conditions in the presence of V, whereas both nitrogenase-1 and -3 are repressed. Finally, nitrogenase-3 is expressed under conditions where both Mo and V are absent (Bishop and Premakumar, 1992). The synthesis of all three nitrogenases is repressed by NH_4^+. Ammonium regulates nitrogenase-2 at the post-transcriptional level, whereas both nitrogenase-1 and -3 are repressed at the transcriptional level (Jacobitz and Bishop, 1992). Because the genes encoding nitrogenase-2 are split into two operons, *vnfHFd* and *vnfDGK*, dinitrogenase reductase-2 and dinitrogenase-2 can be expressed independently. Unlike dinitrogenase-2, dinitrogenase reductase-2 is not only expressed under diazotrophic conditions in the presence of V, but also in the absence of Mo and V, conditions under which nitrogenase-3 is expressed (Bishop and Premakumar, 1992; Premakumar *et al.*, 1989). One reason for this situation is that dinitrogenase reductase-2 is required for transcription of the *anfHDGKOR* operon (Joerger *et al.*, 1991).

The regulatory genes, *nifA*, *vnfA*, and *anfA*, encode transcriptional activators, which are required for the transcription of the *nif*, *vnf*, and *anf* operons, respectively (Bishop and Premakumar, 1992; Joerger *et al.*, 1989a). VnfA is also either directly or indirectly involved in the repression of nitrogenase-1 in cells growing diazotrophically in Mo-deficient medium with or without V. VnfA seems also to be involved, albeit indirectly, in mediating repression of nitrogenase-3 by V (Joerger *et al.*, 1989a; Luque and Pau, 1991; Walmsley *et al.*, 1994). Surprisingly, NifA is required for full transcriptional expression of the *anfHDGKORF* operon (Walmsley *et al.*, 1994). All three transcriptional activators, NifA, VnfA, and AnfA, activate transcription from the *nifB* promoter (Drummond *et al.*, 1996).

The activator proteins have a three-domain structure (Joerger *et al.*, 1989a). An N-terminal domain is thought to interact with a transmitter of environmental signals (Drummond *et al.*, 1986), a central domain is proposed to interact with RNA polymerase, the sigma factor or both, and a C-terminal domain is involved with DNA binding. The highest similarity between these activators is in the C-terminal half of the proteins, where both the potential RNA polymerase-σ^N interaction sites and ATP-binding domains are located. A potential DNA-binding domain (Fischer *et al.*, 1988), which includes a helix-turn-helix motif, is present in both VnfA and AnfA (Joerger *et al.*, 1989a). However, the determinants of promoter specificity lie outside of the helix-turn-helix motif in the C-terminal domain of VnfA and AnfA (Jacob and Drummond, 1993).

An upstream activating sequence (UAS) for *vnfH* and *vnfD* was identified as GTAC-N_6-GTAC and was found to be duplicated with either 17 or 18 bases between each for both *vnfH* and *vnfD* promoters (Woodley *et al.*, 1996). This UAS was hypothesized to be the binding site for VnfA. In a related study, the sequences, 5'-GTACCATGCGGAAC-3' and 5'-GTACCTGCGGGTAC-3', which are located 170 and 140 nucleotides upstream of the *vnfH* promoter, were found to be required for *vnfH* expression and are thought to be the binding site for VnfA (Bageshwar *et al.*, 1998). The N-terminal domains of AnfA and VnfA differ from the N-terminus of NifA in that they have a cysteine cluster, reminiscent of the FNR family of redox-sensitive regulatory proteins (Joerger *et al.*, 1989a; Premakumar *et al.*, 1994). In the case of AnfA, both cysteine-21 and -26 are essential for AnfA function (Premakumar *et al.*, 1994). Using chimeric transcriptional activators generated from VnfA and AnfA of *A. vinelandii*, Frise *et al.* (1994) have shown that promoter specificity is determined by the C-terminal region of AnfA and that the requirement of dinitrogenase reductase-2 for transcriptional activation of the *anfHDGKOR* operon (Joerger *et al.*, 1991) by this protein is a function of its N-terminal domain. Thus, it is possible that the requirement of dinitrogenase reductase-2 for transcription of the *anfHDGKOR* operon involves the reduction of a potential redox center in the N-terminal domain of AnfA (Premakumar *et al.*, 1992). Austin and Lambert (1994) demonstrated that an N-terminally truncated AnfA with the central and C-terminal DNA-binding domains intact could activate transcription in an *in vitro* transcription system. They also showed binding of AnfA to sites between 200-300 base pairs upstream of the *anfH* promoter.

Regulation of the *vnf* and *anf* operons by NH_4^+, V, and Mo appears to be mediated through the effect of NH_4^+, V, and Mo on the transcriptional expression of the *vnfA* and *anfA* genes with the exception that *vnfA* expression is not affected by the presence of NH_4^+. Transcription of *vnfA* is not repressed by NH_4^+ but is repressed by Mo, whereas NH_4^+, V and Mo repress expression of *anfA*. ModE, a repressor of the *modABCD* (Mo transport) operon also plays a role in the repression of *anfA* transcription by Mo (Premakumar *et al.*, 1998).

The details of regulation of both nitrogenase-2 and -3 by V and Mo are not well understood, however, mutants that express nitrogenase-3 in the presence of normally repressive concentrations of Mo can be readily isolated by selecting for tolerance to tungsten under N_2-fixing conditions (Bishop *et al.*, 1980; Premakumar *et al.*, 1996). One such mutant, CA6, may have a mutation in a Mo-transport system that is distinct from the well-characterized high-affinity Mo-transport system described for *A. vinelandii* (Bishop *et al.*, 1980; Premakumar *et al.*, 1996). Also, mutations in the genes that encode the high-affinity Mo-transport system (*modABCD* and *modE*) permit expression of the Mo-independent nitrogenases in the presence of low, but not high, concentrations of Mo (Luque *et al.*, 1993; Mouncey *et al.*, 1995). Thus, the genes involved in transport may also play a role in the ability to sense intracellular concentrations of Mo.

7. MOLYBDENUM-INDEPENDENT NITROGENASE SYSTEMS IN OTHER *AZOTOBACTER* SPECIES AND CLOSELY RELATED NATURAL ISOLATES

The presence of Mo-independent nitrogenase systems in *Azotobacter* species is probably a common occurrence because they are found in the four species that have so far been examined. In addition to nitrogenase-1, *A. chroococcum* has nitrogenase-2, *A. paspali* appears to have both nitrogenases-2 and -3, and A. salinestris has nitrogenase-2 (Fallik *et al.*, 1991; Loveless, and Bishop, 1999; Page, and Collinson, 1987). Another closely related bacterium, *Azomonas macrocytogenes*, has nitrogenase-3 in addition to nitrogenase-1 (Loveless, and Bishop, 1999; Page and Collinson, 1987).

Seven diazotrophic bacteria have been isolated from environmental sources using N-free, Mo-deficient medium as the enrichment medium. These isolates appear to be specifically related the fluorescent pseudomonads and are closely related to *Azotobacter* species. All of the isolates have nitrogenases-1 and -2 and four of the isolates also have nitrogenase-3 (Loveless *et al.*, 1999). Diazotrophs with Mo-independent nitrogenases can be isolated from a variety of environments, including those that are known to have Mo concentrations sufficient for Mo-dependent nitrogen fixation, such as a waste water treatment plant (about 90 nM Mo), and salt marshes (approximately 110 nM Mo). These observations suggest that factors other than the concentration of Mo in the macro environment may be important in determining the presence of these diazotrophs.

7.1. What is the Role of Molybdenum-independent Nitrogenases?

One of the remaining unanswered questions is why certain diazotrophs have Mo-independent nitrogenases. These nitrogenases are found in a rather diverse group of diazotrophs, which includes *Clostridium pasteurianum* (Zinoni *et al.*, 1993), *Rhodobacter capsulatus* (Schneider *et al.*, 1991; 1997; Schuddekopf *et al.*, 1993), *Anabaena variabilis* (Kentemich *et al.*, 1988; Thiel, 1993), *Rhodospirillum rubrum* (Davis *et al.*, 1996; Lehman and Roberts, 1991), *Heliobacterium gestii* (Kimble and Madigan, 1992; Loveless and Bishop, 1999), *Rhodopseudomonas palustris* (at *http://www.jgi.doe.gov/JGI_microbial/html/rhodopseudomonas/rhodops_homepage.html*) and *Methanosarcina barkeri* (Chien *et al.*, 2000). An obvious possibility is that these nitrogenases could provide diazotrophs with a selective advantage under conditions of Mo limitation. One such scenario might be in acid soils with high iron oxide contents, where MoO_4^{2-} is tightly bound to the iron oxide and, thus, biologically unavailable. The limitation of Mo-dependent nitrogen fixation due to high concentrations of sulfate in oxic marine environments has been hypothesized to account for the chronic nitrogen depletion observed in these environments (Howarth and Cole, 1985), leading to the suggestion that Mo-independent nitrogenases might be advantageous for diazotrophic growth in high sulfate environments. However, diazotrophic growth by *A. vinelandii* was Mo dependent even in media containing extremely high ratios of SO_4^{2-}-to-MoO_4^{2-} (Paulsen *et al.*, 1991). Another possibility is that Mo-independent nitrogenases may function more efficiently than Mo-dependent nitrogenase-1 under environmental conditions unrelated to metal availability. For example, there is evidence that nitrogenase-2 may function more efficiently at low temperatures than nitrogenase-1 (Miller and Eady, 1988). Also, it appears that Mo uptake may be inhibited during growth at low temperatures (Walmsley and Kennedy, 1991).

Finally, Mo-independent nitrogenases may provide an advantage to diazotrophs in situations where living organisms could cause local depletions of Mo, such as in plant rhizospheres, in canopies of tropical rain forests, and in some symbiotic relationships (Bishop, 1993). For example, Mo uptake by plants could result in a local depletion of Mo in the rhizosphere. In this case, heterotrophic diazotrophs having Mo-independent nitrogenases should have a selective advantage in colonizing such potentially carbon-rich environments. This logic can also be extended to microbial communities, such as microbial mats and biofilms, where the microorganisms themselves might create Mo-depleted micro zones. One hypothesis is that Mo-deficient micro zones are formed during growth of *A. vinelandii* on a solid substrate surface due to the organism's powerful Mo-accumulation system. Consistent with this hypothesis, wild-type *A. vinelandii* out-competes a mutant lacking nitrogenase-3 during diazotrophic growth on Mo-sufficient agar medium (Maynard *et al.*, 1994). Furthermore, *A. vinelandii* may gain a competitive advantage over other diazotrophs that lack Mo-independent nitrogenases because of its ability to remove relatively large amounts of molybdenum from its immediate environment. Thus, by generating Mo-deficient microenvironments, this organism could create an ecological niche that excludes diazotrophs that do not have Mo-independent nitrogenases. If this hypothesis applies to other diazotrophs that have Mo-independent nitrogenases, one would predict that such diazotrophs can be isolated from macro environments that are known to have sufficient concentrations of Mo for Mo-dependent N_2 fixation.

ACKNOWLEDGEMENTS

C.K. thanks the United States of America Department of Agriculture (USDA), National Research Initiative (NRI) for funding her research during the past eleven years (1992 - 2003) for studies of *Azotobacter vinelandii* genes involved in ammonium sensing and gene regulation and genomic analysis (award numbers 92-

37305-7888, 95-37305-2067, 2001-35-3191-0013). C.K. also thanks the United States Department of Energy (DOE) for DNA sequencing of the *A. vinelandii* genome at the Joint Genome Institute (JGI) in Walnut Creek, CA, its annotation at the DOE Oak Ridge National Laboratories, and for support of the 'Workshop for Annotation of the *Azotobacter vinelandii* Genome', University of Arizona, Tucson, AZ (July 2002) (DOE award number defg0302erb3417). P.B. thanks USDA, NRI for funding (award number 96-35305-3554) his research on the alternative nitrogenases of *Azotobacter vinelandii* and other organisms.

REFERENCES

Austin, S., and Lamber, J. (1994). Purification and *in vitro* activity of a truncated form of ANFA. Transcriptional activator protein of alternative nitrogenase from *Azotobacter vinelandii. J. Biol. Chem., 269*, 18141-18148.

Badran, H., Sohoni, R., Venkatesh, T. V., and Das, H. K. (1999). Construction of a *recF* deletion mutant of *Azotobacter vinelandii* and its characterization. *FEMS Microbiol. Lett., 174*, 363-369.

Bageshwar, U. K., Raina, R., Choudhury, N. R., and Das, H. K. (1998). Analysis of upstream activation of the *vnfH* promoter of *Azotobacter vinelandii. Can. J. Microbiol., 44*, 405-415.

Balajee, S., and Mahadevan, A. (1989). Evidence for a dissimilatory plasmid in *Azotobacter chroococcum. FEMS Microbiol. Lett., 53(1-2),* 223-227.

Bali, A., Blanco, G., Hill, S., and Kennedy, C. (1992). Excretion of ammonium by a *nifL* mutant of nitrogen fixing *Azotobacter vinelandii. Appl. Environ. Microbiol., 58* , 1711-1718.

Becking, G. C. (1992). Methodology in neurotoxicology--activities within the World Health Organization and International Programme on Chemical Safety. *Toxicol Lett., 64-65,* 203-208.

Beijerinck, M. W. (1901). Über oligonitrophile mikroben. Zentr. *Bakteriol. Parasitenk. Abt. II, 7,* 190-217.

Benemann, J. R., McKenna, C. E., Lie, R. F., Traylor, T. G., and Kamen, M. D. (1972). The vanadium effect in nitrogen fixation by *Azotobacter. Biochim. Biophys. Acta, 264*, 25-38.

Bennett, L. T., Cannon, F. C., and Dean, D. R. (1988). Nucleotide sequence and mutagenesis of the *nifA* gene from *Azotobacter vinelandii. Mol. Microbiol., 2,* 315-321.

Bennett, L. T., Jacobson, M. R., and Dean, D. R. (1988). Isolation, sequencing, and mutagenesis of the nifF gene encoding flavodoxin from *Azotobacter vinelandii. J.Biol.Chem., 263,*1364-1369.

Beynon, J., Ally, A., Cannon, M., Cannon, F., Jacobson, M., Cash, V., and Dean, D. (1987). Comparative organization of nitrogen fixation-specific genes from *Azotobacter vinelandii* and *Klebsiella pneumoniae*: DNA sequence of the *nifUSV* genes. *J. Bacteriol., 169,* 4024-4029.

Beynon, J., Cannon, M., Buchanan-Wollaston, V., and Cannon, F. (1983). The *nif* promoters of *Klebsiella pneumoniae* have a characteristic primary structure. *Cell, 34*, 665-671.

Bishop, P. E., and Brill, W. J. (1977). Genetic analysis of *Azotobacter vinelandii* mutant strains unable to fix nitrogen. *J. Bacteriol., 130*, 954-956.

Bishop, P. E., Rizzo, T. M., and Bott, K. F. (1985). Molecular cloning of *nif* DNA from *Azotobacter vinelandii. J. Bacteriol., 162*, 21-28.

Bishop, P. E. (1993). Three genetically distinct nitrogenase systems in *Azotobacter vinelandii.* In L. L. Barton, and B. C. Hemming (Eds.), *Iron chelation in plants and soil microorganisms* (pp. 301-324). New York: Academic Press.

Bishop, P. E., and Premakumar, R. (1992). Alternative nitrogen fixation systems. In G. Stacey, R. H. Burris, H. J. Evans (Eds.), *Biological nitrogen fixation* (pp. 736-762). New York: Chapman and Hall.

Bishop, P. E., Hawkins, M. E., and Eady, R. R. (1986). Nitrogen fixation in Mo-deficient continuous culture by a strain of *Azotobacter vinelandii* carrying a deletion of the structural genes for nitrogenase (*nifHDK*). *Biochem. J., 238*, 437-442.

Bishop, P. E., Jarlenski, D. M. L., and Hetherington, D. R. (1980). Evidence for an alternative nitrogen fixation system in *Azotobacter vinelandii. Proc. Natl. Acad. Sci. USA, 77*, 7342-7346.

Bishop, P. E., Jarlenski, D. M. L., and Hetherington, D. R. (1982). Expression of an alternative nitrogen fixation system in *Azotobacter vinelandii. J. Bacteriol., 150*, 1244-1251.

Bishop, P. E., Premakumar, R., Dean, D. R., Jacobson, M. R., Chisnell, J. R., Rizzo, T. M., et al. (1986). Nitrogen fixation by *Azotobacter vinelandii* strains having deletions in structural genes for nitrogenase. *Science, 232*, 92-94.

Blanco, G., Ramos, F., Medina, J. R., and Tortolero, M. (1990). A chromosomal linkage map of *Azotobacter vinelandii*. *Mol. Gen. Genet., 224*, 241-247.

Blanco, G., Drummond, M., Woodley, P., and Kennedy, C. (1993). Sequence and molecular analysis of the *nifL* gene of *Azotobacter vinelandii*. *Mol. Microbiol., 9,* 869–879.

Bortels, H. (1930). Molybdän als katalysator bei der biolgischen stickstoffbindung. *Arch. Mikrobiol., 1*, 333-342.

Bortels, H. (1936). Weitere untersuchungen über die bedeutung von molybdän, vanadium, wolfram und andere erdaschenstoffe für stickstoffbindende und andere mikroorganismen. *Zentralbl. Bakteriol. Parasitenkd. Infektionskr. Abt.2, 95*, 193-218.

Brewin, B., Woodley, P., and Drummond, M. (1999). The basis of ammonium release in nifL mutants of *Azotobacter vinelandii*. *J.Bacteriol., 181,* 7356-7362.

Brigle K. E., and Dean, D. R. (1988). Revised nucleotide sequence of the *Azotobacter vinelandii nifE* gene. *Nucleic Acids Res., 16(11),* 5214.

Brigle, K. E., Newton, W. E., and Dean, D. R. (1985). Complete nucleotide sequence of the *Azotobacter vinelandii* nitrogenase structural gene cluster. *Gene, 37,* 37-44.

Burns, R. C., Fuchsman, W. H., and Hardy, R. W. F. (1971). Nitrogenase from vanadium-grown *Azotobacter*: isolation, characteristics, and mechanistic implications. *Biochem. Biophys. Res. Commun., 42*, 353-358.

Chatterjee, R., Ludden, P. W., and Shah, V. K. (1997). Characterization of VNFG, the delta subunit of the *vnf*-encoded apodinitrogenase from *Azotobacter vinelandii*. Implications for its role in the formation of functional dinitrogenase 2. *J. Biol. Chem., 272*, 3758-3765.

Chien, Y.-T., Auerbuch, V., Brabban, A., and Zinder, S. H. (2000). Analysis of genes encoding an alternative nitrogenase in the archaeon *Methanosarcina barkeri* 227. *J. Bacteriol., 182*, 3247-3253.

Chisnell, J. R., Premakumar, R., and Bishop, P. E. (1988). Purification of a second alternative nitrogenase from a *nifHDK* deletion strain of *Azotobacter vinelandii*. *J. Bacteriol., 170,* 27-33.

Colnaghi, R., Rudnick, P., He, L., Green, A., Yan, D., Larson, E., and Kennedy, C. (2001). Lethality of *glnD* null mutations in *Azotobacter vinelandii* is suppressible by prevention of glutamine synthetase adenylylation. *Microbiol., 147*, 1267-1276.

Contreras, A., and Casadesus, J. (1987). Tn*10* mutagenesis in *Azotobacter vinelandii*. *Mol. Gen. Genet., 209*, 276-282.

Contreras, A., Maldonado, R., and Casadesus, J. (1991). Tn*5* mutagenesis and insertion replacement in *Azotobacter vinelandii*. *Plasmid, 25*, 76-80.

Davis, R., Lehman, L., Petrovich, R., Shah, V. K., Roberts, G. P., and Ludden, P. W. (1996). Purification and characterizaton of the alternative nitrogenase from the photosynthetic bacterium *Rhodospirillum rurum*. *J. Bacteriol., 178*, 1445-1450.

Dilworth, M. J., Eady, R. R., and Eldridge, M. E. (1988). The vanadium nitrogenase of *Azotobacter chroococcum*. *Biochem. J., 249*, 745-751.

Dixon, R. O. (1978). Nitrogenase-hydrogenase interrelationships in rhizobia. *Biochimie, 60(3),* 233-236.

Dixon, R. (1998). The oxygen-responsive NIFL-NIFA complex, a novel two-component regulatory system controlling nitrogenase synthesis in gamma-proteobacteria, *Arch. Microbiol., 169*, 371-380.

Do Nascimento, G. F., and Tavares, F. A. (1987). Transformation with *nif* genes in *Azotobacter paspali*. *Rev. Microbiol., 18*, 41-45.

Doran, J. L., Bingle, W. H., Roy, K. L., Hiratsuka, K., and Page, W. J. (1987). Plasmid transformation of *Azotobacter vinelandii* OP. *J. Gen. Microbiol., 133*, 2059-2072.

Drummond, M. H., and Wootton, J. C. (1987). Sequence of *nifL* from *Klebsiella pneumoniae*: Mode of action and relationship to two families of regulatory proteins. *Mol. Microbiol., 1,* 37-44.

Drummond, M., Walmsley, J., and Kennedy, C. (1996). Expression from the *nifB* promoter of *Azotobacter vinelandii* can be activated by NifA, VnfA, or AnfA transcriptional activators. *J. Bacteriol., 178*, 788-792.

Drummond, M., Whitty, P., and Wootton, J. (1986). Sequence and domain relationships of *ntrC* and *nifA* from *Klebsiella pneumoniae*: Homologies to other regulatory proteins. *EMBO J., 5*, 441-447.

Eady, R. R., Issack, R., Kennedy, C., Postgate, J. R., and Ratcliffe, H. D. (1978). Nitrogenase synthesis in *Klebsiella pneumoniae*: Comparison of ammonium and oxygen regulation. *Gen. Microbiol., 104(2),* 277-285.

Eady, R. R., Richardson, T. H., Miller, R. W., Hawkins, M., and Lowe, D. J. (1988). The vanadium nitrogenase of *Azotobacter chroococcum*: Purification and properties of the Fe protein. *Biochem. J., 256,* 189-196.
Eady, R. R., Robson, R. L., Richardson, T. H., Miller, R. W., and Hawkins, M. (1987). The vanadium nitrogenase of *Azotobacter chroococcum*: Purification and properties of the VFe protein. *Biochem. J., 244,* 197-207.
Enon, A. L., Mortenson, L. E., and Robson, R. L. (1992). Nucleotide sequences and genetic analysis of hydrogen oxidation (*hox*) genes in *Azotobacter vinelandii. J. Bacteriol., 174,* 4549-4557.
Evans D, Jones, R., Woodley, P, and Robson, R. (1988). Further analysis of nitrogen fixation (*nif*) genes in *Azotobacter chroococcum*, identification and expression in *Klebsiella pneumoniae* of *nifS, nifV, nifM* and *nifB* genes and localization of *nifE/N-, nifU-, nifA-* and *fixABC*-like genes. *J. Gen. Microbiol., 134,* 931-942.
Fallik, E, Hartel, P. G., and Robson R. L. (1993). Presence of a vanadium nitrogenase in *Azotobacter paspali. Appl. Environ. Microbiol., 59,* 1883-1886.
Fallik, E., Hartel, P. G., and Robson, R. L. (1991). Detection of alternative nitrogenases in aerobic gram-negative nitrogen-fixing bacteria. *J. Bacteriol., 173,* 365-371.
Fischer, M.-H., Bruderer, T., and Hennecke, H. (1988). Essential and non-essential domains in the *Braydrhizobium japonicum* NifA protein: Identification of indispensable cysteine residues involved in redox reactivity and/or metal binding. *Nucleic Acids Res., 16,* 2207-2224.
Fisher, R. J., and Brill, W. J. (1969). Mutants of *Azotobacter vinelandii* unable to fix nitrogen. *Biochim Biophys Acta. 184(1),* 99-105.
Frise, E., Green, A., and Drummond, M. (1994). Chimeric trascriptional activators generated in vivo from VnfA and AnfA of *Azotobacter vinelandii*: N-terminal domain of AnfA is responsible for dependence on nitrogenase Fe-protein. *J. Bacteriol., 176,* 6545-6549.
Gangeswaran R, Lowe, D. J., and Eady, R. R. (1993). Purification and characterization of the assimilatory nitrate reductase of *Azotobacter vinelandii. Biochem. J.,* 289, 335-342.
Garg, R. P., Menon, A. L., Jacobs, K., Robson, R. M., and Robson, R. L. (1994). The *hypE* gene completes the *Gene* cluster for H_2-oxidation in *Azotobacter vinelandii. J. Mol. Biol., 236,* 390-396.
Glick, B. R., Butler, B. J., Mayfield, C. I., and Pasternak, J. J. (1989). Effect of transformation of *Azotobacter vinelandii* with the low copy number plasmid pRK290. *Current Microbiol., 19,* 143-146.
Glick, B. R., Brooks, H. E., and Pasternak, J. J. (1986). Physiological effects of plasmid DNA transformation on *Azotobacter vinelandii. Can. J. Microbiol., 32,* 145-148.
Hales, B. J., Case, E. E., Morningstar, J. E., Dezda, M. F., and Mauterer, L. A. (1986). Isolation of a new vanadium-containing nitrogenase from *Azotobacter vinelandii. Biochemistry, 25,* 7251-7255.
Hales, B. J., Langosch, D. J., and Case, E. E. (1986). Isolation and characterization of a second nitrogenase Fe-protein from *Azotobacter vinelandii. J. Biol. Chem., 261,* 15301-15306.
Howarth, R. W., and Cole, J. J. (1985). Molybdenum availability, nitrogen limitation, and phytoplankton growth in natural waters. *Science, 229,* 653-655.
Jacob, J., and Drummond, M. (1993). Construction of chimeric proteins from the sigma N-associated transcriptional activators VnfA and AnfA of *Azotobacter vinelandii* shows that the determinants of promoter specificity lie outside the 'recognition' helix of the HTH motif in the C-terminal domain. *Mol. Microbiol., 10,* 813-821.
Jacobitz, S., and Bishop, P. E. (1992). Regulation of nitrogenase-2 in *Azotobacter vinelandii* by ammonium, molybdenum, and vanadium. *J. Bacteriol., 174,* 3884-3888.
Jacobson, M. R., Brigle, K. E., Bennett, L., Setterquist, R. A., Wilson, R. A., Cash, V. L., Beynon, J., Newton, W. E., and Dean, D. R. (1989). Physical and *Gene*tic map of the major *nif* gene cluster from *Azotobacter vinelandii. J. Bacteriol., 171,* 1017-1027.
Jacobson, M. R., Premakumar, R. and Bishop, P. E. (1986). Transcriptional regulation of nitrogen fixation by molybdenum in *Azotobacter vinelandii. J. Bacteriol., 167,* 480-486.
Joerger, R. D., and Bishop, P. E. (1988). Nucleotide sequence and genetic analysis of the *nifB-nifQ* region from *Azotobacter vinelandii. J. Bacteriol., 170,* 1475-1487.
Joerger, R. D., Jacobson, M. R., and Bishop, P. E. (1989a). Two *nifA*-like genes required for expression of alternative nitrogenases by *Azotobacter vinelandii. J. Bacteriol., 171,* 3258-3267.
Joerger, R. D., Jacobson, M. R., Premakumar, R., Wolfinger, E. D., and Bishop, P. E. (1989b). Nucleotide sequence and mutational analysis of the structural genes (*anfHDGK*) for the second alternative nitrogenase from *Azotobacter vinelandii. J. Bacteriol., 171,* 1075-1086.

Joerger, R. D., Loveless, T. M., Pau, R. N., Mitchenall, L. A., Simon, B. H., and Bishop, P. E. (1990). Nucleotide sequence and mutational analysis of the structural genes for nitrogenase 2 of *Azotobacter vinelandii*. *J. Bacteriol., 172*, 3400-3408.

Joerger, R. D., Wolfinger, E. D., and Bishop, P. E. (1991). The gene encoding dinitrogenase reductase 2 is required for expression of the second alternative nitrogenase from *Azotobacter vinelandii*. *J. Bacteriol., 173*, 4440-4446.

Jones, R., Woodley, P., and Robson, R. (1984). Cloning and organisation of some genes for nitrogen fixation from *Azotobacter chroococcum* and their expression in *Klebsiella pneumoniae*. *Mol. Gen. Genet., 197*, 318-327.

Jung, Y. S., Gao-Sheridan, H. S., Christiansen, J., Dean, D. R., and Burgess, B. K. (1999). Purification and biophysical characterization of a new [2Fe-2S] ferredoxin from *Azotobacter vinelandii,* a putative [Fe-S] cluster assembly/repair protein. *J. Biol .Chem., 274*, 32402-2410.

Kelly, M. J. S., Poole, R. K., Yates, M. G., and Kennedy, C. (1990). Cloning and muta*Genes*is of genes encoding the cytochrome bd terminal oxidase complex in *Azotobacter vinelandii*, mutants deficient in the cytochrome d complex are unable to fix nitrogen in air. *J. Bacteriol., 172*, 6010-6019.

Kennedy, C., and Robson, R. (1983). Activation of *nif* gene expression in *Azotobacter* by the *nifA* gene product of *Klebsiella pneumoniae*. *Nature, 301*, 626-628.

Kennedy, C., and Toukdarian, A. (1987). Genetics of A*zotobacters*: Applications to nitrogen fixation and related aspects of metabolism. *Annu. Rev. Microbiol., 41*, 227-258.

Kennedy, C., and Dean, D. (1992). The *nifU*, *nifS*, and *nifV* gene products are required for activity of all three nitrogenases of *Azotobacter vinelandii*. *Mol. Gen. Genet., 231*, 494-498.

Kennedy, C., Gamal, R., Humphrey, R., Ramos, J., Brigle, K., and Dean, D. (1986). The *nifH*, *nifM*, and *nifN* genes of *Azotobacter vinelandii*: Characterization by Tn*5* mutagenesis and isolation from pLAFR1 gene banks. *Mol. Gen. Genet., 205*, 318-325.

Kennedy, C., Rudnick, P., MacDonald, M., and Melton, T. (2003) Genus *Azotobacter*. In G. Garrity *et al.* (Eds.), *Bergey's manual of systematic bacteriology, Volume II, The proteobacteria* (in press). New York: Springer.

Kentemich, T., Danneberg, G., Hundeshagen, B., and Bothe, H. (1988). Evidence for the occurrence of the alternative , vanadium-containing nitrogenase in the cyanobacterium *Anabaena variabilis*. *FEMS Microbiol. Lett., 51*, 19-24.

Lehman, L. J., and Roberts, G. P. (1991). Identification of an alternative nitrogenase system in *Rhodospirillum rubrum*. *J. Bacteriol., 173*, 5705-5711.

Loveless, T. M., and Bishop, P. E. (1999). Identification of genes unique to Mo-independent nitrogenase systems in diverse diazotrophs. *Can. J. Microbiol., 45*, 312-317.

Loveless, T. M., Saah, J. R., and Bishop, P. E. (1999). Isolation of nitrogen-fixing bacteria containing molybdenum-independent nitrogenases from natural environments. *Appl. Environ. Microbiol., 65*, 4223-4226.

Luque, F., and Pau, R. N. (1991). Transcriptional regulation by metals of structural genes from *Azotobacter vinelandii* nitrogenases. *Mol. Gen. Genet., 227*, 481-487.

Luque, R. F., Mitchenall, L. A., Chapman, M., Christine, R., and Pau, R. N. (1993). Characterization of genes involved in molybdenum transport in *Azotobacter vinelandii*. *Mol. Microbiol., 7*, 447-459.

Maier, R. J., and Moshiri, F. (2000). Role of the *Azotobacter vinelandii* nitrogenase-protective Shethna protein in preventing oxygen-mediated cell death. *J. Bacteriol., 182*, 3854-3857.

Maldonado, R., Jiminez, J., and Casadesus, J. (1994). Changes of ploidy during the *Azotobacter vinelandii* growth cycle. *J. Bacteriol., 176*, 3911-3919.

Manna, A. C., and Das, H. K. (1994). The size of the chromosome of *Azotobacter chroococcum*. *Microbiology, 140*, 1237-1239.

Manna, A. C., and Das, H. K. (1997). Characterization and mutagenesis of the leucine biosynthetic genes of *Azotobacter vinelandii*, an analysis of the rarity of amino acid auxotrophs. *Mol. Gen. Genet., 254*, 207-217.

Maynard, R. H., Premakumar, R., and Bishop, P. E. (1994). Mo-independent nitrogenase 3 is advantageous for diazotrophic growth of *Azotobacter vinelandii* on solid medium containing molybdenum. *J. Bacteriol., 176*, 5583-5586.

McKenna, C. E., Benemann, J. R., and Traylor, T. G. (1970). A vanadium-containing nitrogenase preparation: Implications for the role of molybdenum in nitrogen fixation. *Biochem. Biophys. Res. Commun., 41*, 1501-1508.

Meletzus, D., Rudnick, P., Doetsch, N., Green, A., and Kennedy, C. (1998). Characterization of the *glnK amtB* operon of *Azotobacter vinelandii*. *J. Bacteriol., 180*, 3260-3264.

Merrick, M, Filser, M., Kennedy, C., and Dixon, R. (1978). Polarity of mutations induced by insertion of transposons Tn5, Tn7 and Tn10 into the *nif* gene cluster of *Klebsiella pneumoniae*. *Mol. Gen. Genet., 165(1),* 103-111.

Miller, R. W., and Eady, R. R. (1988). Molybdenum and vanadium nitrogenases of *Azotobacter chroococcum*: Low temperature favours N_2 reduction by vanadium nitrogenase. *Biochem., J., 256*, 429-432.

Mishra, A. K., and Wyss, O. (1968). Induced mutations in *Azotobacter* and isolation of an adenine requiring mutant. *Nucleus, 111*, 96-105.

Mishra, A. K., Roy, B., Prasad, R., and Das, H. K. (1991). Amino acid transport in *Azotobacter vinelandii*, implications of non-availability of amino acid auxotrophs. *FEMS Microbiol. Lett., 79*, 41-44.

Moshiri, F., Kim, J. W., Fu, C., and Maier, R. J. (1994). The FeSII protein of *Azotobacter vinelandii* is not essential for aerobic nitrogen fixation, but confers significant protection to oxygen-mediated inactivation of nitrogenase *in vitro* and *in vivo*. *Mol. Microbiol., 14*, 101-114.

Mouncey, N. J., Mitchenall, L. A., and Pau, R. N. (1995). Mutational analysis of genes of the *mod* locus involved in molybdenum transport, homeostasis, and processing in *Azotobacter vinelandii*. *J. Bacteriol., 177*, 5294-5302.

Mylona, P. V., Premakumar, R., Pau, R. N., and Bishop, P. E. (1996). Characteristics of *orf1orf2* in the *anfHDGK* genomic region encoding nitrogenase 3 of *Azotobacter vinelandii*. *J. Bacteriol., 178*, 204-208.

Nagatani, H. H., Shah, V. K., and Brill, W. J. (1974). Activation of inactive nitrogenase by acid-treated component I. *J. Bacteriol., 120(2),* 697-701.

Owen, D. J., and Ward, A. C. (1985). Transfer of transposable drug-resistance elements Tn*5*, Tn*7*, and Tn*76* to *Azotobacter beijerinckii*, Use of plasmid RP4,Tn*76* as a suicide vector. *Plasmid, 14*, 162-166.

Page, W. J. (1985). Genetic transformation of molybdenum-starved *Azotobacter vinelandii*, increased transformation frequency and recipient range. *Can. J. Microbiol., 31*, 659-662.

Page, W. J., and Grant, G. A. (1987). Effect of mineral iron on the development of transformation competence in *Azotobacter vinelandii*. *FEMS Microbiol. Lett., 41*, 257-262.

Page, W. J., and Collinson, S. K. (1982). Molybdenum enhancement of nitrogen fixation in a Mo-starved *Azotobacter vinelandii* Nif⁻ mutant. *Can. J. Microbiol., 28*, 1173-1180.

Page, W. J., and Collinson, S. K. (1987). Characterization of *Azomonas macrocytogenes* strains isolated from Alberta soils. *Can. J. Microbiol., 33*, 830-833.

Pau, R. N., Eldridge, M. E., Lowe, D. J., Michenall, L. A., and Eady, R. R. (1993). Molybdenum-independent nitrogenases of *Azotobacter vinelandii*: A functional species of alternative nitrogenase-3 isolated from a molybdenum-tolerant strain contains an iron-molybdenum cofactor. *Biochem. J., 293*, 101-107.

Paulsen, D. M., Paerl, H. W., and Bishop, P. E. (1991). Evidence that molybdenum-dependent nitrogen fixation is not limited by high sulfate concentrations in marine environments. *Limnol. Oceanogr., 36*, 1325-1334.

Peters J. W., Fisher, K., and Dean, D. R. (1995). Nitrogenase structure and function: A biochemical-genetic perspective. *Annu Rev Microbiol., 49*, 335-366.

Premakumar, R., Jacobson, M. R., Loveless, T. M., and Bishop, P. E. (1992). Characterization of transcripts expressed from nitrogenase-3 structural genes of *Azotobacter vinelandii*. *Can. J. Microbiol., 38*, 929-936.

Premakumar, R., Chisnell, J. R., and Bishop, P. E. (1989). A comparison of the three dinitrogenase reductases expressed by *Azotobacter vinelandii*. *Can. J. Microbiol., 35*, 344-348.

Premakumar, R., Jacobitz, S., Ricke, S. C., and Bishop, P. E. (1996). Phenotypic characterization of a tungsten-tolerant mutant of *Azotobacter vinelandii*. *J. Bacteriol., 178*, 691-696.

Premakumar, R., Jacobson, M. R., Loveless, T. M., and Bishop, P. E. (1992). Characterization of transcripts expressed from nitrogenase-3 structural genes of *Azotobacter vinelandii*. *Can. J. Microbiol., 38*, 929-936.

Premakumar, R., Lemos, E. M., and Bishop, P. E. (1984). Evidence for two dinitrogenase reductases under regulatory control by molybdenum in *Azotobacter vinelandii*. *Biochim. Biophys. Acta, 797*, 64-70.

Premakumar, R., Loveless, T., and Bishop, P. E. (1992). Effect of amino acid substitutions in a potential metal-binding site of AnfA on expression of the *anfHDGK* operon in *Azotobacter vinelandii*. In R.

Palacios, J. Mora, W. E. Newton (Eds.), *New horizons in nitrogen fixation* (p. 488). Dordrecht: Kluwer Academic Publishers.

Premakumar, R., Loveless, T. M., and Bishop, P. E. (1994). Effect of amino acid substitutions in a potential metal-binding site of AnfA on expression from the *anfH* promoter in *Azotobacter vinelandii*. *J. Bacteriol., 176*, 6139-6142.

Premakumar, R., Pau, R. N., Mitchenall, L. A., Easo, M., and Bishop, P. E. (1998). Regulation of the transcriptional activators AnfA and VnfA by metals and ammonium in *Azotobacter vinelandii*. *FEMS Microbiol. Lett., 164*, 63-68.

Raina, R., Bageshwar, U. K., and Das, H. K. (1993). The ORF encoding a putative ferredoxin-like protein downstream of the *vnfH* gene in *Azotobacter vinelandii* is involved in the vanadium-dependent alternative pathway of nitrogen fixation. *Mol. Gen. Genet., 236*, 459-462.

Raina, R., Reddy, M. A., Ghosal, D., and Das, H. K. (1988). Characterization of the gene for the Fe-protein of the vanadium dependent alternative nitrogenase of *Azotobacter vinelandii* and construction of a Tn*5* mutant. *Mol. Gen. Genet., 214*, 121-127.

Ramos, F., Blanco, G., Gutierrez, J. C., Luque, F., and Tortolero, M. (1993). Identification of an operon involved in the assimilatory nitrate reducing system of *Azotobacter vinelandii*. *Mol. Microbiol., 8*, 1145-1153.

Ramos, J., and Robson, R. L. (1987). Cloning of the gene for phosphoenolpyruvate carboxylase from *Azotobacter chroococcum*, an enzyme important in aerobic nitrogen fixation. *Mol. Gen. Genet., 208*, 481-484.

Rangaraj, P., Rüttiman-Johnson, C., Shah, V. K., and Ludden, P. (2000). Biosynthesis of the iron-molybdenum and iron-vanadium cofactors of the *nif*- and *vnf*-encoded nitrogenases. In E. Triplett (Ed.), *Prokaryotic nitrogen fixation* (pp55-80). Wymondham, UK: Horizon Scientific Press.

Roberts, G. P., MacNeil, T., MacNeil, D., and Brill, W. J. (1978). Regulation and characterization of protein products coded by the *nif* (nitrogen fixation) genes of *Klebsiella pneumoniae*. *J. Bacteriol., 136,* 267-279.

Roberts, G. P., and Brill, W. J. (1980). Gene-product relationships of the *nif* regulon of *Klebsiella pneumoniae*. *J. Bacteriol., 144,* 210-216.

Robson, R. L., Chesshyre, J. A., Wheeler, C., Jones, R., Woodley, P. R., and Postgate, J. R. (1984). Genome size and complexity in *Azotobacter chroococcum*. *J. Gen. Microbiol., 130*, 1603-1612.

Robson, R. L., and Postgate, J. R. (1980). Oxygen and hydrogen in biological nitrogen fixation. *Annu. Rev. Microbiol., 34,* 183-207.

Robson, R. L. (1986). Nitrogen fixation in strains of *Azotobacter chroococcum* bearing deletions of a cluster of genes coding for nitrogenase. *Arch. Microbiol., 146*, 74-79.

Robson, R .L., Eady, R. R., Richardson, T. H., Miller, R.W., Hawkins, M., and Postgate, J. R. (1986). The alternative nitrogenase of *Azotobacter chroococcum* is a vanadium enzyme. *Nature, 322*, 388-390.

Robson, R. L., Woodley, P. R., and Jones, R. (1986). Second gene (*nifH**) coding for a nitrogenase iron-protein in *Azotobacter chroococcum* is adjacent to a gene coding for a ferredoxin-like protein. *EMBO J., 5*, 1159-1163.

Robson, R. L., Woodley, P. R., Pau, R. N., and Eady, R. R. (1989). Structural genes for the vanadium nitrogenase from *Azotobacter chroococcum*. *EMBO J., 8*, 1217-1224.

Rodriguez-Quinones, F., Bosch, R., and Imperial, J. (1993). Expression of the *nifBfdxNnifOQ* region of *Azotobacter vinelandii* and its role in nitrogenase activity. *J. Bacteriol., 175,* 2926-2935.

Rüttimann-Johnson, C., Rubio, L. M., Dean, D. R., and Ludden, P. W. (2003). VnfY is required for full activity of the vanadium-containing dinitrogenase in *Azotobacter vinelandii*. *J. Bacteriol., 185*, 2383-2386.

Rüttimann-Johnson, C., Staples, C. R., Rangaraj, P., Shah, V. K., and Ludden, P. W. (1999). A vanadium and iron cluster accumulates on VnfX during iron-vanadium-cofactor synthesis for the vandium nitrogenase in *Azotobacter vinelandii*. *J. Biol. Chem., 274*, 18087-18092.

Sadoff, H. L., Shimel, B., and Ellis, S. (1979). Characterization of *Azotobacter vinelandii* deoxyribonucleic acid and folded chromosomes. *J. Bacteriol., 138*, 871-877.

Schuddekopf, K., Hennecke, S., Liese, U., Kutsche, M., and Klipp, W. (1993). Characterization of *anf* genes specific for the alternative nitrogenase and identification of *nif* genes required for both nitrogenases in *Rhodobacter capsulatus*. *Mol. Microbiol., 8*, 673-684.

Schneider, K., Gollan, U., Drottboom, M., Selsemeir-Voigt, S., and Müller, A. (1997). Comparative biochemical characterization of the iron-only nitrogenase and the molybdenum nitrogenase from *Rhodobacter capsulatus*. *Eur. J. Biochem., 15*, 789-800.

Schneider, K., Müller, A., Schramm, U., and Klipp, W. (1991). Demonstration of a molybdenum- and vanadium-independent nitrogenase in a *nifHDK*-deletion mutant of *Rhodobacter capsulatus*. *Eur. J. Biochem., 195*, 653-661.

Shah, V. K., Pate, J. L., and Brill, W.J. (1973). Protection of nitrogenase in *Azotobacter vinelandii*. *J. Bacteriol., 115(1),* 15-17.

Singh, R. A., Choudhury, N. R., and Das, H. K. (2000). The replication origin of *Azotobacter vinelandii*. *Mol. Gen. Genet., 262*, 1070-1080.

Smith, B. E., Eady, R. R., Lowe, D. J., and Gormal, C. (1988). The vanadium-iron protein of vanadium nitrogenase from *A. chroococcum* contains an iron-vanadium cofactor. *Biochem. J., 250*, 299-302.

Thiel, T. (1993). Characterization of genes for an alternative nitrogenase in the cyanobacterium *Anabaena variabilis*. *J. Bacteriol., 175*, 6276-6286.

Thompson, J. P., and Skerman, V. B. D. (1979). *Azotobactereaceae, the taxonomy and ecology of the aerobic nitrogen fixing bacteria* (417p.). London: Academic Press.

Tibelius, K. H., Du L., Tito, D., and Stejskal, F. (1993). The *Azotobacter chroococcum* hydrogenase gene cluster, Sequences and genetic analysis of four accessory genes, *hupA, hupB, hupY* and *hupC*. *Gene, 127*, 53-61.

Toukdarian, A., Saunders, G., Selman-Sosa, G., Santero, E., Woodley, P., and Kennedy, C. (1990). Molecular analysis of the *Azotobacter vinelandii glnA* gene encoding glutamine synthetase. *J. Bacteriol., 172*, 6529-6539.

Venkatesh, T. V., and Das, H. K. (1992). The *Azotobacter vinelandii recA* gene - sequence analysis and regulation of expression. *Gene, 113*, 47-53.

Walmsley, J., and Kennedy, C. (1991). Temperature-dependent regulation by molybdenum and vanadium of expression of the structural genes encoding three nitrogenases in *Azotobacter vinelandii*. *Appl. Environ. Microbiol., 57*, 622-624.

Walmsley, J., Toukdarian, A., and Kennedy, C. (1994). The role of regulatory genes *nifA*, *vnfA*, *nfrX*, *ntrC* and *rpoN* in expression of genes encoding the three nitrogenases of *Azotobacter vinelandii*. *Arch. Microbiol., 162*, 422-429.

Waugh, S. I., Paulsen, D. M., Mylona, P. V., Maynard, R. H., Premakumar, R., and Bishop, P. E. (1995). The genes encoding the delta subunits of dinitrogneases 2 and 3 are required for Mo-independent diazotrophic growth by *Azotobacter vinelandii*. *J. Bacteriol., 177*, 1505-1510.

Wolfinger, E. D., and Bishop, P. E. (1991). Nucleotide sequence and mutational analysis of the *vnfENX* region of *Azotobacter vinelandii*. *J. Bacteriol., 173*, 7565-7572.

Woodley, P., Buck, M., and Kennedy, C. (1996). Identification of sequences important for recognition of *vnf* genes by the VnfA transcriptional activator in *Azotobacter vinelandii*. *FEMS Microbiol. Lett., 135*, 213-221.

Wyss, O., and Wyss, M. B. (1950). Mutants of *Azotobacter* that do not fix nitrogen. *J. Bacteriol.,* 59, 287-291.

Wu, G., Hill, S., Kelly, M. J., Sawers, G., and Poole, R. K. (1997). The *cydR* gene product, required for regulation of cytochrome *bd* expression in the obligate aerobe *Azotobacter vinelandii*, is an Fnr-like protein. *Microbiology, 143*, 2197-2207.

Yates, M. G., and Campbell, F. (1989). The effect of nutrient limitation on the competition between an H_2-uptake, hydrogenase-positive recombinant strain of *Azotobacter chroococcum* and the Hup- mutant parent in mixed populations. *J. Gen. Microbiol., 135*, 221-226.

Zheng, L., Cash, V. L., Flint, D. H., and Dean, D. R. (1998). Assembly of iron-sulfur clusters. Identification of an *iscSUA-hscBA-fdx* gene cluster from *Azotobacter vinelandii*. *J. Biol. Chem., 273*, 13264-13272.

Zinoni, F., Robson, R. M., and Robson, R. L. (1993). Organization of potential alternative nitrogenase genes from *Clostridium pasteurianum*. *Biochim. Biophys. Acta, 1174*, 83-86.

Chapter 3

NITROGEN FIXATION IN THE CLOSTRIDIA

J.-S. CHEN
Department of Biochemistry, Virginia Polytechnic Institute and State University, Blacksburg, Virginia 24061, USA

1. INTRODUCTION

Among nitrogen-fixing organisms, the genus *Clostridium* occupies a very special place. The species *Clostridium pasteurianum* is the first known free-living nitrogen-fixing bacterium, and it has been studied in the laboratory ever since its isolation by S. Winogradsky over 100 years ago (Winogradsky, 1895). After the ^{15}N tracer technique became available, a number of *Clostridium* species, including *C. acetobutylicum*, *C. beijerinckii*, and *C. butyricum* were also found to be capable of nitrogen fixation (Rosenblum and Wilson, 1949).

Biochemical studies conducted with *C. pasteurianum* between the late 1950s and early 1970s led to pivotal findings that are the foundation of our current knowledge of the enzymology of biological nitrogen fixation. These crucial events include the preparation of a consistently active nitrogen-fixing cell-free extract (Carnahan *et al.*, 1960a, 1960b), the discovery of the low-redox potential iron-sulfur protein ferredoxin as the natural electron donor for nitrogenase (Mortenson *et al.*, 1962, 1963; Mortenson, 1963), the discovery that ATP, in addition to ferredoxin, is required for nitrogenase activity (Mortenson, 1964), and the separation of nitrogenase into two metalloproteins (Mortenson, 1965, 1966).

In retrospect, the use of an obligate anaerobe in early nitrogen-fixation research had a profound effect on the progress of this field. Because of the O_2-sensitivity of the source organism, anaerobic conditions were used in the preparation of cell-free extracts and pyruvate was used in the assay to support the *in vitro* nitrogen-fixing activity. This combination of circumstances facilitated the realization that nitrogenase is extremely O_2-sensitive and that reduced ferredoxin and ATP, both of which are produced during

W. Klipp, B. Masepohl, J. R. Gallon and W. E. Newton (eds.), Genetics and Regulation of Nitrogen Fixation in Free-Living Bacteria, 53-64.

the "phosphoroclastic" cleavage of pyruvate, are necessary for nitrogenase-catalyzed reduction of dinitrogen (N_2) to ammonia.

Between the early 1970s and the 1980s, many biochemical and biophysical studies were performed on nitrogenase component proteins purified from *Azotobacter vinelandii*, *C. pasteurianum*, *Klebsiella pneumoniae*, and other diazotrophs, and these studies established the fact that nitrogenase is a highly conserved enzyme. Nevertheless, the nitrogenase of *C. pasteurinaum* also exhibits unique properties, particularly the inability of its component proteins to form an active nitrogenase in heterologous combinations with a component protein from either *A. vinelandii* or *K. pneumoniae* (Emerich *et al.*, 1981). Since the 1980s, the characterization of the nitrogen-fixation (*nif*) genes of the clostridia has revealed characteristic structural features in the Fe and MoFe proteins (Chen and Johnson, 1993) and the presence of novel *nif* genes, including the fused *nifN-B* gene and the split *nifV*ω and *nifV*α genes, in *C. acetobutylicum*, *C. beijerinckii* and *C. pasteurianum* but not in any other diazotrophs (see Genomic Aspects of Nitrogen Fixation in the Clostridia, in the volume "*Genomes and Genomics of Nitrogen-Fixing Organisms*" of this series).

2. THE NITROGEN-FIXING CLOSTRIDIA

The genus *Clostridium* is a diverse collection of rod-shaped, spore-forming, obligately anaerobic bacteria that do not carry out dissimilatory sulfate reduction and usually stain Gram positive in young cultures (Collins *et al.*, 1994). At present, there are over 170 named species of *Clostridium* (J. P. Euzéby: List of Bacterial Names with Standing in Nomenclature; www.bacterio.cict.fr/allnames.html; update of March 14, 2003), and they encompass a broad range of physiological types, differing in ranges of carbon and nitrogenous substrates fermented, major end products, and pathogenicity. On the basis of phenotypic criteria and the results of phylogenetic analyses (based on 16S rRNA sequences), it has been proposed that the traditional genus *Clostridium* be rearranged into different genera (Collins *et al.*, 1994). The proposed scheme will separate the currently known nitrogen-fixing clostridia into different genera. However, the better known nitrogen-fixing species, *C. acetobutylicum*, *C. beijerinckii*, *C. butyricum*, and *C. pasteurianum*, will remain in the redefined genus *Clostridium*, which is reserved for species belonging to the rRNA group I of Johnson and Francis (1975).

As mentioned above, *C. pasteurianum* is the first free-living diazotroph that was isolated. Winogradsky, in 1895, reported the isolation and initial characterization of this nitrogen-fixing anaerobe; he named it "*Clostridium pasteurianum*" in honor of Louis Pasteur (Winogradsky, 1895; p. 330):

> "Nous l´appellerons en honneur du grand savant, créateur de notre science: *Clostridium Pasteurianum*."

In a subsequent report, Winogradsky presented a careful sketch of *C. pasteurianum* cells undergoing the sporulation and germination processes (Winogradsky, 1902a), and photomicrographs were used to illustrate the characteristic shape of sporulating *C. pasteurianum* cells (Winogradsky, 1902b). It is of interest that the name "*Clostridium Pastorianum*" was used in the 1902 papers (Winogradsky, 1902a, 1902b). The culture of *C. pasteurianum* from Winogradsky was subjected to single-cell isolation in the McCoy laboratory at the University of Wisconsin, and this further purification did not alter the nitrogen-fixing activity or efficiency of the culture (McCoy *et al.*, 1928). *C. pasteurianum* strain W5 (= ATCC 6013) from the McCoy laboratory represents the organism isolated by Winogradsky. Among the nitrogen-fixing clostridia, *C. pasteurianum* is the most thoroughly characterized in terms of nitrogen fixation.

Rosenblum and Wilson (1949) reported the fixation of $^{15}N_2$ by several clostridia, including *C. aceticum*, *C. acetobutylicum*, *C. beijerinckii*, "*C. butylicum*", *C. butyricum*, *C. felsineum*, *C. kluyverii*, "*C. lactoacetophilum*", "*C. madisoni*", *C. pasteurianum*, *C. pectinovorum*, and *C. tetanomorphum*. Among these species, "*C. butylicum*" (George *et al.*, 1983), "*C. lactoacetophilum*" (Cummins and Johnson, 1971), and "*C. madisoni*" (Keis et al., 2001) have been re-classified as *C. beijerinckii*. Nitrogen fixation by *C. acetobutylicum* (Chen *et al.*, 2001), *C. beijerinckii* (Chen *et al.*, 2001), *C. butyricum* (Kanamori *et al.*, 1989), and *C. kluyverii* (Kanamori *et al.*, 1989) has been confirmed. *C. aceticum* did not grow in a yeast extract-containing medium that supports nitrogen-fixing growth of *C. pasteurianum* (Szech *et al.*, 1989). It should be noted that the medium used by Rosenblum and Wilson (1949) contained low concentrations of yeast extract and tryptone (carried from the inoculum). In addition to these mesophilic and saccharolytic species, the nitrogen-fixing clostridia include species that belong to other physiological groups; the acetogenic *C. formicoaceticum* (Bogdahn *et al.*, 1983), the cellulytic *C. hungatei* (Monserrate *et al.*, 2001), and the acid-tolerant *C. akagii* and *C. acidisoli* (Kuhner *et al.*, 2000). The nitrogen-fixing thermophile "*C. thermosaccharolyticum*" (Bogdahn and Kleiner, 1986) has now been renamed as *Thermoanaerobacterium thermosaccharolyticum* (Collins *et al.*, 1994).

The genus *Paenibacillus* contains nitrogen-fixing species that form spores under anaerobic conditions (Rosado *et al.*, 1997). The former "*Clostridium durum*" (or "*Paenibacillus durum*"), the dominant organism found in a sediment core from the Black Sea, has been reclassified as a member of the species *Paenibacillus azotofixans* (Rosado *et al.*, 1997). Criteria used in the reclassification include the DNA relatedness at the genome level as measured by the DNA-DNA reassociation technique. Thus, some of the spore-forming, nitrogen-fixing rods do not belong to the genus *Clostridium*.

3. DISTINCTIVE FEATURES OF THE *NIF* GENES OF THE CLOSTRIDIA

The nucleotide sequences of the *nif* cluster of *C. acetobutylicum*, *C. beijerinckii*, and *C. pasteurianum* have been determined (see Genomic Aspects of Nitrogen Fixation in the Clostridia, in the volume "*Genomes and Genomics of Nitrogen-Fixing Organisms*" of this series). The organization of the *nif* genes is conserved in these three species (Figure 1). The *nif* cluster of *C. acetobutylicum* is the most concise as it consists of only nine genes: three genes (*nifHDK*) encoding the nitrogenase polypeptides, four genes (*nifE*, *nifN-B*, *nifVω* and *nifVα*) encoding the proteins for the synthesis of the iron-molybdenum cofactor, and two putative genes (*nifI-1* and *nifI-2*) encoding either the P_{II} or GlnB-like proteins for the regulation of nitrogenase activity in response to the availability of fixed nitrogen.

[A] > ***nifH*** > [B] > ***nifD*** > ***nifK*** > [C] > ***nifE*** > ***nifN-B*** > [D] > ***nifVω*** > ***nifVα*** >

Species	Region			
	A	B	C	D
C. acetobutylicum	None	*nifI-1, nifI-2*	None (96 bp)	None
C. beijerinckii	None	*nifI-1, nifI-2*	None (1,1223 bp)	*fdxA, nirJ-1, nirJ-2, nirD, nirH*
C. pasteurianum	nifH2	None	None (284 bp)	*modA, modB*

Figure 1. Organization of the nif *gene cluster in* C. acetobutylicum, C. beijerinckii, *and* C. pasteurianum. *The* nif *genes and open-reading frames of the gene cluster have the same direction of transcription. Arrowheads indicate the direction of transcription. The* nif *cluster of these three species has distinctive features in the four regions marked as A, B, C, and D, and the distinctive features of each region are listed.*

Outside the *nif* cluster, an open-reading frame (CAC2229), which corresponds to the *nifJ* gene (encoding pyruvate:ferredoxin/flavodoxin oxidoreductase) of *K. pneumoniae*, is found on the *C. acetobutylicum* ATCC824 chromosome. The ferredoxin/flavodoxin-linked pyruvate dehydrogenase is a hallmark of the saccharolytic clostridia and is not specific for nitrogen fixation. In addition, two *nifS*-like open-reading frames (CAC2234, CAC 2972) are present on the chromosome of *C. acetobutylicum* ATCC824 (J. Toth and J.-S. Chen, unpublished results). The *nifS* and the related *iscS* genes encode a cysteine desulfurase that is required for the formation of iron-sulfur clusters as well as the synthesis of other compounds (Mihara and Esaki, 2002). Open-reading frames similar to *nifS* are also present in the genomes of the two non-nitrogen-fixing clostridia, *C. perfringens* and *C. tetani* (J. Toth and J.-S. Chen, unpublished results). It suggests that the *nifS*-like genes in these *Clostridium* species are not specific for nitrogen fixation. There are no open-reading frames in the genome of *C. acetobutylicum* ATCC 824 that correspond to the other *nif* genes of *K. pneumoniae*.

The *nif* genes of *C. acetobutylicum*, *C. beijerinckii*, and *C. pasteurianum* have several distinctive features: (1) the *nifHDK* genes encode polypeptides that do not have the residues identified for component interaction in the *A. vinelandii* nitrogenase complex; (2) the fused *nifN-B* gene corresponds to two unlinked genes (*nifN* and *nifB*) in other diazotrophs; and (3) the separate *nifVω* and *nifVα* genes encode polypeptides corresponding to, respectively, the C-terminal half and the N-terminal half of the *nifV*-encoded homocitrate synthase of other diazotrophs.

The Fe protein and MoFe protein from several organisms can form active, heterologous nitrogenase complexes (Emerich *et al.*, 1981). The nitrogenase component proteins of *C. pasteurianum*, however, are exceptional in their inability to form such heterologous complexes. The nitrogenase polypeptides of *C. pasteurianum* are either shorter (NifH and NifK) or longer (NifD) than the corresponding proteins of other bacteria (Chen *et al.*, 1986; Wang *et al.*, 1988b). The NifH, NifD, and NifK sequences of *C. acetobutylicum*, *C. beijerinckii*, and *C. pasteurianum* are highly conserved in that the distinctive features found in *C. pasteurianum* are also found in the other two species (J. Toth and J.-S. Chen, unpublished results). That NifD is larger than NifK is a characteristic of the *Clostridium* nitrogenase. Compared with their counterparts in other organisms, the clostridial NifD (528-533 amino acids) is about 20-to-40 residues longer and has an extra stretch of about 50 amino acids in the 380 to 430 region, whereas clostridial NifK (454-458 amino acids) is about 50 residues shorter in the N-terminal region. It is not known if these differences prevent the formation of active, heterologous nitrogenase complexes. However, a comparison of the nitrogenase component proteins of these three *Clostridium* species with those of *A. vinelandii* has yielded some clue for the incompatibility between the nitrogenase component proteins of *C. pasteurianum* and *A. vinelandii*.

In cross-linking studies performed with the *A. vinelandii* nitrogenase complex, Glu 112 of NifH and Lys 400 of the β-subunit of MoFe protein (numbering of *A. vinelandii* proteins) have been identified as the principal residues contributing to the binding interface (Schmid *et al.*, 2002). Additional residues also provide significant contributions to the binding interface; they include Glu 68, Asp 69, Glu 71, Ile 103, and Glu 111 of NifH and Glu 60, Lys 403, Lys 404, and Ile 423 of NifK. As shown in Figure 2, the nitrogenase component proteins of *Azotobacter* and *Clostridium* differ significantly in these regions. A leucine residue (position 104 or 109 of clostridial proteins) is found in the clostridial NifH in place of Glu 112 of the *Azotobacter* NifH, whereas a Glu (position 344 of Ca NifK), or an Asn (position 345 of Cb NifK) or a Met (position 348 of Cp NifK) is found in the clostridial protein in place of Lys 400 of *Azotobacter* NifK. Therefore, the electrostatic interaction between Glu 112 of NifH and Lys 400 of NifK of *A. vinelandii* is not present in the clostridial nitrogenase complex. Furthermore, the presence of a lysine in the clostridial NifK at the position occupied by Glu 60 of *Azotobacter* NifK as well as the absence of two amino acid residues in the clostridial NifH between the conserved glycine (position 56 or 61 in clostridial proteins) and the three negatively charged residues involved in component interaction in the *Azotobacter* nitrogenase complex also suggest a distinct binding interface in the clostridial nitrogenase complex.

The fused *nifN-B* gene and the split *nifVω* and *nifVα* genes were first observed in *C. pasteurianum* (Chen and Johnson, 1993). These unusual *nif* genes have now been observed in *C. acetobutylicum* and *C. beijerinckii* as well (Chen *et al.*, 2001; J. Toth and J.-S. Chen, unpublished results). The length of the deduced NifN-B polypeptide is 867 (*C. acetobutylicum*), 896 (*C. beijerinckii*) or 929 amino acids (*C. pasteurianum*). Whether or not these unusual features will contribute to a better understanding of the evolution of the nitrogen-fixation genes will await future work. While the *nif* cluster of the clostridia is composed of these distinctive *nif* genes for nitrogenase synthesis, there is a conspicuous absence of a *nifA*-like gene for the transcriptional regulation of the *nif* genes. Not surprisingly, the presumed promoter regions for the *nif* genes of *C. pasteurianum* do not possess the motif of the *nif* promoters of the proteobacteria (Wang *et al.*, 1988a). Despite the lack of a gene in the *nif* cluster for the transcriptional regulation of *nif* genes of the clostridia, the synthesis of nitrogenase in *C. pasteurianum* is subject to the availability of ammonia in the growth medium (see Section 5).

Fe Protein

						*
Av NifH	(65)	GTV**ED**L**E**L	(99)	GRGV**I**T	(110)	E**EE**GAY
Ca NifH	(56)	G--EDVDL	(91)	GRGIIT	(102)	EQLGAY
Cb NifH	(61)	G--DDIEL	(96)	GRGIIT	(107)	EQLGAY
Cp NifH1	(61)	G--EDVEL	(96)	GRGIIT	(107)	EQLGAY

β-Subunit of MoFe Protein

				*		
Av NifK	(59)	R**E**AL	(398)	GN**K**RW**KK**	(422)	Y**I**G
Ca NifK	(12)	RKAL	(342)	KGEKFEK	(364)	KSF
Cb NifK	(12)	RKAL	(343)	PSNVFEK	(365)	KQD
Cp NifK	(12)	RKAL	(346)	PGMKFQK	(368)	KVE

Figure 2. Amino acid residues around the cross-linking site between the Fe protein and the β-subunit of the MoFe protein of A. vinelandii *and the corresponding regions in the clostridial nitrogenase components.* Av, A. vinelandii*;* Ca, C. acetobutylicum *ATCC 824;* Cb, C. beijerinckii *NRRL B593;* Cp, C. pasteurianum *W5. The number in parentheses is the residual number for the first amino acid of the segment.* Av *NifH Glu 112 and* Av *NifK Lys 400 (asterisked) provide the cross-linking site. Glu 68, Asp 69, Glu 71, Ile 103, and Glu 111 of* Av *NifH and Glu 60, Lys 403, Lys 404, and Ile 423 of* Av *NifK (all in bold letters) are the other residues providing the principal contributions to the binding interface.*

4. GENES FOR AMMONIA ASSIMILATION

When nitrogen-fixing cells of *C. pasteurianum* were supplied with $^{15}N_2$ for a short time, they excreted ammonia with an extremely high ^{15}N concentration into the medium (Zelitch *et al.*, 1951a). Using ^{15}N-enriched N_2 or NH_4^+ as a tracer for 45 to 60 min, the nitrogen isotope was found to accumulate in glutamic acid of cellular materials (Zelitch *et al.*, 1951b). Results of these early studies support the proposals that ammonia is a key intermediate in biological nitrogen fixation and that glutamic acid is the primary product of ammonia assimilation. Proof of this path would come after nitrogenase was purified in the 1960s (see Section 1) and after the discovery of the enzyme glutamate synthase in 1970 (see below).

Under nitrogen-limited growth conditions, bacteria assimilate ammonia by a route not involving glutamate dehydrogenase (Tempest *et al.*, 1973). Instead, they use the more efficient glutamine synthetase (GS) for the utilization of the low concentration of ammonia that is available, and the enzyme glutamate synthase ("GOGAT") catalyzes the reductive formation of two glutamate molecules from one glutamine and one 2-oxoglutarate, with NADH, NADPH, or ferredoxin as the electron donor (Vanoni and

Curti, 1999). Activity of glutamate synthase was measured in nitrogen-fixing cells of *C. pasteurianum* soon after this enzyme had been discovered in *Klebsiella aerogenes* (Nagatani *et al.*, 1971; Dainty, 1972).

In the genome of *C. acetobutylicum* (Nölling *et al.*, 2001), putative genes for glutamine synthetase (CAC2658), glutamate synthase small (or β) subunit (CAC1674, CAC0764), glutamate synthase large subunit (CAC1673), and glutamate dehydrogenase (CAC0737) have been identified (annotated genome sequence; GenBank accession number AE001437). An open-reading frame that could encode the small subunit of glutamate synthase has been identified in *C. beijerinckii* NRRL B593 (K. Ishizuka, J. Toth, and J.-S. Chen, unpublished data). Two glutamate synthase genes (GenBank Accession numbers BAB82167 and BAB80960) are present in *Clostridium perfringens* (not a nitrogen-fixing bacterium). Both *C. perfringens* genes encode polypeptides related to the small (or β) subunit of *C. acetobutylicum* glutamate synthase.

Genes for ammonia assimilation in *Clostridium saccharobutylicum* NCP 262 (formerly *C. acetobutylicum* NCP 262) have been studied, although it is unknown whether or not *C. saccharobutylicum* is a diazotroph. From *C. saccharobutylicum* NCP 262, the *glnA* gene for glutamine synthetase (Usdin *et al.*, 1986) and the genes for the large subunit (GenBank accession number AAD41675) and small subunit (GenBank accession numbers AAD41676 and AAP06761) of glutamate synthase have been cloned and sequenced. The *glnA* gene of *C. saccharobutylicum* NCP 262 is not homologous to the *E. coli glnA* gene at either the DNA or the protein level. However, it enables *E. coli glnA* deletion mutants to utilize ammonium sulfate as a sole source of nitrogen. The cloned *glnA* gene from *C. saccharobutylicum* NCP 262 was expressed in *E. coli*, and the clostridial glutamine synthetase was purified from *E. coli*.

Purified bacterial glutamate synthase generally consists of a large subunit and a small subunit, and the enzyme is an iron-sulfur flavoprotein (Vanoni and Curti, 1999). A glutamate synthase with a native M_r of 590,000 has been purified from *C. pasteurianum*, but the enzyme gave five different subunits, with M_r values of 91,000, 86,000, 68,000, 31,000, and 17,500 on SDS-PAGE, and it does not contain iron or flavin (Singhal *et al.*, 1989). Genes or amino acid sequences for the subunits of the *C. pasteurianum* glutamate synthase have not been reported.

5. REGULATION OF NITROGEN FIXATION AND AMMONIA ASSIMILATION

5.1. Nitrogen Fixation

Although the genetic element(s) for the transcriptional control of the *nif* genes have not yet been found in the clostridia, there is evidence that the expression of the *nif* genes is tightly controlled at the transcriptional level. When a sufficient amount of ammonia is present in the growth medium, nitrogenase is not synthesized by *C. pasteurianum* (Mortenson *et al.*, 1967), "*C. thermosaccharolyticum*" (Bogdahn and Kleiner, 1986), or *C. formicoaceticum* (Bogdahn *et al.*, 1983). When provided with a limiting amount of ammonia, a culture of *C. pasteurianum* switches from ammonia-dependent growth to nitrogen-fixing growth, after a diauxic lag following the exhaustion of ammonia in the growth medium (Mortenson, 1978).

When ammonia is added to a nitrogen-fixing culture of *C. pasteurianum*, synthesis of nitrogenase is immediately stopped (Daesch and Mortenson, 1972; Kasap, 2002; Kleiner, 1979). The pre-formed nitrogenase, however, remains active through several cell doublings, which results in a decrease in nitrogenase activity per cell due to dilution of the enzyme. It is interesting to note that the addition of ammonia had different effects on the *in vivo* activity of pre-formed nitrogenase in *C. pasteurianum* and in *C. beijerinckii* (Kasap, 2002). Whereas addition of ammonia does not affect the *in vivo* activity of pre-formed nitrogenase in *C. pasteurianum*, similar additions caused an immediate drop in *in vivo* nitrogenase activity in *C. beijerinckii* (Kasap, 2002). However, there was no similar drop in *in vitro* nitrogenase activity following addition of ammonia to a nitrogen-fixing culture of *C. beijerinckii*. It may be speculated that products of the *nifI-1* and *nifI-2* genes play a role in regulating the activity of nitrogenase in response to the concentration of ammonia in *C. beijerinckii*, assuming *nifI*-like genes are not present in the genome (outside the *nif* cluster) of *C. pasteurianum*.

When sporulation was induced in batch cultures of nitrogen-fixing *C. pasteurianum*, either by the addition of calcium acetate or by an increase in pH, partial inactivation of nitrogenase and excretion of ammonia were observed (Vallespinós and Kleiner, 1980). A drastic drop (to less than 10%) in *in vivo* nitrogenase activity occurred immediately after the culture pH was changed from 6.0 to 8.4; however, about 40% of the activity recovered in 2 h when the pH was maintained at 8.4. At the time of the pH change, the *in vitro* nitrogenase activity fell about 65%, and the activity remained at that level for at least 2 h. It appears that the pH change stopped nitrogenase synthesis, caused partial inactivation of pre-formed nitrogenase, and generated an initial physiological condition that did not support full nitrogenase activity.

To investigate the mechanism of transcriptional control of the *nif* genes, the transcription start site for six *nif* transcription units (including those of *nifH*-like genes) has been determined in *C. pasteurianum* (Wang *et al.*, 1988a). Transcription of *nifH1*, which encodes the characterized iron protein, starts 62 base pairs preceding the translation start site. An analysis of the upstream sequences revealed similarities in the –130 region of these transcription units, and the consensus sequence (ATCAATAT–N_{6-10}–ATGGATTC) is different from the TGT–N_{10}–ACA consensus that is present in the –100 region of the *nif* transcription units of *K. pneumoniae* and *A. vinelandii*. Because a *nifA*-like gene is not present in the *nif* clusters of *C. acetobutylicum*, *C. beijerinckii*, and *C. pasteurianum* or elsewhere in the genome of *C. acetobutylicum*, the promoter and upstream sequences for the *nif* genes of these *Clostridium* species can be expected to differ from those involving NifA.

5.2. Ammonia Assimilation

Regulation of ammonia assimilation has been studied in several nitrogen-fixing clostridia. These studies examined the levels of glutamine synthetase (GS), glutamate synthase (GOGAT), and glutamate dehydrogenase (GDH) in ammonia-grown and nitrogen-fixing cells. *C. pasteurianum* has GS and GDH, whose *in vitro* activities do not change between ammonia-grown and nitrogen-fixing conditions; however, its *in vitro* GOGAT activity, which is higher than the activity of GDH in a nitrogen-fixing cell extract, increases during a period when ammonia is added to the nitrogen-fixing culture (Kleiner, 1979). In earlier studies, it was also shown that GOGAT activity is about two-fold higher in ammonia-grown than nitrogen-fixing cells (Dainty, 1972; Nagatani *et al.*, 1971).

The NAD(P)H-linked spectrophotometric assays for GDH and GOGAT are prone to interference because of residual O_2 present in the assay system and because extracts of *Clostridium* species have active NAD(P)H-oxidizing activities. This technical problem can be circumvented by the use of ^{15}N nuclear magnetic resonance (NMR) spectroscopy with appropriately labeled substrates and assay solutions containing excess NAD(P)H or an NADH-regenerating system (Kanamori *et al.*, 1989). Using ^{15}N NMR spectroscopy, it was shown that nitrogen-fixing *C. kluyverii* uses GS and GOGAT for ammonia assimilation, whereas in ammonia-grown cells, GDH is important for ammonia assimilation (Kanamori *et al.*, 1989). *C. butyricum* has undetectable GDH activity when grown on glucose-salt medium with either ammonia or N_2 as the nitrogen source; therefore, the GS-GOGAT pathway is responsible for ammonia assimilation under both growth conditions. However, high activities of GDH are present when *C. butyricum* is grown on complex nitrogen and carbon sources (Kanamori *et al.*, 1989).

In *C. formicoaceticum*, GS is repressed and alanine dehydrogenase is induced by ammonia, whereas the synthesis of the other ammonia-assimilating enzymes (NADH- and NADPH-dependent GOGAT, NADH- and NADPH-dependent GDH, and NH_4^+-dependent asparagine synthetase) is not influenced by the extracellular concentration of ammonia (Bogdahn *et al.*, 1983). In "*C. thermosaccharolyticum*", now known as *Thermoanaerobacterium thermosaccharolyticum*, GS is partially repressed by ammonia, whereas the GOGAT and GDH activities are both greater under ammonia-grown than under nitrogen-fixing growth conditions (Bogdahn and Kleiner, 1986).

In *C. saccharobutylicum*, GS is sensitive to Mg^{2+} under all growth conditions and treatment with snake venom phosphodiesterase did not either affect this Mg^{2+} sensitivity or alter the mobility of the purified GS (from transformed *E. coli* cells) in SDS-PAGE, suggesting that the GS of *C. saccharobutylicum* is not regulated by an adenylylation-deadenylylation system (Usdin *et al.*, 1986). There is no evidence for a global *ntr* system (Fierro-Monti *et al.*, 1992; Janssen *et al.*, 1988). Regulation of GS activity in *C. saccharobutylicum* may involve an antisense RNA (Fierro-Monti *et al.*, 1992).

The results of these studies suggest that the clostridia use a variety of pathways for the assimilation of ammonia. At present, little is known about how the transcription of genes for GS, GDH and GOGAT is regulated in the nitrogen-fixing clostridia. Future research will reveal whether or not a common underlying mechanism (or hierarchy) is employed by the nitrogen-fixing clostridia to optimize the efficiency of ammonia assimilation.

6. CONCLUDING REMARKS

The nitrogen-fixing clostridia are characterized by the use of a relatively small number of *nif*-specific genes for the fixation of N_2. At present, the *nif* cluster (nine genes) of *C. acetobutylicum* ATCC 824 appears to represent the most concise *nif* cluster of known diazotrophs. To understand the regulation of expression of *nif* genes in the clostridia and to elucidate the precise function of the *nifB* gene product in relation to the *nifE* and *nifN* gene products, the *nif* system of *C. acetobutylicum* is attractive as a model system, despite the fact that the biochemical properties of the nitrogen-fixing system of *C. pasteurianum* have been more thoroughly investigated.

In addition to the simplicity of the *nif* cluster of *C. acetobutylicum* ATCC 824, the genome of this organism has been sequenced (Nölling *et al.*, 2001). This strain is amenable to a range of genetic manipulations, which include gene inactivation through homologous replacement. Useful reporter genes, including those encoding the *E. coli* β-glucuronidase, a clostridial β-galactosidase, and a secondary alcohol dehydrogenase, are now available for the study of promoter activities. The application of sophisticated genetic tools to the study of the *nif* genes of the clostridia can be expected to yield useful new information on several frontiers of nitrogen-fixation research.

REFERENCES

Bogdahn, M., and Kleiner, D. (1986). N_2 fixation and NH_4^+ assimilation in the thermophilic anaerobes *Clostridium thermosaccharolyticum* and *Clostridium thermoautotrophicum*. *Arch. Microbiol.*, *144*, 102-104.

Bogdahn, M., Andreesen, J. R., and Kleiner, D. (1983) Pathways and regulation of N_2, ammonium and glutamate assimilation by *Clostridium formicoaceticum*. *Arch. Microbiol.*, *134*, 167-169.

Carnahan, J. E., Mortenson, L. E., Mower, H. F., and Castle, J. E. (1960a) Nitrogen fixation in cell-free extracts of *Clostridium pasteurianum*. *Biochim. Biophys. Acta*, *38*, 188-189.

Carnahan, J. E., Mortenson, L. E., Mower, H. F., and Castle, J. E. (1960b) Nitrogen fixation in cell-free extracts of *Clostridium pasteurianum*. *Biochim. Biophys. Acta*, *44*, 520-535.

Chen, J.-S., and Johnson, J. L. (1993). Molecular biology of nitrogen fixation in the clostridia. In D. R. Woods (Ed.), *The clostridia and biotechnology* (pp. 371-392). Boston: Butterwoth-Heinemann.

Chen, K. C.-K., Chen, J.-S., and Johnson, J. L. (1986) Structural features of multiple *nifH*-like sequences and very biased codon usage in nitrogenase genes of *Clostridium pasteurianum*. *J. Bacteriol.*, *166*, 162-172.

Chen, J.-S., Toth, J., and Kasap, M. (2001). Nitrogen-fixation genes and nitrogenase activity in *Clostridium acetobutylicum* and *Clostridium beijerinckii*. *J. Ind. Microbiol. Biotechnol.*, *27*, 281-286.

Collins, M. D., Lawson, P. A., Willems, A., Cordoba, J. J., Fernandez-Garayzabal, J., Garcia, P. et al. (1994) The phylogeny of the genus *Clostridium*: proposal of five new genera and eleven new species combinations. *Int. J. Syst. Microbiol.*, *44*, 812-826.

Cummins, C., and Johnson, J. L. (1971) Taxonomy of the clostridia: wall composition and DNA homologies in *Clostridium butyricum* and other butyric acid-producing clostridia. *J. Gen. Microbiol.*, *67*, 33-46.

Daesch, G., and Mortenson, L. E. (1972) Effect of ammonia on the synthesis and function of the N_2-fixing enzyme system in *Clostridium pasteurianum*. *J. Bacteriol.*, *110*, 103-109.

Dainty, R. H. (1972) Glutamate biosynthesis in *Clostridium pasteurianum* and its significance in nitrogen metabolism. *Biochem. J.*, *126,* 1055-1056.

Emerich, D. W., Hageman, R. V., and Burris, R. H. (1981) Interactions of dinitrogenase and dinitrogenase reductase. *Adv. Enzymol.*, *52*, 1-22.

Fierro-Monti, I. P., Reid, S. J. and Woods, D. R. (1992) Differential expression of a *Clostridium acetobutylicum* antisense RNA: implications for regulation of glutamine synthetase. *J. Bacteriol.*, *174*, 7642-7647.

George, H. A., Johnson, J. L., Moore, W. E. C., Holdeman, L. V., and Chen, J.-S. (1983) Acetone, isopropanol, and butanol production by *Clostridium beijerinckii* (syn. *Clostridium butylicum*) and *Clostridium aurantibutyricum*. *Appl. Environ. Microbiol.*, *45*, 1160-1163.

Janssen, P. L., Jones, W. A., Jones, D. T., and Woods, D. R. (1988) Molecular analsysis and regulation of the *glnA* gene of the Gram-positive anaerobe *Clostridium acetobutylicum*. *J. Bacteriol.*, *170*, 400-408.

Johnson, J. L., and Francis, B. S. (1975) Taxonomy of the clostridia: ribosomal ribonucleic acid homologies among the species. *J. Gen. Microbiol.*, *88*, 229-244.

Kanamori, K., Weiss, R. L., and Roberts, J. D. (1989) Ammonia assimilation pathways in nitrogen-fixing *Clostridium kluyverii* and *Clostridium butyricum*. *J. Bacteriol.*, *171*, 2148-2154.

Keis, S., Shaheen, R., and Jones, D. T. (2001) Emended description of *Clostridium acetobutylicum*, *Clostridium beijerinckii*, and descriptions of *Clostridium saccharobutylacetonicum* sp. nov. and *Clostridium saccharobutylicum* sp. nov. *Int. J. Syst. Evol. Microbiol.*, *51*, 2095-2103.

Kleiner, D. (1979) Regulation of ammonium uptake and metabolism by nitrogen fixing bacteria III. *Clostridium pasteurianum*. *Arch. Microbiol.*, *120*, 263-270.

Kuhner, C. H., Matthies, C., Acker, G., Schmittroth, M., Gößner, A. S.,and Drake, H. L. (2000) *Clostridium akagii* sp. nov. and *Clostridium acidisoli* sp. nov.: acid-tolerant, N_2-fixing clostridia isolated from acidic forest soil and litter. *Int. J. Syst. Evol. Microbiol.*, *50*, 873-881.

Kasap, M. (2002) *Nitrogen metabolism and solvent production in Clostridium beijerinckii NRRL B593.* Ph.D. dissertation. Virginia Polytechnic Institute and State University, Blacksburg, Virginia, U.S.A.

McCoy, E., Higby, W. M., and Fred, E. B. (1928) The assimilation of nitrogen by pure cultures of Clostridium pasteurianum and related organisms. *Centralbl. Bakteriol. Parasitenk. Infektionskr. Abt. II*, *76*, 314-320.

Mihara, H., and Esaki, N. (2002) Bacterial cysteine desulfurase: their function and mechanisms. *Appl. Microbiol. Biotechnol.*, *60*, 12-23.

Monserrate, E., Leschine, S. B., and Canale-Parola, E. (2001) *Clostridium hungatei* sp. nov., a mesophilic, N_2-fixing cellulytic bacterium isolated from soil. *Int. J. Syst. Evol. Microbiol.*, *51*, 123-132.

Mortenson, L. E. (1963) Ferredoxin requirement for nitrogen fixation by extracts of *Clostridium pasteurianum*. *Biochim. Biophys. Acta*, *81*, 473-478

Mortenson, L. E. (1964) Ferredoxin and ATP, requirements for nitrogen fixation in cell-free extracts of *Clostridium pasteurianum*. *Proc. Natl. Acad. Sci. U.S.A.*, *52*, 272-279.

Mortenson, L. E. (1965) Nitrogen fixation in extracts of *Clostridium pasteurianum*. In A. San Pietro (Ed.) *Non-heme iron proteins: Role in energy conversion* (pp. 243-259). Yellow Springs (Ohio): The Antioch Press.

Mortenson, L. E. (1966) Components of cell-free extracts of *Clostridium pasteurianum* required for ATP-dependent H_2 evolution from dithionite and for N_2 fixation. *Biochim. Biophys. Acta*, 127, 18-25.

Mortenson, L. E. (1978) Regulation of nitrogen fixation. *Curr. Topics Cell. Regulation, 13*, 179-232.

Mortenson, L. E., Morris, J., and Jeng, D. Y. (1967) Purification, metal composition and properties of molybdoferredoxin and azoferredoxin, two of the components of the nitrogen-fixing system of *Clostridium pasteurianum*. *Biochim. Biophys. Acta, 141,* 516-522.

Mortenson, L. E., Valentine. R. C., and Carnahan, J. E. (1962) An electron transport factor from *Clostridium pasteurianum*. *Biochem. Biophys. Res. Commun.*, *7*, 448-452.

Mortenson, L. E., Valentine, R. C., and Carnahan, J. E. (1963) Ferredoxin in the phosphoroclastic reaction of pyruvic acid and its relation to nitrogen fixation in *Clostridium pasteurianum*. *J. Biol. Chem.*, *238*, 794-800.

Nagatani, H., Shimizu, M., and Valentine, R. C. (1971) The mechanism of ammonia assimilation in nitrogen fixing bacteria. *Arch. Mikrobiol., 79,* 164-175.

Nölling, J., Breton, G., Omelchenko, V., Makarova, K. S., Zeng, Q., Gibson, R. *et al.* (2001) Genome sequence and comparative analysis of the solvent-producing bacterium *Clostridium acetobutylicum*. *J. Bacteriol.*, *183*, 4823-4838.

Rosado, A. S., van Elsas, J. D., and Seldin, L. (1997) Reclassification of *Paenibacillus durun* (formerly *Clostridium durum* Smith and Cato 1974) Collins et al. 1994 as a member of the species *P. azotofixans* (formerly *Bacillus azotofixans* Seldin et al. 1984) Ash et al. 1994. *Int. J. Syst. Bacteriol.*, *47*, 569-572.

Rosenblum, E. D., and Wilson, P. W. (1949) Fixation of isotopic nitrogen by Clostridium. *J. Bacteriol.*, *57*, 413-414.

Schmid, B., Einsle, O., Chiu, H.-J., Willing, A., Yoshida, M., Howard, J B., and Rees, D. C. (2002) Biochemical and structural characterization of the cross-linked complex of nitrogenase: Comparison to the ADP-AlF_4^- stabilized structure. *Biochem. 41*, 15557-15565.

Singhal, R. K., Krishnan, I. S., and Dua, R. D. (1989) Stabilization, purification, and characterization of glutamate synthase from *Clostridium pasteurianum*. *Biochem., 28,* 7928-7935.

Szech, U., Braun, M., and Kleiner, D. (1989) Uptake and excretion of amino acids by saccharolytic clostridia. *FEMS Microbiol. Lett., 58,* 11-14.

Tempest, D. W., Meers, J. L., and Brown, C. M. (1973) Glutamate synthetase (GOGAT): a key enzyme in the assimilation of ammonia by prokaryotic organisms. In S. Prusiner and E. R. Stadtman (Eds.) *The enzymes of glutamine metabolism* (pp. 167-182). New York: Academic Press.

Usdin, K. P., Zappa, H., Jones, D. T., and Woods, D. R. (1986) Cloning, expression, and purification of glutamate synthetase from *Clostridium acetobutylicum*. *Appl. Environ. Microbiol., 52,* 413-419.

Vallespinós, F., and Kleiner, D. (1980) Selective inactivation of nitrogenase and glutamate synthase during sporulation of *Clostridium pasteurianum*. *J. Gen. Microbiol., 117,* 543-545.

Vanoni, M. A., and Curti, B. (1999) Glutamate synthase: A complex iron-sulfur flavoprotein. *Cell. Mol. Life Sci., 55,* 617-638.

Wang, S.-Z., Chen, J.-S., and Johnson, J. L. (1988a) The presence of five *nifH*-like sequences in *Clostridium pasteurianum*: Sequence divergence and transcription properties. *Nucleic Acid Res., 16,* 439-454.

Wang, S.-Z., Chen, J.-S., and Johnson, J. L. (1988b) Distinct structural features of the α and β subunits of nitrogenase molybdenum-iron protein of *Clostridium pasteurianum*: an analysis of amino acid sequences. *Biochem., 27,* 2800-2810.

Winogradsky, S. (1895). Recherches sur l'assimilation de l'azote libre de l'atmosphère par les microbes. *Arch. Sci. Biol. (St. Petersburg), 3,* 297-352.

Winogradsky, S. (1902a). *Clostridium Pastorianum*, seine Morphologie und seine Eigenschaften als Buttersäuereferment. *Zentralbl. Bakteriol. Parasitenk. Infektionskr. Hyg. Abt. II, 9,* 43-54.

Winogradsky, S. (1902b). *Clostridium Pastorianum*, seine Morphologie und seine Eigenschaften als Buttersäuereferment. *Zentralbl. Bakteriol. Parasitenk. Infektionskr. Hyg. Abt. II, 9,* 107-112.

Zelitch, I., Rosenblum, E. D., Burris, R. H., and Wilson, P. W. (1951a) Isolation of the key intermediate in biological nitrogen fixation by *Clostridium*. *J. Biol. Chem., 191,* 295-298.

Zelitch, I., Rosenblum, E. D., Burris, R. H., and Wilson, P. W. (1951b) Comparison of the metabolism of ammonia and molecular nitrogen in *Clostridium*. *J. Bacteriol., 62,* 747-752.

Chapter 4

REGULATION OF NITROGEN FIXATION IN METHANOGENIC ARCHAEA

J. A. LEIGH
Department of Microbiology, University of Washington, Seattle, WA 98195, USA

1. INTRODUCTION

The biochemical mechanism of nitrogen fixation in the domain Archaea is evidently similar to that in the well-studied Bacteria. Proteins that are genetically and biochemically homologous to bacterial dinitrogenase (MoFe protein) and dinitrogenase reductase (Fe protein) are involved, as are either molybdenum-containing or alternative cofactors. A high expenditure of ATP is presumably also a requirement as in Bacteria. Hence, it is not surprising that nitrogen fixation is highly regulated. As in many bacterial diazotrophs, this regulation takes place at two levels: transcriptional control and control of nitrogenase enzyme activity.

In contrast to this biochemical similarity, the regulatory mechanisms of nitrogen fixation differ in the Archaea from those that have been elucidated in many Bacteria, in particular the Proteobacteria. In *Methanococcus maripaludis*, transcriptional regulation of *nif* gene expression is mediated by a repressor rather than an activator, and appears to be independent of any known nitrogen sensor/transducer protein. The mechanism for control of nitrogenase activity is unknown, but is clearly different from the mechanism known in Proteobacteria. Our current knowledge of these regulatory systems in the diazotrophic Archaea is the subject of this chapter.

2. HISTORY AND BACKGROUND

Nitrogen fixation was discovered in the Archaea in 1984, in both *Methanosarcina barkeri* (Murray and Zinder, 1984) and *Methanococcus thermolithotrophicus* (Belay *et al.*, 1984). Known nitrogen-fixing species among the Archaea are restricted to

W. Klipp, B. Masepohl, J. R. Gallon and W. E. Newton (eds.), Genetics and Regulation of Nitrogen Fixation in Free-Living Bacteria, 65-71.

the methanogenic Euryarchaeota. Early studies of *nif* genetics showed that diazotrophic methanogens possess a *nifH* gene, followed in order by two genes encoding P_{II} nitrogen sensor/transducer homologs, and *nifD* and *nifK* (Sibold *et al.*, 1991; Souillard *et al.*, 1988; Souillard and Sibold, 1989). Later studies in *M. barkeri* (Chien *et al.*, 1998; Chien and Zinder, 1994, 1996) and *M. maripaludis* (Kessler *et al.*, 1998), and recent genome sequences (see J. A. Leigh, Genomics of Diazotrophic Archea, in the volume "*Genomes and Genomics of Nitrogen-Fixing Organisms*" of this series), show a common arrangement in which the above genes are followed by *nifE*, *nifN*, and sometimes *nifX*. This gene order resembles that found in many bacterial diazotrophs. However, in contrast to Bacteria, the known *nif* genes in archaeal diazotrophs are distinguished by their inclusion within a single operon, at least in *M. maripaludis* (Kessler *et al.*, 1998).

3. TRANSCRIPTIONAL REGULATION

Nitrogen regulation of *nif* mRNA levels has been observed in *M. thermolithotrophicus* (Souillard and Sibold, 1989), *M. maripaludis* (Kessler *et al.*, 1998), and *M. barkeri* (Chien *et al.*, 1998; Chien and Zinder, 1996). As with all Archaea, transcription in these organisms is carried out by an apparatus that resembles a simplified eukaryotic system. A TATA box-binding protein binds to a TATA box promoter sequence, followed by recruitment of transcription factor B and a multi-subunit RNA polymerase (Bell and Jackson, 2001). Nevertheless, *nif* regulation in *M. maripaludis* and *M. thermolithotrophicus* resembles the bacterial repression paradigm. The regulation of *nif* in *M. barkeri* occurs by a different mechanism that has not yet been thoroughly characterized (Chien *et al.*, 1998).

3.1. Nitrogen Operators

Due to the availability of genetic tools, the mechanism of archaeal *nif* regulation has been studied mainly in *M. maripaludis*. As originally noted in *M. thermolithotrophicus* (Souillard and Sibold, 1989), the promoter region preceding the *nif* operon in *M. maripaludis* contains two similar inverted repeats (Figure 1A). Genetic tests showed that the first of these repeats, but not the second repeat, was required for repression of *nif* transcription when cells were grown on ammonia (Cohen-Kupiec *et al.*, 1997). In addition, a role in repression for both repeats was observed when cells were grown on alanine as a nitrogen source (Lie and Leigh, 2002). Growth on ammonia generally results in marked down-regulation of nitrogen assimilatory functions in *M. maripaludis*, whereas growth on alanine causes partial down-regulation. Hence, the first repeat (*nif* operator 1) is the primary operator for *nif* repression, whereas the second repeat (*nif* operator 2) is required for repression when cells are only partially nitrogen-depleted. A model, involving cooperative binding of a tetrameric repressor protein to the two operators, is shown in Figure 2. In this model, two operators downstream of the transcription start site (Figure 1A) mediate marked repression.

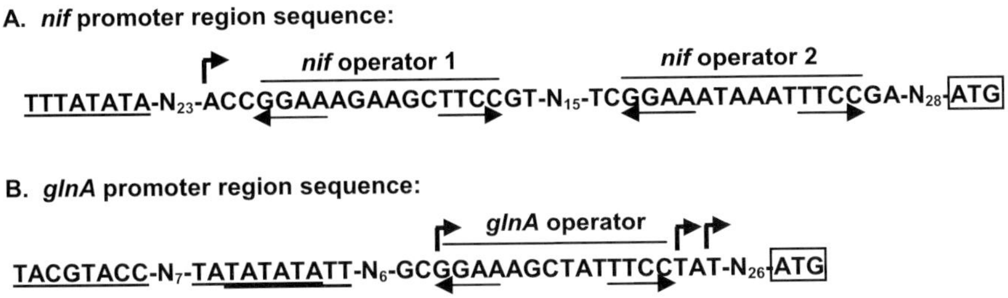

Figure 1. *Promoter region sequences. Underscores indicate TATA boxes. Transcription starts are marked with bent arrows. Horizontal arrows indicate inverted repeats. Start codons are boxed. A,* nif *promoter region. B,* glnA *promoter region; the upstream start site is constitutive but the two downstream start sites are regulated similarly by nitrogen.*

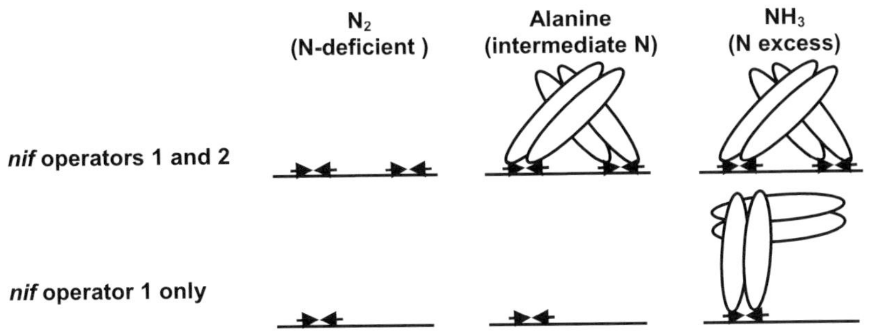

Figure 2. *Model for binding of the nitrogen repressor to* nif *operators. The nitrogen sources N_2, alanine, and NH_3 result in deficient, intermediate, and excess nitrogen conditions, respectively. With operators 1 and 2 intact, repression occurs even under intermediate nitrogen conditions. With only* nif *operator 1 (*nif *operator 2 mutationally altered), repression occurs only under conditions of nitrogen excess.*

Genes that are regulated by nitrogen in *M. maripaludis* (and certain other methanogens) appear to form a regulon, which is governed by a common repression mechanism that involves a consensus operator sequence (Kessler and Leigh, 1999). The *glnA* gene (encoding glutamine synthetase), as well as *nif* in *M. maripaludis*, belongs to this regulon and, like *nif*, the function of the operator has been demonstrated genetically (Cohen-Kupiec *et al.*, 1999). However, the two operons offer an interesting contrast with regard to promoter region architecture and regulatory outcome. Instead of two operators in a promoter region with a single transcription start site, the *glnA* promoter region contains one operator and three transcription start sites (Figure 1B) (Cohen-Kupiec *et al.*, 1999). Several differences in the regulatory outcome may be noted. First, transcription from the furthest upstream *glnA* start site is constitutive. Evidently, the single operator immediately downstream of the start site is ineffective in this case. Second, both downstream start sites are regulated by alanine as well as ammonia (Lie and Leigh, 2002). Therefore, the single operator upstream of these sites mediates complete repression. The *nif* and *glnA* promoter regions each appears to be suitably designed with respect to the function regulated. High-affinity binding of a repressor to two *nif* operators is well suited to the rigorous regulation needed for an expensive process such as nitrogen fixation. Binding of a repressor to a single *glnA* operator achieves the necessary regulation of ammonia assimilation at the same time that constitutive expression from the upstream start site allows glutamine synthesis for incorporation into proteins.

3.2. The Nitrogen Repressor

The protein that binds to nitrogen operator sequences and represses gene expression has been identified in *M. maripaludis* (Lie and Leigh, 2003). This protein is designated NrpR. Molecular weight measurements on the native and denatured protein are consistent with the homotetrameric structure proposed in Figure 2. NrpR contains a helix-turn-helix DNA-binding motif and two mutually homologous domains whose functions were previously unknown (Figure 3). One function of these novel domains is presumably either to sense the nitrogen state of the cell or to interact with other proteins that do so. Homologs of NrpR are found in some species of Euryarchaeota but are not present in the Crenarchaeota or the Bacteria. Some homologs lack the second of the mutually homologous domains. Homologs in *Methanosarcina* species include proteins that also lack the helix-turn-helix motif, suggesting that, whatever the function of the remaining domain, it is not always directly connected with DNA binding.

3.3. The Nitrogen Sensor

Nitrogen transcriptional regulators in Proteobacteria (such as the activator systems NtrBC and NifLA) receive information on the nitrogen state of the cell from members of the P_{II} family of proteins. P_{II} in *E. coli* responds to the levels of metabolites thought to indicate nitrogen excess or deficiency. In Archaea no protein has yet been identified that senses the nitrogen state of the cell and affects transcriptional regulation. The *M. maripaludis* genome encodes five P_{II} homologs. Work in my laboratory produced null mutations in all five P_{II} genes in a single mutant, but found no discernible effect on transcriptional regulation (unpublished data). Whether NrpR itself senses nitrogen or interacts with an unknown protein that does so is currently unknown.

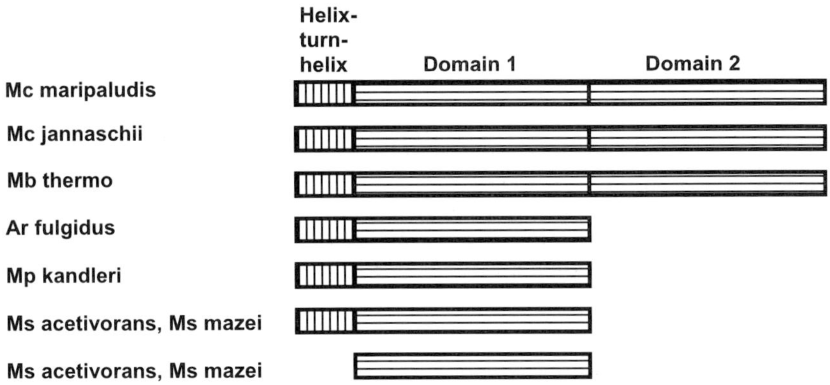

Figure 3. The NrpR repressor protein. Homologs are found in (top to bottom) Methanococcus maripaludis, Methanococcus jannaschii, Methanothermobacter thermoautotrophicus, Archaeoglobus fulgidus, Methanopyrus kandleri, Methanosarcina acetivorans, *and* Methanosarcina mazei. *The latter two species each contains two NrpR homologs as shown.*

4. REGULATION OF NITROGENASE ACTIVITY

In addition to the regulation of *nif* gene expression, nitrogenase enzyme activity is regulated in many diazotrophic Bacteria. This mode of regulation, often termed "switch-off", prevents existing nitrogenase proteins from unnecessarily using ATP after a favorable nitrogen source is encountered. Among the diazotrophic Archaea, switch-off has been observed in *M. barkeri* (Lobo and Zinder, 1988; 1990) and *M. maripaludis* (Kessler *et al.*, 2001; Kessler and Leigh, 1999). As is the case for transcriptional regulation, the phenomenon is similar but the mechanism is different compared to that in the Proteobacteria. Similarities include the observations that the process is rapid, reversible, and important for accelerated growth upon the switch from N_2 to ammonia (Kessler *et al.*, 2001). The major difference is that, unlike certain Proteobacteria, detectable covalent modification of dinitrogenase reductase does not occur (Kessler *et al.*, 2001; Lobo and Zinder, 1988, 1990). The mechanism could involve covalent modification of another nitrogenase protein or some sort of non-covalent interaction.

Although we do not know what component of the nitrogenase machinery is affected by switch-off, we do know that members of the P_{II} nitrogen sensor/transducer proteins are involved. Among the five P_{II} proteins of *M. maripaludis* mentioned above, two belong to subfamilies designated $NifI_1$ and $NifI_2$ (Arcondeguy *et al.*, 2001; Kessler *et al.*, 2001). These proteins are encoded within the *nif* operon of diazotrophic methanogens and contain distinctive amino-acid sequences within a domain called the T loop (see J. A. Leigh, Genomics of Diazotrophic Archea, in the volume "*Genomes and Genomics of Nitrogen-Fixing Organisms*" of this series). In *M. maripaludis* both $NifI_1$ and $NifI_2$ are required for switch-off (Kessler *et al.*, 2001). $NifI_1$ and $NifI_2$ could sense the nitrogen state of the cell and interact with a component of nitrogenase or interact with some other protein that does so.

5. SUMMARY

Although regulatory strategies in diazotrophic Bacteria and Archaea are based on common goals, mechanisms of regulation are different. Transcriptional regulation of *nif* gene expression and regulation of nitrogenase activity by switch-off occur in both domains. A model for regulation in *M. maripaludis* is shown in Figure 4.

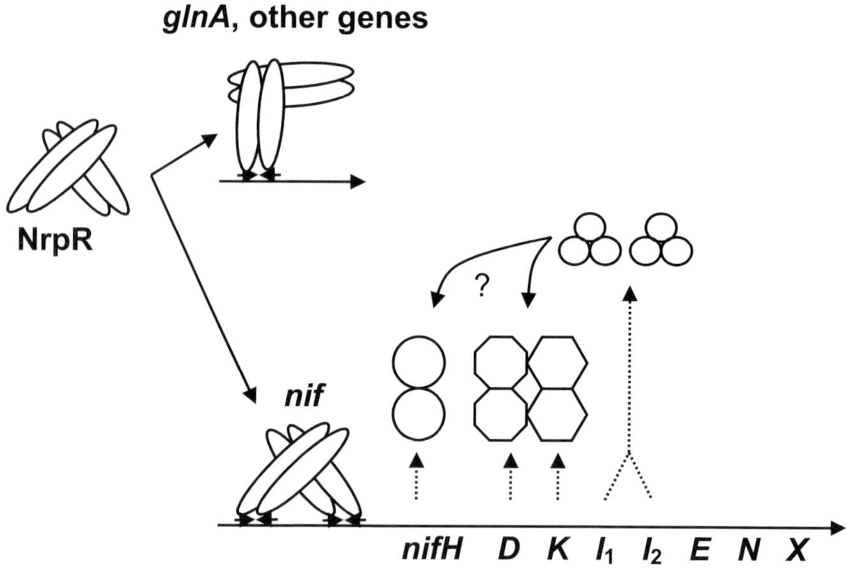

Figure 4. Model for regulation of nitrogen fixation in Methanococcus maripaludis. *The repressor NrpR regulates transcription of the* nif *operon,* glnA, *and other genes by binding to inverted repeat operators.* $NifI_1$ *and* $NifI_2$, P_{II} *nitrogen sensor subfamilies encoded in the* nif *operon, control nitrogenase activity by an unknown mechanism. Proteins encoded by genes are represented by shapes above vertical dotted arrows.*

Transcriptional regulation by nitrogen in *M. maripaludis* and related methanogens operates by a repression mechanism that involves a novel repressor protein, NrpR. Expression of *nif* and *glnA* respond differently to repression due to differences in promoter region architecture. The nitrogen-sensing mechanism for transcriptional regulation is unknown. Switch-off in diazotrophic methanogens occurs by an unknown mechanism, which differs from that known to occur in

Proteobacteria. Nitrogen sensing for switch-off appears to involve P_{II} protein homologs belonging to distinct subfamilies, $NifI_1$ and $NifI_2$.

REFERENCES

Arcondeguy, T., Jack, R., and Merrick, M. (2001). PII signal transduction proteins, pivotal players in microbial nitrogen control. *Microbiol. Mol. Biol. Rev., 65*, 80-105.

Belay, N., Sparling, R., and Daniels, L. (1984). Dinitrogen fixation by a thermophilic methanogenic bacterium. *Nature, 312*, 286-288.

Bell, S. D., and Jackson, S. P. (2001). Mechanism and regulation of transcription in Archaea. *Curr. Opinion Microbiol., 4*, 208-213.

Chien, Y., Helmann, J. D., and Zinder, S. H. (1998). Interactions between the promoter regions of nitrogenase structural genes (*nifHDK2*) and DNA-binding proteins from N_2- and ammonium-grown cells of the archaeon *Methanosarcina barkeri* 227. *J. Bacteriol., 180*, 2723-2728.

Chien, Y. T., and Zinder, S. H. (1994). Cloning, DNA sequencing, and characterization of a *nifD*-homologous gene from the archaeon *Methanosarcina barkeri* 227 which resembles *nifD1* from the eubacterium *Clostridium pasteurianum*. *J. Bacteriol., 176*, 6590-6598.

Chien, Y. T., and Zinder, S. H. (1996). Cloning, functional organization, transcript studies, and phylogenetic analysis of the complete nitrogenase structural genes (*nifHDK2*) and associated genes in the archaeon *Methanosarcina barkeri* 227. *J. Bacteriol., 178*, 143-148.

Cohen-Kupiec, R., Blank, C., and Leigh, J. A. (1997). Transcriptional regulation in Archaea: in vivo demonstration of a repressor binding site in a methanogen. *Proc. Natl. Acad. Sci. U.S.A., 94*, 1316-1320.

Cohen-Kupiec, R., Marx, C. J., and Leigh, J. A. (1999). Function and regulation of *glnA* in the methanogenic archaeon *Methanococcus maripaludis*. *J. Bacteriol., 181*, 256-261.

Kessler, P. S., Blank, C., and Leigh, J. A. (1998). The *nif* gene operon of the methanogenic archaeon *Methanococcus maripaludis*. *J. Bacteriol., 180*, 1504-1511.

Kessler, P. S., Daniel, C., and Leigh, J. A. (2001). Ammonia switch-off of nitrogen fixation in the methanogenic archaeon *Methanococcus maripaludis*: mechanistic features and requirement for the novel GlnB homologues, $NifI_1$ and $NIfI_2$. *J. Bacteriol., 183*, 882-889.

Kessler, P. S., and Leigh, J. A. (1999). Genetics of nitrogen regulation in *Methanococcus maripaludis*. *Genetics, 152*, 1343-1351.

Lie, T. J., and Leigh, J. A. (2002). Regulatory response of *Methanococcus maripaludis* to alanine, an intermediate nitrogen source. *J. Bacteriol., 184*, 5301-5306.

Lie, T. J., and Leigh, J. A. (2003). A novel repressor of *nif* and *glnA* expression in the methanogenic archaeon, *Methanococcus maripaludis*. *Mol. Microbiol., 47*, 235-246.

Lobo, A. L., and Zinder, S. H. (1988). Diazotrophy and nitrogenase activity in the archaebacterium *Methanosarcina barkeri* 227. *Appl. Environ. Microbiol., 54*, 1656-1661.

Lobo, A. L., and Zinder, S. H. (1990). Nitrogenase in the archaebacterium *Methanosarcina barkeri* 227. *J. Bacteriol., 172*, 6789-6796.

Murray, P. A., and Zinder, S. H. (1984). Nitrogen fixation by a methanogenic archaebacterium. *Nature, 312*, 284-286.

Sibold, L., Henriquet, M., Possot, O., and Aubert, J. P. (1991). Nucleotide sequence of *nifH* regions from *Methanobacterium ivanovii* and *Methanosarcina barkeri* 227 and characterization of *glnB*-like genes. *Res. Microbiol., 142*, 5-12.

Souillard, N., Magot, M., Possot, O., and Sibold, L. (1988). Nucleotide sequence of regions homologous to *nifH* (nitrogenase Fe protein) from the nitrogen-fixing archaebacteria *Methanococcus thermolithotrophicus* and *Methanobacterium ivanovii*: evolutionary implications. *J. Mol. Evol., 27*, 65-76.

Souillard, N., and Sibold, L. (1989). Primary structure, functional organization and expression of nitrogenase structural genes of the thermophilic archaebacterium *Methanococcus thermolithotrophicus*. *Mol. Microbiol., 3*, 541-551.

Chapter 5

NITROGEN FIXATION IN HETEROCYST-FORMING CYANOBACTERIA

T. THIEL
Department of Biology, University of Missouri-St. Louis
8801 Natural Bridge Rd., St. Louis, MO 63121, USA

1. INTRODUCTION

Cyanobacteria are prokaryotes that obtain their energy from photosynthesis, using the same photosystem that is used by plants. They live in diverse environments that are often low in essential nutrients, especially fixed nitrogen. They are the only organisms that are capable of both O_2-evolving photosynthesis and nitrogen fixation. Fossils indicate that cyanobacteria were responsible for the production of O_2 on earth via photosynthesis and thus they evolved mechanisms to protect nitrogenase from O_2 very early in evolutionary history. This chapter focuses on cyanobacterial strains that produce differentiated cells called heterocysts, which maintain a microaerobic environment to support nitrogenase activity. Strains of *Anabaena* spp, and *Nostoc* spp., particularly, *Anabaena* sp. PCC 7120, are the best studied of the heterocyst-forming cyanobacteria. Some other genera that produce heterocysts that have been studied are *Anabaenopsis*, *Aphanizomenon*, *Nodularia*, *Cylindrospermum*, *Cylindrospermopsis*, *Scytonema*, *Calothrix*, *Chlorogloeopsis*, and *Fischerella* (Rippka *et al.*, 1979). Most of these strains fix nitrogen exclusively in heterocysts that form only in the absence of external fixed nitrogen. Heterocysts protect nitrogenase from O_2 that is produced by photosystem II in adjacent vegetative cells and from O_2 that enters the cell from the environment. The differentiation of heterocysts from vegetative cells in the filament is a complex, regulated process that involves changes in structure and function. Heterocysts form in the cyanobacterial filament in a semi-regular pattern suggesting that there is cell-to-cell communication in the filament that leads to this patterning of differentiation.

W. Klipp, B. Masepohl, J. R. Gallon and W. E. Newton (eds.), Genetics and Regulation of Nitrogen Fixation in Free-Living Bacteria, 73-110.

Wolk (2000) suggests that 100-200 genes are likely to be dedicated to the process of heterocyst differentiation and nitrogen fixation in *Anabaena* sp. PCC 7120. With the sequence of the genome now available (Kaneko *et al.*, 2001) and genetic tools for this strain at hand, all the genes that are responsible for heterocyst differentiation and nitrogen fixation may soon be elucidated.

2. STRUCTURE OF HETEROCYSTS

2.1. Heterocyst Cell Layers

Heterocysts are terminally differentiated cells that form in a semi-regular pattern in filaments starved for a source of fixed nitrogen. These cells, occurring about every 10-15 cells in a filament, are evident by differences in structure compared to vegetative cells. They are typically somewhat larger than vegetative cells with a thicker cell envelope (Figure 1). Cyanobacterial cells, including heterocysts, have a

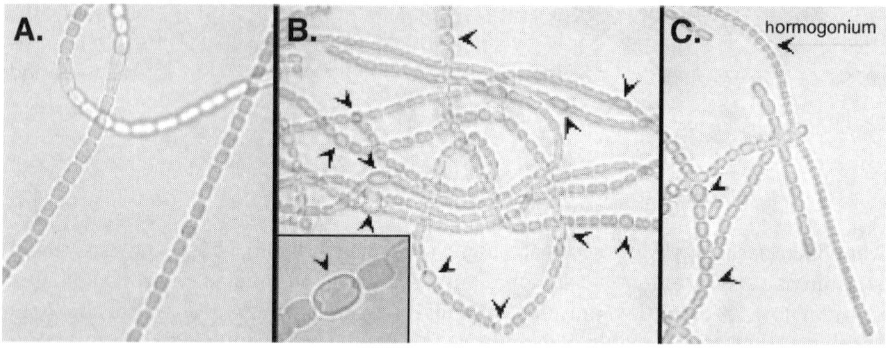

Figure 1. Light micrographs of Anabaena variabilis *ATCC 29413 grown (A) with ammonium, or (B and C) without a source of fixed nitrogen. Heterocysts are indicated by arrowheads. The inset to panel B shows the distinctive polar regions of the heterocyst. Panel C shows filaments with heterocysts and a type of differentiated filament called a hormogonium. A hormogonium comprises small vegetative cells that result from cell division in the absence of cell growth. They are motile and, in the absence of fixed nitrogen, differentiate heterocysts prior to resumption of growth. Rippka* et al. *(1979) use the differentiation of hormogonia to distinguish between the genera* Nostoc *(with hormogonia) and* Anabaena *(without), so it appears that this strain is actually a* Nostoc.

Gram-negative peptidoglycan cell wall with an outer bilayer membrane. However, the peptidoglycan layer in cyanobacteria is thicker than is typical of Gram-negative bacteria and has the cross-linking and covalently linked polysaccharides that are characteristic of the walls of Gram-positive cells (Weckesser and Jürgens, 1988).

Heterocysts have additional layers external to the cell wall, not present in vegetative cells, that limit the diffusion of gases into these cells (Murry and Wolk, 1989; Walsby, 1985). There is an inner laminated layer comprising heterocyst-specific glycolipids that are derivatives of hexoses with long-chain polyhydroxy-alcohols (Bryce *et al.*, 1972; Lambein and Wolk, 1973; Winkenbach *et al.*, 1972; Wolk, 1973; Wolk *et al.*, 1994). The outermost layers of the heterocyst are made of specific polysaccharides whose composition has been described in detail (Cardemil and Wolk, 1976; 1979; 1981a; 1981b). This polysaccharide layer can be further subdivided into an inner homogeneous layer and an external fibrous layer, which is typically the first layer that is deposited during the differentiation of a heterocyst (Adams, 1999).

2.1.1. Glycolipid Layer

The glycolipids of heterocysts are typically long-chained either di- or tri-hydroxy saturated fatty acid chains generally with an ether link to glucose (Gambacorta *et al.*, 1995; Soriente *et al.*, 1993; 1995; Wolk *et al.*, 1994; Wolk and Quine, 1975). The glycolipid layer appears to be critical in the commitment stage of heterocyst formation, after which these cells are no longer able to regress to vegetative cells. This layer is essential for O_2 protection of nitrogenase; mutants lacking the glycolipid layer have a Fox^- phenotype, meaning that they are able to fix N_2 only under anaerobic conditions (Black *et al.*, 1995; Ernst *et al.*, 1992). Synthesis of the glycolipid layer requires O_2; therefore, cells grown under anaerobic conditions cannot synthesize a complete glycolipid layer and cannot fix N_2 in air (Rippka and Stanier, 1978).

The biochemical pathway for synthesis of the heterocyst glycolipid is not yet well understood. Genes required for either formation or deposition of the heterocyst glycolipid layer include *hglC*, *hglD*, *hetM* (aka *hglB*), *hglE*, *hglK*, and *devBCA* (Bauer *et al.*, 1997; Black and Wolk, 1994; Campbell *et al.*, 1997; Fiedler *et al.*, 1998). In *Anabaena* sp. PCC 7120, *hglE*, *hglD*, *hglC*, *hetM*, *hetN* and *hetI* (in that order, although *hetI* is in the opposite orientation to the other genes) are clustered on the *Anabaena* sp. PCC 7120 genome with genes of unknown function separating *hglD* from *hglE* and *hglC* from *hetM*, *hetN* and *hetI* (Kaneko *et al.*, 2001; Cyanobase: http://www.kazusa.or.jp/cyanobase/). The clustering of these genes suggests that they may all be involved in some aspect of glycolipid layer formation. The proteins encoded by *hglE*, *hglD*, *hglC*, *hetM*, and *hetN* all show similarity to enzymes involved in fatty acid or polyketide biosynthesis.

Based on the known structure of the hydrocarbon chains of heterocyst glycolipids, it appears that these glycolipids may be more like polyketides than like fatty acids. Polyketides and fatty acids differ in several characteristics. Polyketides have longer carbon chains than fatty acids and they often form complex cyclic structures. Synthesis of heterocyst glycolipids appears to require a multienzyme complex, which is also characteristic of polyketide synthesis (Campbell *et al.* 1997). A *hetM* mutant has a Fox^- phenotype (incapable of aerobic nitrogen fixation) and is deficient in heterocyst glycolipids (Bauer *et al.*, 1997; Black and Wolk, 1994; Ernst *et al.*, 1992), indicating that the glycolipids help to protect nitrogenase from O_2. Although the next gene, *hetN*, is part of the cluster of genes involved in glycolipid synthesis, it does not appear to affect glycolipid synthesis directly (Callahan and Buikema, 2001). A mutant in this gene results in a phenotype of multiple contiguous heterocysts (Bauer *et al.*, 1997; Black and Wolk, 1994) and its role (see Section 6.1.3) is apparently in maintenance of heterocyst spacing (Callahan and Buikema, 2001). The product of *hetI* is most like a family of phosphopantetheinyl transferases, which are responsible for activation of polyketide and fatty acid synthases, and hence, may be essential for activation of several such enzymes (Wolk, 2000). Black and Wolk (1994) were unable to obtain a *hetI* mutant, suggesting that this gene is essential in *Anabaena* sp. PCC 7120. The product of *hglE* is most similar to polyketide synthases and a mutant in *hglE* in *Nostoc punctiforme* has a Fox$^-$ phenotype and is deficient in heterocyst glycolipids (Campbell *et al.* 1997). There are two *hglE*-like genes in *Anabaena* sp. PCC 7120; the one most similar to *hglE* in *N. punctiforme* is part of the cluster that contains *hglD* and *hglC*. The other copy has no genes of known function nearby (Kaneko *et al.*, 2001; Cyanobase: http://www.kazusa.or.jp/cyanobase/).

Another gene involved in glycolipid synthesis is *hglK*, which encodes a protein that may be involved in the extracellular assembly of the glycolipid; glycolipids are made in an *hglK* mutant, but the extracellular layer is not formed (Black *et al.* 1995). Similarly, the products of the ABC transporter genes, *devBCA*, are also required for assembly of glycolipids outside the cell (Fiedler *et al.*, 1998a; 1998b; 2001; Maldener *et al.*, 1994). See Section 5.7 for more information on *devBCA*. The product of *devR*, which is also required for normal glycolipid deposition in *N. punctiforme*, is similar to a class of response regulators that lack a C-terminal effector domain (Campbell *et al.*, 1996). The protein is phosphorylated *in vivo* and is thought to be a phosphotransferase involved in a phosphorelay with as yet unidentified proteins (Hagen and Meeks, 1999). In *Anabaena* sp. PCC 7120, DevR is involved in polysaccharide synthesis (see section 2.1.2).

2.1.2. Polysaccharide Layer

Very little is known about the biosynthesis and assembly of heterocyst polysaccharides. Genes that are involved in heterocyst polysaccharide synthesis include *hepA*, *hepB*, *hepC*, *hepK*, *apb2*, and *apb3*. HepA, which is essential for synthesis of heterocyst polysaccharides, is an ATP-binding cassette inner-membrane transport protein that is induced 4.5-7 h after nitrogen step-down in developing heterocysts. Its expression is dependent on HetR (Black *et al.*, 1993; Holland and Wolk, 1990; Khudyakov and Wolk, 1997; Leganés, 1994). A mutant in *hepB* was identified as defective in heterocyst maturation and the gene product appears to be involved in polysaccharide biosynthesis (Wolk *at al.*, 1999; see also GenBank accession AJ250131; ASU68036). HepC, which is similar to UDP-galactose-lipid carrier transferases, is required for synthesis of the heterocyst envelope polysaccharide, possibly for the synthesis of the side branches that contain galactosyl residues (Zhu *et al.*, 1998). Interestingly, HepC also appears to have some regulatory activity. It represses transcription of *hepA*, although this effect may be indirect (Zhu *et al.*, 1998). In contrast, two DNA-binding proteins (Apb2 and Apb3), which bind upstream of *hepC*, appear to regulate directly expression of *hepC* and, hence, *hepA*. Mutants in *apb2* and *apb3* fail to induce expression of *hepC* and *hepA*, preventing maturation of heterocysts and, thus, nitrogen fixation under aerobic conditions (Koksharova and Wolk, 2002).

A mutation in a gene, *hcwA,* also blocks transcription of *hepA*. HcwA is most similar to an autolysin (N-acetylmuramoyl-L-alanine amidase) of *Bacillus subtilis* (Zhu *et al.*, 2001). HcwA, when expressed in *E. coli*, shows cell-wall lysis activity. In *Anabaena* sp. PCC 7120, transcription of *hcwA* is constitutive, possibly resulting in degradation of peptidoglycan as vegetative cells develop into heterocysts. This degradation may increase the permeability of the peptidoglycan layer in developing heterocysts, thereby facilitating transfer of glycolipids and polysaccharides to the outside of the cell wall. A mutant in the *hcwA* gene may prevent an external regulatory signal from reaching and acting on *hepA*, or, possibly, heterocyst envelope components may accumulate in cells of a *hcwA* mutant, thereby repressing transcription of *hepA* (Zhu *et al.*, 2001).

The product of *hepK* is also required for polysaccharide synthesis. A mutation in that gene prevents induction of *hepA* (Zhu *et al.*, 1998). HepK is an autokinase and DevR is its cognate response regulator. Together, these proteins comprise a two-component system that regulates biosynthesis of a heterocyst envelope polysaccharide (Zhou and Wolk, 2003). A strain containing a *hepK::gfp* transcriptional fusion expresses GFP primarily in proheterocysts, which is consistent with its role in deposition of polysaccharide in developing heterocysts (Zhou and Wolk, 2003).

2.1.3. Peptidoglycan/LPS Layer

Penicillin-binding proteins are involved in synthesis of the peptidoglycan layer of the cell wall. A gene coding for a putative penicillin-binding protein, *pbpB*, is required for aerobic nitrogen fixation in *Anabaena* sp. strain PCC 7120. A mutation in this gene results in a Fox^- phenotype with filaments showing abnormal heterocysts (Lázaro *et al.*, 2001). In addition, mutations, which affect synthesis of the lipopolysaccharide (LPS) layer of *Anabaena* sp. PCC 7120, lead to phage resistance and to the loss of heterocyst function under aerobic conditions, suggesting that the LPS layer has a role in formation of the heterocyst envelope and, hence, in protection of nitrogenase from O_2 (Xu *et al.*, 1997).

2.2. Polar Regions of Heterocysts and Cell Junctions

Heterocysts can be distinguished morphologically from vegetative cells by distinct regions at the polar regions of the cell, where they join a vegetative cell (Figure 1 B, inset). These refractile regions are rich in a nitrogen-storage polymer, cyanophycin, comprising a polymer that is a multi-L-arginyl-poly-L-aspartate (Sherman *et al.*, 2000; Simon, 1971). A mutant of *A. variabilis* lacking the enzyme for synthesis of cyanophycin is deficient in the polar granules normally present in heterocysts but is not impaired in diazotrophic growth (Ziegler *et al.*, 2001), suggesting that this reserve material is not required for heterocyst function. The polar region also contains honeycomb membranes with high respiratory activity for scavenging O_2 that enters heterocysts from adjacent vegetative cells (Murry *et al.*, 1981).

The septum or pore junction between a heterocyst and a vegetative cell is very narrow compared to the junctions between vegetative cells (Fay, 1992). The very small channels, called microplasmodesmata, which span the cytoplasmic membrane separating adjacent cells, increase in number at junctions between heterocysts and vegetative cells (Fay, 1992; Giddings and Staehelin, 1978). These narrow channels are thought to be conduits for transport of metabolites between cells in a filament. There is also evidence that these structures may be essential for heterocyst differentiation. In plants, the movement protein of tobacco mosaic virus modifies the plasmodesmata (the plasmalemma junctions that connect adjacent plant cells) allowing the viral nucleic acid to move from cell to cell (Deom *et al.*, 1990). Expression of the tobacco mosaic virus movement protein in *Anabaena* sp. PCC 7120 results in accumulation of the movement protein in the cell envelope layer and also blocks heterocyst differentiation (Heinlein *et al.*, 1998; Zahalak *et al.*, 1995).

Compounds may also move between cells in a filament through the periplasm. There is evidence that the outer cell membrane continuously surrounds the cells along a filament and does not surround each cell individually, suggesting that the periplasm is common to all cells in a filament (Drews and Weckesser 1982). This provides a potential mechanism for cell-to-cell transfer either of metabolites made in heterocysts or of differentiation signals that control pattern formation.

3. NITROGENASE GENES

3.1. Conventional Nitrogenase

The genes involved in nitrogen fixation have been characterized best in *Klebsiella*, where they form a contiguous cluster of twenty genes (Arnold *et al.*, 1988; Dean and Jacobson, 1992). As is the case with other diazotrophic bacteria, the genes required for nitrogenase synthesis in cyanobacteria are highly conserved. The structural genes for nitrogenase, *nifH*, *nifD* and *nifK,* as well as those for the scaffolding proteins, encoded by *nifE* and *nifN*, form two operons in heterocystous cyanobacteria as well as in non-heterocystous cyanobacteria (Buikema and Haselkorn, 1993; Flores and Herrero, 1994; Thiel *et al.*, 1997; 1998). The organization and order of a large cluster of *nif* and *nif*-related genes is highly conserved among cyanobacteria (Buikema and Haselkorn 1993; Huang *et al.*, 1999; Thiel *et al.*, 1997; 1998) (see Figure 2). Upstream of *nifH* in both *N. commune* and *N. punctiforme*, there is a gene that encodes a hemoglobin-like protein, called cyanoglobin (*glbN*) (Angeloni and Potts, 1994; Hill *et al.*, 1996; Meeks *et al.*, 2001; Potts *et al.*, 1992), whose function is not known. In the genome of *Anabaena* sp. PCC 7120, the *nifV1* (homocitrate synthase), *nifP* (serine O-acetyltransferase), *nifZ*, and *nifT* genes are about 17 kb downstream from the end of the major *nif* cluster, whereas in *N. punctiforme*, these same genes are about 10 kb upstream of *nifB* (Kaneko *et al.*, 2001; Cyanobase: http://www.kazusa.or.jp/cyanobase/; JGI: http://www.jgi.doe.gov/JGI_microbial/html/nostoc/nostoc_homepage.html).

There are few mutants in cyanobacterial *nif* genes. Either a mutation in *nifB* or a deletion in *nifKE* in *A. variabilis* abolished nitrogenase activity, whereas mutations in either *nifS* or *nifU* had no effect (Lyons and Thiel, 1995). A mutation in *nifV1* did not affect diazotrophic growth of *Anabaena* sp. PCC 7120 (Stricker *et al.*, 1997), whereas a mutant in *fdxH* demonstrated that this gene is required only for optimum nitrogenase activity (Masepohl *et al.*, 1997). There is no homolog of *nifM* in either *Anabaena* sp. PCC 7120 or *N. punctiforme*. However, the *nifM* genes of *K. pneumoniae* and *Azotobacter vinelandii* show very low sequence identity, which suggests that a functionally equivalent protein may exist in cyanobacteria. Other genes not identified in cyanobacterial genomes include the regulatory proteins encoded by *nifL* and *nifA*. The gene encoding the flavodoxin of *Anabaena* sp. is similar to the *nifF* gene of *K. pneumoniae*, however, a true homologue of *nifF* is missing. Also absent *in Anabaena* sp. *are nifY*, which in other organisms shares significant homology in the C-terminal regions with *nifX*, and the *nifQ* gene. Both NifQ and NifY appear involved in FeMo cofactor biosynthesis (Allen *et al.*, 1994)

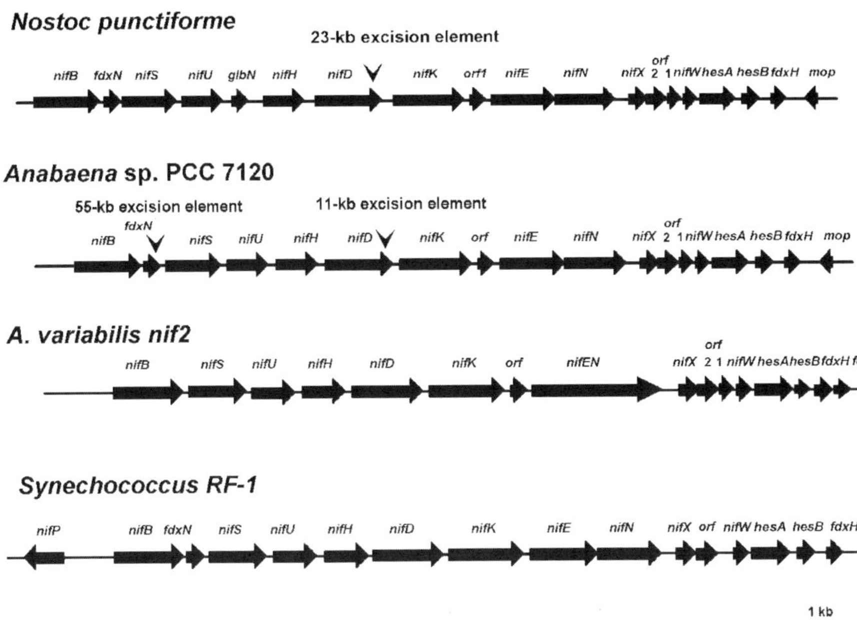

Figure 2. Organization of nif *genes in cyanobacteria. The maps of genes and orfs were constructed from GenBank sequence data based on published information (Huang* et al., *1999; Thiel* et al., *1997; 1998) and on whole genome analysis (Cyanobase: http://www.kazusa.or.jp/cyanobase/; JGI: http://www.jgi.doe.gov/JGI_microbial/html/nostoc/nostoc_homepage.html).*

3.2. V-nitrogenase

The *nif* genes described above encode the Mo-dependent nitrogenase that functions exclusively in heterocysts. *A. variabilis*, and a few closely related cyanobacterial strains (but not *Anabaena* sp. PCC 7120), has two alternative nitrogenases (Thiel *et al.*, 1998; Thiel and Pratte, 2001). One is a V-dependent nitrogenase, encoded by *vnfH*, *vnfDG*, *vnfK*, *vnfE* and *vnfN*, which functions only in the absence of Mo (Lyons and Thiel, 1995; Thiel, 1993; 1996). The *vnfDGK* genes are most similar to their homologs in certain species of *Azotobacter* (Bishop and Joerger, 1990; Bishop and Premakur 1992). The *vnfEN* genes of *A. variabilis* show very weak similarity to other *nifEN*/*vnfEN* genes but show greater similarity to the *vnfDK* genes just upstream than to the *A. vinelandii vnfEN* genes (Thiel, 1996).

A mutant with a *vnfN* insertion lacks V-nitrogenase activity; thus, the putative VnfEN scaffolding protein is essential for the V-nitrogenase system in *A. variabilis*. In *A. vinelandii*, NifEN can substitute for VnfEN (Wolfinger and Bishop, 1991). Unlike in *Azotobacter*, there is no *vnfX* gene downstream of *vnfN* in *A. variabilis* (Thiel, 1996). A mutation in *nifB1*, which prevented diazotrophic growth with Mo under aerobic conditions, also prevented growth using the V-nitrogenase, suggesting that production of an active V-nitrogenase requires some of the *nif1* genes (Lyons and Thiel, 1995). The *vnf* genes are expressed only under conditions in which heterocysts form and their expression is apparently controlled by the same developmental factors that control the Mo-dependent *nif1* genes. The *vnf* genes are repressed by Mo, but do not require vanadium for transcription. In mutants that fail to transport molybdate and are, thus, always starved for Mo, the *vnf* genes are constitutively expressed in cells grown in the absence of fixed nitrogen. Because such mutants lack Mo, they fix N_2 using V-nitrogenase (T. Thiel, unpublished).

3.3. Second Mo-nitrogenase (Nif2)

A 13-kb region of the chromosome of *A. variabilis* contains most of the *nif* genes of the *nif2* gene cluster (GenBank U49859; see Figure 2). Also associated with this cluster are ferredoxin genes downstream of *nifW*. The second Mo-nitrogenase in *A. variabilis* is encoded by these genes and functions in vegetative cells under strictly anaerobic conditions (Thiel *et al.* 1995; 1997). Like the heterocyst-specific nitrogenases (Nif1 and Vnf), the Nif2 nitrogenase requires the global nitrogen regulatory protein, NtcA, for expression; however, there is no recognizable consensus NtcA-binding site (see Section 5.1) upstream of the transcriptional start site of *nifH2* (Thiel and Pratte, 2001). The genes for the second Mo-nitrogenase are most similar to the *nif* genes of *Plectonema boryanum* (Thiel *et al.* 1997), a nitrogen-fixing cyanobacterium that lacks heterocysts and fixes N_2 only under anaerobic conditions.

3.4. Additional Copies of nif *Genes*

It is not unusual for cyanobacteria to have multiple copies of *nifH*. *Anabaena* sp. PCC 7120 has two *nifH* genes and *A. variabilis* has four copies. Three of the four *nifH* genes in *A. variabilis* encode dinitrogenase reductases (Fe proteins), one for each of the Nif1, Nif2, and Vnf nitrogenases, whereas the fourth is a homolog of the second copy that is present in *Anabaena* sp. PCC 7120 and has no known function. *N. punctiforme* has only one complete set of *nif* genes and these appear to encode the heterocyst-specific Mo-nitrogenase (Meeks *et al.*, 2001), but has two additional copies of *nifH* and one additional copy of *nifE* and *nifN*. One of the extra copies of *nifH* is immediately upstream of the extra copy of *nifEN* forming a contiguous cluster of three genes. The third copy of *nifH* is not near any other *nif* gene. The functions of these extra copies are as yet unknown. Although all *nifH* genes have some sequence similarity to the gene encoding protochlorophyllide reductase (*chlL*), phylogenetic analysis of *nifH* genes indicates these genes are much more closely related to each other than any of them are to *chlL*. All the cyanobacterial *nifH* genes cluster in a single monophyletic group (T. Thiel, unpublished data).

3.5. nif-*Gene Excision Elements*

The *nifD* genes of *Anabaena* sp. PCC 7120 and *A. variabilis* are interrupted close to the 3' end by an 11-kb excision element (Brusca *et al.*, 1989; Golden *et al.*, 1985). In contrast, the *nifD* gene of *N. punctiforme* has a 24-kb insertion at exactly the same site (Meeks *et al.*, 2001). The two excision elements are almost completely different, except for the highly conserved excisase gene (*xisA*) that removes the element during heterocyst differentiation (Golden and Wiest, 1988) and a small *orf* of unknown function. The excision element in *nifD* in *N. punctiforme* contains homologs of two genes of unknown function present in *Synechocystis* sp. PCC 6803 and one homolog of a gene of *Anabaena* sp. called protein X (Sato, 1994). The *N. punctiforme* excision element contains a bacterial retron-like gene. *Anabaena* sp. PCC 7120 also has a gene that appears to be a retron, however, it is not located in the excision element and it is not closely related phylogenetically to the gene in *N. punctiforme*. In *Anabaena* sp. PCC 7120, but not in either *N. punctiforme* or *A. variabilis*, there is a 55-kb excision element in the *fdxN* gene (Brusca *et al.*, 1989; Golden *et al.*, 1988; Masepohl *et al.*, 1997; Meeks *et al.*, 1988; Ramaswamy *et al.*, 1997). The rearrangement of the *fdxN* element in *Anabaena* sp. PCC 7120 requires three genes encoded within the 55-kb element, *xisF*, *xisH*, and *xisI*. Expression of *xisH* and *xisI* on a plasmid results in excision of the element in vegetative cells. However, loss of the 55-kb element has no adverse effect on the strain suggesting that, as is the case with the 11-kb element, this element is not essential for diazotrophic growth of *Anabaena* sp. PCC 7120 (Ramaswamy *et al.*, 1997).

4. HETEROCYST METABOLISM

4.1. Respiratory Electron Transport and Nitrogen Fixation

High amounts of respiration are characteristic of heterocysts (Fay and Walsby, 1966). Respiration has two functions related to nitrogen fixation; removal of O_2 to maintain an anaerobic environment for nitrogenase, and synthesis of ATP for N_2 reduction (Fay, 1992). The respiratory chain in cyanobacteria is part of the photosynthetic electron-transport chain, except for the terminal oxidases, which are key enzymes in respiration (for a review, see Schmetterer, 1994). The cyanobacterial terminal oxidases are similar to the aa3-type cytochrome c oxidases found in mitochondria. There are three sets of *coxCBA* genes encoding terminal oxidases in both *Anabaena* sp. PCC 7120 and *N. punctiforme* (Jones and Haselkorn, 2002; Cyanobase: http://www.kazusa.or.jp/cyanobase/; JGI: http://www.jgi.doe.gov/JGI_microbial/html/nostoc/nostoc_homepage.html). One copy is expressed exclusively in proheterocysts and heterocysts of *Anabaena* sp. PCC 7120, although a mutant lacking that aa3-type oxidase is still capable of diazotrophic growth, presumably using terminal oxidases encoded by other *cox* genes (Jones and Haselkorn, 2002). In *A. variabilis*, a mutation in a different copy of *coxCBA*, expressed constitutively with or without fixed nitrogen, resulted in a strain unable to grow chemoheterotrophically in the dark using fructose as sole carbon source. This cytochrome oxidase mutant was not impaired in diazotrophic growth (Schmetterer *et al.*, 2001). The redundancy of cytochrome oxidases in heterocyst-forming cyanobacteria apparently ensures respiratory support for nitrogen fixation.

4.2. Transport of Fixed Nitrogen

Heterocyst differentiation is inhibited in cells grown with a source of fixed nitrogen. Concentrations of ammonium in the range of 3-7 μM repress heterocyst differentiation in *Anabaena* sp. PCC 7120 (Meeks *et al.*, 1983). The most common sources of fixed nitrogen for cyanobacteria are ammonium, nitrate, nitrite and urea (Flores and Herrero, 1994). Transport systems for all these compounds have been identified in cyanobacteria. In *Anabaena* sp. PCC 7120 (Cai and Wolk, 1997; Frías *et al.* 1997), the genes involved in the uptake and reduction of nitrate and nitrite comprise a cluster of *nirA* (ferredoxin nitrite reductase)-*nrtA-nrtB-nrtC-nrtD* (four genes encoding an ABC transporter)-*narB* (ferredoxin nitrate reductase). In *N. punctiforme*, the genes between *nirA* and *narB* are replaced by a single gene that is similar to a nitrate/nitrite transporter belonging to the major facilitator superfamily found in some unicellular cyanobacteria called either *nrtP* (Sakamoto *et al.* 1999) or *napA* (Wang *et al.*, 2000). Elsewhere in the genome, there are genes very similar to *nrtABCD*, suggesting that this strain has two nitrate/nitrite transport systems (Meeks *et al.*, 2001). The genes in the *nirA-narB* cluster are strongly expressed in the absence of ammonium and expression is somewhat increased in the presence of nitrate or nitrite (Kikuchi *et al.*, 1996; Frías *et al.*, 1997). These genes require the transcriptional activator, NtcA. In *Anabaena* sp. PCC 7120, the *ntcB* gene, which is upstream of the *nirA-narB* cluster, is also required for activation of these genes and expression of *ntcB* is repressed by ammonium (Frías *et al.*, 2000). NtcB is a member of the LysR family of transcriptional regulators. Thus, ammonium represses transcription of the *nirA-narB* gene cluster by repressing synthesis of the activator NtcB and by reducing the amount of NtcA, both of which are required for activation of those genes (Frías *et al.*, 2000). When ammonium is removed from the medium, cells of *Anabaena* sp. PCC 7120 respond within about 30 min by greatly increasing the transcription of the *nirA-narB* genes (Cai and Wolk, 1997).

Concentrations of ammonium in the μM range are efficiently transported in unicellular cyanobacteria by a permease that is induced in the absence of ammonium (Montesinos *et al.*, 1998; Vazquez-Bermudez *et al.*, 2002). The products of the *amt* genes of *Synechococcus* sp. and *Synechocystis* sp. are responsible for ammonium/methylammonium transport (Montesinos *et al.*, 1998; Vazquez-Bermudez *et al.*, 2002) as is, presumably, the *amt1* homologue present in the *Anabaena* sp. PCC 7120 genome (Kaneko *et al.*, 2001; Cyanobase: http://www.kazusa.or.jp/cyanobase/). Urea is also transported in cyanobacteria by a specific permease. In *Anabaena* sp. PCC 7120, there is a cluster of genes, *urtABCDE*, that encodes an ABC-type transporter for urea. These genes are expressed under nitrogen-limiting conditions under the control of the global nitrogen regulatory protein NtcA (Valladares *et al.*, 2002; also see Section 5.1).

4.3. Assimilation of Ammonium by GS/GOGAT

In cyanobacteria, the major pathway for assimilation of fixed nitrogen is via glutamine synthase (GS) and ferredoxin-dependent glutamine:2-oxoglutarate aminotransferase (GOGAT). GS makes glutamine from glutamate and ammonium, whereas GOGAT makes glutamate from glutamine and oxoglutarate (Figure 3). Together, these enzymes make glutamate from oxoglutarate and ammonium. The importance of this pathway for assimilation of N_2 by nitrogen fixation in *Anabaena cylindrica* was demonstrated by the conversion of $^{13}N_2$ first to $^{13}NH4^+$ by nitrogenase and then, through the activities of GS and then GOGAT, to ^{13}N-glutamine followed by ^{13}N-glutamate (Meeks *et al.*, 1977; 1978; Thomas *et al.*, 1977; Wolk *et al.*, 1976). The GS inhibitor, methionine sulfoximine (MSX), inhibited synthesis of both ^{13}N-glutamine and ^{13}N-glutamate. GOGAT activity is very low or absent in heterocysts (Thomas *et al.*, 1977), suggesting that glutamate is supplied to heterocysts from adjacent vegetative cells. GS activity is increased after nitrogen step-down, presumably as a result of the increased transcription of *glnA* from a heterocyst-specific promoter (Tumer *et al.*, 1983). In heterocysts, the ammonium made by nitrogen fixation is assimilated by GS and transported to vegetative cells where the glutamine is converted to glutamate via GOGAT.

There is a single *glnA* gene encoding a type I GS in *Anabaena* sp. PCC 7120. The only GS mutants are those that make an MSX-resistant enzyme; however, such mutants still have some GS activity (Chapman and Meeks, 1983; Spiller *et al.*, 1986). There is a mutant in *glnA* in the unicellular cyanobacterium, *Synechocystis* sp. PCC 6803, however, that strain (but not *Anabaena* sp. PCC 7120) has a second Type III GS, encoded by *glnN* (Reyes and Florencio, 1994). In addition to GS/GOGAT, ammonia can be assimilated directly into glutamate via glutamate dehydrogenase (*gdhA*), which is present in both *Anabaena* sp. PCC 7120 and *N. punctiforme* (Cyanobase: http://www.kazusa.or.jp/cyanobase/; Meeks *et al.*, 2001).

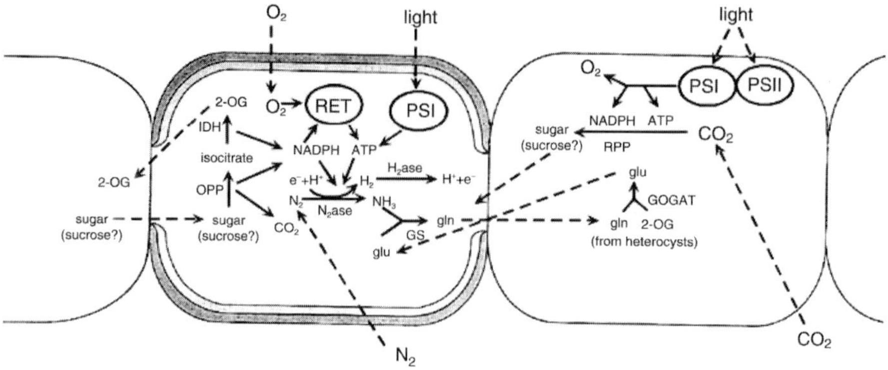

Figure 3. Major metabolic pathways in heterocysts and vegetative cells related to nitrogen fixation. The heterocyst (depicted with a multilayer envelope) reduces N_2 to ammonium using nitrogenase, which produces H_2 as a byproduct. Hydrogenase breaks down the H_2 to regenerate electrons. The reductant required for nitrogenase is produced by metabolism via the oxidative pentose phosphate (OPP) pathway in heterocysts of a sugar, provided by vegetative cells, producing 2-oxoglutarate (2-OG), which is transported to vegetative cells. The ATP for nitrogenase comes from cyclic photophosphorylation using PSI and from the respiratory electron transport chain, which also consumes O_2 that enters the heterocyst. The ammonium that is produced by nitrogenase is assimilated first to glutamine (gln) by glutamine synthetase (GS) and is then transported to vegetative cells where, with 2-oxoglutarate, it is converted to glutamate (glu) by glutamine:2-oxoglutarate aminotransferase (GOGAT). Glu returns to the heterocyst for the assimilation of ammonium. The sugars that support cyanobacterial growth and nitrogen fixation are made in vegetative cells by O_2-evolving photosynthesis and CO_2 reduction via the reductive pentose phosphate pathway (RPP). IDH, isocitrate dehydrogenase; RET, respiratory electron transport; PSI and PSII, photosystems I and II.

The assimilation of ammonium requires 2-oxoglutarate, made by isocitrate dehydrogenase (encoded by *icd*). In *Anabaena* sp. PCC 7120, expression of *icd* is highest under diazotrophic growth conditions and the gene appears to be essential because fully segregated mutants in *icd* could not be obtained and partially segregated mutants required oxoglutarate (Muro-Pastor and Florencio, 1994; Muro-Pastor *et al.*, 1996). Because it is a key carbon compound in nitrogen assimilation and is regulated by nitrogen availability, 2-oxoglutarate is a candidate for a sensor of the carbon/nitrogen balance in the cell (Herrero *et al.*, 2001; Meeks and Elhai, 2002; see Section 5.1).

4.4. Proteolysis and Nitrogen Fixation

Heterocyst differentiation requires degradation of some existing proteins and synthesis of new proteins required for heterocyst functions. New proteins made in developing heterocysts must be made from amino acids derived from existing proteins because nitrogen fixation and, hence, amino-acid biosynthesis occur only in mature heterocysts. Heterocysts are typically deficient in the photosystem II light-harvesting pigments known as biliproteins, such as phycocyanin, and their degradation is associated with nitrogen starvation and heterocyst differentiation. Loss of phycocyanin results from its degradation (Foulds and Carr, 1977; Thiel, 1990; Wood and Haselkorn, 1978; 1980) and from lack of expression of the genes for this protein in heterocysts (Johnson *et al.*, 1988; Wealand *et al.*, 1989). Phycobilisome degradation is mediated by a small protein encoded by the *nblA* gene (Collier and Grossman, 1994). In the unicellular cyanobacterium, *Synechocystis* sp. PCC 6803, there are two tandem copies of *nblA* genes that are cotranscribed under conditions of nitrogen starvation (Baier *et al.*, 2001; Richaud *et al.*, 2001). Mutations in either copy of *nblA* lead to the same phenotype. Under conditions of nitrogen starvation, there is little or no growth of the mutant and there is little degradation of phycocyanin compared to wild-type cells (Baier *et al.*, 2001; Richaud *et al.*, 2001).

Loss of phycocyanin fluorescence is a characteristic of mature heterocysts (Figure 4B). If loss of biliproteins occurs preferentially in differentiating heterocysts, then the loss of fluorescence in spaced vegetative cells in a filament after nitrogen step-down could potentially identify cells destined to become heterocysts. However, such dimly fluorescent, spaced cells are not evident in filaments of *A. variabilis* subjected to nitrogen step-down. Some biliprotein degradation is evident at the stage of proheterocysts; however, even then, not all proheterocysts have degraded their biliproteins significantly (Figure 4A). In this strain, heterocyst differentiation can be reversed by the addition of ammonium up to about 8 h after nitrogen step-down (Thiel and Pratte, 2001) and there are no clear morphological differences among cells in a filament until about 12 h, when proheterocysts are evident. By 12 h after step-down, there is loss of phycocyanin fluorescence providing visible evidence of biliprotein degradation; however, this loss is quite variable within and between filaments. It is notable that similar amounts of fluorescence are often seen in adjacent vegetative cells, or even in contiguous clusters of four or more cells, suggesting that sibling cells had initiated biliprotein degradation at about the same time. The loss of phycocyanin fluorescence in proheterocysts shows no consistent pattern. Some proheterocysts have lost most of their phycocyanin fluorescence whereas others are strikingly bright. The presence of many relatively highly fluorescent proheterocysts at 12 h after nitrogen step-down indicates that degradation of these proteins is not strongly correlated with the differentiation process, but rather occurs variably in cells in the filament in response to nitrogen starvation (Figure 4). In contrast, by 24 h after nitrogen step-down, when heterocysts are mature and fixing N_2 at a high rate, vegetative cells are more uniformly fluorescent and heterocysts are more consistently dimly fluorescent (Figure 4B). This situation is consistent with the inability of mature heterocysts to synthesize new biliproteins (Johnson *et al.*, 1988; Wealand *et al.*, 1989).

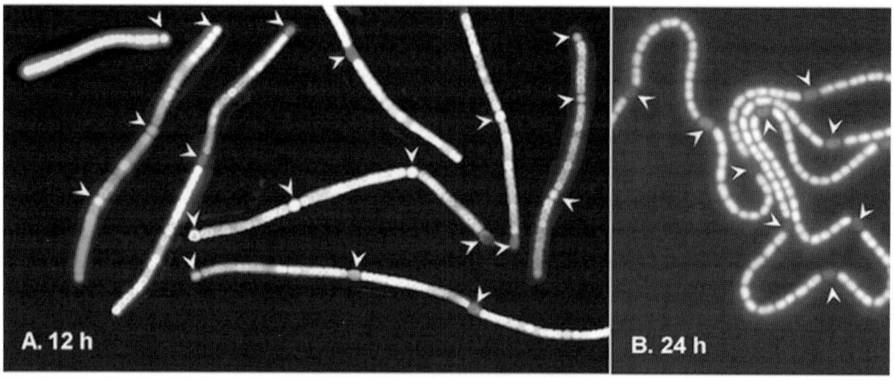

Figure 4. Fluorescence images of differentiating filaments of A. variabilis *ATCC 29413. Cells were shifted from a medium containing 5 mM NH_4Cl to a medium lacking a source of fixed nitrogen for 24 h. At various times, samples were viewed by epifluorescence microscopy with a Texas red filter set that allows visualization of phycocyanin fluorescence. Images were captured with a cooled CCD camera. No visible signs of heterocysts were evident at 2, 4, 6, 8, or 10 h after nitrogen shift. Proheterocysts were first seen about 12 h after nitrogen shift (panel A, arrowheads). Mature heterocysts at 24 h are shown in panel B.*

4.5. Hydrogenases

Nitrogenase produces H_2 during the reduction of N_2, thus potentially wasting large amounts of reductant. Cyanobacteria have two types of hydrogenases, both are NiFe enzymes. One is an uptake hydrogenase with a large subunit encoded by *hupL* and a small subunit encoded by *hupS*. These genes have been characterized in *Anabaena* sp. PCC 7120 (Carrasco *et al.*, 1995), *A. variabilis* (Boison *et al.*, 2000; Carrasco *et al.*, 1995; Happe *et al.*, 2000) and *Nostoc* sp. PCC 73102 (Oxelfelt *et al.*, 1998). In *Anabaena* sp. PCC 7120, but not in either *A. variabilis* or *Nostoc* sp. PCC 73102, the *hupL* gene is interrupted by a 10.5-kb excision element that is excised in response to nitrogen step-down by a site-specific excisase encoded by *xisC*, a gene near the right border of the element (Carrasco *et al.*, 1995). The excision of the 10.5-kb element and expression of *hupL* are very late events in heterocyst formation, occurring around the same time as expression of *nifH* (Carrasco *et al.*, 1995). The uptake hydrogenase is made primarily under diazotrophic growth conditions and acts only after nitrogen fixation begins (Happe *et al.*, 2000). It is found in the thylakoid membranes of heterocysts and functions to recycle the H_2 produced by nitrogenase into the electron transport chain for production of ATP. Mutants either in *hupSL* in *A. variabilis* or in *hupL* in *Anabaena* sp. PCC 7120 generate much more H_2 after nitrogen fixation begins than do the wild-type strains, suggesting that such mutants may be useful for photobiological H_2 production based on nitrogenase (Happe *et al.*, 2000; Masukawa *et al.*, 2002).

The second hydrogenase, a bi-directional, $NAD(P)^+$-reducing hydrogenase, can both evolve and take up H_2. The diaphorase (dihydrolipoamide:NAD oxido-reductase) component is encoded by *hox(E)FU* and the hydrogenase component is encoded by *hoxHY*. This enzyme is not present in all cyanobacteria and is not associated with nitrogen fixation, although some diazotrophic strains, such as *Anabaena* sp. PCC 7120 and *A. variabilis*, have this enzyme (Axelsson and Lindblad, 2002; Axelsson *et al.*, 1999; Boison *et al.*, 2000; Schmitz and Bothe, 1996; Schmitz *et al.*, 2002). This enzyme is also found in unicellular cyanobacteria and in both vegetative cells and heterocysts of filamentous strains; however its function is as yet unclear (Boison *et al.*, 2000).

4.6. *Photosynthesis and Carbon Metabolism in Heterocysts*

Heterocysts lack photosystem II activity and ribulose bisphosphate carboxylase, hence, they cannot photoreduce CO_2 using the reductive pentose phosphate pathway (RPP) to provide carbon skeletons for assimilation of fixed nitrogen (reviewed by Wolk, 1982; Wolk *et al.*, 1994). Photosystem I activity, capable of producing ATP by cyclic photophosphorylation, is present in heterocysts, providing energy but not reductant for nitrogen fixation (Figure 2). Despite the lack of carbon fixation in heterocysts, about 43% of the dry weight of heterocysts is carbohydrate, most of it in the heterocyst envelope (Wolk, 1973, 1982). Heterocysts metabolize carbohydrates, primarily through the oxidative pentose phosphate (OPP) pathway, which produces NADPH. The activity of glucose-6-phosphate dehydrogenase (G6PD), the enzyme that controls carbon flow into the OPP pathway, is much higher in heterocysts than in vegetative cells (reviewed by Wolk, 1982). G6PD is subject to complex regulation by key metabolites and by the reduction potential in the cell, possibly preventing futile cycling of the oxidative and reductive pentose phosphate pathways (Cossar *et al.*, 1984; Udvardy *et al.*, 1984). A mutation in the gene encoding G6PD (*zwf*) in *N. punctiforme* resulted in loss of nitrogenase activity, presumably because of insufficient reductant for nitrogen fixation from the OPP pathway (Summers *et al.*, 1995).

Because heterocysts cannot fix carbon, it must be supplied to heterocysts from vegetative cells. In free-living cyanobacteria, the maximum frequency of heterocysts is about 20-25%, suggesting that about 4-5 vegetative cells are required to support the fixed carbon needs of a heterocyst. The carbon compound(s) supplied to heterocysts are not known, but sucrose is a candidate. *Anabaena* sp. PCC 7120 has three enzymes that are known to be required for sucrose metabolism in plants (Cumino *et al.*, 2001, 2002; Curatti *et al.*, 1998, 2000, 2002): sucrose-phosphate synthase (makes sucrose-6-phosphate from UDP-glucose and fructose-6-phosphate); sucrose-phosphate phosphatase (removes the phosphate from sucrose-6-phosphate); and sucrose synthase (catalyzes the reversible conversion of sucrose and UDP to fructose and UDP-glucose). The genome of *Anabaena* sp. PCC 7120 has two genes identified as sucrose synthases and one identified as sucrose phosphate phosphatase, but there is no gene similar to that for sucrose-phosphate synthase (Kaneko *et al.*, 2001; Cyanobase: http://www.kazusa.or.jp/cyanobase/).

Expression of sucrose synthase is low in cells of *Anabaena* sp. PCC 7119 grown in the absence of combined nitrogen. A mutation in the *susA* gene, encoding sucrose synthase, greatly increases sucrose concentrations, whereas overexpression of the gene leads to undetectable amounts of sucrose, suggesting that this enzyme is important for the cleavage of sucrose (Curatti *et al.*, 2002). The strain that overexpresses *susA*, leading to very low amounts of sucrose, is impaired in nitrogen fixation, indicating that sucrose may be essential in carbon flow in filaments growing diazotrophically (Curatti *et al.*, 2002).

5. GENES IMPORTANT FOR HETEROCYST FORMATION

5.1. Global Regulation by NtcA

Heterocyst differentiation is typically repressed in cells grown with fixed nitrogen sources, such as ammonium or nitrate. Induction of heterocysts by removal of fixed nitrogen requires a regulatory protein called NtcA; mutants in *ntcA* are blocked in an early step in heterocyst differentiation. This protein is more generally involved in regulation of nitrogen metabolism in all cyanobacteria. NtcA, a member of the CRP (cAMP receptor protein) family of DNA-binding proteins (Frías *et al.*, 1993, 1994; Vega-Palas *et al.*, 1992, Wei *et al.*, 1993, 1994), is essential for transcription of the *nir-nar* operon for uptake and synthesis of nitrate and nitrite and it acts very early in the differentiation process to activate genes required for heterocyst formation. NtcA binds to a sequence located about 25 nucleotides upstream of the –10 region of the NtcA-activated promoter (Herrero *et al.*, 2001). The optimal sequence for NtcA binding is TGTA–N_8–TACA; the N_8 region of the NtcA-binding site of *glnA* of *Anabaena* sp. PCC 7120 provides the best binding (Fanyi Jiang *et al.*, 2000). The sequence GT–N_{10}–AC is essential for any NtcA binding (Vazquez-Bermudez *et al.*, 2002a). However, there are enormous differences in binding affinity among known NtcA-binding sites, suggesting that such differences affect the regulation of various genes by NtcA (Jiang *et al.*, 2000; Vazquez-Bermudez *et al.*, 2002a).

NtcA directly regulates *nirA*, *hetC*, *devABC*, *ntcB*, *urt*, and *glnA* in *Anabaena* sp. PCC 7120 via binding to the consensus-binding site (Herrero *et al.*, 2001). There is also evidence that it binds near *xisA*, *nifH*, *petH* (ferredoxin: $NADP^+$ reductase) and *rbcLS*, which encodes Rubisco (Frías *et al.*, 1993; Ramasubramanian *et al.*, 1994; Valladares *et al.*, 1999); however, these genes do not have an NtcA-binding site that corresponds well to the consensus sequence. Although there is no recognizable NtcA-binding site upstream of *ntcA*, NtcA binds upstream of the gene and transcription of *ntcA* requires that protein; thus, it is autoregulatory (Herrero *et al*, 2001; Ramasubramanian *et al.*, 1996; Tanigawa *et al.*, 2002). Similarly, there is no recognizable NtcA-binding site upstream of *hetR*; however, two of the four transcription start sites reported for *hetR* require NtcA, and one of those start sites also requires HetR (Muro-Pastor *et al.*, 2002; see Section 5.3). In *Anabaena* sp. PCC 7120, *ntcA* is expressed in the presence and in the absence of ammonium or nitrate; however, expression of *ntcA* is higher in cultures that are fixing N_2 (Herrero *et al.*, 2001; Muro-Pastor *et al.*, 2002). The increased expression of *ntcA* under conditions in which fixed nitrogen is limited was shown to be under the control of two transcription start sites that required NtcA as well as HetR (Muro-Pastor *et al.*, 2002; see Section 5.3).

Although control of NtcA-regulated genes by that protein is not well understood, recent evidence suggests an important role for the metabolite, 2-oxoglutarate, which serves as the substrate for assimilation of ammonium in cyanobacteria (see Figure 2). 2-Oxoglutarate greatly enhances the binding of NtcA to the *glnA* promoter region in unicellular cyanobacteria and is required for NtcA-dependent transcription of *glnA* and *ntcA* (Tanigawa *et al.*, 2002; Vasquez-Bermudez *et al.*, 2002a). This finding suggests that, in unicellular cyanobacteria, regulation of expression of nitrogen-assimilation genes may be directly related to the concentration of 2-oxoglutarate in the cell. Because 2-oxoglutarate is used only for nitrogen assimilation in cyanobacteria (there is an incomplete tricarboxylic cycle in cyanobacteria, so 2-oxoglutarate is not an intermediate in that pathway), this metabolite may serve as the sensor of the nitrogen status of the cell (Meeks and Elhai, 2002; Muro-Pastor *et al.*, 2001; Tanigawa *et al.*, 2002). When ammonium concentrations in the cell are low and 2-oxoglutarate is high, binding of 2-oxoglutarate to NtcA may result in a change in the conformation of that protein, which increases its binding to DNA and enhances transcription, possibly by interaction with RNA polymerase (Tanigawa *et al.*, 2002).

In *Anabaena* sp. PCC 7120 and *A. variabilis* ATCC 29413, mutants in *ntcA* cannot use nitrate as sole nitrogen source and they cannot differentiate heterocysts to fix N_2; thus, they require ammonium for growth (Frías *et al.*, 1994; Thiel *et al.*, 2002; Wei *et al.*, 1994). The Nif2 nitrogenase of *A. variabilis* ATCC 29413, which functions in vegetative cells under anaerobic conditions, also requires a functional *ntcA* gene; hence, all nitrogen fixation in that strain requires NtcA (Thiel and Pratte, 2001). In contrast, inducible transport of molybdate, a metal that is required as part of the cofactors for both nitrate reductase and nitrogenase (whose expression requires NtcA), does not require NtcA (Thiel *et al.*, 2002). Although NtcA is most associated with activation of genes involved in nitrogen metabolism, it has also been reported to bind to the promoter region of *gor* (encoding glutathione reductase), where is appears to act as a repressor (Jiang, *et al.*, 1997). DNA binding by NtcA to *gor in vitro* is regulated in a redox-dependent manner involving cysteine residues in the NtcA protein, suggesting that regulation of genes by NtcA may respond to the redox state of the cell as well as the nitrogen status. Such changes in redox state may be particularly important in the developing heterocyst as it becomes microaerobic (Jiang *et al.*, 1997).

5.2. PII Protein in Nitrogen Regulation

In addition to NtcA, the PII protein, encoded by *glnB*, is a key regulator of nitrogen metabolism, acting as a sensor of 2-oxoglutarate. In *E. coli*, cells that are ammonium-depleted have a high ratio of 2-oxoglutarate-to-glutamine and modify PII by the addition of UMP at Tyr51 (reviewed by Arcondeguy *et al.*, 2001). The unmodified PII indirectly, via NtrB/NtrC, inhibits transcription of genes, such as *glnA*, which are required in ammonium-limited cells. The modified form of PII no longer inhibits transcription of these genes, thus, helping to activate their expression and facilitating growth under nitrogen-limiting conditions.

In unicellular cyanobacteria, PII is involved in the regulation of assimilation of both nitrate/nitrite and bicarbonate (Forchhammer and Tandeau de Marsac, 1995a; Hisbergues *et al.*, 1999; Lee *et al.*, 1998; 1999). In *Synechococcus elongatus* PCC 7942, under conditions of ammonium starvation, ATP and 2-oxoglutarate bind to PII synergistically to phosphorylate Ser49 using the PII kinase (Forchhammer and Hedler, 1997; Forchhammer and Tandeau de Marsac, 1995b). Phosphorylation of PII, leading to the inactive form, is highest when 2-oxoglutarate is abundant, corresponding to low concentrations of ammonium. Dephosphorylation, leading to the active form of PII that inhibits genes involved in nitrate/nitrite assimilation, is highest when 2-oxoglutarate is scarce, corresponding to high concentrations of ammonium. Dephosphorylation of PII occurs by the action of a phosphatase that belongs to the PP2C family (Irmler *et al.*, 1997; Irmler and Forchhammer, 2001; Ruppert *et al.*, 2002). In *Synechocystis* sp. PCC 6803, dephosphorylation of PII is sensitive to both 2-oxoglutarate and oxaloacetate concentrations as well as to the energy status of the cell (Ruppert *et al.*, 2002).

The *glnB* gene of *S. elongatus* has two transcription start sites; one is expressed constitutively and the other requires NtcA (Lee *et al.*, 1999). The latter promoter is highly expressed under conditions of starvation for fixed nitrogen, irrespective of CO_2 concentration, and in the presence of nitrate when concentrations of CO_2 are high (Lee *et al.*, 1999). Phosphorylation of PII is also NtcA dependent, whereas dephosphorylation is not. Thus, NtcA modulates the phosphorylation state of PII in response to both the nitrogen and carbon status of the cell, possibly via the interaction of NtcA with 2-oxoglutarate (Lee *et al.*, 1999).

A *glnB* mutant of *S. elongatus* PCC 7942 constitutively expresses glutamine synthetase and takes up nitrate and nitrite even in the presence of ammonium, although transcriptional control by ammonium of *glnA* and *nir-nrtABCD-narB* is unaffected (Forchhammer and Tandeau de Marsac, 1995a; Lee *et al.*, 1998). Further, in the *glnB* mutant, nitrate uptake is no longer dependent on CO_2 fixation (Forchhammer and Tandeau de Marsac, 1995a). The unphosphorylated form of PII appears to be responsible for the short-term inhibition by ammonium of nitrate and nitrite transport in the wild-type strain. Neither a *glnB* mutant nor an *ntcA* mutant is able to repress transcription of *rbcL* in response to nitrogen starvation, as occurs in the wild-type strain of *S. elongatus* (Fadi Aldehni *et al.*, 2003). A *glnB* mutant with a fusion of the NtcA-regulated *glnB* promoter to the *luxAB* reporter produced low amounts of luciferase in the presence of ammonium, indicating that PII is not required for NtcA-mediated inhibition of *glnB* by ammonium. However, after nitrogen starvation, activation of *glnB* transcription from the NtcA-dependent promoter was impaired in the *glnB* mutant, suggesting that PII is required for regulation of NtcA-dependent genes under nitrogen starvation conditions (Fadi Aldehni *et al.*, 2003). The PII-independent repression of NtcA-dependent promoters by NtcA probably involves 2-oxoglutarate. When ammonium concentrations are high and 2-oxoglutarate is low, NtcA might bind its target DNA site in the promoter; however, without 2-oxoglutarate, it may act as a repressor instead of an activator (Fadi Aldehni *et al.*, 2003).

Control by PII of nitrogen metabolism in heterocyst-forming cyanobacteria has not been studied. In *N. punctiforme*, the *glnB* gene could not be mutated unless a second copy was provided *in trans*, suggesting that the gene is essential in this strain, and surprisingly, there was no evidence of phosphorylation of PII *in vivo*, although phosphorylation was demonstrated *in vitro* (Hanson, *et al.*, 1998).

5.3. Heterocyst-specific Regulatory Gene: hetR

An early gene that is important for heterocyst differentiation is *hetR* (Black *et al.*, 1993; Buikema and Haselkorn, 1991; 2001). Mutations in *hetR* completely block heterocyst differentiation and transcription of *hetR* is detectable within about 30 min of nitrogen step-down, the earliest known heterocyst-specific gene to be transcribed (Buikema and Haselkorn, 2001). Extra copies of *hetR* on a plasmid result in multiple contiguous heterocysts under nitrogen-fixing conditions and result in heterocyst development in the presence of nitrate (Buikema and Haselkorn, 1991). Expression of *hetR* from a copper-inducible promoter by the addition of copper to the medium resulted in a very high frequency of heterocysts (up to 30%) under nitrogen-fixing conditions, with high frequencies of non-fixing heterocysts even in the presence of either ammonia or nitrate (Buikema and Haselkorn, 2001).

After nitrogen step-down, a strain containing a *hetR*::*luxAB* reporter showed expression of *luxAB* in a pattern suggesting that these were cells destined to become heterocysts. However, even with combined nitrogen, there was a low amount of *hetR* expression (Black *et al.*, 1993; Buikema and Haselkorn, 1991; 2001). Expression of green fluorescent protein (GFP) under the control of the *hetR* promoter was strongest in differentiated heterocysts 25 h after nitrogen step-down (Haselkorn, 1998). Transcription of *hetR* requires a functional copy of *hetR*, indicating that HetR positively activates *hetR* gene transcription (Black *et al.*, 1993).

HetR is a serine protease that shows autodegradation (Dong *et al.*, 2000; Zhou *et al.*, 1998a, 1998b). The protease inhibitor, dansyl fluoride, binds the ser152 and a mutant altered at ser152 residue lacked autodegradation and failed to differentiate heterocysts; thus, ser152 is at the active site of HetR (Dong *et al*, 2000). Mutations at ser142 affected neither heterocyst differentiation nor autodegradation, whereas the ser179 mutation prevented heterocyst differentiation and autodegradation (Dong *et al.*, 2000; Zhou *et al.*, 1998b). After nitrogen step-down, the HetR protein is made in the wild-type strain and in mutants at ser142 (S142A), ser152 (S152A), and ser179 (S179N); however, the two mutants that failed to differentiate heterocysts (at ser152 and ser179) continued to synthesize the protein long after accumulation of HetR ceased in both the wild-type strain and the ser 142 mutant, which makes normal heterocysts (Dong *et al.*, 2000).

It has been suggested that the autoproteolysis of HetR regulates the differential accumulation of this protein in different cell types. This degradation may be responsible for loss of the protein in vegetative cells, whereas inhibition of that activity results in its accumulation in heterocysts (Dong *et al.*, 2000). Studies using a *hetR* promoter fused to GFP in *N. punctiforme* showed that *hetR* is transcribed in both cell types, although at a higher amount in heterocysts; however a functional HetR-GFP fusion protein was not visible in vegetative cells, but was present in heterocysts (Wong and Meeks, 2001). This finding supports the model that HetR is made in all cells, but is degraded by autoproteolysis in vegetative cells and not in heterocysts. The non-functional *hetR* mutant at ser179 cannot autodegrade the protein, which resulted in increased transcription and accumulation of HetR in all cell types (Dong *et al.*, 2000). A *hetR* promoter fused to GFP, when expressed in the ser179 mutant strain, also resulted in increased synthesis of GFP in all cells after nitrogen step-down (Haselkorn, 1998). These results along with those of Dong *et al.* (2002) suggest that accumulation of HetR as a consequence of the loss of autodegradation results in loss of ability to differentiate heterocysts.

5.4. Dual Regulation of Transcription of ntcA *and* hetR *by NtcA and HetR*

There are four transcription start sites for *hetR*. The 1.4-kb transcript is made from the transcription start site 184 nucleotides upstream of the translation start site for the protein (designated as –184); the 1.5-kb transcript is made from the transcription start site 271 nucleotides upstream of the translation start site (designated as –271); and the two transcripts of about 1.9-kb are made from transcription start sites at 696 and 728 nucleotides upstream of the translation start site (designated as either –696 or –728) (Buikema and Haselkorn, 2001; Muro-Pastor *et al.*, 2002). Although primer extension experiments indicated that the promoter, which produces one of the 1.9-kb transcripts, was the predominant transcript, Northern blot analysis showed that the smallest transcript (1.4 kb) was the first to be induced at about 30 min after nitrogen step-down and was (possibly along with the 1.5-kb transcript) the dominant transcript in wild-type cells (Buikema and Haselkorn, 2001). The smallest transcript is constitutively expressed, even in the presence of ammonium, with or without either NtcA or HetR present (Muro-Pastor *et al.*, 2002). Transcription from the other three promoters is induced by starvation of the cells for fixed nitrogen.

Transcription from the site at –271 (to give the 1.5-kb transcript) requires HetR and NtcA, whereas the site at –728 (for the larger 1.9 kb transcript) requires only NtcA. Thus, only the site at –271 appears to be autoregulatory (Buikema and Haselkorn, 2001; Muro-Pastor *et al.*, 2002). Interestingly, the ser179 mutant of *hetR* shows induction of the 1.4-kb and the 1.9-kb transcripts, corroborating the findings of Dong *et al.* (2000) that the mutant with ser179 in HetR does not prevent activation of transcription of *hetR*. Transcription of the smaller of the two 1.9-kb transcripts, from the start site at –696, requires neither NtcA nor HetR. Despite the evidence for a requirement for NtcA, based on transcripts of *hetR* in wild-type versus *ntcA* mutant strains, there is no recognizable NtcA-binding site upstream of the NtcA-regulated start sites of *hetR* and purified NtcA does not bind to these promoter regions, suggesting that regulation of these promoters by NtcA might be indirect (Buikema and Haselkorn, 2001; Muro-Pastor *et al.*, 2002).

Northern blots indicate that there are three *ntcA* transcripts of about 0.8, 1.1 and 1.4 kb (Muro-Pastor *et al.*, 2002). In nitrogen step-down experiments, *ntcA* transcripts increased about 3-fold by 6 h after the shift and increased as much as 5-fold by 9 h after the shift. By 24 h after step-down, the amount of transcription decreased to the amount seen at about 6 h after shift. In a *hetR* mutant, there was a lower amount of *ntcA* transcription, which increased slightly in response to the shift from ammonium to N_2. This result suggests that the induction of *ntcA* depends on HetR, which is made by about 3 h, and that higher amounts of *ntcA* expression are not required for the early stages of heterocyst differentiation but only for the later stages after *hetR* is induced (Muro-Pastor *et al.*, 2002).

Primer extension products showed that the three *ntcA* transcripts in *Anabaena* sp. PCC 7120 begin at 49, 136, and 180 bp upstream of the translation start site (designated as –49, –136, –180) (Muro-Pastor *et al.*, 2002). The –49 and the –180 regions can function as promoters for the *lacZ* reporter gene (Ramasubramanian *et al.*, 1996). Complementation of a *ntcA* mutant by *ntcA* genes provided *in trans* on a plasmid required the –180/190 region for growth of the complemented mutant on N_2, whereas only the –49 region was required for growth on nitrate (Ramasubramanian *et al.*, 1996). This observation suggests that the –180 start site is required for the high amount of *ntcA* expression seen after nitrogen step-down and probably for heterocyst differentiation, whereas the –49 site is used in mature heterocysts. The –180 transcript was transiently expressed only after nitrogen step-down and its transcription required both NtcA and HetR. The transcript at –136 was expressed constitutively and initiation at this site required neither NtcA nor HetR. The –49 transcript was expressed at a low constitutive amount but its transcription increased after nitrogen shift and that increase required NtcA and HetR. The major transcript in heterocysts is from the –49 start site. It appears that both the –180 and the –49 transcription start sites are regulated by both NtcA and HetR, but the –136 transcript is independent of both factors (Muro-Pastor *et al.*, 2002).

The induction of *ntcA*, after nitrogen step-down, peaks later than that of *hetR*, suggesting that the lower concentrations of NtcA early in the differentiation process are sufficient to induce *hetR* expression. Subsequently, HetR induces increased transcription of *ntcA*. This finding implies that high concentrations of NtcA are not required for the induction of heterocyst formation, but rather for the formation and maturation of heterocysts. Evidence for a joint role for HetR and NtcA also comes from the requirement of genes activated later during development, such as *devBCA* and *devH*, for both HetR and NtcA, whereas expression of the early gene, *ntcA*, is dependent only on NtcA. Similarly, other NtcA-regulated genes that are not involved in diazotrophic growth, such as *nir-nar*, do not require HetR. One function of HetR appears to be to increase the concentration of NtcA during heterocyst differentiation to allow expression of later heterocyst-specific genes required for maturation (Muro-Pastor *et al.*, 2002).

5.5. *The* hetC *Gene and its Product*

Another gene that is important for the early steps of heterocyst formation is *hetC*. The product of this gene appears most similar to ATP-binding cassette (ABC) membrane transport proteins, in particular, to those that export toxins (Khuduyakov and Wolk, 1997; Xu and Wolk, 2001). The *hetC* gene is activated directly by NtcA. However, HetC is not required for the activation of *hetR* by NtcA and HetR is not required for the activation of *hetC* (Herrero *et al.*, 2001; Muro-Pastor *et al.*, 1999; 2002). A mutant in *hetC* fails to differentiate heterocysts. However, a pattern of spaced vegetative cells showing low phycocyanin fluorescence (characteristic of heterocysts) along with expression of GFP from a *hetR::gfp* fusion (Xu and Wolk, 2001) is evident after 48 h, suggesting that these cells are in an early stage of heterocyst formation. The weakly fluorescent cells in the *hetC* mutant are capable of cell division, but produce very small progeny cells. A *hetR/hetC* double mutant does not produce these weakly fluorescent cells and addition of the PatS pentapeptide, which represses heterocyst formation (see Section 6.1.1), prevents formation of these spaced weakly fluorescent cells (Xu and Wolk, 2001; Yoon and Golden 1998), suggesting that both HetR and the PatS pentapeptide act at a step in differentiation earlier than HetC. The *hetC* mutation has no effect on the transcription of *hetR*, but blocks transcription of *hepC* (required for heterocyst polysaccharide formation) and is also required for its own activation.

The *hetC* gene is first transcribed about 3-4 h after nitrogen step-down and expression of this gene is strongest in proheterocysts and heterocysts. The transcription start site is –571 from the translation start site (Muro-Pastor *et al.*, 1999, 2002). A *hetC* mutant apparently does not allow the developing heterocyst to block cell division, suggesting that HetC may be involved in the export of a factor, which normally inhibits this change in the heterocyst to a non-dividing state (Xu and Wolk, 2001). The *hetP* gene, located about 1 kb from *hetC*, shows similarity to *hetC*, both in the early time of its expression and the phenotype of the *hetP* mutant, *i.e.*, a pattern of spaced vegetative cells with low phycocyanin fluorescence. The predicted protein product of *hetP* shows no similarity to existing proteins and its function is not yet known (Fernández-Piñas *et al.*, 1994).

5.6. The hetF *Gene and its Product*

A gene called *hetF*, which has been identified in *N. punctiforme* (it is also present in the genome of *Anabaena* sp. PCC 7120), has characteristics similar to *hetR*. It is expressed early after nitrogen step-down, is essential for heterocyst formation, and leads to multiple contiguous heterocysts when it is overexpressed (Wong and Meeks, 2001). Unlike *hetR*, which is found in filamentous cyanobacterial strains that do not make heterocysts, *hetF* has been identified only in heterocyst-forming strains. In a *hetF* mutant, the normal increase in *hetR* transcription that occurs 3-6 h after nitrogen step-down was delayed and the increased transcription of *hetR* was seen in all cells in a strain containing a *hetR* promoter fused to GFP. A functional HetR-GFP fusion protein accumulated specifically in heterocysts in the *hetF*$^+$ strain, but non-specifically in all cells in the *hetF* mutant. HetF is thought to work in conjunction with HetR in activation of heterocyst formation, perhaps by either increasing *hetR* transcription or aiding in the autoproteolysis of HetR, which is associated with loss of HetR in vegetative cells (Wong and Meeks, 2001).

5.7. The devBCA *and* devH *Genes and their Products*

The product of the *devBCA* genes is an ABC-type transporter that is required for deposition of the glycolipid layer of heterocysts. It is thought to function as an exporter either of heterocyst-specific glycolipids or, possibly, of components required for assembly of these glycolipids (Fiedler *et al.*, 1998). A mutant in *devA* is blocked in late-stage heterocyst development (Maldener *et al.*, 1994) and DevA may be involved in activation of *hetM*, which is essential for glycolipid formation (Cai and Wolk, 1997). There is a transcription start site over 700 nucleotides upstream of *devB* that is induced within about 1 h after nitrogen step-down; however, high amounts of *devBCA* transcription do not occur until about 7 h after the shift. Upstream of this transcription start site is a canonical NtcA-binding site, which is conserved in several heterocyst-forming strains. The promoter region for this transcript binds NtcA, thus, expression of *devBCA* is directly dependent on NtcA. Transcription of this operon also requires HetR (Fiedler *et al.*, 2001). This finding suggests that transcription of *devBCA* occurs early after nitrogen step-down using the initial low concentrations of NtcA, but that later high transcription requires the higher concentrations of NtcA, which result from HetR activation of *ntcA* transcription (Muro-Pastor *et al.*, 2002). The higher amounts of *devBCA* expression are probably required for formation of the glycolipid layer of the heterocyst.

DevH belongs to the CRP family of DNA-binding proteins, as does NtcA. The *devH* gene is weakly expressed in vegetative cells and is induced by 6 h after nitrogen step-down, suggesting that it is important for nitrogen fixation. Like many other heterocyst-specific genes, *devH* transcription depends on NtcA and HetR. Mutants in *devH* make heterocysts that do not function under aerobic growth conditions (Hebbar and Curtis, 2000).

6. HETEROCYST PATTERN FORMATION

6.1. Genes Involved in Heterocyst Pattern Formation

6.1.1. The patS *Gene and its Product*

The *patS* gene encodes a peptide that is essential for normal heterocyst pattern formation. Mutants in *patS* produce multiple contiguous heterocysts, whereas either extra copies or overexpression of *patS* results in suppression of heterocysts (Yoon and Golden, 1998; 2001). Thus, the PatS peptide acts as an inhibitor of heterocyst differentiation. A synthetic peptide comprising only the C-terminal pentapeptide of PatS (PatS-5) suppresses heterocyst formation when added to the growth medium at very low concentrations and restores a normal number of heterocysts when added to cells of a *patS* mutant. However, the spacing of heterocysts is not normal in a complemented *patS* mutant unless the complementing *patS* gene is under the control of the heterocyst-specific *hepA* promoter. Thus, the normal spacing of heterocysts depends not only on PatS but also on its expression in heterocysts at the appropriate time during differentiation. The expression of *patS* is initially found in small groups of cells about 8-10 h after nitrogen step-down but, by 12-14 h, its expression is typically seen only in single cells, followed a few hours later by expression only in proheterocysts. These results are consistent with a model in which PatS is produced in differentiating cells and is itself, or perhaps as its C-terminal pentapeptide product, exported from those cells and diffuses through the filament, producing a gradient of the inhibitor that controls heterocyst spacing. In a strain that overexpresses *patS*, thus producing no heterocysts, overexpression of another gene, *hetL*, restores heterocyst formation, however, the mechanism of the interaction of these two gene products is not yet understood (Liu and Golden, 2002).

6.1.2. The patA *Gene and its Product*

One type of pattern mutant in *Anabaena* sp. PCC 7120, leading to heterocyst formation only at the ends at filaments, is in the *patA* gene (Liang *et al.*, 1992). Overexpression of *hetR*$^+$ (either from a multi-copy plasmid or by addition of copper to cells containing the *hetR*$^+$ gene under the control of the *petE* copper-inducible promoter) in a *patA* mutant results in the same phenotype as the *patA* mutant alone. When a *hetR* promoter, driving synthesis of GFP, is introduced into a *patS* mutant, expression from this promoter occurs only in terminal cells. Therefore, the *patA* mutation in some way represses the induction of spaced heterocysts within a filament and represses the multiple contiguous heterocysts found in a strain overexpressing *hetR*, possibly by a direct or indirect effect of PatA on *hetR* (Buikema and Haselkorn, 2001).

6.1.3. The hetN *Gene and its Product*

In *Anabaena* sp. PCC 7120, the *hetN*-gene product is important in heterocyst spacing, probably for the maintenance of spacing. HetN acts as a suppressor of heterocyst formation. When the normal promoter for *hetN* is replaced by the copper inducible promoter from *petE*, expression of *hetN* in the presence of high concentrations of copper results in complete suppression of heterocyst formation (Callahan and Buikema, 2001). Repression of *hetN* by removal of copper, in the absence of fixed nitrogen, leads initially (24 h) to a normal heterocyst pattern that later (48 h) changes to a pattern of multiple contiguous heterocysts. Overexpression of *hetN* blocks expression of *hetR* (Li *et al.*, 2002) and prevents the normal spaced pattern of expression of GFP from a *hetR*::*gfp* fusion present on a multi-copy plasmid (Callahan and Buikema, 2001), suggesting that HetN may act by either directly or indirectly inhibiting *hetR* expression.

In a strain containing a *hetR*::*gfp* fusion, GFP was expressed in multiple contiguous cells when *hetN* expression was repressed by lack of copper. Even in the presence of high concentrations of HetR (expressed from a plasmid under the control of the copper-inducible promoter), which normally leads to 20% heterocysts, high concentrations of HetN repress heterocyst formation (Callahan and Buikema, 2001). In the absence of copper, when *hetN* was not expressed, normal amounts of heterocyst glycolipids were made, suggesting that HetN is not required for their synthesis. In a strain containing a *hetN*::*gfp* fusion, expression of GFP was seen primarily in developing heterocysts about 17 h after induction (Callahan and Buikema, 2001). Li *et al.* (2002) determined that *hetN* is expressed in vegetative cells; however, after nitrogen step-down, expression of *hetN* increases and it is localized to mature heterocysts, particularly in the membrane fractions. These results suggest that HetN is required for production of a signal from mature heterocysts that suppresses heterocyst differentiation. Thus, HetN is thought to be responsible for the maintenance, but not the establishment, of the spacing of new heterocysts in a filament (Callahan and Buikema, 2001).

6.2. Commitment to Heterocyst Differentiation

During the early stages of heterocyst differentiation, cells are not committed to develop into heterocysts, but may regress to normal vegetative cells (Adams and Carr, 1989). It appears from the expression pattern of *pats*, using a *patS-gfp* fusion, that the clusters of cells, which begin to differentiate soon after nitrogen step-down, resolve into a single proheterocyst by about 12-14 h (Yoon and Golden, 2001). When are proheterocysts committed to complete differentiation? Addition of either the PatS-5 pentapeptide or nitrate, each of which suppresses heterocyst formation, to *Anabaena* sp. PCC 7120 at times up to about 8 h after nitrogen step-down, results in very low frequencies of heterocysts, consistent with the ability of proheterocysts at this stage to reverse the differentiation process. By 8-10 h after nitrogen step-down, the time when clusters of PatS-expressing cells are resolving, some of the proheterocysts are committed to differentiate even in the presence of either nitrate or PatS-5. By 12-14 h, when clusters have already resolved, most proheterocysts have committed to differentiate, suggesting that resolution of clusters by PatS represents the commitment to differentiation (Yoon and Golden, 2001). Thiel and Pratte (2001) found a very similar pattern of timing of commitment in *A. variabilis* ATCC 29413.

6.3. Model for Initiation of Heterocyst Differentiation

In filaments grown with fixed nitrogen, the vegetative cells are indistinguishable and there is no evidence that any particular cell is destined to become a heterocyst. The selection of certain cells for differentiation among all the apparently identical cells may be influenced by the timing of the cell cycle (Adams and Carr 1989; Meeks and Elhai, 2002). Heterocyst differentiation requires DNA replication (Adams and Carr 1989; Meeks and Elhai, 2002). Evidence for this includes the inhibition of differentiation by DNA-replication inhibitors and the requirement for the histone-like protein HU for differentiation (Adams and Carr 1989; Khudyakov and Wolk, 1996). The requirement for DNA replication suggests that the cell cycle may be important for heterocyst formation, *i.e.*, only cells in a certain stage of the cell cycle are competent to undergo differentiation (Adams and Carr, 1989). However, sibling cells are typically near the same stage of the cell cycle, suggesting that either pairs of cells or possibly even four adjacent cells may all be at a stage that would allow heterocyst formation. This situation suggests then that several adjacent cells at the same stage of the cell cycle may all begin to differentiate.

The first description of the differentiation of strings of cells was by Wilcox *et al.* (1973a), who also suggested that competition leads to resolution of the cluster into a single cell that completes the differentiation process. This suggestion indicates that there is competition among adjacent cells, with all but one cell regressing to vegetative cells (Meeks and Elhai, 2002). The development of multiple contiguous heterocysts in some mutants (Black and Wolk, 1994; Wong and Meeks, 2001; Yoon and Golden, 1998), either in the presence of excess HetR (Buikema and Wolk, 1991; Callahan and Buikema, 2001), in the presence of 7-azatryptophan (Adams, 1992; Wilcox *et al.*, 1973b) or with intense illumination (Adams and Carr, 1981), suggests that these conditions prevent the normal regression of clusters of developing heterocysts.

A two-stage model for heterocyst differentiation has been proposed (Meeks and Elhai, 2002). Stage one is initiation, in which only those nitrogen-starved cells at a certain stage in the cell cycle (generally contiguous sibling cells) are competent to begin differentiation with the transcription of *hetR*. In the second stage, HetR induces synthesis of an inhibitor, possibly PatS or PatS-5, which is released and transported to nearby cells causing them to regress to vegetative cells. Autoregulation of *hetR* by HetR results in high concentrations of HetR, which promote heterocyst formation, whereas high concentrations of PatS promote regression. In those cells in which HetR is most abundant, more PatS will be made and exported, causing adjacent cells to regress. According to this model, initiation of heterocyst formation is biased by the position of cells in the cell cycle and then resolution to a single committed cell, by competition among these early differentiating cells, leads to patterned heterocyst formation (Meeks and Elhai, 2002). Either mutations or agents, which interfere with resolution, lead to the phenotype of multiple contiguous heterocysts, whereas those that overstimulate resolution may completely inhibit heterocyst formation. A model of heterocyst differentiation is presented in Figure 5.

7. REGULATION

7.1. Developmental versus *Environmental Control of Nitrogen Fixation*

In most diazotrophic bacteria, expression of nitrogenase genes is controlled by environmental factors, such as either the availability of fixed nitrogen or by the O_2 tension. The nitrogenase genes in heterocysts are expressed only after heterocysts form, therefore, it appears that their expression is not controlled directly by the environment outside the cell, but by the development of the heterocyst, which provides the microaerobic environment that is deficient in fixed nitrogen as required for nitrogenase activity. This situation was elegantly demonstrated in *Anabaena* sp. PCC 7120, where a luciferase reporter gene under the control of the *nifH* promoter was expressed only in heterocysts under either aerobic or anaerobic conditions (Elhai and Wolk, 1990). Similar results in *A. variabilis*, using a *lacZ* reporter under the control of the *nifH1* promoter, also confirmed that the *nif1* genes in that strain were expressed only in heterocysts. In contrast, *lacZ* under control of the *nifH2* promoter was expressed in vegetative cells but only in an anaerobic environment that supports nitrogen fixation, suggesting that these genes were regulated exclusively by environmental factors (Thiel *et al.*, 1995).

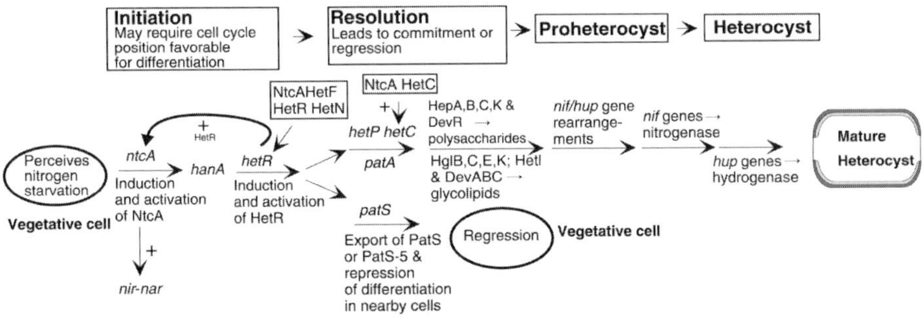

Figure 5. Model for the differentiation of heterocysts. This model presents a simplified view of some of the genes known to be involved in heterocyst formation. The major temporal divisions correspond to (1) initiation of differentiation in one or more contiguous cells, (2) resolution to only one cell committed to differentiate, (3) continuation of differentiation of that cell to a morphologically distinguishable proheterocyst, and (4) maturation to a heterocyst. The horizontal pathway with arrows showing major genes or gene products corresponds to a timeline, with expression of earlier genes, such as ntcA *and* hetR, *preceding middle genes, such as* hetC, *followed by expression of genes for the heterocyst envelope, and, finally, the expression of the genes for nitrogenase and then hydrogenase. Arrows with a + sign indicate proteins that are known to activate transcription of the specified genes. For simplicity, not all known regulatory relationships are shown. For further details concerning the genes and proteins in this model please consult the text.*

7.2. Sigma Factors

In many bacteria, alternative sigma factors provide a mechanism for differential control of coordinately regulated genes. Six sigma factors have been identified in *Anabaena* sp. PCC 7120, including several that are differentially expressed in either the presence or absence of fixed nitrogen (Golden and Yoon, 1998). The principal sigma factor is encoded by *sigA*, whereas *sigB* and *sigC* encode proteins that belong to the sigma 70 group and are primarily expressed under conditions of nitrogen starvation (Brahamsha and Haselkorn, 1991; 1992). However, neither of these latter genes is essential for diazotrophic growth (Brahamsha and Haselkorn, 1992). Mutations in *sigD*, *sigE*, or *sigF* led to strains that were able to grow diazotrophically, although mutations in either *sigD* or *sigE* delayed growth and a *sigD/sigE* double mutant failed to grow diazotrophically because the filaments fragmented (Khudyakov and Golden, 1991). The *Anabaena* sp. PCC 7120 genome reveals 12 potential sigma factor genes including the six described previously. Three of them, including *sigB*, are on the alpha and beta plasmids of this strain

(Kaneko *et al.*, 2001). However, the role of these sigma factors in heterocyst differentiation is as yet unknown.

7.3. Role of Fixed Nitrogen in Heterocyst Development and Pattern Formation

Expression of the Nif2 nitrogenase system in vegetative cells under anaerobic diazotrophic conditions supports good growth of a *nif1* mutant. However, these cells still form heterocysts, as do wild-type cells grown under the same conditions (Thiel and Pratte, 2001). Thus, nitrogen fixation in vegetative cells of a filament does not repress heterocyst formation, suggesting that, under these conditions, the products of nitrogen fixation made in vegetative cells do not repress the differentiation of new heterocysts. Exogenously added fixed nitrogen does suppress heterocyst formation, suggesting that cells have a mechanism to distinguish between internally fixed nitrogen, which does not suppress heterocysts, and externally supplied fixed nitrogen, which does suppress new heterocyst formation (Thiel and Pratte, 2001). However, under some conditions, even externally supplied fixed nitrogen does not suppress heterocyst differentiation. For example, mutants lacking *patS* (Yoon and Golden, 2001) and strains that overexpress *hetR* (Buikema and Haselkorn, 1991, 2001) produce heterocysts in the presence of either nitrate or ammonium. In addition, glutamine supports good growth of a strain of *A. variabilis*, which takes up glutamine well, and represses nitrogenase, but it does not suppress heterocyst formation (Thiel and Leone, 1986). All of this evidence suggests that gradients of fixed nitrogen are not primarily responsible for the differentiation of spaced heterocysts. Probably the most compelling argument that internally fixed nitrogen plays no role in heterocyst pattern formation is the fact that the *de novo* pattern of heterocysts formed initially after nitrogen step-down occurs in the absence of nitrogen fixation. While there might be completely different mechanisms to account for the initial spaced pattern versus the maintained spaced pattern, the simplest model suggests that the products of nitrogen fixation do not control pattern formation in either case.

Yoon and Golden (2001) argue that there is a relatively high threshold concentration of available fixed nitrogen that both supports good growth while also allowing heterocysts to form, and that internally fixed nitrogen (such as that produced by Nif2 in *A. variabilis* under anaerobic conditions) falls far below that high threshold. In fact, an external concentration of ammonium as low as 3-7 μM represses heterocyst formation in *Anabaena* sp. PCC 7120 (Meeks *et al.*, 1983). That strain transports nitrate and ammonium very efficiently and the internal concentrations of these molecules would be expected to remain fairly high until the external concentration falls below the K_m for transport, at which time the cells would begin to sense nitrogen deprivation. One early response to low external fixed nitrogen is the induction of the *nir-nar* operon for the transport of nitrate and nitrite, producing more transport protein and allowing scavenging of some of the remaining fixed nitrogen (Cai and Wolk, 1997). However, at the same time the cells are increasing transport of nitrate, they also induce heterocyst formation, which is reversible for several hours after induction. This observation suggests that the filaments do not begin the expensive process of heterocyst formation until the external concentration of fixed nitrogen has fallen to a concentration that does not allow adequate amounts of fixed nitrogen to be transported to maintain normal growth.

Further evidence that differentiating filaments are quite deficient in fixed nitrogen is the fact that they degrade cellular protein. This expensive process surely would not occur if the threshold concentration of fixed nitrogen were high enough to support continued growth until nitrogen fixation could begin. In contrast to the view expressed by Yoon and Golden (2001), this observation suggests that the threshold concentration of fixed nitrogen, which leads to perception of starvation, is very low and that this concentration is sensed by lack of transport of either nitrate or ammonium into the cell. Thus, possibly a major difference in the response of the cells to internally fixed nitrogen (which does not repress heterocyst differentiation) and externally supplied fixed nitrogen (which normally does) is that internally fixed nitrogen does not repress expression of *ntcA*, whereas fixed nitrogen transported into the cell does, resulting in repression of heterocyst formation.

The availability of the genome sequences for two heterocyst-forming strains (as well as the sequence of the genome of *A. variabilis* ATCC 29413 that should be completed soon) should lead to an increased understanding of the genetic program for heterocyst differentiation as well as the patterning of heterocysts in a cyanobacterial filament.

ACKNOWLEDGEMENTS

The research in the Thiel laboratory was supported by National Science Foundation grant MCB-9723754 and USDA grants 97-35305-4970 and 99-35100-7582.

REFERENCES

Adams, D. G. (1992). The effect of Dl-7 azatryptophan on heterocyst development in the cyanobacterium *Anabaena cylindrica*. *J. Gen. Microbiol., 138,* 355-362.

Adams, D. G., and Carr, N. G. (1981). Heterocyst differentiation and cell division in the cyanobacterium *Anabaena cylindrica*: effect of high light intensity. *J. Cell Sci., 49*, 341-352.

Adams, D. G., and Carr, N. G. (1989). Control of heterocyst development in the cyanobacterium *Anabaena cylindrica*. *J. Gen. Microbiol., 135,* 839-850.

Adams, D. G. and Duggan, P. S. (1999). Heterocyst and akinete differentiation in cyanobacteria. *New Phytol., 144*, 3-33.

Allen, R. M., Madden, M. S., Ludden, P. W., and Shah, V. K. (1994). Biosynthesis of the iron-molybdenum cofactor of nitrogenase. *Crit. Rev. Biotech., 14*, 225-249.

Angeloni, S., and Potts, M. (1994). Analysis of the sequences within and flanking the cyanoglobin-encoding gene, *glbN*, of the cyanobacterium *Nostoc commune* UTEX 584. *Gene, 146*, 133-134.

Arcondeguy, T., Jack, R., and Merrick, M. (2001). PII signal transduction proteins: pivotal players in microbial nitrogen control. *Microbiol. Mol. Biol. Rev., 65*, 80-105.

Arnold, W., Rump, A., Klipp, W., Priefer, U. B., and Pühler, A. (1988). Nucleotide sequence of a 24,206-base-pair DNA fragment carrying the entire nitrogen fixation gene cluster of *Klebsiella pneumoniae*. *J. Mol. Biol., 203*, 715-738.

Axelsson, R., and Lindblad, P. (2002). Transcriptional regulation of *Nostoc* hydrogenases: Effects of oxygen, hydrogen, and nickel. *Appl. Environ. Microbiol., 68*, 444-447.

Axelsson, R., Oxelfelt, F., and Lindblad, P. (1999). Transcriptional regulation of *Nostoc* uptake hydrogenase. *FEMS Microbiol. Lett., 170*, 77-81.

Baier, K., Nicklisch, S., Grundner, C., Reinecke, J., and Lockau, W. (2001). Expression of two *nblA*-homologous genes is required for phycobilisome degradation in nitrogen-starved Synechocystis sp. PCC6803. *FEMS Microbiol. Lett., 195*, 35-39.

Bauer, C. C., Ramaswamy, K. S., Endley, S., Scappino, L. A., Golden, J. W., and Haselkorn, R. (1997). Suppression of heterocyst differentiation in *Anabaena* PCC 7120 by a cosmid carrying wild-type genes encoding enzymes for fatty acid synthesis. *FEMS Microbiol. Lett., 151*, 23-30.

Bishop, P. E., and Joerger, R. D. (1990). Genetics and molecular biology of alternative nitrogen fixation systems. *Ann. Rev. Plant Physiol. Plant Mol. Biol., 41*, 109-125.

Bishop, P. E., and Premakumar, R. (1992). Alternative nitrogen fixation systems. In G. Stacey, R. H. Burris and H. J. Evans (Eds.), *Biological nitrogen fixation* (pp. 736-762). New York: Chapman and Hall.

Black, K., Buikema, W., and Haselkorn, R. (1995). The *hglK* gene is required for localization of heterocyst-specific glycolipids in the cyanobacterium *Anabaena* sp. strain PCC 7120. *J. Bacteriol., 177*, 6440-6448.

Black, T. A., Cai, Y., and Wolk, C. P. (1993). Spatial expression and autoregulation of *hetR*, a gene involved in the control of heterocyst development in *Anabaena*. *Mol. Microbiol., 9*, 77-84.

Black, T. A., and Wolk, C. P. (1994). Analysis of a Het$^-$ mutation in *Anabaena* sp. strain PCC 7120 implicates a secondary metabolite in the regulation of heterocyst spacing. *J. Bacteriol., 176*, 2282-2292.

Boison, G., Bothe, H., and Schmitz, O. (2000). Transcriptional analysis of hydrogenase genes in the cyanobacteria *Anacystis nidulans* and *Anabaena variabilis* monitored by RT-PCR. *Curr. Microbiol., 40*, 315-21.

Brahamsha, B., and Haselkorn, R. (1991). Isolation and characterization of the gene encoding the principal sigma factor of the vegetative cell RNA polymerase from the cyanobacterium *Anabaena* sp. strain PCC 7120. *J. Bacteriol., 173,* 2442-2450.

Brahamsha, B., and Haselkorn, R. (1992). Identification of multiple RNA polymerase sigma factor homologs in the cyanobacterium *Anabaena* sp. strain PCC 7120: cloning, expression, and inactivation of the *sigB* and *sigC* genes. *J. Bacteriol., 174,* 7273-7282.

Brusca, J. S., Hale, M. A., Carrasco, C. D., and Golden, J. W. (1989). Excision of an 11-kilobase-pair DNA element from within the *nifD* gene in *Anabaena variabilis* heterocysts. *J. Bacteriol., 171*, 4138-4145.

Bryce, T., Welti, D., Walsby, A., and Nichols, B. (1972). Monohexoside derivatives of long-chain polyhydroxy alcohols; a novel class of glycolipids specific to heterocystous algae. *Phytochem., 11*, 295-302.

Buikema, W. J., and Haselkorn, R. (1991). Characterization of a gene controlling heterocyst differentiation in the cyanobacterium *Anabaena* 7120. *Genes Develop., 5*, 321-330.

Buikema, W. J., and Haselkorn, R. (1993). Molecular genetics of cyanobacterial development. *Ann. Rev. Plant Physiol., 44*, 33-52.

Buikema, W. J., and Haselkorn, R. (2001). Expression of the *Anabaena hetR* gene from a copper-regulated promoter leads to heterocyst differentiation under repressing conditions. *Proc. Natl. Acad. Sci. U.S.A., 98*, 2729-2734.

Cai, Y., and Wolk, C. P. (1997). Nitrogen deprivation of *Anabaena* sp. strain PCC 7120 elicits rapid activation of a gene cluster that is essential for uptake and utilization of nitrate. *J. Bacteriol., 179*, 258-266.

Callahan, S. M., and Buikema, W. J. (2001). The role of HetN in maintenance of the heterocyst pattern in *Anabaena* sp. PCC 7120. *Mol. Microbiol., 40*, 941-950.

Campbell, E. L., Cohen, M. F., and Meeks, J. C. (1997). A polyketide-synthase-like gene is involved in the synthesis of heterocyst glycolipids in *Nostoc punctiforme* strain ATCC 29133. *Arch. Microbiol., 167*, 251-258.

Campbell, E. L., Hagen, K. D., Cohen, M. F., Summers, M. L., and Meeks, J. C. (1996). The *devR* gene product is characteristic of receivers of two-component regulatory systems and is essential for heterocyst development in the filamentous cyanobacterium *Nostoc* sp. strain ATCC 29133. *J. Bacteriol., 178,* 2037-2043.

Cardemil, L., and Wolk, C. P. (1976). The polysaccharides from heterocyst and spore envelopes of a blue green alga. Methylation analysis and structure of the backbones. *J. Biol. Chem., 251*, 2967-2975.

Cardemil, L., and Wolk, C. P. (1979). The polysaccharides from heterocyst and spore envelopes of a blue-green alga. Structure of the basic repeating unit. *J. Biol. Chem., 254*, 736-741.

Cardemil, L., and Wolk, C. P. (1981a). Isolated heterocysts of *Anabaena variabilis* synthesize envelope polysaccharide. *Biochim. Biophys. Acta, 674*, 265-276.

Cardemil, L., and Wolk, C. P. (1981b). Polysaccharides from the envelopes of heterocysts and spores of the blue- green algae *Anabaena variabilis* and *Cylindrospermum licheniforme*. *J. Phycol., 17*, 234-240.

Carrasco, C. D., Buettner, J. A., and Golden, J. W. (1995). Programmed DNA rearrangement of a cyanobacterial *hupL* gene in heterocysts. *Proc. Natl. Acad. Sci. U.S.A., 92,* 791-795.

Chapman, J. S., and Meeks, J. C. (1983). Glutamine and glutamate transport by *Anabaena variabilis*. *J. Bacteriol., 156,* 122-129.

Collier, J. L., and Grossman, A. R. (1994). A small polypeptide triggers complete degradation of light-harvesting phycobiliproteins in nutrient-deprived cyanobacteria. *EMBO J., 13*, 1039-1047.

Cossar, J. D., Rowell, P., and Stewart, W. D. P. (1984). Thioredoxin as a modulator of glucose-6-phosphate dehydrogenase in a N_2- fixing cyanobacterium. *J. Gen. Microbiol., 130*, 991-998.

Cumino, A., Curatti, L., Giarrocco, L., and Salerno, G. L. (2002). Sucrose metabolism: *Anabaena* sucrose-phosphate synthase and sucrose-phosphate phosphatase define minimal functional domains shuffled during evolution. *FEBS Lett., 517*, 19-23.

Cumino, A., Ekeroth, C., and Salerno, G. L. (2001). Sucrose-phosphate phosphatase from *Anabaena* sp. strain PCC 7120: isolation of the protein and gene revealed significant structural differences from the higher-plant enzyme. *Planta (Berlin), 214*, 250-256.

Curatti, L., Flores, E., and Salerno, G. (2002). Sucrose is involved in the diazotrophic metabolism of the heterocysts-forming cyanobacterium *Anabaena* sp. *FEBS Lett., 513*, 175-178.

Curatti, L., Folco, E., Desplats, P., Abratti, G., Limones, V., Herrera-Estrella, L., *et al.* (1998). Sucrose-phosphate synthase from *Synechocystis* sp. strain PCC 6803: identification of the *spsA* gene and characterization of the enzyme expressed in *Escherichia coli*. *J. Bacteriol., 180*, 6776-6779.

Curatti, L., Porchia, A. C., Herrera-Estrella, L., and Salerno, G. L. (2000). A prokaryotic sucrose synthase gene (*susA*) isolated from a filamentous nitrogen-fixing cyanobacterium encodes a protein similar to those of plants. *Planta (Berlin), 211*, 729-735.

Dean, D. R., and Jacobson, M. R. (1992). Biochemical genetics of nitrogenase. In G. Stacey, R. H. Burris and H. J. Evans (Eds.), *Biological nitrogen fixation* (pp. 763-834). New York: Chapman and Hall.

Deom, C. M., Schubert, K. R., Wolf, S., Holt, C. A., Lucas, W. J., and Beachy, R. N. (1990). Molecular characterization and biological function of the movement protein of tobacco mosaic virus in transgenic plants. *Proc. Natl. Acad. Sci. U.S.A., 87,* 3284-3288.

Dong, Y., Huang, X., Wu, X.-Y., and Zhao, J. (2000). Identification of the active site of HetR protease and its requirement for heterocyst differentiation in the cyanobacterium *Anabaena* sp. strain PCC 7120. *J. Bacteriol., 182,* 1575-1579.

Drews, G., and Weckesser, J. (1982). Function, structure and composition of cell walls and external layers. In N. G. Carr and B. A. Whitton (Eds.), *The biology of cyanobacteria* (pp. 333-357). Blackwell Scientific Publications, Ltd., Oxford, United Kingdom.

Elhai, J., and Wolk, C. P. (1990). Developmental regulation and spatial pattern of expression of the structural genes for nitrogenase in the cyanobacterium *Anabaena*. *EMBO J., 9*, 3379-3388.

Ernst, A., Black, T. A., Cai, Y. P., Panoff, J. M., Tiwari, D. N., and Wolk, C. P. (1992). Synthesis of nitrogenase in mutants of the cyanobacterium *Anabaena* sp. strain PCC 7120 affected in heterocyst development or metabolism. *J. Bacteriol., 174*, 6025-6032.

Fadi Aldehni, M., Sauer, J., Spielhaupter, C., Schmid, R., and Forchhammer, K. (2003). Signal transduction protein P(II) is required for NtcA-regulated gene expression during nitrogen deprivation in the cyanobacterium *Synechococcus elongatus* strain PCC 7942. *J. Bacteriol., 185*, 2582-2591.

Fay, P. (1992). Oxygen relations of nitrogen fixation in cyanobacteria. *Microbiol. Rev., 56*, 340-373.

Fay, P., and Walsby, A. E. (1966). Metabolic activities of isolated heterocysts of the blue-green alga *Anabaena cylindrica*. *Nature*, 209, 94-95.

Fernández-Piñas, F., Leganés, F., and Wolk, C. P. (1994). A third genetic locus required for the formation of heterocysts in *Anabaena* sp. strain PCC 7120. *J. Bacteriol., 176*, 5277-5283.

Fiedler, G., Arnold, M., Hannus, S., and Maldener, I. (1998a). The DevBCA exporter is essential for envelope formation in heterocysts of the cyanobacterium *Anabaena* sp. strain PCC 7120. *Mol. Microbiol., 27*, 1193-1202.

Fiedler, G., Arnold, M., and Maldener, I. (1998b). Sequence and mutational analysis of the *devBCA* gene cluster encoding a putative ABC transporter in the cyanobacterium *Anabaena variabilis* ATCC 29413. *Biochim. Biophys. Acta, 1375*, 140-143.

Fiedler, G., Muro-Pastor, A. M., Flores, E., and Maldener, I. (2001). NtcA-dependent expression of the *devBCA* operon, encoding a heterocyst-specific ATP-binding cassette transporter in *Anabaena* spp. *J. Bacteriol., 183*, 3795-3799.

Flores, E., and Herrero, A. (1994). Assimilatory nitrogen metabolism and its regulation. In D. Bryant (Ed.), *The molecular biology of cyanobacteria* (pp. 487-517). Dordrecht, Netherlands: Kluwer Academic Publishers.

Forchhammer, K., and Hedler, A. (1997). Phosphoprotein PII from cyanobacteria: analysis of functional conservation with the PII signal-transduction protein from *Escherichia coli. Eur. J. Biochem., 244,* 869-875.

Forchhammer, K., and Tandeau de Marsac, N. (1995a). Functional analysis of the phosphoprotein PII (*glnB* gene product) in the cyanobacterium *Synechococcus* sp. strain PCC 7942. *J. Bacteriol., 177,* 2033-2040.

Forchhammer, K., and Tandeau de Marsac, N. (1995b). Phosphorylation of the PII protein (*glnB* gene product) in the cyanobacterium *Synechococcus* sp. strain PCC 7942: analysis of *in vitro* kinase activity. *J. Bacteriol., 177*, 5812-5817.

Foulds, I. J., and Carr, N. G. (1977). A proteolytic enzyme degrading phycocyanin in the cyanobacterium *Anabaena cylindrica. FEMS Microbiol. Lett., 2*, 117-119.

Frías, J. E., Flores, E., and Herrero, A. (1994). Requirement of the regulatory protein NtcA for the expression of nitrogen assimilation and heterocyst development genes in the cyanobacterium *Anabaena* sp. PCC 7120. *Mol. Microbiol., 14*, 823-832.

Frías, J. E., Flores, E., and Herrero, A. (1997). Nitrate assimilation gene cluster from the heterocyst-forming cyanobacterium *Anabaena* sp. strain PCC 7120. *J. Bacteriol., 179*, 477-486.

Frías, J. E., Flores, E., and Herrero, A. (2000). Activation of the *Anabaena nir* operon promoter requires both NtcA (CAP family) and NtcB (LysR family) transcription factors. *Mol. Microbiol., 38*, 613-625.

Frías, J. E., Merida, A., Herrero, A., Martin-Nieto, J., and Flores, E. (1993). General distribution of the nitrogen control gene *ntcA* in cyanobacteria. *J. Bacteriol., 175*, 5710-5713.

Gambacorta, A., Soriente, A., Trincone, A., and Sodano, G. (1995). Biosynthesis of the heterocyst glycolipids in the cyanobacterium *Anabaena cylindrica. Phytochem., 39*, 771-774.

Giddings, T. H., Jr., and Staehelin, L. A. (1978). Plasma membrane architecture of *Anabaena cylindrica*: occurrence of microplasmodesmata and changes associated with heterocyst development and the cell cycle. *Cytobiol., 16*, 235-249.

Golden, J. W. (2002). PatS peptide inhibits heterocyst development and influences pattern formation in *Anabaena. Biol. Environ.-Proc. Royal Irish Acad., 102B*, 57-59.

Golden, J. W., Carrasco, C. D., Mulligan, M. E., Schneider, G. J., and Haselkorn, R. (1988). Deletion of a 55-kilobase-pair DNA element from the chromosome during heterocyst differentiation of *Anabaena* sp. strain PCC 7120. *J. Bacteriol., 170*, 5034-5041.

Golden, J. W., Robinson, S. J., and Haselkorn, R. (1985). Rearrangement of nitrogen fixation genes during heterocyst differentiation in the cyanobacterium *Anabaena. Nature, 314,* 419-423.

Golden, J. W., and Wiest, D. R. (1988). Genome rearrangement and nitrogen fixation in *Anabaena* blocked by inactivation of *xisA* gene. *Science, 242*, 1421-1423.

Golden, J. W., and Yoon, H.-S. (1998). Heterocyst formation in *Anabaena. Curr. Opin. Microbiol., 1,* 623-629.

Hagen, K. D., and Meeks, J. C. (1999). Biochemical and genetic evidence for participation of DevR in a phosphorelay signal transduction pathway essential for heterocyst maturation in *Nostoc punctiforme* ATCC 29133. *J. Bacteriol., 181,* 4430-4434.

Hanson, T. E., Forchhammer, K., Tandeau de Marsac, N., and Meeks, J. C. (1998). Characterization of the *glnB* gene product of *Nostoc punctiforme* strain ATCC 29133: *glnB* or the PII protein may be essential. *Microbiol., 144*, 1537-1547.

Happe, T., Schütz, K., and Böhme, H. (2000). Transcriptional and mutational analysis of the uptake hydrogenase of the filamentous cyanobacterium *Anabaena variabilis* ATCC 29413. *J. Bacteriol., 182,*1624-1631.

Haselkorn, R. (1998). How cyanobacteria count to 10. *Science, 283*, 891-892.

Hebbar, P. B., and Curtis, S. E. (2000). Characterization of *devH*, a gene encoding a putative DNA binding protein required for heterocyst function in *Anabaena* sp. strain PCC 7120. *J. Bacteriol., 182*, 3572-3581.

Heinlein, M., Wood, M. R., Thiel, T., and Beachy, R. N. (1998). Targeting and modification of prokaryotic cell-cell junctions by tobacco mosaic virus cell-to-cell movement protein. *Plant J., 14*, 345-351.

Herrero, A., Muro-Pastor, A. M., and Flores, E. (2001). Nitrogen control in cyanobacteria. *J. Bacteriol., 183*, 411-425.

Hill, D. R., Belbin, T. J., Thorsteinsson, M. V., Bassam, D., Brass, S., Ernst, A., *et al.* (1996). GlbN (cyanoglobin) is a peripheral membrane protein that is restricted to certain *Nostoc* spp. *J. Bacteriol., 178*, 6587-6598.

Hisbergues, M., Jeanjean, R., Joset, F., Tandeau de Marsac, N., and Bedu, S. (1999). Protein PII regulates both inorganic carbon and nitrate uptake and is modified by a redox signal in *Synechocystis* PCC 6803. *FEBS Lett., 463,* 216-220.

Holland, D., and Wolk, C. P. (1990). Identification and characterization of *hetA*, a gene that acts early in the process of morphological differentiation of heterocysts. *J. Bacteriol., 172,* 3131-3137.

Huang, T.-C., Lin, R.-F., Chu, M.-K., and Chen, H.-M. (1999). Organization and expression of nitrogen-fixation genes in the aerobic nitrogen-fixing unicellular cyanobacterium *Synechococcus* sp. strain RF-1. *Microbiol., 145*, 743-753.

Irmler, A., and Forchhammer, K. (2001). A PP2C-type phosphatase dephosphorylates the PII signaling protein in the cyanobacterium *Synechocystis* PCC 6803. *Proc. Natl. Acad. Sci. U.S.A., 98,* 12978-12983.

Irmler, A., Sanner, S., Dierks, H., and Forchhammer, K. (1997). Dephosphorylation of the phosphoprotein PII in *Synechococcus* PCC 7942: identification of an ATP and 2-oxoglutarate-regulated phosphatase activity. *Mol. Microbiol., 26,* 81-90.

Jiang, F., Mannervik, B., and Bergman, B. (1997). Evidence for redox regulation of the transcription factor NtcA, acting both as an activator and a repressor, in the cyanobacterium *Anabaena* PCC 7120. *Biochem. J., 327*, 513-517.

Jiang, F., Wisen, S., Widersten, M., Bergman, B., and Mannervik, B. (2000). Examination of the transcription factor NtcA-binding motif by in vitro selection of DNA sequences from a random library. *J. Mol. Biol., 301*, 783-793.

Johnson, T. R., Haynes, J. I., Wealand, J. L., Yarbrough, L. R., and Hirschberg, R. (1988). Structure and regulation of genes encoding phycocyanin and allophycocyanin from *Anabaena variabilis* ATCC 29413. *J. Bacteriol., 170*, 1858-1865.

Jones, K. M., and Haselkorn, R. (2002). Newly identified cytochrome c oxidase operon in the nitrogen-fixing cyanobacterium *Anabaena* sp. strain PCC 7120 specifically induced in heterocysts. *J. Bacteriol., 184*, 2491-2499.

Kaneko, T., Nakamura, Y., Wolk, C. P., Kuritz, T., Sasamoto, S., Watanabe, A., *et al.* (2001). Complete genomic sequence of the filamentous nitrogen-fixing cyanobacterium *Anabaena* sp. strain PCC 7120. *DNA Res., 8*, 205-213; 227-253.

Khudyakov, I. Y., and Golden, J. W. (2001). Identification and inactivation of three group 2 sigma factor genes in *Anabaena* sp. strain PCC 7120. *J. Bacteriol., 183*, 6667-6675.

Khudyakov, I. Y., and Wolk, C. P. (1996). Evidence that the *hanA* gene coding for HU protein is essential for heterocyst differentiation in, and cyanophage A-4(L) sensitivity of, *Anabaena* sp. strain PCC 7120. *J. Bacteriol., 178*, 3572-3577.

Khudyakov, I. Y., and Wolk, C. P. (1997). *hetC*, a gene coding for a protein similar to bacterial ABC protein exporters, is involved in early regulation of heterocyst differentiation in *Anabaena* sp. strain PCC 7120. *J. Bacteriol., 179*, 6971-6978.

Kikuchi, H., Aichi, M., Suzuki, I., and Omata, T. (1996). Positive regulation by nitrite of the nitrate assimilation operon in the cyanobacteria *Synechococcus* sp. strain PCC 7942 and *Plectonema boryanum*. *J. Bacteriol., 178*, 5822-5825.

Koksharova, O. A., and Wolk, C. P. (2002). Novel DNA-binding proteins in the cyanobacterium *Anabaena* sp. strain PCC 7120. *J. Bacteriol., 184*, 3931-3940.

Lambein, F., and Wolk, C. P. (1973). Structural studies on the glycolipids from the envelope of the heterocyst of *Anabaena cylindrica*. *Biochemistry, 12*, 791-798.

Lázaro, S., Fernández-Piñas, F., Fernández-Valiente, E., Blanco-Rivero, A., and Leganés, F. (2001). *pbpB*, a gene coding for a putative penicillin-binding protein, is required for aerobic nitrogen fixation in the cyanobacterium *Anabaena* sp. strain PCC 7120. *J. Bacteriol., 183*, 628-636.

Lee, H. M., Flores, E., Forchhammer, K., Herrero, A., and Tandeau de Marsac, N. (2000). Phosphorylation of the signal transducer PII protein and an additional effector are required for the PII-mediated regulation of nitrate and nitrite uptake in the cyanobacterium *Synechococcus* sp. PCC 7942. *Eur. J. Biochem., 267,* 591- 600.

Lee, H. M., Flores, E., Herrero, A., Houmard, J., and Tandeau de Marsac, N. (1998). A role for the signal transduction protein PII in the control of nitrate/nitrite uptake in a cyanobacterium. *FEBS Lett., 427,* 291-295.

Lee, H. M., Vázquez-Bermúdez, M. F., and Tandeau de Marsac, N. (1999). The global nitrogen regulator NtcA regulates transcription of the signal transducer PII (GlnB) and influences its phosphorylation level in response to nitrogen and carbon supplies in the cyanobacterium *Synechococcus* sp. strain PCC 7942. *J. Bacteriol., 181,* 2697-2702.

Leganés, F. (1994). Genetic evidence that *hepA* gene is involved in the normal deposition of the envelope of both heterocysts and akinetes in *Anabaena variabilis* ATCC 29413. *FEMS Microbiol. Lett., 123*, 63-67.

Li, B., Huang, X., and Zhao, J. (2002). Expression of *hetN* during heterocyst differentiation and its inhibition of *hetR* up-regulation in the cyanobacterium *Anabaena* sp. PCC 7120. *FEBS Lett., 517*, 87-91.

Liang, J. H., Scappino, L., and Haselkorn, R. (1992). The *patA* gene product, which contains a region similar to *cheY* of Escherichia coli, controls heterocyst pattern formation in the cyanobacterium *Anabaena* 7120. *Proc. Natl. Acad. Sci. U.S.A., 89*, 5655-5659.

Liu, D., and Golden, J. W. (2002). *hetL* overexpression stimulates heterocyst formation in *Anabaena* sp. strain PCC 7120. *J. Bacteriol., 184*, 6873-6881.

Lockau, W., Massalsky, B., and Dirmeier, A. (1988). Purification and partial characterization of a calcium-stimulated protease from the cyanobacterium, *Anabaena variabilis*. *Eur J. Biochem., 172*, 433-438.

Lyons, E. M. and Thiel, T. (1995). Characterization of *nifB*, *nifS*, and *nifU* genes in the cyanobacterium *Anabaena* variabilis: NifB is required for the vanadium-dependent nitrogenase. *J. Bacteriol., 177*, 1570-1575.

Maldener, I., Fiedler, G., Ernst, A., Fernandez-Pinas, F., and Wolk, C. P. (1994). Characterization of *devA*, a gene required for the maturation of proheterocysts in the cyanobacterium *Anabaena* sp. strain PCC 7120. *J. Bacteriol., 176*, 7543-7549.

Maldener, I., Lockau, W., Cai, Y. P., and Wolk, C. P. (1991). Calcium-dependent protease of the cyanobacterium *Anabaena*: molecular cloning and expression of the gene in *Escherichia coli*, sequencing and site-directed mutagenesis. *Mol. Gen. Genet., 225*, 113-120.

Masepohl, B., Schölisch, K., Görlitz, K., Kutzki, C., and Böhme, H. (1997). The heterocyst-specific *fdxH* gene product of the cyanobacterium *Anabaena* sp. PCC 7120 is important but not essential for nitrogen fixation. *Mol. Gen. Genet., 253,* 770-776.

Masukawa, H., Mochimaru, M., and Sakurai, H. (2002). Disruption of the uptake hydrogenase gene, but not of the bidirectional hydrogenase gene, leads to enhanced photobiological hydrogen production by the nitrogen-fixing cyanobacterium *Anabaena* sp. PCC 7120. *Appl. Microbiol. Biotech., 58*, 618-624.

Meeks, J. C, and Elhai, J. (2002). Regulation of cellular differentiation in filamentous cyanobacteria in free-living and plant-associated symbiotic growth states. *Microbiol. Mol. Biol. Rev., 66,* 94-121.

Meeks, J. C., Elhai, J., Thiel, T., Potts, M., Larimer, F., Lamerdin, J., *et al.* (2001). An overview of the genome of *Nostoc punctiforme*, a multicellular, symbiotic cyanobacterium. *Photosyn. Res., 70*, 85-106.

Meeks, J. C., Joseph, C. M., and Haselkorn, R. (1988). Organization of the *nif* genes in cyanobacteria in symbiotic association with Azolla and Anthoceros. *Arch. Microbiol., 150*, 61-71.

Meeks, J. C., Wolk, C. P., Lockau, W., Schilling, N., Shaffer, P. W., and Chien, W.-S. (1978). Pathways of assimilation of $^{13}NH_4^+$ by cyanobacteria with and without heterocysts. *J. Bacteriol., 134*, 125-130.

Meeks, J. C., Wolk, C. P., Thomas, J., Lockau, W., Shaffer, P. W., Austin, S. M., *et al.* (1977). The pathways of assimilation of $^{13}NH_4^+$ by the cyanobacterium, *Anabaena cylindrica*. *J. Biol. Chem., 252*, 7894-7900.

Meeks, J. C., Wycoff, K. L., Chapman, J. S., and Enderlin, C. S. (1983). Regulation of expression of nitrate and dinitrogen assimilation by *Anabaena* species. *Appl. Environ. Microbiol., 45,* 1351-1359.

Montesinos, M. L., Muro-Pastor, A. M., Herrero, A. and Flores, E. (1998). Ammonium/methyl-ammonium permeases of a cyanobacterium. Identification and analysis of three nitrogen-regulated *amt* genes in *Synechocystis* sp. PCC 6803. *J. Biol. Chem., 273*, 31463-31470.

Muro-Pastor, A. M., and Florencio, F. J. (1994). $NADP^+$-isocitrate dehydrogenase from the cyano-bacterium *Anabaena* sp. strain PCC 7120: purification and characterization of the enzyme and cloning, sequencing, and disruption of the *icd* gene. *J. Bacteriol., 176*, 2718-2726.

Muro-Pastor, A. M., Reyes, J. C., and Florencio, F. J. (1996). The NADP+-isocitrate dehydrogenase gene (*icd*) is nitrogen regulated in cyanobacteria. *J. Bacteriol., 178*, 4070-4076.

Muro-Pastor, A. M., Reyes, J. C., and Florencio, F. J. (2001). Cyanobacteria perceive nitrogen status by sensing intracellular 2-oxoglutarate levels. *J. Biol. Chem., 276*, 38320-38328.

Muro-Pastor, A. M., Valladares, A., Flores, E., and Herrero, A. (1999). The *hetC* gene is a direct target of the NtcA transcriptional regulator in cyanobacterial heterocyst development. *J. Bacteriol., 181*, 6664-6669.

Muro-Pastor, A. M., Valladares, A., Flores, E., and Herrero, A. (2002). Mutual dependence of the expression of the cell differentiation regulatory protein HetR and the global nitrogen regulator NtcA during heterocyst development. *Mol. Microbiol., 44*, 1377-1385.

Murry, M. A., Olafsen, A. G., and Benemann, J. R. (1981). Oxidation of diaminobenzidine in the hetero-cysts of *Anabaena cylindrica. Curr Microbiol., 6*, 201-206.

Murry, M. A., and Wolk, C. P. (1989). Evidence that the barrier to the penetration of oxygen into heterocysts depends upon two layers of the cell envelope. *Arch. Microbiol., 151*, 469-474.

Oxelfelt, F., Tamagnini, P., and Lindblad, P. (1998). Hydrogen uptake in *Nostoc* sp. strain PCC 73102. Cloning and characterization of *hupSL* homologue. *Arch. Microbiol., 169*, 267-274.

Potts, M., Angeloni, S. V., Ebel, R. E., and Bassam, D. (1992). Myoglobin in a cyanobacterium. *Science, 256*, 1690-1692.

Ramasubramanian, T. S., Wei, T.-F., and Golden, J. W. (1994). Two *Anabaena* sp. strain PCC 7120 DNA-binding factors interact with vegetative cell- and heterocyst-specific genes. *J. Bacteriol., 176,*1214-1223.

Ramasubramanian, T. S., Wei, T.-F., Oldham, A. K., and Golden, J. W. (1996). Transcription of the *Anabaena* sp. strain PCC 7120 *ntcA* gene: multiple transcripts and NtcA binding. *J. Bacteriol., 178*, 922-926.

Ramaswamy, K. S., Carrasco, C. D., Fatma, T., and Golden, J. W. (1997). Cell-type specificity of the *Anabaena fdxN*-element rearrangement requires *xisH* and *xisI*. *Mol. Microbiol., 23*, 1241-9.

Reyes, J. C., and Florencio, F. J. (1994). A new type of glutamine synthetase in cyanobacteria - the protein encoded by the *glnN* gene supports nitrogen assimilation in *Synechocystis* sp. strain PCC 6803. *J. Bacteriol., 176*, 1260-1267.

Richaud, C., Zabulon, G., Joder, A., and Thomas, J.-C. (2001). Nitrogen or sulfur starvation differentially affects phycobilisome degradation and expression of the *nblA* gene in *Synechocystis* strain PCC 6803. *J. Bacteriol., 183*, 2989-2994.

Rippka, R., Deruelles, J., Waterbury, J. B., Herdman, M., and Stanier, R. Y. (1979). Generic assignments, strain histories and properties of pure cultures of cyanobacteria. *J. Gen. Microbiol., 111*, 1-61.

Rippka, R., and Stanier, R. Y. (1978). The effects of anaerobiosis on nitrogenase synthesis and heterocyst development by Nostocacean cyanobacteria. *J. Gen. Microbiol., 105*, 83-94.

Ruppert, U., Irmler, A., Kloft, N., and Forchhammer, K. (2002). The novel protein phosphatase PphA from *Synechocystis* PCC 6803 controls dephosphorylation of the signalling protein PII. *Mol. Microbiol., 44*, 855-864.

Sakamoto T., Inoue-Sakamoto, K., and Bryant D. A. (1999). A novel nitrate/nitrite permease in the marine cyanobacterium *Synechococcus* sp. strain PCC 7002. *J. Bacteriol., 181*, 7363-7372.

Sato, N. (1994). A cold-regulated cyanobacterial gene cluster encodes RNA-binding protein and ribosomal protein S21. *Plant Mol. Biol., 24*, 819-823.

Schmetterer, G. (1994). Cyanobacterial respiration. In D. Bryant (Ed.), *The molecular biology of cyanobacteria* (pp. 409-435). Dordrecht, Netherlands: Kluwer Academic Publishers.

Schmetterer, G., Valladares, A., Pils, D., Steinbach, S., Pacher, M., Muro-Pastor, A. M., *et al.* (2001). The *coxBAC* operon encodes a cytochrome c oxidase required for heterotrophic growth in the cyanobacterium *Anabaena variabilis* strain ATCC 29413. *J. Bacteriol., 183*, 6429-6434.

Schmitz, O., Boison, G., Salzmann, H., Bothe, H., Schuetz, K., Wang, S.-H., *et al.* (2002). HoxE: a subunit specific for the pentameric bidirectional hydrogenase complex (HoxEFUYH) of cyanobacteria. *Biochim. Biophys. Acta, 1554*, 66-74.

Schmitz, O., and Bothe, H. (1996). The diaphorase subunit HoxU of the bidirectional hydrogenase as electron transferring protein in cyanobacterial respiration? *Naturwiss., 83*, 525-527.

Sherman, D. M., Tucker, D., and Sherman, L. A. (2000). Heterocyst development and localization of cyanophycin in N_2-fixing cultures of *Anabaena* sp. PCC 7120 (cyanobacteria). *J. Phycol., 36*, 932-941.

Simon, R. (1971). Cyanophycin granules from the blue-green alga *Anabaena cylindrica*: a reserve material consisting of copolymers of aspartic acid and arginine. *Proc. Natl. Acad. Sci. U.S.A., 68*, 265-267.

Soriente, A., Bisogno, T., Gambacorta, A., Romano, I., Sili, C., Trincone, A., *et al.* (1995). Reinvestigation of heterocyst glycolipids from the cyanobacterium, *Anabaena cylindrica. Phytochem., 38*, 641-645.

Soriente, A., Gambacorta, A., Trincone, A., Sili, C., Vincenzini, M., and Sodano, G. (1993). Heterocyst glycolipids of the cyanobacterium *Cyanospira rippkae. Phytochem.*, 393-396.

Spiller, H., Latorre, C., Hassan, M. E., and Shanmugam, K. T. (1986). Isolation and characterization of nitrogenase-derepressed mutant strains of cyanobacterium *Anabaena variabilis. J. Bacteriol., 165*, 412-419.

Stricker, O, Masepohl, B., Klipp, W., and Böhme, H. (1997). Identification and characterization of the *nifV-nifZ-nifT* gene region from the filamentous cyanobacterium *Anabaena* sp. strain PCC 7120. *J. Bacteriol., 179*, 2930-2937.

Summers, M. L., Wallis, J. G., Campbell, E. L., and Meeks, J. C. (1995). Genetic evidence of a major role for glucose-6-phosphate dehydrogenase in nitrogen fixation and dark growth of the cyanobacterium *Nostoc* sp. strain ATCC 29133. *J. Bacteriol., 177,* 6184-6194.

Tamagnini, P., Costa, J.-L., Almeida, L., Oliveira, M.-J., Salema, R., and Lindblad, P. (2000). Diversity of cyanobacterial hydrogenases, a molecular approach. *Curr Microbiol., 40*, 356-361.

Tandeau de Marsac, N., and Houmard, J. (1993). Adaptation of cyanobacteria to environmental stimuli - new steps towards molecular mechanisms. *FEMS Microbiol. Rev., 104*, 119-189.

Tandeau de Marsac, N., Lee, H. M., Hisbergues, M., Castets, A. M., and Bedu, S. (2001). Control of nitrogen and carbon metabolism in cyanobacteria. *J. Appl. Phycol., 13*, 287-292.

Tanigawa, R., Shirokane, M., Maeda, S.-I., Omata, T., Tanaka, K., and Takahashi, H. (2002). Transcriptional activation of NtcA-dependent promoters of *Synechococcus* sp. PCC 7942 by 2-oxoglutarate in vitro. *Proc. Natl. Acad. Sci. U.S.A., 99*, 4251-4255.

Thiel, T. (1990). Protein turnover and heterocyst differentiation in the cyanobacterium *Anabaena variabilis. J. Phycol., 26*, 50-54.

Thiel, T. (1993). Characterization of genes for an alternative nitrogenase in the cyanobacterium *Anabaena variabilis. J. Bacteriol., 175*, 6276-6286.

Thiel, T. (1996). Isolation and characterization of the *vnfEN* genes of the cyanobacterium *Anabaena variabilis. J. Bacteriol., 178*, 4493-4499.

Thiel, T., and Leone, M. (1986) Effect of glutamine on growth and heterocyst differentiation in the cyanobacterium *Anabaena variabilis. J. Bacteriol., 168*, 769-774.

Thiel, T., Lyons, E. M., and Erker, J. E. (1997). Characterization of genes for a second Mo-dependent nitrogenase in the cyanobacterium *Anabaena variabilis. J. Bacteriol., 179*, 5222-5225.

Thiel, T., Lyons, E. M., Erker, J. E., and Ernst, A. (1995). A second nitrogenase in vegetative cells of a heterocyst-forming cyanobacterium. *Proc. Natl. Acad. Sci. U.S.A., 92*, 9358-9362.

Thiel, T., Lyons, E. M., and Thielemier, J. (1998). Organization and regulation of two clusters of *nif* genes in the cyanobacterium *Anabaena variabilis*. In L. W. Peschek, G. Schmetterer (Ed.), *Phototrophic prokaryotes* (pp. 517-521). New York: Plenum Press.

Thiel, T., and Pratte, B. (2001). Effect on heterocyst differentiation of nitrogen fixation in vegetative cells of the cyanobacterium *Anabaena variabilis* ATCC 29413. *J. Bacteriol., 183*, 280-286.

Thiel, T., Pratte, B., and Zahalak, M. (2002). Transport of molybdate in the cyanobacterium *Anabaena variabilis* ATCC 29413. *Arch. Microbiol., 179*, 50-56.

Thomas, J., Meeks, J. C., Wolk, C. P., Shaffer, P. W., and Austin, S. M. (1977). Formation of glutamine from $[^{13}N]$ammonia, $[^{13}N]$dinitrogen, and $[^{14}C]$glutamate by heterocysts isolated from *Anabaena cylindrica. J. Bacteriol., 129*, 1545-1555.

Tumer, N. E., Robinson, S. J., and Haselkorn, R. (1983). Different promoters for the *Anabaena* glutamine synthetase gene during growth using molecular or fixed nitrogen. *Nature, 306*, 337-342.

Udvardy, J., Borbëly, G., Juhäsz, A., and Farkas, G. L. (1984). Thioredoxins and the redox modulation of glucose-6-phosphate dehydrogenase in *Anabaena* sp. strain PCC 7120 vegetative cells and heterocysts. *J. Bacteriol., 157*, 681-683.

Valladares, A., Montesinos, M. L., Herrero, A., and Flores, E. (2002). An ABC-type, high-affinity urea permease identified in cyanobacteria. *Mol. Microbiol., 43*, 703-715.

Valladares, A., Muro-Pastor, A., Fillat, M., Herrero, A., and Flores, E. (1999). Constitutive and nitrogen-regulated promoters of the *petH* gene encoding ferredoxin:$NADP^+$ reductase in the heterocyst-forming cyanobacterium *Anabaena* sp. *FEBS Lett., 449*, 159-164.

Vazquez-Bermudez, M. F., Herrero, A., and Flores, E. (2002a). 2-Oxoglutarate increases the binding affinity of the NtcA (nitrogen control) transcription factor for the *Synechococcus glnA* promoter. *FEBS Lett., 512*, 71-74.

Vazquez-Bermudez, M. F., Paz-Yepes, J., Herrero, A., and Flores, E. (2002b). The NtcA-activated *amt1* gene encodes a permease required for uptake of low concentrations of ammonium in the cyanobacterium *Synechococcus* sp. PCC 7942. *Microbiol., 148*, 861-869.

Vega-Palas, M. A., Flores, E., and Herrero, A. (1992). NtcA, a global nitrogen regulator from the cyanobacterium *Synechococcus* that belongs to the Crp family of bacterial regulators. *Mol. Microbiol., 6*, 1853-1859.

Walsby, A. (1985). The permeability of heterocysts to the gases nitrogen and oxygen. *Proc. Royal Soc. (London) B, 226*, 345-366.

Wang, Q., Li, H., and Post, A. F. (2000). Nitrate assimilation genes of the marine diazotrophic, filamentous cyanobacterium *Trichodesmium* sp. strain WH9601. *J. Bacteriol., 182,* 1764-1767.

Wealand, J. L., Myers, J. A., and Hirschberg, R. (1989). Changes in gene expression during nitrogen starvation in *Anabaena variabilis* ATCC 29413. *J. Bacteriol., 171*, 1309-1313.

Weckesser, J., and Jürgens U. J. (1988). Cell walls and external layers. *Meth. Enzymol., 167,* 173-188.

Wei, T.-F., Ramasubramanian, T. S., and Golden, J. W. (1994). *Anabaena* sp. strain PCC 7120 *ntcA* gene required for growth on nitrate and heterocyst development. *J. Bacteriol., 176,* 4473-4482.

Wei, T.-F., Ramasubramanian, T. S., Pu, F., and Golden, J. W. (1993). *Anabaena* sp. strain PCC 7120 *bifA* gene encoding a sequence-specific DNA-binding protein cloned by in vivo transcriptional interference selection. *J. Bacteriol., 175,* 4025-4035.

Wilcox, M., Mitchison, G. J., and Smith, R. J. (1973a). Pattern formation in the blue-green alga, *Anabaena*: I. Basic mechanism. *J. Cell Sci., 12*, 707-725.

Wilcox, M., Mitchison, G. J., and Smith, R. J. (1973b). Pattern formation in the blue-green alga, *Anabaena*: II. Controlled proheterocyst regression. *J. Cell Sci., 13*, 637-649.

Winkenbach, F., Wolk, C. P. , and Jost, M. (1972). Lipids of membranes and of the cell envelope in heterocysts of a blue-green alga. *Planta, 107*, 69-80.

Wolfinger, E. D., and Bishop, P. E. (1991). Nucleotide sequence and mutational analysis of the *vnfENX* region of *Azotobacter vinelandii*. *J. Bacteriol., 173*, 7565-7572.

Wolk, C. P. (1973). Physiology and cytological chemistry blue-green algae. *Bacteriol. Rev., 37*, 32-101.

Wolk, C. P. (1982). Heterocysts. In N. G. Carr and B. A. Whitton (Eds.), *The biology of cyanobacteria* (pp. 359-386). London: Blackwell Scientific Publishers.

Wolk, C. P. (2000). Heterocyst formation in *Anabaena*, In Y. Brun, and L. Shimkets (Eds.), *Prokaryotic development* (pp. 83-104). Washington, DC: American Society for Microbiology.

Wolk, C. P., Ernst, A., and Elhai, J. (1994). Heterocyst metabolism and development. In D. Bryant (Ed.), *The molecular biology of cyanobacteria* (pp. 769-823). Dordrecht: Kluwer Academic Publishers.

Wolk, C. P., and Quine, M. P. (1975). Formation of one-dimensional patterns by stochastic processes and by filamentous blue-green algae. *Dev. Biol., 46*, 370-382.

Wolk, C. P., Thomas, J., Shaffer, P. W., Austin, S. M., and Galonsky, A. (1976). Pathway of nitrogen metabolism after fixation of ^{13}N-labeled nitrogen gas by the cyanobacterium, *Anabaena cylindrica. J. Biol. Chem., 251*, 5027-5034.

Wolk, C. P., Zhu, J., and Kong, R. (1999). Genetic analysis of heterocyst formation. In G. A. Peschek, W. Löffelhardt, and G. Schmetterer (Eds.), *The phototrophic prokaryotes.* (pp. 509-515). Kluwer Academic Publishers, New York, N.Y.

Wong, F. C. Y., and Meeks, J. C. (2001). The *hetF* gene product is essential to heterocyst differentiation and affects HetR function in the cyanobacterium *Nostoc punctiforme*. *J. Bacteriol., 183*, 2654-2661.

Wood, N. B., and Haselkorn, R. (1978). Proteinase activity during heterocyst differentiation in nitrogen-fixing cyanobacteria. In H. Holzer and G. Cohen (Eds.), *Limited proteolysis in microorganisms: Biological function, use in protein structural and functional studies*. (pp. 159-166). Bethesda, MD: Fogarty Intl. Center Publ.

Wood, N. B., and Haselkorn, R. (1980). Control of phycobiliprotein proteolysis and heterocyst differentiation in *Anabaena*. *J. Bacteriol., 141*, 1375-1385.

Xu, X., and Wolk, C. P. (2001). Role for *hetC* in the transition to a nondividing state during heterocyst differentiation in *Anabaena* sp. *J. Bacteriol., 183*, 393-396.

Xu, X., Khudyakov, I., and Wolk, C. P. (1997). Lipopolysaccharide dependence of cyanophage sensitivity and aerobic nitrogen fixation in *Anabaena* sp. strain PCC 7120. *J. Bacteriol., 179*, 2884-2891.

Yoon, H.-S., and Golden, J. W. (1998). Heterocyst pattern formation controlled by a diffusible peptide. *Science, 283*, 935-938.

Yoon, H.-S., and Golden, J. W. (2001). PatS and products of nitrogen fixation control heterocyst pattern. *J. Bacteriol., 183*, 2605-2613.

Zahalak, M., Beachy, R. N., and Thiel, T. (1995). Expression of the movement protein of tobacco mosaic virus in the cyanobacterium *Anabaena* sp strain PCC 7120. *Mol. Plant Microbe Interact., 8*, 192-199.

Zhou, R., Cao, Z., and Zhao, J. (1998a). Characterization of HetR protein turnover in *Anabaena* sp. PCC 7120. *Arch. Microbiol., 169*, 417-423.

Zhou, R., Wei, X., Jiang, N., Li, H., Dong, Y., Hsi, K., *et al.* (1998b). Evidence that HetR protein is an unusual serine-type protease. *Proc. Natl. Acad. Sci. U.S.A., 95*, 4959-4963.

Zhou, R., and Wolk, C. P. (2003). A two-component system mediates developmental regulation of biosynthesis of a heterocyst polysaccharide. *J. Biol. Chem,. 278*, 19939-19946.

Zhu, J., Jäger, K., Black, T., Zarka, K., Koksharova, O., and Wolk, C. P. (2001). HcwA, an autolysin, is required for heterocyst maturation in *Anabaena* sp. strain PCC 7120. *J. Bacteriol., 183*, 6841-6851.

Zhu, J., Kong, R., and Wolk, C. P. (1998). Regulation of *hepA* of *Anabaena* sp. strain PCC 7120 by elements 5' from the gene and by *hepK*. *J. Bacteriol., 180*, 4233-4242.

Ziegler, K., Stephan, D. P., Pistorius, E. K., Ruppel, H. G., and Lockau, W. (2001). A mutant of the cyanobacterium *Anabaena variabilis* ATCC 29413 lacking cyanophycin synthetase: growth properties and ultrastructural aspects. *FEMS Microbiol. Lett., 196*, 13-18.

Chapter 6

N_2 FIXATION BY NON-HETEROCYSTOUS CYANOBACTERIA

J. R. GALLON
Biochemistry Research Group, School of Biological Sciences, University of Wales Swansea, Singleton Park, Swansea, SA2 8PP, U. K.

1. INTRODUCTION

1.1. Interactions between Photosynthesis and N_2 Fixation: Benefits and Problems

In order to fix N_2, diazotrophs need to synthesize nitrogenase, to maintain the enzyme in a functionally active form and to supply it with the ATP and reductant needed for catalytic activity. They also need to provide carbon skeletons for assimilation of newly fixed nitrogen. In phototrophic diazotrophs, these processes can be maintained at the expense of solar energy which, in most environments, is freely available. Among phototrophs, cyanobacteria have a particular advantage because photosynthesis in these organisms uses water, rather than less readily available compounds, as the ultimate electron donor. The other side of the coin, however, is that cyanobacterial photosynthesis produces O_2, a potent inhibitor of nitrogenase. Cyanobacteria, therefore, face the unique problem of balancing two essential, but incompatible, cellular processes: oxygenic photosynthesis and O_2-sensitive N_2 fixation. Much of the research into cyanobacterial N_2 fixation has been driven by the desire to understand how cyanobacteria deal with this problem.

1.2. N_2 Fixation and O_2

O_2 affects N_2 fixation in various ways (for a review, see Gallon, 1992). *In vitro*, it acts as a reversible, uncompetitive inhibitor of nitrogenase (Burris, 1979), probably as a result of interactions with one or more forms of the Fe protein (Thorneley and Ashby, 1989). More significantly, O_2 also irreversibly inactivates nitrogenase either directly,

W. Klipp, B. Masepohl, J. R. Gallon and W. E. Newton (eds.), Genetics and Regulation of Nitrogen Fixation in Free-Living Bacteria, 111-139.

as O_2 itself, or indirectly, as a consequence of the formation of reactive oxygen species, such as singlet O_2, superoxide (O_2^{-}), hydrogen peroxide (H_2O_2) and hydroxyl radicals ($OH^{·}$). *In vivo*, O_2 can inhibit N_2 fixation simply by diverting electrons away from N_2 reduction. For example, respiratory chains that terminate in O_2 may compete with nitrogenase for reduced ferredoxin or flavodoxin. In most diazotrophs, O_2 also inhibits *nif* gene transcription.

Even though cyanobacteria produce O_2, their nitrogenases show no particular tolerance to this inhibitor. For example, isolated nitrogenase from the unicellular cyanobacterium *Gloeothece* is inhibited by 50% after only 10 min exposure to O_2 at a concentration of 0.34% (v/v). *In vivo*, however, the enzyme is much less susceptible to inhibition by O_2: it takes 60 min at 30% (v/v) O_2 (a concentration higher than that in air) to inhibit N_2 fixation by 50% (see Figure 1). *Gloeothece* and other cyanobacteria, therefore, appear to be well-equipped with mechanisms for limiting contact between O_2 and nitrogenase and for replacing O_2-damaged nitrogenase where contact does

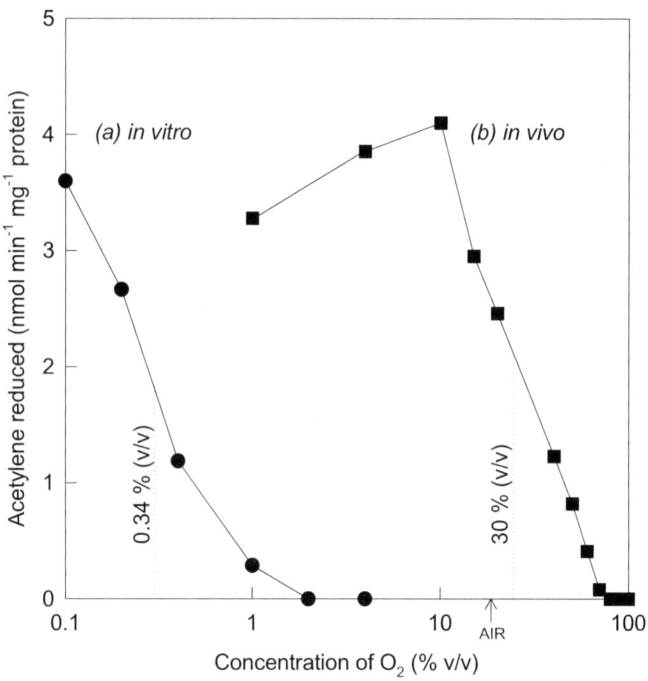

Figure 1. Effect of O_2 on Gloeothece *nitrogenase (a)* in vitro *and (b)* in vivo. *Activity was measured as acetylene reduction after10 min exposure* (in vitro) *or 60 min exposure* (in vivo) *to an atmosphere containing O_2 at the concentration indicated. For methodological details, see Gallon* et al. *(1993).*

occur. These mechanisms have been reviewed by Bergman *et al.* (1997), Fay (1992) and Gallon (1992). In the case of *Gloeothece*, however, the maximum rates of nitrogenase activity occur *in vivo* at 10% O_2 (Figure 1), suggesting that N_2 fixation actually requires O_2. This observation emphasizes that the interrelations between N_2 fixation and O_2 can be very complex in cyanobacteria.

1.3. Heterocystous and Non-heterocystous Cyanobacteria

N_2-fixing cyanobacteria may be conveniently divided into heterocystous and non-heterocystous types. All heterocystous cyanobacteria can fix N_2 aerobically. In such aerobically grown cultures, nitrogenase is found only in fully differentiated heterocysts. However, in anaerobic conditions, certain heterocystous cyanobacteria also produce nitrogenase in vegetative cells (see Thiel, this volume). Because heterocysts do not photo-evolve O_2, photosynthetic O_2 production (in vegetative cells) is spatially separated from the O_2-sensitive nitrogenase (in heterocysts).

In contrast, the majority of non-heterocystous cyanobacteria either cannot fix N_2 at all or can do so only when O_2 is either absent or present at concentrations lower than those in air (referred to as microaerobic conditions). Nevertheless, a few strains can fix N_2 aerobically. Although not representative of non-heterocystous cyanobacteria in general, it is these strains that have been most extensively studied because they seem especially well adapted to protect nitrogenase from inactivation by O_2.

Non-heterocystous cyanobacteria may be either unicellular or filamentous but, in both types, nitrogenase occurs in cells that appear to be both morphologically undifferentiated and, unlike heterocysts, fully competent photosynthetically, *i.e.*, O_2-producing (Bergman *et al.*, 1997). The strategy used by most non-heterocystous cyanobacteria is to effect a temporal separation between photosynthesis and N_2 fixation. It is found that, when grown under alternating periods of 12 h of light and 12 h of darkness, N_2 fixation is confined to the dark period. In this way, the two incompatible processes of N_2 fixation and photosynthetic O_2 production are separated in time. In contrast, heterocystous cyanobacteria fix N_2 mainly, and often exclusively, during the light phase of a cycle of alternating light and darkness (Khamees *et al.*, 1987). As a first approximation then, an alternative method of dividing N_2-fixing cyanobacteria is into two groups: those that separate N_2 fixation and oxygenic photosynthesis in space; and those that separate the two processes in time. Heterocystous cyanobacteria fall into the first group, whereas most non-heterocystous cyanobacteria fall into the second group.

2. NON-HETEROCYSTOUS CYANOBACTERIA

2.1. Historical Perspective

The first suggestions that cyanobacteria might be able to fix N_2 date back to the end of the 19th century (Frank, 1889; Prantl, 1889). However, it was several decades before the ability of cyanobacteria to fix N_2 was unequivocally demonstrated; firstly, using pure cultures (Drewes, 1928) and purified gases (Fogg, 1942) and, secondly, using $^{15}N_2$ (Burris *et al.*, 1943). Until the 1960s, it was believed that only heterocystous cyanobacteria were capable of N_2 fixation. In 1961, however, Dugdale *et al.* (1961)

used $^{15}N_2$ to demonstrate that natural populations of the non-heterocystous filamentous marine cyanobacterium, *Trichodesmium*, could fix N_2. A few years later, with the advent of the acetylene-reduction technique, laboratory cultures of the unicellular cyanobacterium *Gloeothece* (then called *Gloeocapsa*) were shown to fix N_2 aerobically (Wyatt and Silvey, 1969), whereas cultures of the filamentous cyanobacterium, *Plectonema* (*Leptolyngbya*) *boryanum*, could only fix N_2 under microaerobic conditions (Stewart and Lex, 1970). Subsequently, many other non-heterocystous cyanobacteria have been shown to fix N_2, several under aerobic conditions (Bergman *et al.*, 1997; Castenholz and Waterbury, 1989; Lundgren *et al.*, 2001; Zehr *et al.*, 2001).

2.2. Significance of Non-heterocystous Cyanobacteria to Studies on N_2 Fixation

2.2.1. Contribution to Global N_2 Fixation

The recent demonstration of N_2 fixation by nanoplanktonic marine unicellular cyanobacteria at rates (Zehr *et al.*, 2001) comparable with previous estimates of N_2 fixation by *Trichodesmium* (Gallon, 2001) means that non-heterocystous cyanobacteria may be responsible for more than 50% of the annual rate of N_2 fixation on Earth. The contribution that these organisms make to the global nitrogen cycle in general, and to oceanic productivity in particular, is therefore immense.

2.2.2. Biotechnological Exploitation of N_2-fixing Non-heterocystous Cyanobacteria

Cyanobacteria use light energy to sustain metabolism, including N_2 fixation. Because they can fix CO_2 as well as N_2, they can literally live on fresh air, requiring only a few dissolved inorganic nutrients for their healthy growth. They, therefore, represent a cheap source of biomass. They are also useful organisms for solar energy conversion whether as a means of generating fixed nitrogen for agriculture (Musgrave *et al.*, 1982), for nitrogenase-catalysed industrial production of H_2 (Phlips and Mitsui, 1986), or as components of photosynthetic solar cells (Ochiai *et al.*, 1983). Although photosynthetic bacteria can also couple photosynthesis to N_2 fixation, the fact that cyanobacteria can sustain these processes under aerobic conditions makes them easier to handle. Similarly, while heterocystous cyanobacteria are no less suitable than non-heterocystous species for industrial exploitation, in some circumstances, the special properties of the latter can be an advantage. For example, because of the temporal separation of N_2 fixation and photosynthesis, non-heterocystous cyanobacteria would sustain H_2 production during the night, fuelled by the previous day's sunshine. In concert with heterocystous cyanobacteria, these organisms could provide continuous fuel production under natural illumination (alternating day and night).

Further, unicellular non-heterocystous cyanobacteria are an ideal group of organisms with which to probe the complex interrelations between N_2 fixation and O_2; something that needs to be understood if we are ever to introduce the ability to fix N_2 into the leaves of agriculturally important crop plants. Despite some encouraging progress in this area (Dixon *et al.*, 1997), there remain many problems yet to be solved.

2.2.3. Nuisance Organisms

In certain situations, N_2-fixing non-heterocystous cyanobacteria can be a problem. The marine cyanobacterium *Trichodesmium*, for example, forms extensive and unsightly surface blooms that, when washed ashore, can severely damage the recreational potential of beaches, at least in the short term. *Trichodesmium* also contains a potent neurotoxin (Hawser *et al.*, 1991) and has been implicated in the deaths of fish and shellfish. Though no human fatalities have so far been reported, this cyanobacterium nevertheless represents a potential human health hazard both through direct contact, while swimming, and through ingestion of poisoned shellfish.

Several species of non-heterocystous cyanobacteria occur naturally on limestone cliffs, in marble quarries, and in caves. Their ability to photosynthesise and to fix N_2 sustains a complex epilithic biofilm that can form nuisance growths on historically important buildings, sculptures, and frescos, especially where the prevailing light intensity is low. Not only are these biofilms unsightly, they can actually destroy the substrate on which they grow (Albertano and Urzi, 1999).

2.3. Classification of Non-heterocystous Cyanobacteria

The classical taxonomy of cyanobacteria divides these organisms into five 'subsections' or orders, three for non-heterocystous types and two for heterocystous types (Castenholz, 2001; Castenholz and Waterbury, 1989). The non-heterocystous cyanobacteria comprise Subsection I (Chroococcales), which are unicellular cyanobacteria that reproduce by binary fission; Subsection II (Pleurocapsales) are unicellular cyanobacteria that produce daughter cells smaller than the parent; and Subsection III (Oscillatoriales) consists of cyanobacteria that produce filaments of cells known as trichomes. All three subsections have N_2-fixing representatives (Bergman *et al.*, 1997).

For the student of N_2 fixation, however, a classification based on behavioural features is more useful (see Table 1). Initially, this system divides cyanobacteria into: (A) those that can fix N_2 aerobically; and (B) those that cannot. The former category is then subdivided into Group A1, which contains those organisms that separate N_2 fixation from photosynthetically-produced O_2 in terms of space, and Group A2, consisting of organisms that separate the two processes in terms of time, at least under alternating light and darkness. Because aerobically grown heterocystous cyanobacteria confine N_2 fixation to the differentiated heterocysts, whereas photosynthetic production of O_2 occurs in the undifferentiated vegetative cells, these cyanobacteria are all placed in Group A1. In contrast, almost all of those non-heterocystous cyanobacteria that can fix N_2 aerobically effect a temporal separation between N_2 fixation and photosynthetically-generated O_2 and are assigned to Group A2. Although the result has superficial similarity with the conventional heterocystous *versus* non-heterocystous classification system, there are, at least, two non-heterocystous cyanobacteria that apparently fall into Group A1. These are the related marine genera, *Trichodesmium* (Capone *et al.*, 1997) and *Katagnymene* (Lundgren *et al.*, 2001). Both are filamentous, non-heterocystous cyanobacteria but, unlike other non-heterocystous cyanobacteria, they fix N_2 during the light period of a cycle of alternating light and darkness. Though non-heterocystous in nature, these cyanobacteria, therefore, behave like heterocystous cyanobacteria. However, as discussed below (see Section 2.5.6.), *Trichodesmium* may also effect a partial, at least, temporal separation between photosynthetic O_2 production and N_2 fixation. This cyanobacterium does not, therefore, sit comfortably in either Group A1 or Group A2, so Group A3 (see Table 1) has been created to contain those non-heterocystous cyanobacteria, like *Trichodesmium*, that effect both a spatial and a temporal separation between maximum rates of photosynthetic O_2 evolution and maximum rates of N_2 fixation.

Table 1. A Classification of N_2-fixing Cyanobacteria based on Behaviour

A. Cyanobacteria that can fix N_2 aerobically
A1. Cyanobacteria that separate N_2 fixation from oxygenic photosynthesis in space. Includes heterocystous genera, for example, *Anabaena*.
A2. Cyanobacteria that separate N_2 fixation from oxygenic photosynthesis in time. Includes non-heterocystous genera, such as *Gloeothece*, *Cyanothece* and *Lyngbya*
A3. Cyanobacteria that separate N_2 fixation from oxygenic photosynthesis both in space and in time. Includes non-heterocystous genera, such as *Trichodesmium* and *Katagnymene*
B. Cyanobacteria that can fix only N_2 either anaerobically or microaerobically
Many non-heterocystous cyanobacteria, for example, *Plectonema boryanum*.

An overall relatedness among diazotrophic non-heterocystous cyanobacteria can be seen in phylogenetic trees based on *nifH* (Lundgren *et al.*, 2001; Orcutt *et al.*, 2002; Zehr *et al.*, 2001). The non-heterocystous cyanobacteria cluster in a small number of groups that are themselves closely related and similar to some heterocystous cyanobacteria. The behavioural differences referred to above are not, therefore, reflected in the nucleotide sequence of *nifH*. Interestingly, phylogenetic studies on *Cyanothece* strains, based on small subunit ribosomal RNA sequence data, suggest that, during evolution, these cyanobacteria may have experienced multiple gains and losses of the ability to fix N_2 (Turner *et al.*, 2001).

The either spatial or temporal separation between N_2 fixation and oxygenic photosynthesis in cyanobacteria appears to solve the incompatibility problem that exists between these processes, and there is little doubt that these separations are important in natural populations (Bergman *et al.*, 1997). However, a closer examination of the behaviour of certain cyanobacteria reveals that the situation is rather more complex than appears at first sight. For example, most diazotrophic non-heterocystous cyanobacteria can grow and fix N_2 in the laboratory under continuous illumination (Gallon *et al.*, 1991; Huang and Chow, 1986; Mullineaux *et al.*, 1981a; Stal and Krumbein, 1981). Thus, their ability to fix N_2 does not absolutely depend upon their experiencing periods of darkness. In addition, during anaerobic incubation under N_2 (with photosynthetic O_2 production inhibited by addition of 10 μM 3-(2',4'-dichlorophenyl)-1,1-dimethyl urea), the heterocystous cyanobacterium, *Anabaena variabilis* ATCC 29413, produces a nitrogenase in vegetative cells (Thiel *et al.*, 1995; see Thiel, this volume). Active nitrogenase is not, therefore, always confined to the heterocysts of heterocystous cyanobacteria. It should also be noted that both natural populations and laboratory cultures of the non-heterocystous *Trichodesmium* fix N_2 during the day (see, for example, Bergman *et al.*, 1997; Capone *et al.*, 1990; Chen *et al.*, 1998; Gallon *et al.*, 1996), when photo-evolution of O_2 is also occurring.

The ability of non-heterocystous cyanobacteria to fix N_2 under continuous illumination and the ability of certain strains to fix N_2 during the light phase of a cycle of alternating light and darkness suggest that these organisms can maintain an active nitrogenase while simultaneously photo-evolving O_2. It is, however, far from certain that the N_2-fixing cells in these cultures actually show net O_2 production under these conditions. For example, O_2-consuming reactions in these cells may more than compensate for any photosynthetic production of O_2 (Gallon, 2001).

2.4. Distribution of Non-heterocystous Cyanobacteria

Diazotrophic cyanobacteria are found everywhere, from the tropics to the poles, in soils, in fresh waters, in the oceans, in hot springs, in clouds, and in such unlikely environments as the hind-gut of dragonflies (Stewart and Schlichting, 1966). The only factors that seem to limit their distribution are their requirement for light and their preference for either neutral or alkaline environments. Non-heterocystous representatives occur in tropical rice paddies (Huang and Chow, 1988) and polar lakes (Fernandez-Valiente *et al.*, 2001), temperate (Rippka *et al.*, 1979) and polar (Pandey *et al.*, 2000) bogs, in caves (Griffiths *et al.*, 1987), in desert soils both in the tropics (Friedmann and Galun, 1974; Issa *et al.*, 2001) and at the poles (Friedmann and Ocampo, 1976), on stones, rocks and bark (Coute *et al.*, 1999), endolithically (Weber *et al.*, 1996), in frozen ice (Laamanen, 1996) and in air (Gregory *et al.*, 1955). However, it is in the marine environment that non-heterocystous cyanobacteria make their major contribution to global N_2 fixation; in salt marshes (Griffiths *et al.*, 1987), in microbial mats on the littoral fringes of the oceans (Stal, 1995), and in tropical open waters, where the filamentous organism *Trichodesmium* (Capone *et al.*, 1997) and a range of unicellular N_2-fixing cyanobacteria (Zehr *et al.*, 2001) are widespread.

2.5. Specific Organisms

2.5.1. Plectonema (Leptolyngbya) boryanum

Plectonema boryanum, a species that has been assigned by Anagnostidis and Komárek (1988) to the genus *Leptolyngbya*, is a filamentous non-heterocystous cyanobacterium that was first shown to fix N_2 by Stewart and Lex in 1970. Although it is capable of fixing N_2 under either anaerobic or microaerobic conditions, *P. boryanum* cannot fix N_2 aerobically. It is, therefore, a representative of Group B in the classification system shown in Table 1. The need to exclude O_2 rigorously means that studies on organisms such as *P. boryanum* are more difficult to perform than are studies on aerotolerant non-heterocystous diazotrophs. In consequence, rather less is known about the physiology of N_2 fixation in *P. boryanum* and the other non-heterocystous cyanobacteria that can fix N_2 only anaerobically or microaerobically.

Although *P. boryanum* can fix N_2 under an O_2-free atmosphere when incubated under continuous illumination, it cannot fix N_2 under alternating light and darkness (Misra, 1999). Nevertheless, during growth under continuous illumination, it effects a temporal separation between photosynthetic O_2 production and N_2 fixation, with peaks of N_2 fixation corresponding to troughs in photosynthetic O_2 production and *vice versa*. Successive peaks in both N_2 fixation and photosynthesis are separated by a period of 24 h and these fluctuations in activity seem to be controlled by regulation of gene transcription. The genes involved include (for photosynthesis) those encoding phycocyanin (*cpcBA*), the chlorophyll *a*-binding protein of Photosystem II (*psbC*), and the D1 reaction centre protein of Photosystem II (*psbA*), along with (for N_2 fixation) *nifH*, which encodes the Fe-protein of nitrogenase (Misra and Tuli, 2000). In addition, during the 'photosynthetic' phase, light energy is preferentially transferred to the O_2-evolving Photosystem II, whereas, during the 'N_2-fixing' phase, light energy is preferentially transferred to Photosystem I (Misra and Tuli, 1994), which does not generate O_2. This situation would allow N_2 fixation to be supported by cyclic photophosphorylation and a light-mediated input of electrons from donors other than water, in a manner similar to that reported for heterocysts (Wolk *et al.*, 1994).

2.5.2. Gloeothece

Gloeothece was the first unicellular cyanobacterium shown to fix N_2 (Wyatt and Silvey, 1969). It can fix N_2 aerobically and is characterised by an extensive polysaccharide slime capsule that surrounds individual cells. Because *Gloeothece* grows epilithically in limestone caves and in rock crevices, the slime acts to protect it from desiccation, but it may also function as an extracellular vacuole for the temporary storage of the amino acid products of N_2 fixation (Flynn and Gallon, 1990).

Early studies on batch cultures of *Gloeothece* showed that, when grown under constant illumination, this cyanobacterium effected a temporal separation between maximum rates of oxygenic photosynthesis and N_2 fixation (Gallon *et al.*, 1974). This separation of N_2 fixation and photosynthetic O_2 production was much more obvious, however, when cultures were grown under alternating 12-h light and 12-h darkness. Photosynthesis occurred only during the light period, whereas nitrogenase activity, measured either as acetylene reduction (Mullineaux *et al.*, 1981a) or as N_2 reduction (Gallon, 1992), occurred predominantly during the hours of darkness. Although this pattern can be modified in the laboratory by altering environmental conditions, (Du and Gallon, 1993; Ortega-Calvo and Stal, 1991), natural populations fix N_2 at night (Griffiths *et al.*, 1987). It is, therefore, assumed that this constitutes 'natural' behaviour and *Gloeothece* is, accordingly, assigned to Group A2 (see Table 1).

The behaviour of cultures of *Gloeothece* during incubation under alternating light and darkness has been extensively investigated and a summary of some of the major fluctuations that occur in N_2 fixation and related metabolism is presented in Figure 2.

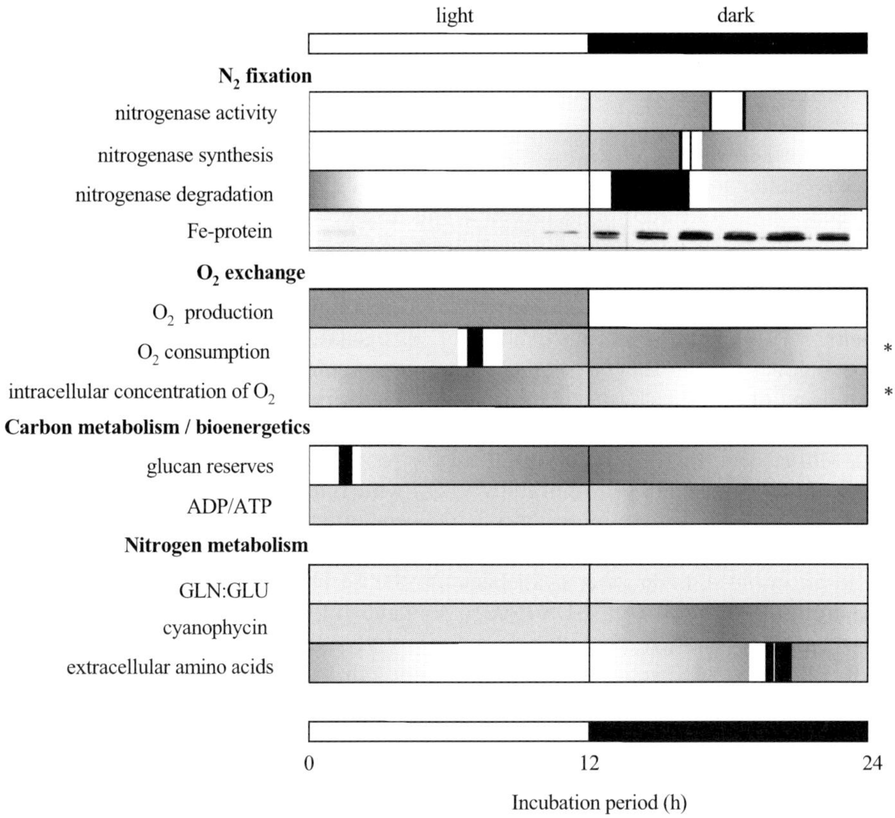

Figure 2. Fluctuations in N_2 fixation, O_2 exchange, carbon metabolism and nitrogen metabolism during growth of Gloeothece *under alternating 12-h light and 12-h darkness. For each set of measurements, the darker the scale, the greater the activity or concentration; white indicates a value of zero. The asterisks indicate hitherto unpublished data. For details, see text and Flynn and Gallon (1990); Gallon (1988, 1990); Reade* et al. *(1999).*

In *Gloeothece*, the pattern of nitrogenase activity is reflected in the behaviour of the nitrogenase proteins. During the dark period, when N_2 fixation is occurring, both the Fe protein and the MoFe protein of nitrogenase can be detected in the cells. However, at the end of the dark period, nitrogenase disappears from the cells only to reappear about 2 h before cultures enter the next dark phase. Synthesis of nitrogenase is regulated at the level of transcription and commences about 2 h before the onset of darkness, abruptly ceasing about 6 h into the dark phase (Gallon *et al.*, 1988). Loss of nitrogenase activity during the latter half of the dark phase is caused mainly by inactivation of pre-existing enzyme by atmospheric O_2 (Mullineaux *et al.*, 1981b). However, after synthesis of nitrogenase has ceased, the enzyme proteins disappear from the cells (Reade *et al.*, 1999). *Gloeothece*, therefore, undergoes a diurnal round not only of nitrogenase synthesis but also of degradation.

In the dark, synthesis and activity of nitrogenase is supported by the breakdown of glucan reserves that accumulate during the period of illumination. Coincident with the cessation of nitrogenase synthesis, there is an abrupt decrease in the rate of both glucan breakdown (Gallon *et al.*, 1988) and of respiratory O_2 consumption (Figure 2; Yunes, Gallon and Chaplin, unpublished data), along with a dramatic increase in the intracellular ratio of ADP to ATP. It seems likely, therefore, that ATP limitation, resulting from the depletion of glucan reserves, prevents any further synthesis of nitrogenase. The concomitant decline in respiration results in an increase in the intracellular concentration of O_2 (Figure 2; Du and Gallon, unpublished data), which would, in turn, exacerbate inactivation of nitrogenase. Several lines of evidence suggest that carbon-supported metabolic consumption of O_2 is an important component of the mechanisms that minimize the adverse effect of both photosynthetic and atmospheric O_2 on *Gloeothece* nitrogenase (Gallon, 2001; Gallon *et al.*, 1993) and, furthermore, that Ca^{2+} is essential for the proper operation of these mechanisms (Gallon and Hamadi, 1984). Intriguingly, O_2, which inhibits nitrogenase synthesis in most diazotrophs, has no more than a transient effect on nitrogenase synthesis in *Gloeothece* (Maryan *et al.*, 1986; Page and Gallon, 1992).

In *Gloeothece*, as in other cyanobacteria, the Fe protein of nitrogenase exists in two forms that appear, on SDS/PAGE, to have different molecular weights (for details, see Section 3.2.). These two forms are interconvertible, even in the presence of inhibitors of protein synthesis, and it is assumed that the larger form (M_r = 40 000) represents a modified form of the smaller form (M_r = 38 500). In cultures growing under alternating light and darkness, newly synthesized Fe protein always first appears in its higher molecular-weight form. This form may be catalytically inactive (at least under aerobic conditions) and it is possible that the conversion of this form to the lower molecular-weight, presumed active, form of the Fe protein is a factor in explaining the increase in nitrogenase activity that occurs during the first few hours of the dark period (Du and Gallon, 1993; Reade *et al.*, 1999). However, there is no corresponding change to the higher molecular-weight form coincident with the loss of activity towards the end of the dark period. Rather, this activity loss appears to be related to the complete disappearance of the protein.

Nitrogenase breakdown often commences before *Gloeothece* cells enter the light phase, but becomes very rapid once cultures become illuminated. It seems likely that synthesis of a critical protein component of a nitrogenase-degrading system is induced about 8 h into the dark phase (Reade *et al.*, 1999). Degradation of nitrogenase *in vivo* occurs very rapidly and gives no obvious breakdown products. In contrast, degradation of the native Fe protein (M_r 40 000 and 38 500) *in vitro*, using extracts of *Gloeothece* prepared from cultures sampled 10 h into the dark period, gave a number of products, predominant among which were two polypeptides of M_r 34 700 and 32 900 respectively (Dougherty *et al.*, 1996a). Breakdown of nitrogenase occurred more rapidly when extracts were exposed to air than when they were incubated anaerobically, implying that at least part of the degradative process is stimulated by O_2. However, exposure of nitrogenase *in vitro* to air for 15 min, which completely destroyed activity, did not greatly increase subsequent degradation of the Fe protein in the absence of O_2, suggesting that the more rapid breakdown observed in the presence of O_2 probably reflects O_2-stimulated proteolysis rather than specific targeting of O_2-damaged protein by the degradative system.

Extracts of *Gloeothece* sampled from cells at the beginning of the light period showed elevated proteolytic activity *in vitro* when compared to extracts taken at the end of the dark period. However, the slower rate of degradation of Fe protein in the latter extracts could be stimulated in the presence of dithiothreitol (Dougherty *et al.*, 1996a), a compound that mimics the action of thioredoxin in activating other enzymes that show light stimulation. The nitrogenase-specific proteolytic system in *Gloeothece* may, therefore, respond to illumination through the light-mediated reduction of thioredoxin. However, because addition of $MgATP^{2-}$ also stimulated degradation of nitrogenase in extracts from cells sampled at the end of the dark period, the increased proteolytic activity that occurs when cultures of *Gloeothece* enter the light period could equally reflect increased ATP synthesis, which would also follow illumination. Most probably, the stimulation of nitrogenase breakdown in the light is due to an increase in three factors linked to photosynthesis: O_2, reductant (i.e. reduced thioredoxin), and ATP.

The diurnal rhythm of nitrogenase activity is, therefore, imposed upon *Gloeothece* by the illumination cycle, with glucan reserves functioning as the primary link between the pattern of illumination and N_2 fixation. About half way through the dark period, glucan reserves are depleted to the extent that neither synthesis nor activity of nitrogenase can be maintained. Nitrogenase then becomes more susceptible to inactivation by atmospheric O_2 but is also degraded by a specific proteolytic system, one or more components of which are synthesised towards the end of the dark period. This proteolytic system is stimulated by O_2 and by light. The onset of illumination initially accelerates the degradation of what is, by this stage, largely inactive nitrogenase but it also triggers the events (most probably replenishment of glucan reserves) that allow nitrogenase synthesis to recommence after a lag of about 10 h.

The intracellular ratio of glutamine to glutamate is very low in cultures of *Gloeothece* grown under N_2-fixing conditions. This finding implies that the cells are permanently nitrogen-stressed and that it is through carbon metabolism, rather than nitrogen metabolism, that nitrogenase activity is regulated in response to the pattern of illumination. Nevertheless, during the period of N_2 fixation, newly fixed nitrogen accumulates as amino acids, which are released into the slime capsule that surrounds cells of *Gloeothece* (Flynn and Gallon, 1990), and as cyanophycin (Gallon *et al.*, 1990). These compounds then provide nitrogen for metabolic processes during the following light period.

2.5.3. Cyanothece (Synechococcus*)*

Like *Gloeothece*, *Cyanothece* is a unicellular cyanobacterium that can fix N_2 aerobically. However, it lacks the extensive slime capsule that surrounds *Gloeothece*. Taxonomic assignation of unicellular cyanobacteria is not easy and it is now considered that certain other aerotolerant N_2-fixing unicellular cyanobacteria, previously identified as either *Aphanothece* or *Synechococcus*, may more accurately be described as *Cyanothece* (Bergman *et al.*, 1997). However, even the distinction between *Cyanothece* and *Gloeothece* can be blurred because it is fairly easy to isolate capsule-free cells from cultures of *Gloeothece* (Du and Gallon, 1993; Rippka and Cohen-Bazire, 1983). Nevertheless, although several *Cyanothece*/*Synechococcus* strains behave similarly at the physiological level, they all behave differently from *Gloeothece*.

When grown under alternating periods of light and darkness, *Cyanothece* ATCC 51142, like *Gloeothece*, shows a diurnal rhythm in N_2 fixation, confining nitrogenase activity to the period of darkness. Also, as with *Gloeothece*, newly fixed nitrogen is stored as cyanophycin (Li *et al.*, 2001) and the nitrogenase proteins are completely turned over every 24 h (Sherman *et al.*, 1998). *Cyanothece* ATCC 51142 differs from *Gloeothece*, however, in that the 24-h pattern of N_2 fixation, which is seen in cultures incubated under alternating light and darkness, persists when cultures are transferred to constant illumination (Colón-López *et al.*, 1997). In contrast, the 24-h pattern seen with cultures of *Gloeothece*, when incubated under alternating light and darkness, is rapidly 'forgotten' once the signal of a period of darkness is removed (Gallon, 2001), converting to a 40-h cycle, as seen in cultures that were grown for many generations under constant illumination (Mullineaux *et al.*, 1981a). Thus, both *Cyanothece* and *Gloeothece* display an endogenous rhythm of N_2 fixation under constant illumination but, whereas the pattern with *Cyanothece* is circadian (see Section 3.4.), that with *Gloeothece* operates on a 40-h cycle.

In *Synechococcus* (*Cyanothece*) BG 43511, the endogenous 24-h rhythm in N_2 fixation has been shown to be related to the synchronisation of cell division during incubation under alternating light and darkness (Mitsui *et al.*, 1986). *Synechococcus* BG 43511 has a doubling time of about 24 h and, in synchronous cultures, N_2 fixation is confined to a specific phase in the cell cycle. This is not the case in *Gloeothece,* where the endogenous 40-h rhythm in N_2 fixation does not correspond very well to the doubling time of the organism under constant illumination, which varied from 50-80 h (Flynn and Gallon, 1990; Gallon *et al.*, 1988; Mullineaux *et al.*, 1981a).

In both *Cyanothece* ATCC 51142 and *Gloeothece*, the endogenous rhythm in N_2 fixation is paralleled by a rhythm in respiratory O_2 consumption (supported by carbohydrate breakdown) and opposed by a rhythm in photosynthetic O_2 production. In *Cyanothece*, modulation of photosynthesis appears to have both short-term and long-term components. The long-term effects are based on regulation of gene expression and synthesis of several of the proteins involved in photosynthetic O_2 production is regulated in a cyclic manner (Sherman *et al.*, 1998). The short-term effects relate to reorganizations in the structure of the reaction centers of Photosystems I and II (Meunier *et al.*, 1997, 1998; Tucker and Sherman, 2000). In *Cyanothece*, the photosynthetic apparatus can exist in two states that are relevant to N_2 fixation. State 2 favours cyclic electron flow around Photosystem I, which does not result in any O_2 evolution, and consists of trimeric Photosystem I complexes and monomeric Photosystem II complexes. Under these circumstances, the phycobilisomes are primarily attached to Photosystem I, so light energy absorbed by the accessory phycobiliproteins is preferentially directed to Photosystem I. State 1 favours linear electron flow from water, which results in O_2 evolution, to NADPH and has a dimeric Photosystem II and a monomeric Photosystem I. In this state, phycobilisomes can more readily attach to Photosystem II. The dimeric form of Photosystem II can exist in two forms, 1 and 2, that reflect different forms of the D1 protein. Form 2 is an adaptation to high levels of illumination and permits high rates of non-cyclic electron flow when light intensities are very high.

In *Cyanothece* growing under alternating 12-h light and 12-h darkness, the organisation of the photosystems is such that, during the dark period, when cells are fixing N_2 maximally, State 2 predominates (Figure 3). There is a significant jump to State 1 at 6 h into the dark period as nitrogenase activity declines rapidly. During most of the light period, State 1 dominates, with Form 2 of the D1 protein accumulating towards the end of the light period (Meunier *et al.*, 1997). These transitions are probably mediated through the redox state of the plastoquinone pool (Sherman *et al.*, 1998) and persist in cultures transferred to constant light so, during the period of N_2 fixation, there is little photosynthetic O_2 production, even in the light, because the photosynthetic apparatus is in State 2.

In addition to these state changes within Photosystems I and II, the structure of the manganese-stabilising protein (MSP) in *Cyanothece* suggests that this protein too may be involved in the observed fluctuations seen in photosynthetic O_2 production (Tucker *et al.*, 2001). MSP functions to protect the O_2-evolving Mn centers of Photosystem II from inactivation. At its N-terminus, cyanobacterial MSP contains cysteine residues that may be important for structural stability. These residues may respond to redox status by forming intramolecular disulfide bridges. This, in turn, may induce conformational changes in MSP and result, during certain phases of the alternating light/dark cycle, in altered stability within the O_2-evolving Mn complex. As yet, however, there is no direct evidence for this suggestion.

In the longer term, the rhythmicity of N_2 fixation and photosynthesis in *Cyanothece* is regulated at the genetic level (Sherman *et al.*, 1998; Tucker *et al.*, 2001), with certain genes being transcribed at specific times during the light/dark cycle. The genes, *nifH* and *nifDK*, which encode the Fe and MoFe proteins of nitrogenase, are transcribed during the early part of the dark phase, when State 2 of the photosynthetic apparatus predominates (Figure 3). In contrast, *psaA* and *psaB*, which encode the reaction center proteins of Photosystem I, and the *psbA* genes, which in *Cyanothece* give rise to Forms 1 and 2 of the D1 protein of Photosystem II, are transcribed during the light phase. The genes *psbDI* and *psbDII*, which encode the structural D2 protein of Photosystem II, and *psbO*, which encodes MSP, are also transcribed at specific times during the light/dark cycle, coincident with the predominance of State 1. These characteristic patterns of expression, along with the organisational regulatory features described above, persist in cultures transferred from alternating light and darkness to constant illumination. They, therefore, constitute an entrained circadian rhythm.

In *Synechococcus* (*Cyanothece*) RF-1, a whole family of *nif* and *nif*-related genes undergo cyclic expression both under alternating light and darkness and under constant illumination (Huang *et al.*, 1999). In the latter case, they are all expressed during the subjective 'dark' phase. In cultures of *Cyanothece* ATCC 51142 incubated under constant illumination, the Fe protein of nitrogenase disappears during the subjective 'light' phase, just as it does during the light phase of a light/dark cycle. The MoFe protein, however, does not (Colón-López *et al.*, 1997). The reason for this is not clear, although it implies that the system that degrades the MoFe protein requires an external, dark-modulated signal to be activated and may not, therefore, be part of an endogenous circadian response.

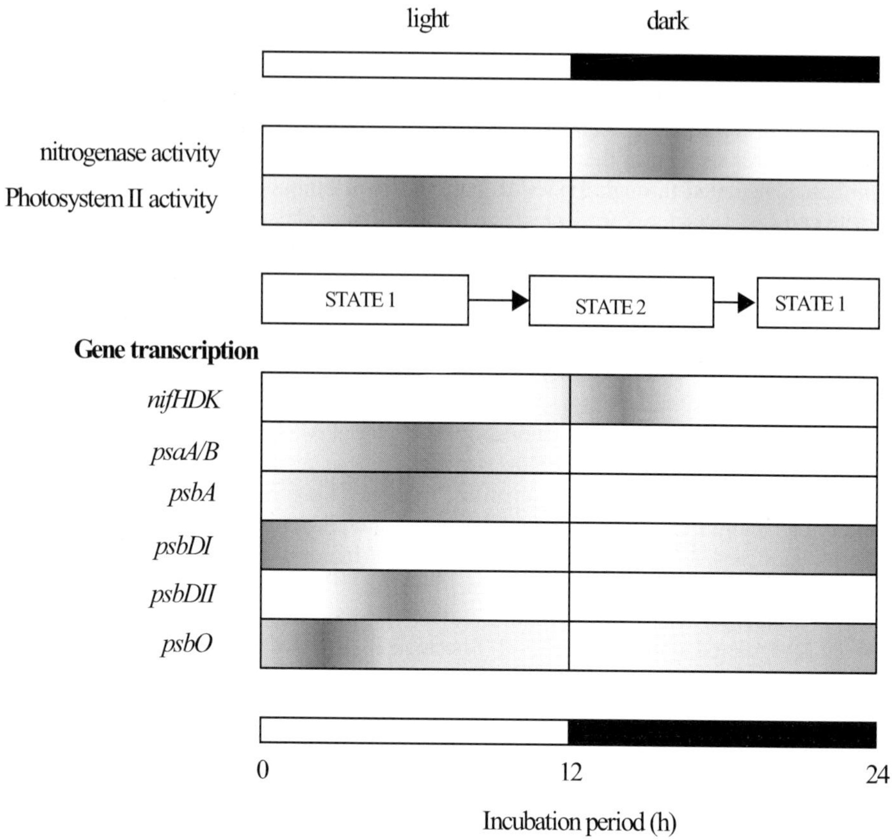

Figure 3. Modulation of photosynthetic O_2 production in cultures of Cyanothece *ATCC 51142 incubated under alternating 12-h light and 12-h darkness (based on Sherman* et al.*, 1998 and Tucker* et al.*, 2001). For each set of measurements, the darker the scale, the greater the activity or concentration; white indicates a value of zero. See text for further details.*

2.5.4. Lyngbya (Oscillatoria)

Lyngbya, some strains of which have previously been classified as *Oscillatoria* (Bergman *et al.*, 1997), is a filamentous non-heterocystous cyanobacterium. Several strains of *Lyngbya* can fix N_2 aerobically. Like the unicellular strains described above, when grown under alternating periods of light and darkness, *Oscillatoria limosa* (probably a strain of *Lyngbya aestuarii)* fixes N_2 in the dark. Moreover, as in *Cyanothece*, the circadian rhythm of N_2 fixation in *O. limosa* appears to be endogenous (Stal and Krumbein, 1985). However, unlike in either *Gloeothece* or

Cyanothece, nitrogenase does not disappear from cultures of *O. limosa* during the light period. Instead, during the period when no activity is seen, the Fe protein of nitrogenase is present exclusively in its larger modified form, which is presumed to be catalytically inactive under aerobic growth conditions (Stal and Bergman, 1990). *O. limosa* also differs from other non-heterocystous cyanobacteria in that it is capable of N_2 fixation in the dark under strictly anaerobic conditions (Stal and Heyer, 1987), whereas both *Gloeothece* (Du and Gallon, 1993) and *Synechococcus* (*Cyanothece*) BG 43511 (Ikemoto and Mitsui, 1994) fixed N_2 only during the light phase of a cycle of alternating light and darkness under anaerobic conditions.

Two strains of *Lyngbya*, one designated *L. aestuarii* (Paerl *et al.*, 1991) and apparently distinct from *O. limosa*, and one designated *L. majuscula* (Jones, 1990), have been reported to fix N_2 during the light phase of a cycle of alternating light and darkness. In this respect, they resemble heterocystous cyanobacteria (see Thiel, this volume) and *Trichodesmium* (see below). In the case of *L. majuscula*, however, recent studies (Lundgren, 2001) have failed to confirm the initial observation, so the status of these organisms remains uncertain.

2.5.5. Trichodesmium

Trichodesmium is one of the most significant N_2-fixing organisms on Earth, accounting, on its own, for the fixation of 87.4 Tg of N per year (as calculated by Gallon, 2001). It is a filamentous cyanobacterium that is widely distributed in the tropical ocean and is unusual, if not unique, among non-heterocystous cyanobacteria in fixing N_2 during the light phase of a cycle of alternating light and darkness. In not effecting a temporal separation between photosynthetic O_2 production and N_2 fixation, *Trichodesmium* behaves more like a heterocystous cyanobacterium than the other non-heterocystous strains. This resemblance between *Trichodesmium* and heterocystous cyanobacteria is strengthened by the observation, based on immuno-localization studies, that nitrogenase is confined to small numbers of contiguous cells along individual filaments of *Trichodesmium* (Bergman and Carpenter, 1991; Fredriksson and Bergman, 1995; Janson *et al.*, 1994; Lin *et al.*, 1998). These cells, therefore, resemble heterocysts, even though they are not morphologically differentiated and retain the ability to divide.

It is not certain, however, whether there is a true spatial separation between photosynthetic O_2 production and N_2 fixation in *Trichodesmium*. For example, there has been much debate about whether the N_2-fixing cells of *Trichodesmium* are fully photosynthetically active (*i.e.*, O_2-evolving). Unlike heterocysts, they do not appear to lack critical components of the O_2-evolving machinery of photosynthesis (Siddiqui *et al.*, 1992) and they even retain the ability to fix CO_2 (Lin *et al.*, 1998). However, several lines of evidence suggest that these N_2-fixing cells may consume, rather than produce, O_2 (Berman-Frank *et al.*, 2001; Gallon, 2001), so it appears that any photosynthetic O_2 production within these cells is more than compensated for by reactions that consume O_2. In some natural populations and in laboratory cultures of *Trichodesmium*, maximum rates of N_2 fixation occurred during the middle of the light period, coincident with a transient, but marked, decrease in the rate of photosynthetic CO_2 fixation (Berman-Frank *et al.*, 2001). During the same period, cultures showed a net consumption of O_2, implying that respiratory O_2 uptake exceeded photosynthetic O_2 production. This behaviour would, therefore, create a temporal separation between the periods of maximum rates of N_2 fixation and photosynthetic O_2 production, even though the separation is less complete than that seen in other non-heterocystous cyanobacteria.

High resolution measurements of *in vivo* chlorophyll fluorescence in *Trichodesmium* showed that the quantum yield of the O_2-evolving Photosystem II was lowest during periods of maximum N_2 fixation (Berman-Frank *et al.*, 2001). Modulation of Photosystem II activity was rapid and occurred in all cells, not just those that were fixing N_2. Moreover, under N_2-fixing conditions, some cells showed considerably higher fluorescence than others, implying the presence of zones of relative photosynthetic inactivity. Unfortunately, however, these zones were randomly distributed along individual filaments of *Trichodesmium* and did not coincide with the nitrogenase-containing cells, which occurred only in central regions of filaments. Nevertheless, taken together, all of these findings suggest that *Trichodesmium* may have the capacity to separate N_2 fixation and photosynthetic O_2 production in both space and time. It, therefore, appears in Group A3 in Table 1. Neither separation may be very efficient on its own, but together they allow *Trichodesmium* to fix N_2 during the day, an unusual behaviour for a non-heterocystous cyanobacterium. As in *Gloeothece*, transcription of *nifHDK* in *Trichodesmium* appears to be more tolerant of O_2 than is the case in most diazotrophs (Dominic *et al.*, 1998).

Two other non-heterocystous filamentous cyanobacteria behave similarly to *Trichodesmium*. *Symploca* PCC 8002 (originally classified as *Microcoleus chthonoplastes*) shows a pattern of N_2 fixation where activity increases steadily in the light and declines in the dark (Fredriksson *et al.*, 1998) whereas, in *Katagnymene*, which is phylogenetically closely related to *Trichodesmium*, N_2 fixation occurred exclusively in the light (Lundgren *et al.*, 2001). In both cases, nitrogenase is confined to a subset of cells, although, in *Katagnymene*, individual filaments often contain more than one zone of nitrogenase-containing cells. It should be noted, however, that another strain of *Symploca*, designated S84, has been reported to fix N_2 mainly during the dark period of an alternating 12-h light and 12-h dark cycle (Kumazawa *et al.*, 2001).

3. PATTERNS OF N_2 FIXATION

3.1. Alternating Light and Darkness

With the exception of *Trichodesmium*, *Katagnymene* and *Symploca,* which all confine nitrogenase to specialised, although undifferentiated, cells, non-heterocystous cyanobacteria fix N_2 during the dark phase of a cycle of alternating light and darkness. However, the molecular mechanisms underpinning this behaviour can vary from organism to organism. For example, in *Gloeothece* and *Cyanothece*, nitrogenase disappears from the cells during the light phase, implying regulation of nitrogenase synthesis and degradation. In *O. limosa*, however, nitrogenase does not disappear in the light but, rather, is converted to a modified form that is catalytically inactive, at least under aerobic conditions.

When inhibitors of proteases, such as pepstatin A, are added to cultures of *Gloeothece*, which are 6 h into the dark phase, nitrogenase persists into the light phase, though it remains catalytically inactive. Under these conditions, the Fe protein is present during the light phase exclusively in its modified form (Reade *et al.*, 1999). Thus, transfer from darkness to light stimulates modification of the Fe protein in *Gloeothece* just as occurs in *O. limosa*. It, therefore, appears that, unlike *Gloeothece*, *O. limosa* lacks a system for degrading nitrogenase.

In *Trichodesmium* (Capone *et al.*, 1990; Ohki *et al.*, 1992) and *Katagnymene* (Lundgren *et al.*, 2001), nitrogenase proteins can be detected throughout the period of alternating light and darkness, although their intracellular concentrations may be lower during the period when N_2 fixation is not occurring. In this respect, these cyanobacteria more closely resemble heterocystous cyanobacteria (Ernst *et al.*, 1990; Evans *et al.*, 2000; Gallon *et al.*, 2002) and *O. limosa* (Stal and Bergman, 1990) than either *Gloeothece* or *Cyanothece*.

3.2. Covalent Modification of Nitrogenase

In a number of cyanobacteria, although not *P. boryanum* (Rai *et al.*, 1992), the Fe protein of nitrogenase undergoes covalent modification (Bergman *et al.*, 1997). The extent of this modification is affected by O_2 (Du and Gallon, 1993), fixed nitrogen (Cheng *et al.*, 1999), and the pattern of illumination (Reade *et al.*, 1999). Unlike the situation in certain anoxygenic photosynthetic bacteria (see Nordlund and Ludden, this volume), modification of the Fe protein in cyanobacteria does not involve ADP-ribosylation. In *Gloeothece*, there is evidence that the modified form of the Fe protein is palmitoylated (Gallon *et al.*, 2000). However, a single modification of this type would be insufficient to explain the mass difference of either 1500 (as estimated by SDS/PAGE) or 2361 (as measured by MALDI-TOF mass spectrometry of the immunoprecipitated *Gloeothece* Fe-protein) (Dougherty *et al.*, 1996b) between the unmodified and modified Fe protein. Such a mass difference is equivalent to 7-10 palmitoyl groups per molecule of modified Fe protein, which seems most unlikely.

The functional significance of palmitoylation of the Fe protein is not clear. It does not, for example, appear either to stabilize the protein or to target it for degradation (Gallon *et al.*, 2000). However, it may be significant that other palmitoylated proteins interact with membranes. Examples include Ras, which shares other features in common with the Fe protein of nitrogenase (Howard and Rees, 1994). Perhaps modification of the Fe protein brings it into closer contact with membranes in order to allow greater interaction with membrane-bound electron transport systems, to minimise O_2 damage to the enzyme (O_2 is less soluble in lipids than in water) or, extending the analogy with Ras even further, for some role in signal transduction.

Based on the nucleotide sequence of *nifH*, the Fe protein of nitrogenase in *Cyanothece* ATCC 51142 (M_r = 35 822) is larger than that from *Synechococcus* (*Cyanothece*) RF-1 (32 304; Huang *et al.*, 1999), *P. boryanum* (32 333; Fujita *et al.*, 1991) or *Trichodesmium* (32 276; Sroga *et al.*, 1996; 32 230; Zehr *et al.*, 1997). Its larger size is due to an extension of about 37 amino acid residues at its N-terminus (Colón-López *et al.*, 1999). *Cyanothece* Fe protein may, however, be similar to the Fe protein of *Gloeothece* nitrogenase, which, in its unmodified form, has an M_r of 36 098, as determined directly by MALDI-TOF mass spectrometry (Gallon *et al.*, 2000). Because the extended N-terminal region of *Cyanothece* Fe-protein resembles the PEST-like domains found in proteins that undergo rapid degradation, it has been suggested that this extension may target the protein for proteolytic attack (Colón-López *et al.*, 1999). It may be significant that, in cultures of *Cyanothece* and *Gloeothece* growing under alternating light and darkness, but not in *Trichodesmium*, the Fe protein disappears rapidly from cultures during the period when they are not fixing N_2. It may also be significant that the proteolytic system that degrades *Gloeothece* Fe protein was not active with the Fe protein from *Klebsiella pneumoniae* (Gallon *et al.*, 1995), although it did act on the Fe protein from *Rhodospirillum rubrum* (Reade *et al.*, 1999). On the other hand, the Fe protein of nitrogenase in *Synechococcus* (*Cyanothece*) RF-1 disappears during the light phase of a cycle of light and darkness (Chow and Tabita, 1994) and yet this protein lacks the N-terminal sequence seen in *Cyanothece* ATCC 51142. In *Synechococcus* (*Cyanothece*) RF-1, it has been suggested that it is covalent modification of the Fe protein that targets it for degradation (Chow and Tabita, 1994). This is not, however, the case in *Gloeothece*, in which both modified and unmodified Fe protein appear equally susceptible to proteolytic degradation (Cheng *et al.*, 1999; Dougherty *et al.*, 1996a; Du and Gallon, 1993). The mechanism by which Fe protein is targeted for proteolytic degradation seems, therefore, to be rather complex and may vary among different non-heterocystous cyanobacteria.

3.3. Regulation of N_2 Fixation by Fixed Nitrogen

As in other diazotrophs, addition of either ammonium or nitrate to N_2-grown cultures of non-heterocystous cyanobacteria rapidly inhibits N_2 fixation. Addition of ammonium to N_2-fixing cultures of *Gloeothece* decreased nitrogenase synthesis and stimulated both degradation and covalent modification of the Fe protein (Cheng *et al.*, 1999) under both aerobic and microaerobic conditions. Inhibition by nitrate depends upon reduction of nitrate to ammonium, but ammonium itself does not directly regulate N_2 fixation. Rather, the observed effects of ammonium on N_2 fixation are exerted through changes in the intracellular concentrations of organic compounds related to ammonium assimilation. A likely candidate is glutamine (Thomas *et al.*, 1982), the immediate organic product of ammonium assimilation.

Regulation of nitrogen metabolism in non-diazotrophic unicellular cyanobacteria is under the control of the global nitrogen regulator NtcA (de Marsac *et al.*, 2001), but it is not clear whether this regulation extends to N_2 fixation. For example, NtcA is more weakly expressed in N_2-fixing cultures of *Cyanothece* than in cultures growing on either ammonium or nitrate (Bradley and Reddy, 1997). Moreover, the effects of certain inhibitors of ammonium assimilation on N_2 fixation in *Gloeothece* appear to conflict with the perceived mechanism of action of NtcA. The binding of NtcA to the promoters of NtcA-regulated genes is stimulated by 2-oxoglutarate (Tanigawa *et al.*, 2002). Because 2-oxoglutarate is the form in which organic carbon enters the glutamine synthetase/glutamate synthase (GOGAT) pathway of ammonium assimilation (see, for example, Lea, 1999), its intracellular concentration would correlate negatively with availability of ammonium. Enhancement of NtcA binding by 2-oxoglutarate may, therefore, explain how NtcA acts to switch on certain genes in response to nitrogen-starvation. Addition of ammonium to a cyanobacterial culture would result in a decrease in the concentration of intracellular 2-oxoglutarate and, therefore, in an inhibition of expression of NtcA-regulated genes. In contrast, inhibitors of glutamine synthetase, such as methionine sulphoximine (MSX), and inhibitors of GOGAT, such as azaserine and albizziine, would block ammonium assimilation and increase the intracellular concentrations of 2-oxoglutarate. These inhibitors might, therefore, be expected to stimulate the expression of NtcA-regulated genes, even in the presence of ammonium.

In *Gloeothece*, however, whereas addition of MSX allowed acetylene reduction to continue, even in the presence of ammonium, addition of either azaserine or albizziine inhibited acetylene reduction whether or not ammonium was also present (Thomas *et al.*, 1981). Although by no means conclusive, these findings do not appear consistent with a role for 2-oxoglutarate in signalling intracellular nitrogen status in *Gloeothece*. Glutamine, whose intracellular concentration increases dramatically following addition of ammonium (Flynn and Gallon, 1990) and, to a lesser extent when GOGAT is inhibited, but which decreases when glutamine synthetase is inhibited, may be a much better candidate.

Nitrite also inhibits N_2 fixation in cultures of *Gloeothece* but by a different mechanism from that of either nitrate or ammonium. For example, the effect of nitrite is not exerted *via* glutamine and does not occur under microaerobic conditions (Cheng *et al.*, 1999).

Cultures of *Trichodesmium* may simultaneously fix N_2 and take up organic and inorganic forms of combined nitrogen, at least when present at low concentrations (Mulholland and Capone, 1999; Mulholland *et al.*, 1999). This behaviour may be important for the transfer of nitrogen between N_2-fixing and non-fixing cells, and also between *Trichodesmium* and associated organisms within the aggregated bundles of individual filaments that commonly occur in natural populations of this cyanobacterium (Mulholland and Capone, 2000).

3.4. Circadian Rhythms in N_2-fixing Cyanobacteria

In most non-heterocystous cyanobacteria, modulation of N_2 fixation is under the control of an endogenous rhythm that, once entrained to a period close to 24 h, persists in cultures transferred to continuous illumination (Bergman *et al.*, 1997) or, in the case of cultures of *Cyanothece* ATCC 51142 grown heterotrophically on exogenous glycerol, to constant darkness (Schneegurt *et al.*, 2000). In contrast, *Gloeothece* shows an endogenous free-running period (FRP) of about 40 h in N_2 fixation (Mullineaux *et al.*, 1981a), photosynthesis, and respiration (Gallon, 2001), whereas the FRP in *Oscillatoria* (*Lyngbya*) UCSB8 is about 120 h (Gallon *et al.*, 1991). Although incubation under alternating light and darkness can impose a daily rhythm of N_2 fixation on these cyanobacteria, this rhythm is rapidly lost once the external signal is removed (Gallon, 2001).

For a rhythm to be classified as 'circadian', it must fulfil the following criteria (see, for example, Rönneberg and Merrow, 2001): (a) it must show sustained rhythmicity under constant conditions; (b) it should have an FRP close to 24 h; (c) it should be entrainable by environmental signals, such as light; and (d) it should be temperature compensated (*i.e.*, the rhythm should be the same at different constant temperatures). Only in *Synechococcus* (*Cyanothece*) RF-1 have all these criteria been fulfilled for N_2 fixation (Grobbelaar and Huang, 1992; Grobbelaar *et al.*, 1986; Huang and Chow, 1990; Huang and Grobbelaar, 1995), but there is no reason to doubt that the same is true for other strains of *Cyanothece* (Colón-López *et al.*, 1997; Mitsui *et al.*, 1987) and other non-heterocystous cyanobacteria, such as *O. limosa* (Stal and Krumbein, 1985), *P. boryanum* (Misra and Tuli, 2000) and *Trichodesmium* (Chen *et al.*, 1998, 1999).

A circadian clock consists of at least three fully integrated components, which are a pacemaker, an input pathway, and an output pathway. The pacemaker generates a rhythm through a combination of positive and negative regulatory circuits (Bell-Pedersen *et al.*, 2001; Miyoshi *et al.*, 2001; Ruoff *et al.*, 2001). This rhythm does not have to be circadian. The input pathway effectively entrains the pacemaker to a 24-h cycle and so represents a link between an external environmental signal (such as light) and the pacemaker. The output pathway generates one or more specific rhythms under the influence of the pacemaker. In order to explain some of the properties of clock mutants in various organisms, it has been further proposed that the clock may have other elements, such as a feedback loop between the pacemaker-generated rhythm and the input pathway, and one or more inputs from non-rhythmic elements (Rönneberg and Merrow, 2001).

A model for circadian regulation of gene expression in cyanobacteria has been proposed for the unicellular cyanobacterium, *Synechococcus elongatus* PCC 7942 (Golden *et al.*, 1997; Johnson and Golden, 1999). Although this particular cyanobacterium does not fix N_2, this model could easily apply to those unicellular cyanobacteria, like *Cyanothece*, that exhibit a circadian rhythm in N_2 fixation. The model assumes the existence of a pacemaker, which is set initially by exposure to alternating light and darkness (presumably *via* a photoreceptor). This pacemaker has an output signal that affects gene transcription within the cell. However, additional pathways exist to adjust both phase (*i.e.*, <u>when</u> during the diurnal cycle a specific gene is transcribed) and amplitude (the <u>magnitude</u> of the daily oscillation). In this way, the single pacemaker controls the oscillations of a variety of biochemical pathways.

In *S. elongatus*, most, if not all, of the genes are transcribed with a circadian rhythm. They are conventionally classified as dusk-peaking (Class 1) and dawn-peaking genes (Class 2). Class 1 genes are expressed maximally about 12 h after transfer from alternating light and darkness to continuous illumination (*i.e.*, as cells would normally be going into the dark), whereas Class 2 genes are transcribed maximally 12 h later. Class 1 genes are relatively common and include *psbA* (see Section 2.5.3.), whereas Class 2 genes (*e.g.*, *opcA*, which encodes an activator of glucose-6-phosphate dehydrogenase) are much rarer. In N_2-fixing strains of *Cyanothece* (Huang *et al.*, 1988), nitrogenase peaks during the early part of the subjective 'dark' period, so the *nif* genes appear to be closer to Class 1 than to Class 2.

Studies using mutants of *S. elongatus* have implicated three genes, *kaiA*, *kaiB* and *kaiC*, in the imposition of circadian rhythms (Figure 4). The gene *kaiA* is transcribed as a monocistronic operon, but *kaiBC* are transcribed together. Both operons show cyclic transcription that is abolished by mutation of any one of the *kai* genes. Continuous overexpression of the *kaiC* gene product represses the promoter of *kaiBC* (negative feedback), whereas overexpression of *kaiA* enhances it (positive feedback). These effects are sufficient on their own to sustain a rhythmic reaction. Transient expression of *kaiC* (lasting no more than a few hours) resets the clock. So, the level of expression of KaiC (the product of *kaiC*) is directly linked to the phase of the oscillation.

Interactions among KaiA, KaiB and KaiC (Figure 4) are important in maintaining both clock function and signal output (Iwasaki *et al.*, 1999; Taniguchi *et al.*, 2001). Precise details of these interactions are not yet clear and it has recently emerged that they may involve another protein, SasA. SasA, which acts as an enhancer of the rhythmic output rather than as an essential component of the clock, is a sensory histidine kinase that interacts with KaiC (Iwasaki *et al.*, 2000). KaiC has been shown to bind ATP and GTP and to undergo autophosphorylation. In this and other structural features, KaiC resembles switch proteins (like Ras) that are involved in signal transduction. During the day, KaiC exists as a complex (perhaps a homo-hexamer) of M_r = 350 000 (Kageyama *et al.*, 2003). KaiA, possibly along with KaiB, occurs as a separate, small (M_r = 60 000) complex. One or both of these complexes may act as a positive regulator of *kaiBC* expression. At night, however, KaiA, KaiB and KaiC probably form part of a single, much larger complex (M_r = 500 000 - 670 000), which also contains SasA and may act as a negative regulator of *kaiBC* expression. The formation and breakdown of this large complex have been proposed as core events in establishing the endogenous circadian rhythm in *S. elongatus* (Kageyama *et al.*, 2003). The *kaiABC* gene cluster has been found in a range of heterocystous and non-heterocystous cyanobacteria, including N_2-fixing genera such as *Lyngbya*, *Leptolyngbya* and *Oscillatoria* (Lorne *et al.*, 2000), so circadian rhythms may be widespread, if not universal, in these organisms.

A candidate for the input regulator is CikA, a bacteriophytochrome, recently shown to occur in *S. elongatus* (Schmitz *et al.*, 2000). CikA has a histidine protein kinase motif and sequence similarity to phytochrome in higher plants (for more information about phytochrome, see Møller, 2002). Although CikA seems to play a key role in conveying signals to the clock, it fails to form a photo-reactive holoprotein with a bilin cofactor in the laboratory and may not be a photoreceptor itself (Mutsuda *et al.*, 2003). On the output side, there is evidence that circadian gene expression is modulated by several group 2 sigma factors (Nair *et al.*, 2002; Tsinoremas *et al.*, 1996). Disruption of individual sigma factor genes does not abolish rhythmicity, but affects period, phase, or amplitude of expression from subsets of genes. Another protein, CpmA (circadian phase modifier), regulates the phase of expression of at least three operons, *psbAI*, *psbAII* and *kaiA*, by up to 10 h (Katayama *et al.*, 1999).

In growth studies under various alternating cycles of light and darkness, using strains of *S. elongatus* with FRPs ranging from 23 h to 30 h, it emerged that the strain possessing the FRP closest to the prevailing pattern of illumination consistently outperformed strains whose FRP was more distant from this pattern (Ouyang *et al.*, 1998). Thus, all other factors being equal (admittedly unlikely), an illumination pattern of 12-h light and 12-h darkness would favour *Cyanothece* (FRP for N_2 fixation = 24 h) over *Gloeothece* (FRP for N_2 fixation = 40 h).

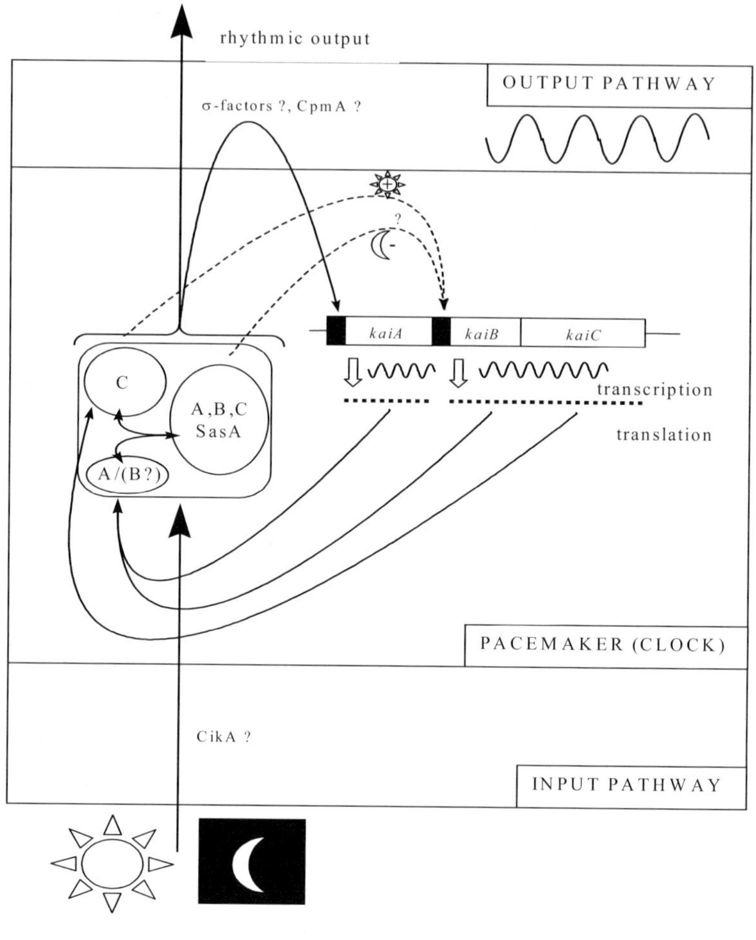

Figure 4. A model for the generation of circadian rhythms in cyanobacteria (based on Johnson and Golden, 1999). The operons kaiA *and* kaiBC *are expressed rhythmically and the latter is subject to positive and, perhaps, negative feedback regulation by complexes involving the gene products KaiA, KaiB and KaiC, along with SasA. These proteins interact in a variety of ways in order to generate the rhythmic signal output. For further details, see text.*

At the time of writing, no genome sequence data are available for any non-heterocystous N_2-fixing cyanobacteria, although sequences are (or will shortly become) available for related organisms, such as the unicellular but non-diazotrophic cyanobacterium *Synechocystis* PCC 6803 (Nakamura *et al.*, 1998), a number of

Synechococcus species (including PCC 7942), *Microcystis aeruginosa*, *Gloeobacter violaceus* PCC 7421, and the heterocystous cyanobacteria, *Anabaena* PCC 7120 (Kaneko *et al.*, 2001) and *Nostoc punctiforme* ATCC 29133 (Meeks *et al.*, 2001). Many of these can be found at http://www.kazusa.or.jp/cyanobase. An active programme to sequence the genome of *Trichodesmium* is, however, under way (see http://genome.ornl.gov/microbial/tery) and, given the increasing ease and decreasing cost with which genomes can be sequenced, it is a reasonable expectation that genome sequences for a number of non-heterocystous, N_2-fixing cyanobacteria will soon become available. These data, coupled with the burgeoning application of proteomics in cyanobacterial research, will then open up the prospect of addressing fundamental problems, such as the identity, and global regulation, of genes that are expressed at specific times during growth under alternating light and darkness, the nature of their gene products, and the probing of circadian clocks in N_2-fixing cyanobacteria.

ACKNOWLEDGEMENTS

I thank Sue Golden for her valuable help and advice with the section on circadian rhythms in cyanobacteria and also for making information available to me prior to publication.

REFERENCES

Albertano, P., and Urzi, C. (1999). Structural interactions among epilithic cyanobacteria and heterotrophic microorganisms in Roman hypogea. *Microbial. Ecol., 38*, 244-252.

Anagnostidis, K., and Komárek, J. (1988). Modern approach to the classification system of cyanophytes. 3. Oscillatoriales. *Arch. Hydrobiol., 80*, 327-472.

Bell-Pedersen, D., Crosthwaite, S. K., Lakin-Thomas, P. L., Merrow, M., and Okland, M. (2001). The *Neurospora* circadian clock: simple or complex? *Phil. Trans. Royal Soc. Lond. B, 356*, 1697-1709.

Bergman, B., and Carpenter, E. J. (1991). Nitrogenase confined to randomly distributed trichomes in the marine cyanobacterium *Trichodesmium*. *J. Phycol., 27*, 158-165.

Bergman, B., Gallon, J. R., Rai, A. N., and Stal, L. J. (1997). N_2 fixation by non-heterocystous cyanobacteria. *FEMS Microbiol. Rev., 19*, 139-185.

Berman-Frank, I., Lundgren, P., Chen, Y.-B., Kupper, H., Kolber, Z., Bergman, B., *et al.* (2001). Segregation of nitrogen fixation and oxygenic photosynthesis in the marine cyanobacterium *Trichodesmium*. *Science, 294*, 1534-1537.

Bradley, R. L., and Reddy, K. J. (1997). Cloning, sequencing, and regulation of the global nitrogen regulator gene *ntcA* in the unicellular diazotrophic cyanobacterium *Cyanothece* sp. strain BH68K. *J. Bacteriol., 179*, 4407-4410.

Burris, R. H. (1979). Inhibition. In R. W. F. Hardy and R. C. Burns (Eds.), *A treatise on dinitrogen fixation* (Vol. 2, pp. 569-604). New York: Wiley.

Burris, R. H., Eppeling, F. J., Wahlin, H. B., and Wilson, P. W. (1943). Detection of nitrogen fixation with isotopic nitrogen. *J. Biol. Chem., 148*, 349-357.

Capone, D. G., O'Neil, J. M., Zehr, J., and Carpenter, E. J. (1990). Basis for diel variation in nitrogenase activity in the marine planktonic cyanobacterium *Trichodesmium thiebautii*. *Appl. Environ. Microbiol., 56*, 3532-3536.

Capone, D. G., Zehr, J. P., Paerl, H. W., Bergman, B., and Carpenter, E. J. (1997). *Trichodesmium*: A globally significant marine cyanobacterium. *Science, 276*, 1221-1229.

Castenholz, R. W. (2001). Phylum BX. Cyanobacteria. Oxygenic photosynthetic Bacteria. In D. R. Boone and R. W. Castenholz (Eds.), *Bergey's manual of systematic bacteriology* (Vol. 1, pp. 473-599). New York: Springer.

Castenholz, R. W., and Waterbury, J. B. (1989). Group 1. Cyanobacteria. In R. G. E. Murray, D. J. Brenner, M. P. Bryant, J. G. Holt, N. R. Krieg, J. W. Moulder, *et al.* (Eds.), *Bergey's manual of systematic bacteriology* (Vol. 3, pp. 1710-1799). Baltimore: Williams and Wilkins.

Chen, Y.-B., Dominic, B., Mellon, M. T., and Zehr, J. P. (1998). Circadian rhythm of nitrogenase gene expression in the diazotrophic filamentous nonheterocystous cyanobacterium *Trichodesmium* sp. strain IMS 101. *J. Bacteriol., 180*, 3598-3605.

Chen, Y.-B., Dominic, B., Zani, S., Mellon, M. T., and Zehr, J. P. (1999). Expression of photosynthesis genes in relation to nitrogen fixation in the diazotrophic filamentous nonheterocystous cyanobacterium *Trichodesmium* sp IMS 101. *Plant Mol. Biol., 41*, 89-104.

Cheng, J., Hipkin, C. R., and Gallon, J. R. (1999). Effects of inorganic nitrogen compounds on the activity and synthesis of nitrogenase in *Gloeothece* (Nägeli) sp. ATCC 27152. *New Phytol., 141*, 61-70.

Chow, T. J., and Tabita, F. R. (1994). Reciprocal light-dark transcriptional control of *nif* and *rbc* expression and light-dependent posttranslational control of nitrogenase activity in *Synechococcus* sp strain RF-1. *J. Bacteriol., 176*, 6281-6285.

Colón-López, M. S., Sherman, D., and Sherman, L. A. (1997). Transcriptional and translational regulation of nitrogenase in light-dark- and continuous-light-grown cultures of the unicellular cyanobacterium *Cyanothece* sp. strain ATCC 51142. *J. Bacteriol., 179*, 4319-4327.

Colón-López, M. S., Tang, H. Y., Tucker, D. L., and Sherman, L. A. (1999). Analysis of the *nifHDK* operon and structure of the NifH protein from the unicellular, diazotrophic cyanobacterium, *Cyanothece* strain sp ATCC 51142. *Biochim. Biophys. Acta, 1473*, 363-375.

Coute, A., Tell, G., and Therezien, Y. (1999). Aerophytic cyanobacteria from New Caledonia. *Cryptogamie Algol., 20*, 301-344.

de Marsac, N. T., Lee, H. M., Hisbergues, M., Castets, A. M., and Bedu, S. (2001). Control of nitrogen and carbon metabolism in cyanobacteria. *J. Appl. Phycol., 13*, 287-292.

Dixon, R. A., Cheng, Q., Shen, G.-F., Day, A., and Dowson-Day, M. (1997). *Nif* gene transfer and expression in chloroplasts: Prospects and problems. *Plant Soil, 194*, 193-203.

Dominic, B., Chen, Y.-B., and Zehr, J. P. (1998). Cloning and transcriptional analysis of the *nifUHDK* genes of *Trichodesmium* sp. IMS101 reveals stable *nifD*, *nifDK* and *nifK* transcripts. *Microbiol. UK, 144*, 3359-3368.

Dougherty, L. J., Brown, E. G., and Gallon, J. R. (1996a). Nitrogenase-specific proteolytic activity in the unicellular cyanobacterium *Gloeothece*. *Biochem. Soc. Trans., 24*, 401S.

Dougherty, L. J., Brown, E. G., Gallon, J. R., Larsen, M., Moertz, E., and Roepstorff, P. (1996b). Modification of the Fe-protein of *Gloeothece* nitrogenase: A mass spectrometric study. *Biochem. Soc. Trans., 24*, 477S.

Drewes, K. (1928). Über die Assimilation des Luftstickstoffs durch Blaualgen. *Zbl. Bakt. Parasit., Abt II, 76*, 88-101.

Du, C., and Gallon, J. R. (1993). Modification of the Fe protein of the nitrogenase of *Gloeothece* (Nägeli) sp. ATCC 27152 during growth under alternating light and darkness. *New Phytol., 125*, 121-129.

Dugdale, R. C., Menzel, D. W., and Ryther, J. H. (1961). Nitrogen fixation in the Sargasso Sea. *Deep-Sea Res., 7*, 297-300.

Ernst, A., Liu, Y.-D., Reich, S., and Böger, P. (1990). Diurnal nitrogenase modification in the cyanobacterium *Anabaena variabilis*. *Bot. Acta, 103*, 183-189.

Evans, A. M., Gallon, J. R., Jones, A., Staal, M., Stal, L. J., Villbrandt, M., *et al.* (2000). Nitrogen fixation by Baltic cyanobacteria is adapted to the prevailing photon flux density. *New Phytol., 147*, 285-297.

Fay, P. (1992). Oxygen relations of nitrogen fixation in cyanobacteria. *Microbiol. Rev., 56*, 340-373.

Fernandez-Valiente, E., Quesada, A., Howard-Williams, C., and Hawes, I. (2001). N_2 fixation in cyanobacterial mats from ponds on the McMurdo Ice Shelf, Antarctica. *Microbial. Ecol., 42*, 338-349.

Flynn, K. J., and Gallon, J. R. (1990). Changes in intracellular and extracellular α-amino acids in *Gloeothece* during N_2 fixation and following addition of ammonium. *Arch. Microbiol., 153*, 574-579.

Fogg, G. E. (1942). Studies on nitrogen fixation by blue-green algae. I. Nitrogen fixation by *Anabaena cylindrica* Lemm. *J. Exp. Bot., 19*, 78-87.

Frank, B. (1889). Über den experimentellen Nachweis der Assimilation freien Stickstoffs durch erdbodenwohnende Algen. *Bericht der Deutschen botanischen Gesellschaft, 7*, 24-42.

Fredriksson, C., and Bergman, B. (1995). Nitrogenase quantity varies diurnally in a subset of cells within colonies of the non-heterocystous cyanobacteria *Trichodesmium* spp. *Microbiol. UK, 141*, 2471-2478.

Fredriksson, C., Malin, G., Siddiqui, P. J. A., and Bergman, B. (1998). Aerobic nitrogen fixation is confined to a subset of cells in the non-heterocystous cyanobacterium *Symploca* PCC 8002. *New Phytol., 140*, 531-538.

Friedmann, E. I., and Galun, M. (1974). Desert algae, lichens and fungi. In G. W. Brown (Ed.), *Desert biology* (Vol. 2, pp. 166-212). New York: Academic Press.

Friedmann, E. I., and Ocampo, R. (1976). Endolithic blue-green algae in the dry valleys: Primary producers in the Antarctic desert ecosystem. *Science, 193*, 1247-1249.

Fujita, Y., Takahashi, Y., Shonai, F., Ogura, Y., and Matsubara, H. (1991). Cloning, nucleotide sequences and differential expression of the *nif H* and *nif H*-like (*frx C*) genes from the filamentous nitrogen-fixing cyanobacterium *Plectonema boryanum*. *Plant Cell Physiol., 32*, 1093-1106.

Gallon, J. R. (1992). Tansley Review No. 44. Reconciling the incompatible: N_2 fixation and O_2. *New Phytol., 122*, 571-609.

Gallon, J. R. (2001). N_2 fixation in phototrophs: Adaptation to a specialized way of life. *Plant Soil, 230*, 39-48.

Gallon, J. R., Cheng, J., Dougherty, L. J., Gallon, V. A., Hilz, H., Pederson, D. M., *et al.* (2000). A novel covalent modification of nitrogenase in a cyanobacterium. *FEBS Lett., 468*, 231-233.

Gallon, J. R., Evans, A. M., Jones, D. A., Albertano, P., Congestri, R., Bergman, B., *et al.* (2002). Maximum rates of N_2 fixation and primary production are out of phase in a developing cyanobacterial bloom in the Baltic Sea. *Limnol. Oceanog., 47*, 1514-1521.

Gallon, J. R., and Hamadi, A. F. (1984). Studies on the effects of oxygen on acetylene reduction (nitrogen fixation) in *Gloeothece* sp. ATCC 27152. *J. Gen. Microbiol., 130*, 495-503.

Gallon, J. R., Hashem, M. A., and Chaplin, A. E. (1991). Nitrogen fixation by *Oscillatoria* spp. under autotrophic and photoheterotrophic conditions. *J. Gen. Microbiol., 137*, 31-39.

Gallon, J. R., Jones, D. A., and Page, T. S. (1996). *Trichodesmium*, the paradoxical diazotroph. *Arch. Hydrobiol., 117 (supplement: Algological Studies 83)*, 215-243.

Gallon, J. R., LaRue, T. A., and Kurz, W. G. W. (1974). Photosynthesis and nitrogenase activity in the blue-green alga *Gloeocapsa*. *Can. J. Microbiol., 20*, 1633-1637.

Gallon, J. R., Pederson, D. M., and Smith, G. D. (1993). The effect of temperature on the sensitivity of nitrogenase to oxygen in the cyanobacteria *Anabaena cylindrica* (Lemmermann) and *Gloeothece* (Nägeli). *New Phytol., 124*, 251-257.

Gallon, J. R., Perry, S. M., and Flynn, K. J. (1990). The immediate fate of N_2 fixed by *Gloeothece* growing under alternating light and darkness. In P. M. Gresshoff, L. E. Roth, G. Stacey and W. E. Newton (Eds.), *Nitrogen fixation: Achievements and objectives* (p. 534). New York: Chapman and Hall.

Gallon, J. R., Perry, S. M., Rajab, T. M. A., Flayeh, K. A. M., Yunes, J. S., and Chaplin, A. E. (1988). Metabolic changes associated with the diurnal pattern of N_2 fixation in *Gloeothece*. *J. Gen. Microbiol., 134*, 3079-3087.

Gallon, J. R., Reade, J. P. H., Dougherty, L. J., Pederson, D. M., and Rogers, L. J. (1995). Degradation of nitrogenase in *Gloeothece*. In I. A. Tikhonovich, N. A. Provorov, V. I. Romanov and W. E. Newton (Eds.), *Nitrogen fixation: Fundamentals and applications* (p. 230). Dordrecht: Kluwer.

Golden, S. S., Ishiura, M., Johnson, C. H., and Kondo, T. (1997). Cyanobacterial circadian rhythms. *Annu. Rev. Plant Physiol. Plant Mol. Biol., 48*, 327-354.

Gregory, P. H., Hamilton, E. D., and Sreeramulu, T. (1955). Occurrence of the alga *Gloeocapsa* in the air. *Nature, 176*, 1270.

Griffiths, M. S. H., Gallon, J. R., and Chaplin, A. E. (1987). The diurnal pattern of dinitrogen fixation by cyanobacteria *in situ*. *New Phytol., 107*, 649-657.

Grobbelaar, N., and Huang, T.-C. (1992). Effect of oxygen and temperature on the induction of a circadian rhythm in *Synechococcus* RF-1. *J. Plant Physiol., 140*, 391-394.

Grobbelaar, N., Huang, T. C., Lin, H. Y., and Chow, T. J. (1986). Dinitrogen-fixing endogenous rhythm in *Synechococcus* RF-1. *FEMS Microbiol. Lett., 37*, 173-177.

Hawser, S. P., Carpenter, E. J., Codd, G. A., and Capone, D. G. (1991). A neurotoxic factor associated with the bloom-forming cyanobacterium *Trichodesmium*. *Toxicon, 29*, 277-278.

Howard, J. B., and Rees, D. C. (1994). Nitrogenase: a nucleotide-dependent molecular switch. *Annu. Rev. Biochem., 63*, 235-264.

Huang, T.-C., and Chow, T.-J. (1986). New type of N_2-fixing unicellular cyanobacterium. *FEMS Microbiol. Lett., 50*, 127-130.

Huang, T.-C., and Chow, T.-J. (1988). Comparative studies of some nitrogen-fixing unicellular cyanobacteria isolated from rice fields. *J. Gen. Microbiol., 134*, 3089-3097.

Huang, T.-C., and Chow, T.-J. (1990). Characterization of the rhythmic nitrogen-fixing activity of *Synechococcus* sp. RF-1 at the transcription level. *Curr. Microbiol., 20*, 23-26.

Huang, T.-C., Chow, T.-J., and Hwang, I.-S. (1988). The cyclic synthesis of the nitrogenase of *Synechococcus* RF-1 and its control at the transcription level. *FEMS Microbiol. Lett., 50*, 127-130.

Huang, T.-C., and Grobbelaar, N. (1995). The circadian clock in the prokaryote *Synechococcus* RF-1. *Microbiol. UK, 141*, 535-540.

Huang, T.-C., Lin, R.-F., Chu, M.-K., and Chen, H.-M. (1999). Organization and expression of nitrogen-fixation genes in the aerobic nitrogen-fixing unicellular cyanobacterium *Synechococcus* sp. strain RF-1. *Microbiol. UK, 145*, 743-753.

Ikemoto, H., and Mitsui, A. (1994). Diazotrophic synchronous growth of a marine unicellular cyanobacterium, *Synechococcus* sp strain Miami BG 043511, under aerobic and microaerobic/anaerobic conditions. *Microbiol. UK, 140*, 2153-2158.

Issa, O. M., Stal, L. J., Defarge, C., Coute, A., and Trichet, J. (2001). Nitrogen fixation by microbial crusts from desiccated Sahelian soils (Niger). *Soil Biol. Biochem., 33*, 1425-1428.

Iwasaki, H., Taniguchi, Y., Ishiura, M., and Kondo, T. (1999). Physical interactions among circadian clock proteins KaiA, KaiB and KaiC in cyanobacteria. *EMBO J., 18*, 1137-1135.

Iwasaki, H., Williams, S. B., Kitayama, Y., Ishiura, M., Golden, S. S., and Kondo, T. (2000). A KaiC-interacting sensory histidine kinase, SasA, necessary to sustain robust circadian oscillation in cyanobacteria. *Cell, 101*, 223-233.

Janson, S., Carpenter, E. J., and Bergman, B. (1994). Compartmentalization of nitrogenase in a nonheterocystous cyanobacterium: *Trichodesmium contortum*. *FEMS Microbiol. Lett., 118*, 9-14.

Johnson, C. H., and Golden, S. S. (1999). Circadian programmes in cyanobacteria. *Annu. Rev. Microbiol., 53*, 389-409.

Jones, K. (1990). Aerobic nitrogen fixation by *Lyngbya* sp., a marine tropical cyanobacterium. *Br. Phycol. J., 25*, 287-289.

Kageyama, H., Kondo, T., and Iwasaki, H. (2003). Circadian formation of clock protein complexes by KaiA, KaiB, KaiC and SasA in cyanobacteria. *J. Biol. Chem., 278*, 2388-2395.

Kaneko, T., Nakamura, Y., Wolk, C. P., Kuritz, T., S., S., Watanabe, A., *et al.* (2001). Complete genome sequence of the filamentous nitrogen-fixing cyanobacterium *Anabaena* sp PCC 7120. *DNA Res., 8*, 205-213.

Katayama, M., Tsinoremas, N. F., Kondo, T., and Golden, S. S. (1999). *CpmA*, a gene involved in an output pathway of the cyanobacterial circadian system. *J. Bacteriol., 181*, 3516-3524.

Khamees, H. S., Gallon, J. R., and Chaplin, A. E. (1987). The pattern of acetylene reduction by cyanobacteria grown under alternating light and darkness. *Br. Phycol. J., 22*, 55-60.

Kumazawa, S., Yumura, S., and Yoshisuji, H. (2001). Photoautotrophic growth of a recently isolated N_2-fixing marine non-heterocystous filamentous cyanobacterium, *Symploca* sp (Cyanobacteria). *J. Phycol., 37*, 482-487.

Laamanen, M. (1996). Cyanoprokaryotes in the Baltic Sea ice and winter plankton. *Arch. Hydrobiol., 117 (supplement: Algological Studies 83)*, 423-433.

Lea, P. J. (1999). Nitrogen metabolism. In P. J. Lea and R. C. Leegood (Eds.), *Plant biochemistry and molecular biology* (2nd ed., pp. 163-191). Chichester: John Wiley.

Li, H., Sherman, D. M., Bao, S. L., and Sherman, L. A. (2001). Pattern of cyanophycin accumulation in nitrogen-fixing and non-nitrogen-fixing cyanobacteria. *Arch. Microbiol., 176*, 9-18.

Lin, S., Henze, S., Lundgren, P., Bergman, B., and Carpenter, E. J. (1998). Whole-cell immunolocalization of nitrogenase in marine diazotrophic cyanobacteria, *Trichodesmium* spp. *Appl. Environ. Microbiol., 64*, 3052-3058.

Lorne, J., Scheffer, J., Lee, A., Painter, M., and Miao, V. P. W. (2000). Genes controlling circadian rhythm are widely distributed in cyanobacteria. *FEMS Microbiol. Lett., 189*, 129-133.

Lundgren, P. (2001). *Marine non-heterocystous cyanobacteria: diazotrophic characterization and molecular phylogeny*. PhD thesis, University of Stockholm, Stockholm.

Lundgren, P., Söderbäck, E., Singer, A., Carpenter, E. J., and Bergman, B. (2001). *Katagnymene*: characterization of a novel marine diazotroph. *J. Phycol., 37*, 10521062.

Maryan, P. S., Eady, R. R., Chaplin, A. E., and Gallon, J. R. (1986). Nitrogen fixation by the unicellular cyanobacterium *Gloeothece*. Nitrogenase synthesis is only transiently repressed by O_2. *FEMS Microbiol. Lett., 34*, 251-255.

Meeks, J. C., Elhai, J., Thiel, T., Potts, M., Larimer, F., Lamerdin, J., *et al.* (2001). An overview of the genome of *Nostoc punctiforme*, a multicellular, symbiotic cyanobacterium. *Photosynth. Res., 70*, 85-106.

Meunier, P. C., Colón-López, M. S., and Sherman, L. A. (1997). Temporal changes in state transitions and photosystem organization in the unicellular, diazotrophic cyanobacterium *Cyanothece* sp. ATCC 51142. *Plant Physiol., 115*, 991-1000.

Meunier, P. C., Colón-López, M. S., and Sherman, L. A. (1998). Photosystem II cyclic heterogeneity and photoactivation in the diazotrophic, unicellular cyanobacterium *Cyanothece* species ATCC 51142. *Plant Physiol., 116*, 1551-1562.

Misra, H. S. (1999). Oxygen implication in the diazotrophic growth of *Plectonema boryanum* in dark-light cycles. *Plant Science, 143*, 135-142.

Misra, H. S., and Tuli, R. (1994). Nitrogen fixation by *Plectonema boryanum*: a photosystem-II independent component. *Microbiol. UK, 140*, 971-976.

Misra, H. S., and Tuli, R. (2000). Differential expression of photosynthesis and nitrogen fixation genes in the cyanobacterium *Plectonema boryanum*. *Plant Physiol., 122*, 731-736.

Mitsui, A., Cao, S., Takahashi, A., and Arai, T. (1987). Growth synchrony and cellular parameters of the unicellular nitrogen-fixing marine cyanobacterium, *Synechococcus* sp. strain Miami BG 043511 under continuous illumination. *Physiol. Plantarum, 69*, 1-8.

Mitsui, A., Kumazawa, S., Takahashi, A., Ikemoto, H., Cao, S., and Arai, T. (1986). Strategy by which nitrogen-fixing unicellular cyanobacteria grow photoautotrophically. *Nature, 323*, 720-722.

Miyoshi, F., Nakayama, F., and Tomita, M. (2001). Estimation of genetic networks of the circadian rhythm in cyanobacterium using the E-CELL system. *Genome Informatics, 12*, 308-309.

Møller, S. G., Ingles, P. J., and Whitelam, G. C. (2002). Tansley Review No. 136. The cell biology of phytochrome signalling. *New Phytol., 154*, 553-590.

Mulholland, M. R., and Capone, D. G. (1999). Nitrogen fixation, uptake and metabolism in natural and cultured populations of *Trichodesmium* spp. *Mar. Ecol. Prog. Ser., 188*, 33-49.

Mulholland, M. R., and Capone, D. G. (2000). The nitrogen physiology of the marine N_2-fixing cyanobacteria *Trichodesmium* spp. *Trends Pl. Sci., 5*, 148-153.

Mulholland, M. R., Ohki, K., and Capone, D. G. (1999). Nitrogen utilization and metabolism relative to patterns of N_2 fixation in cultures of *Trichodesmium* NIBB1067. *J. Phycol., 35*, 977-988.

Mullineaux, P. M., Gallon, J. R., and Chaplin, A. E. (1981a). Acetylene reduction (nitrogen fixation) by cyanobacteria grown under alternating light-dark cycles. *10*, 245-247.

Mullineaux, P. M., Gallon, J. R., and Chaplin, A. E. (1981b). Nitrogen fixation in cultures of the cyanobacterium *Gloeocapsa* (*Gloeothece*) sp. 1430/3 incubated in the dark. *J. Gen. Microbiol., 124*, 141-146.

Musgrave, S. C., Kerby, N. M., Codd, G. A., and Stewart, W. D. P. (1982). Sustained ammonia production by immobilized filaments of the nitrogen-fixing cyanobacterium *Anabaena* 27893. *Biotechnol. Lett., 4*, 647-652.

Mutsuda, M., Michel, K-P., Zhang, X., Montgomery, B. L. and Golden, S. S. (2003). Biochemical properties of CikA, an unusual phytochrome-like histidine protein kinase that resets the circadian clock in *Synechococcus elongatus* PCC 7942. *J. Biol. Chem., 278*, 19102-19110.

Nair, U., Ditty, J. L., Min, H. and Golden, S. S. (2002). Roles for sigma factors in global circadian regulation of the cyanobacterial genome. *J. Bacteriol. 184*, 3530-3538.

Nakamura, Y., Kaneko, T., Hirosawa, M., Miyajima, N., and Tabata, S. (1998). CyanoBase, a WWW database containing the complete nucleotide sequence of the genome of *Synechocystis* sp. strain PCC6803. *Nucl. Acids Res., 26*, 63-67.

Ochiai, H., Shibata, H., Sawa, Y., Shoga, M., and Ohta, S. (1983). Properties of semiconductors coated with living films of cyanobacteria. *Appl. Biochem. Biotech, 8*, 289-303.

Ohki, A., Zehr, J. P., and Fujita, Y. (1992). Regulation of nitrogenase activity in relation to the light-dark regime in the filamentous non-heterocystous cyanobacterium *Trichodesmium* sp. NIBB 1067. *J. Gen. Microbiol., 138*, 2679-2685.

Orcutt, K. M., Rasmussen, U., Webb, E. A., Waterbury, J. B., Gundersen, K., and Bergman, B. (2002). Characterization of *Trichodesmium* spp. by genetic techniques. *Appl. Environ. Microbiol., 68*, 2236-2245.

Ortega-Calvo, J.-J., and Stal, L. J. (1991). Diazotrophic growth of the unicellular cyanobacterium *Gloeothece* sp. PCC 6909 in continuous culture. *J. Gen. Microbiol., 137*, 1789-1797.

Ouyang, Y., Andersson, C. R., Kondo, T., Golden, S. S., and Johnson, C. H. (1998). Resonating circadian clocks enhance fitness in cyanobacteria. *Proc. Natl. Acad. Sci. U.S.A., 95*, 8660-8664.

Paerl, H. W., Prufert, L. E., and Ambrose, W. W. (1991). Contemporaneous N_2 fixation and oxygenic photosynthesis in the non-heterocystous mat-forming cyanobacterium *Lyngbya aestuarii*. *Appl. Environ. Microbiol., 57*, 3086-3092.

Page, T. S., and Gallon, J. R. (1992). DNA topoisomerases in *Gloeothece* (Nägeli) sp. ATCC 27152: a role in N_2 fixation? *New Phytol., 121*, 33-42.

Pandey, K. D., Kashyap, A. K., and Gupta, R. K. (2000). Nitrogen-fixation by non-heterocystous cyanobacteria in an antarctic ecosystem. *Israel J. Pl. Sci., 48*, 267-270.

Phlips, E. J., and Mitsui, A. (1986). Characterization and optimization of hydrogen production by a saltwater blue-green alga *Oscillatoria* sp Miami BG 7. 2. Use of immobilization for enhancement of hydrogen production. *Int. J. Hydrogen Energy, 11*, 83-89.

Prantl, K. A. E. (1889). Die Assimilation freien Stickstoffs und der Parasitismus von *Nostoc*. *Hedwigia, 28*, 135-136.

Rai, A. N., Borthakur, M., and Bergman, B. (1992). Nitrogenase derepression, its regulation and metabolic changes associated with diazotrophy in the non-heterocystous cyanobacterium *Plectonema boryanum* PCC 73110. *J. Gen. Microbiol., 138*, 481-491.

Reade, J. P. H., Dougherty, L. J., Rogers, L. J., and Gallon, J. R. (1999). Synthesis and proteolytic degradation of nitrogenase in cultures of the unicellular cyanobacterium *Gloeothece* strain ATCC 27152. *Microbiol. UK, 145*, 1749-1758.

Rippka, R., and Cohen-Bazire, G. (1983). The Cyanobacteriales: A legitimate order based on the type strain *Cyanobacterium stanieri*? *Ann. Microbiol. (Inst. Pasteur), 134B*, 21-36.

Rippka, R., Deruelles, J., Waterbury, J. B., Herdman, M., and Stanier, R. Y. (1979). Generic assignments, strain histories and properties of pure cultures of cyanobacteria. *J. Gen. Microbiol., 111*, 1-61.

Rönneberg, T., and Merrow, M. (2001). Circadian systems: different levels of complexity. *Phil. Trans. Royal Soc. Lond. B, 356*, 1687-1696.

Ruoff, P., Vinsjevik, M., Monnerjahn, C., and Rensing, L. (2001). The Goodwin model: Simulating the effect of light pulses on the circadian sporulation rhythm of *Neurospora crassa*. *J. Theor. Biol., 209*, 29-42.

Schmitz, O., Katayama, M., Williams, S. B., Kondo, T., and Golden, S. S. (2000). CikA, a bacteriophytochrome that resets the cyanobacterial circadian clock. *Science, 289*, 765-768.

Schneegurt, M. A., Tucker, D. L., Ondr, J. K., Sherman, D. M., and Sherman, L. A. (2000). Metabolic rhythms of a diazotrophic cyanobacterium, *Cyanothece* sp strain ATCC 51142, heterotrophically grown in continuous dark. *J. Phycol., 36*, 107-117.

Sherman, L. A., Meunier, P., and Colón-López, M. S. (1998). Diurnal rhythms in metabolism: a day in the life of a unicellular, diazotrophic cyanobacterium. *Photosynth. Res., 58*, 25-42.

Siddiqui, P. J. A., Carpenter, E. J., and Bergman, B. (1992). Ultrastructure and immunolocalization of phycobiliproteins and ribulose 1,5-bisphosphate carboxylase/oxygenase in the marine cyanobacterium *Trichodesmium thiebautii*. *J. Phycol., 28*, 320-327.

Sroga, G. E., Landegren, U., Bergman, B., and Lagerström-Fermér, M. (1996). Isolation of *nifH* and part of *nifD* by modified capture polymerase chain reaction from a natural population of the marine cyanobacterium *Trichodesmium* sp. *FEMS Microbiol. Lett., 136*, 137-145.

Stal, L. J. (1995). Tansley Review No. 84. Physiological ecology of cyanobacteria in microbial mats and other communities. *New Phytol., 131*, 1-32.

Stal, L. J., and Bergman, B. (1990). Immunological characterization of nitrogenase in the filamentous cyanobacterium *Oscillatoria limosa*. *Planta, 182*, 287-291.

Stal, L. J., and Heyer, H. (1987). Dark anaerobic nitrogen fixation (acetylene reduction) in the cyanobacterium *Oscillatoria*. *FEMS Microbiol. Ecol., 45*, 227-232.

Stal, L. J., and Krumbein, W. E. (1981). Aerobic nitrogen fixation in pure cultures of a benthic marine *Oscillatoria* (cyanobacteria). *FEMS Microbiol. Lett., 11*, 295-298.

Stal, L. J., and Krumbein, W. E. (1985). Nitrogenase activity in the non-heterocystous cyanobacterium *Oscillatoria* sp. grown under alternating light-dark cycles. *Arch. Microbiol., 143*, 67-71.

Stewart, K. W., and Schlichting, H. E. (1966). Dispersal of algae and protozoa by selected aquatic insects. *J. Ecol., 54*, 551-562.

Stewart, W. D. P., and Lex, M. (1970). Nitrogenase activity in the blue-green alga *Plectonema boryanum* strain 594. *Arch. Mikrobiol., 73*, 250-260.

Tanigawa, R., Shirokane, M., Maeda, S., Omata, T., Tanaka, K., Takahashi, H., *et al.* (2002). Transcriptional activation of NtcA-dependent promoters of *Synechococcus* sp PCC 7942 by 2-oxoglutarate *in vitro*. *Proc. Natl. Acad. Sci. U.S.A., 99*, 4251-4255.

Taniguchi, Y., Yamaguchi, A., Hijikata, A., Iwasaki, H., Kamagata, K., Ishiura, M., *et al.* (2001). Two KaiA-binding domains of cyanobacterial circadian clock protein KaiC. *FEBS Lett., 496*, 86-90.

Thiel, T., Lyons, E. M., Erker, J. C., and Ernst, A. (1995). A second nitrogenase in vegetative cells of a heterocyst-forming cyanobacterium. *Proc. Natl. Acad. Sci. U.S.A., 92*, 9358-9362.

Thomas, J. H., Cronshaw, A. D., Gallon, J. R., and Chaplin, A. E. (1981). The effects of ammonia on nitrogen fixation by *Gloeocapsa*. *Biochem. Soc. Trans., 9*, 226P.

Thomas, J. H., Mullineaux, P. M., Cronshaw, A. D., Chaplin, A. E., and Gallon, J. R. (1982). The effects of structural analogues of amino acids on NH_4^+-assimilation and acetylene reduction (nitrogen fixation) in *Gloeocapsa* (*Gloeothece*) sp. CCAP 1430/3. *J. Gen. Microbiol., 128*, 885-893.

Thorneley, R. N. F., and Ashby, G. A. (1989). Oxidation of nitrogenase iron protein by dioxygen without inactivation could contribute to high respiration rates of *Azotobacter* species and facilitate nitrogen fixation in aerobic environments. *Biochem. J., 261*, 181-187.

Tsinoremas, N. F., Ishiura, M., Kondo, T., Anderson, C. R., Tanaka, K., Takahashi, H., *et al.* (1996). A sigma factor that modifies the circadian expression of a subset of genes in cyanobacteria. *EMBO J., 15*, 2488-2495.

Tucker, D. L., Hirsh, K., Li, H., Boardman, B., and Sherman, L. A. (2001). The manganese stabilizing protein (MSP) and the control of O_2 evolution in the unicellular, diazotrophic cyanobacterium, *Cyanothece* sp ATCC 51142. *Biochim. Biophys. Acta, 1504*, 409-422.

Tucker, D. L., and Sherman, L. A. (2000). Analysis of chlorophyll-protein complexes from the cyanobacterium *Cyanothece* sp ATCC 51142 by non-denaturing gel electrophoresis. *Biochim. Biophys. Acta, 1468*, 150-160.

Turner, S., Huang, T.-C., and Chaw, S.-M. (2001). Molecular phylogeny of nitrogen-fixing unicellular cyanobacteria. *Bot. Bull. Acad. Sinica, 42*, 181-186.

Weber, B., Wessels, D. C. J., and Büdel, B. (1996). Biology and ecology of cryptoendolithic cyanobacteria of a sandstone outcrop in the Northern Province, South Africa. *Arch. Hydrobiol., 117 (supplement: Algological Studies 83)*, 565-575.

Wolk, C. P., Ernst, A., and Elhai, J. (1994). Heterocyst metabolism and development. In D. A. Bryant (Ed.), *The molecular biology of cyanobacteria* (pp. 769-823). Dordrecht: Kluwer.

Wyatt, J. T., and Silvey, J. K. G. (1969). Nitrogen fixation by *Gloeocapsa*. *Science, 165*, 908-909.

Zehr, J. P., Harris, D., Dominic, B., and Salerno, J. (1997). Structural analysis of the *Trichodesmium* nitrogenase iron protein: Implications for aerobic nitrogen fixation activity. *FEMS Microbiol. Lett., 153*, 303-309.

Zehr, J. P., Waterbury, J. B., Turner, P. J., Montoya, J. P., Omoregie, E., Steward, G. F., *et al.* (2001). Unicellular cyanobacteria fix N_2 in the subtropical North Pacific Ocean. *Nature, 412*, 635-638.

Chapter 7

NITROGEN FIXATION IN THE PHOTOSYNTHETIC PURPLE BACTERIUM *RHODOBACTER CAPSULATUS*

B. MASEPOHL, T. DREPPER AND W. KLIPP
Ruhr-Universität Bochum, Fakultät für Biologie, Lehrstuhl für Biologie der Mikroorganismen, D-44780 Bochum, Germany

1. INTRODUCTION

The ability to fix dinitrogen is widespread among nonsulfur photosynthetic purple bacteria (*Rhodospirillaceae*), including *Rhodobacter capsulatus*, *Rhodobacter sphaeroides*, *Rhodopseudomonas viridis*, and *Rhodospirillum rubrum* (Madigan, 1995; Madigan *et al.*, 1984). Among all *Rhodospirillaceae* species tested so far, strains of *R. capsulatus* show the highest specific activities for nitrogenase and shortest generation times under diazotrophic growth conditions. Extensive genetic and biochemical studies and the availability of the nearly completed genome sequence make *R. capsulatus* one of the best-characterized diazotrophic organisms (Fonstein *et al.*, 1992; 1998; Masepohl and Klipp, 1996; Overbeek *et al.*, 2000; Schneider *et al.*, 1997; Williams and Taguchi, 1995) and, therefore, this review mainly deals with N_2 fixation in *R. capsulatus* as a representative of purple nonsulfur photosynthetic bacteria. *R. capsulatus* is a ubiquitous bacterium that can be isolated from pond and soil samples by standard enrichment techniques (Marrs, 1974; Weaver *et al.*, 1975). Most studies on *R. capsulatus* (formerly called *Rhodopseudomonas capsulata*; Imhoff *et al.*, 1984) have been performed with strain B10S, a spontaneous streptomycin-resistant derivative of B10 (Klipp *et al.*, 1988), and strain SB1003, a rifampicin-resistant B10 derivative (Cullen *et al.*, 1997; Marrs, 1974).

Phototrophic purple bacteria show remarkable metabolic diversity. *R. capsulatus* is capable of five distinctly different modes of growth, namely photoautotrophic, photoheterotrophic, chemoorganotrophic by aerobic respiration, chemoorganotrophic by fermentation, and chemolithotrophic (with H_2 as electron donor and O_2 as electron acceptor). Diazotrophic growth by *R. capsulatus* is

W. Klipp, B. Masepohl, J. R. Gallon and W. E. Newton (eds.), Genetics and Regulation of Nitrogen Fixation in Free-Living Bacteria, 141-173.

possible under several of these conditions, but growth on N_2 is optimal under photosynthetic conditions (Madigan, 1995). In addition to the "classical" Mo-containing and *nif*-encoded nitrogenase, *R. capsulatus* harbors an alternative iron-only and *anf*-encoded nitrogenase, similar to the heterometal-free nitrogenase from *Azotobacter vinelandii* (Bishop and Premakumar, 1992; Schneider *et al.*, 1991b). In *R. capsulatus*, synthesis and activity of both nitrogenases is controlled by the environmental factors, ammonium, molybdenum, light, and O_2. Several different classes of regulatory proteins, which include the two-component NtrB/NtrC system, the PII signal transduction proteins (GlnB and GlnK), the specific transcriptional activators (NifA1, NifA2, and AnfA), the molybdate-dependent repressor proteins (MopA and MopB), the ammonium transporter (AmtB), the ADP-ribosyl transferring DraT/DraG system, and the histone-like protein (HvrA), form a complex regulatory cascade that integrates these environmental signals.

2. ORGANIZATION OF NITROGEN-FIXATION GENES

2.1. Genes Encoding the Molybdenum Nitrogenase (nif Genes)

The nitrogen-fixation genes in *R. capsulatus* are organized into four gene clusters separated from each other in the chromosome (Fig. 1; Fonstein *et al.*, 1995; 1998).

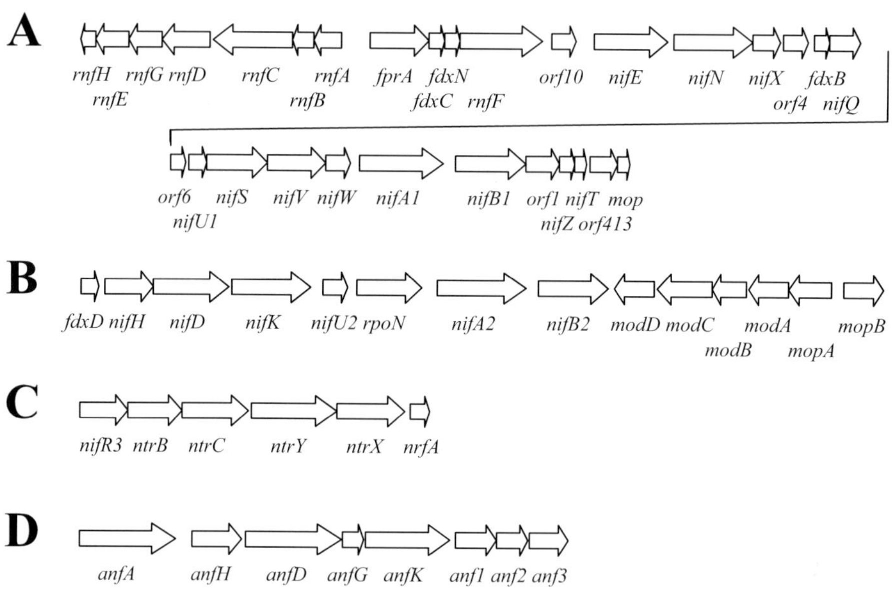

Figure 1. Organization of nitrogen-fixation genes in R. capsulatus. *Arrows indicate sizes and direction of transcription of* nif *and* anf *genes. See text for details.*

Homologues of most of the *Klebsiella pneumoniae nif* genes are found in the so-called *nif* regions A and B (Figures 1A and 1B; Tables 1A and 1B), whereas *nif* region C carries *ntrB*/*ntrC*-like regulatory genes (Figure 1C; Table 1C; Klipp *et al.*, 1988). The fourth nitrogen-fixation gene region consists solely of *anf* genes, which are required for synthesis and activity of the alternative iron-only nitrogenase (Figure 1D; Table 1D; Schüddekopf *et al.*, 1993). In addition to these *nif*, *anf* and *ntr* genes, *R. capsulatus* contains other genes involved in nitrogen fixation and assimilation. These genes are not part of any of the four nitrogen-fixation gene clusters but are scattered in the chromosome (Table 1E).

Table 1A. Nitrogen-fixation genes located in nif *region A*

Gene	*Phenotype*		*Characteristics, function of gene product*
	Nif	Anf	
rnfH	n.t.	n.t.	Soluble protein
rnfE	—	—	Integral membrane protein
rnfG	n.t.	n.t.	Transmembrane segment in N-terminus
rnfD	—	—	Integral membrane protein
rnfC	—	—	Iron-sulfur protein
rnfB	—	+	Iron-sulfur protein
rnfA	—	—	Integral membrane protein
fprA	+	+	FMN-binding protein (Orf14)
fdxC	n.t.	n.t.	[2Fe-2S] ferredoxin
fdxN	+/—	+/—	2x [4Fe-4S] ferredoxin
rnfF	—	—	Membrane protein
orf10	+	+	Soluble protein
nifE	—	+	FeMoco biosynthesis
nifN	—	+	FeMoco biosynthesis
nifX	+	+	FeMoco processing
orf4	+	+	Homologue of *A. vinelandii orf3*
fdxB	+	+	2x [4Fe-4S] ferredoxin
nifQ	q	+	Processing of molybdenum
orf6	+	+	Homologue of *A. vinelandii orf6*
nifU1	2x +	2x +	Truncated *nifU* gene
nifS	—	+	L-Cysteine desulfurase, FeS core formation
nifV	+/—	—	Homocitrate synthase
nifW	+	+	Stabilization of dinitrogenase
nifA1	2x —	2x +	Transcriptional activator of *nif* genes
nifB1	2x —	2x —	FeMoco and FeFeco biosynthesis
orf1	+	+	Periodic protein, homologue of *A. vinelandii orf2*
nifZ	+	+	Accumulation of active MoFe protein
nifT	+	n.t.	Homologue of *B. japonicum fixU*
orf413	+	n.t.	Soluble protein
mop	+	n.t.	Molybdate-binding protein

Abbreviations: n.t. (not tested); Nif^{q} (diazotrophic growth with 1 mM but not with 4 μM molybdate); 2x (duplicated gene; phenotype of double mutant)

Table 1B. Nitrogen-fixation genes located in nif *region B*

Gene	Phenotype		Characteristics, function of gene product
	Nif	Anf	
fdxD	n.t.	n.t.	[2Fe-2S] ferredoxin
nifH	–	+	Dinitrogenase reductase (Fe protein) subunit
nifD	–	+	Dinitrogenase (MoFe protein) α subunit
nifK	–	+	Dinitrogenase (MoFe protein) β subunit
nifU2	2x +	2x +	Truncated *nifU* gene
rpoN	–	–	Alternative sigma factor (σ^{54}, RpoN, NtrA, NifR4)
nifA2	2x –	2x +	Transcriptional activator of *nif* genes
nifB2	2x –	2x –	FeMoco and FeFeco biosynthesis
modD	+	+	Molybdenum transport
modC	m	+	Molybdenum transport, ATP-binding
modB	m	+	Molybdenum transport, membrane protein
modA	m	+	Periplasmic molybdenum-binding protein
mopA	+	+	Molybdate-dependent repressor
mopB	+	+	Molybdate-dependent repressor

Abbreviations: Nif^{m} (diazotrophic growth with 50 μM but not with 100 nM molybdate)

Table 1C. Nitrogen-fixation genes located in nif *region C*

Gene	Phenotype		Characteristics, function of gene product
	Nif	Anf	
nifR3	+	n.t.	Homologue of enterobacterial *yhdG*
ntrB	+/–	n.t.	Sensor kinase NifR2 (2-component regulatory system)
ntrC	–	–	Response regulator NifR1 (2-component regulatory system)
ntrY	+	+	Sensor kinase (2-component regulatory system)
ntrX	n.t.	n.t.	Response regulator (2-component regulatory system)
nrfA	n.t.	n.t.	Homologue of *E. coli hfq* encoding HF-I

Table 1D. Nitrogen-fixation genes located in the anf *gene region*

Gene	Phenotype		Characteristics, function of gene product
	Nif	Anf	
anfA	+	–	Transcriptional activator of *anf* genes
anfH	+	–	Alternative dinitrogenase reductase (Fe protein-3)
anfD	+	–	Alternative dinitrogenase (FeFe protein) α subunit
anfG	+	–	Alternative dinitrogenase (FeFe protein) δ subunit
anfK	+	–	Alternative dinitrogenase (FeFe protein) β subunit
anf1	+	–	Homologue of *A. vinelandii anfO*
anf2	+	–	Homologue of *A. vinelandii anfR*
anf3	+	–	Soluble protein

Table 1E. Nitrogen-fixation and assimilation genes scattered in the chromosome

Gene	*Phenotype*		*Characteristics, function of gene product*
	Nif	Anf	
nifF	f	n.t.	Flavodoxin
draT	+	+	Dinitrogenase reductase ADP-ribosyl transferase
draG	+	+	Dinitrogenase reductase-activating glycohydrolase
glnB	+	+	PII-like signal transduction protein (NifR5)
glnA	Gln	Gln	Glutamine synthetase
glnK	+	+	PII-like signal transduction protein
amtB	+	n.t.	(Methyl)-ammonium transport
amtY	+	n.t.	Ammonium transport
glnD	L	L	Uridylyl transferase / uridylyl-removing enzyme
himA	–	–	IHF (integration host factor) α subunit
hip	n.t.	n.t.	IHF (integration host factor) β subunit
regB	n.t.	n.t.	Sensor kinase (2-component regulatory system)
senC	n.t.	n.t.	Maximal induction from the *puf* and *puh* operons
regA	n.t.	n.t.	Response regulator (2-component regularory system)
hvrA	+	n.t.	Histone-like protein

Abbreviations: Nif^{f} (no diazotrophic growth when iron is omitted from the basic minimal medium); Gln (glutamine auxotroph); L (*glnD* null mutation is lethal)

2.1.1. The Structural Genes of Nitrogenase

As in most N_2-fixing organisms, the structural genes of the molybdenum nitrogenase, *nifHDK*, are organized in a single operon, which gives rise to a 4.4-kb mRNA corresponding to the full-length *nifHDK* transcript and several smaller RNAs, which correspond to *nifH*, *nifHD*, and *nifDK* (Willison *et al.*, 1993). The smaller RNA species seem to result from endonucleolytic cleavage of the full-length *nifHDK* transcript. In contrast to *K. pneumoniae* and *A. vinelandii*, the *R. capsulatus nifHDK* operon does not contain *nifY*- and *nifT*-like genes. Instead, a *nifT*-like gene is located in *nif* region A (Figure 1) as part of the *nifB1-orf1-nifZ-nifT* transcriptional unit, whereas a *nifY*-like gene has neither been identified in any of the four nitrogen-fixation gene regions nor elsewhere in the genomic sequence of *R. capsulatus*. In *K. pneumoniae* and *A. vinelandii*, the *nifM* gene product is required for maturation of NifH, but no *nifM*-like gene has yet been identified in *R. capsulatus*. At present, it uncertain whether *R. capsulatus* NifH does not require such a maturation step or whether maturation involves a non-specific prolyl-peptidyl isomerase.

As a photosynthetic bacterium, *R. capsulatus* contains bacteriochlorophyll *a* as the photochemically active reaction center pigment. Bacteriochlorophyll biosynthesis includes two reduction steps, namely reduction of protochlorophyllide into a chlorin, and reduction of chlorin into a bacteriochlorin (Burke *et al.*, 1993b). Both protochlorophyllide reductase and chlorin reductase are three-subunit enzymes in which each subunit from one reductase shares significant amino acid identity with a subunit of the other. Interestingly, these subunits also show some similarity to the products of the structural genes of nitrogenase, *nifHDK*, with the most-striking similarities between the so-called "chlorophyll iron protein" subunits, encoded by the *bchL* and *bchX* genes with the nitrogenase iron protein, encoded by *nifH* (Burke *et al.*, 1993a).

2.1.2. Cofactor Biosynthesis

In *K. pneumoniae*, the gene products of at least six different *nif* genes (*nifQ*, *nifB*, *nifE*, *nifN*, *nifV*, and *nifH*) are known to be involved in the biosynthesis of the iron-molybdenum cofactor (FeMoco) of nitrogenase (Rangaraj *et al.*, 2000). All six genes are also present in *R. capsulatus*, indicating that FeMoco biosynthesis follows a similar pathway in both organisms. As in *K. pneumoniae* and *A. vinelandii*, the *R. capsulatus nifEN* genes are closely associated, which is in agreement with the hypothesis that *nifEN* may have evolved by duplication of an ancestral *nifDK* pair (Fani *et al.*, 2000). The *R. capsulatus nifEN* genes are part of a *nifENX-orf4-fdxB-nifQ* operon (Moreno-Vivian *et al.*, 1989a). *R. capsulatus nifE* and *nifN* mutants exhibit a Nif⁻ phenotype but can grow diazotrophically via the alternative nitrogenase (Schüddekopf *et al.*, 1993). In contrast to other diazotrophic bacteria, the *nifE* gene product is not absolutely essential for biosynthesis of FeMoco, although the presence of NifE/NifN guarantees formation of a much higher intracellular concentration of MoFe proteins containing FeMoco (Siemann *et al.*, 2001). Despite the requirement of NifX for *in vitro* biosynthesis of FeMoco in *A. vinelandii* (Rangaraj *et al.*, 2001; Shah *et al.*, 1999), *nifX* mutations in *A. vinelandii*, *K. pneumoniae* or *R. capsulatus* have little effect on nitrogenase activity (Gosink *et al.*, 1990; Jacobson *et al.*, 1989; Moreno-Vivian *et al.*, 1989b).

In contrast to the situation in other diazotrophic bacteria, the *nifB* gene of *R. capsulatus* is duplicated (*nifB1*, *nifB2*), and either copy is sufficient for diazotrophic growth (Klipp *et al.*, 1988; Masepohl *et al.*, 1988). Moreover, in addition to *nifB*, the regulatory *nifA* gene (*nifA1*, *nifA2*), a truncated *nifU* gene (*nifU1*, *nifU2*), the genes encoding molybdate-dependent repressor proteins (*mopA*, *mopB*), and the regulatory *ntrBC* genes (*ntrB*/*ntrC*, *ntrY*/*ntrX*) are duplicated in *R. capsulatus* (Figure 1; Masepohl and Klipp, 1996), indicating that duplication of nitrogen-fixation genes seems to be a peculiarity of *R. capsulatus*. As in many other diazotrophic proteobacteria, the *nifA* and *nifB* genes are closely associated in *R. capsulatus* (*nifA1-nifB1*, *nifA2-nifB2*), although these genes do not belong to the same transcriptional units.

Studies on the *nifS* and *nifU* gene products from *A. vinelandii* revealed that both NifS (a cysteine desulfurase supplying inorganic sulfide) and NifU (involved in mobilization of Fe) are required to supply the enlarged demand of Fe_2S_2 clusters for formation of other FeS clusters in the nitrogenase component proteins (Zheng *et al.*, 1994; Yuvaniyama *et al.*, 2000). Housekeeping homologues of both NifU and NifS, respectively designated IscU and IscS, are found not only in *A. vinelandii* and other diazotrophs but also in non-nitrogen fixing organisms. The organization of the genes in a *nifU-nifS-nifV-nifW* transcriptional unit is conserved among many diazotrophic bacteria, including *R. capsulatus* (Masepohl *et al.*, 1993a). As mentioned above, *R. capsulatus* harbors two copies of a truncated *nifU* gene (*nifU1*, *nifU2*) within *nif* regions A and B (Figure 1). These truncated *nifU* genes encode proteins similar to the C-terminal domain of the highly modular NifU protein from *A. vinelandii*. Because NifU1 and NifU2 do not contain homologues of those domains of the *A. vinelandii* NifU that coordinate either the permanent or a transient FeS cluster, it seems unlikely that *R. capsulatus* NifU1 and/or NifU2 are involved in Fe mobilization. Accordingly, mutations in these genes do not alter the activities of either the molybdenum nitrogenase or the alternative nitrogenase. However, no "full-length" copy of *nifU* has been identified in the *R. capsulatus* genome sequence. The *R. capsulatus nifS* gene product is essential for activity of molybdenum nitrogenase, but it is dispensable for the alternative nitrogenase. In contrast, the NifS protein from *A. vinelandii* is required for the activity of all its nitrogenases, the molybdenum nitrogenase, the vanadium nitrogenase, and the heterometal-free alternative nitrogenase. *R. capsulatus* contains three additional copies of *nifS*-like genes, but the roles of the respective gene products in FeS formation remain unknown.

2.1.3. Electron Transport to Nitrogenase

Two types of electron-carrier proteins, ferredoxins and flavodoxins, are known to serve as electron donors to nitrogenase. At least under iron-replete conditions, electron transfer to nitrogenase in *R. capsulatus* is mediated by the products of the *rnf* (***R****hodobacter* **n**itrogen **f**ixation) and *fdxN* genes located in *nif* region A (Schmehl *et al.*, 1993). The *rnf* gene products are thought to form a membrane complex analogous to the predicted ion channel that is formed by the NqrB, NqrD, and NqrE proteins in *Vibrio alginolyticus* (Jouanneau *et al.*, 1998; Kumagai *et al.*, 1997; Rich *et al.*, 1995). The proton-motive force or membrane potential in chromatophores might be coupled by the Rnf protein complex to achieve the endergonic reduction of the two-[4Fe-4S]-containing ferredoxin, FdxN (FdI), by NADH. The Rnf and FdxN proteins form the primary electron-donor pathway to nitrogenase under photosynthetic conditions, thereby linking nitrogen fixation to the cyclic electron-transport system. In addition, the *rnf* and *fdxN* gene products also play an essential role in nitrogen fixation during anaerobic DMSO-dependent growth in the dark, indicating that linkage to the photosynthetic electron transport is not a prerequisite for the function of the Rnf proteins (Saeki and Kumagai, 1998). The supply of reductant through the Rnf complex may be rate limiting for nitrogenase activity *in vivo* as demonstrated by the observation that overexpression of the *rnfABCDGEH* operon leads to enhanced nitrogenase activity (Jeong and Jouanneau, 2000).

In addition to the *fdxN* gene, five ferredoxin-encoding genes have been identified in *R. capsulatus*. The *fdxC* and *fdxB* genes are located in *nif* region A (Figure 1) and encode the [2Fe-2S] ferredoxin, FdIV, and the two-[4Fe-4S]-containing ferredoxin, FdIII, respectively (Grabau *et al.*, 1991; Jouanneau *et al.*, 1993; Moreno-Vivian *et al.*, 1989a; Saeki *et al.*, 1990, 1991, 1993; Schmehl *et al.*, 1993). FdIV may serve as the *in vivo* electron donor to a flavoprotein encoded by the *fprA* gene, which is part of the *fprA-fdxC-fdxN-rnfF* transcriptional unit (Figure 1; Jouanneau *et al.*, 2000; Schmehl *et al.*, 1993). *R. capsulatus* mutants lacking either FprA or FdIV have no detectable phenotype. The *fdxD* gene is found upstream of the structural genes of nitrogenase, *nifHDK*, in *nif* region B (Figure 1) and encodes the [2Fe-2S] ferredoxin, FdV (Armengaud *et al.*, 1994; Willison *et al.*, 1993). The *fdxA* and the *fdxE* genes, which encode the [3Fe-4S]/[4Fe-4S]-containing ferredoxin, FdII, and the [2Fe-2S] ferredoxin, FdVI, respectively, are not located within the known nitrogen-fixation gene regions. In contrast to *fdxN*, *fdxC*, *fdxB*, and *fdxD*, which are transcribed only under nitrogenase-derepressing conditions, the *fdxA* and *fdxE* gene products are synthesized under both fixed-nitrogen-limiting and fixed-nitrogen-sufficient conditions (Armengaud *et al.*, 1997; Duport *et al.*, 1990, 1992; Jouanneau *et al.*, 1990, 1992; Naud *et al.*, 1994; Suetsugu *et al.*, 1991). Beside FdxN, only the *fdxA* gene product can donate electrons to nitrogenase *in vitro*.

In addition to the above mentioned ferredoxins, a flavodoxin similar to *K. pneumoniae* NifF has been purified from *R. capsulatus* (Yakunin *et al.*, 1993). The synthesis of *R. capsulatus* NifF is under dual control by iron and fixed nitrogen availability. Like FdxN, the *nifF* gene product can, *in vitro*, efficiently donate electrons to nitrogenase. NifF also contributes significantly to nitrogenase activity *in vivo* under N_2-fixing conditions in the presence of iron, but is absolutely required for nitrogen fixation under iron limitation (Gennaro *et al.*, 1996). The monocistronic *nifF* gene is not located within any of the four nitrogen-fixation gene clusters shown in Figure 1, but its expression is under the control of the central *nif* gene activator, NifA. Taken together, under iron-sufficient conditions, electron transfer to *R. capsulatus* nitrogenase occurs (mainly) via the Rnf-FdxN system, whereas, under iron-limiting conditions, NifF is the immediate (and single) electron donor to nitrogenase reductase. In *K. pneumoniae*, electron transfer to nitrogenase involves a pyruvate:flavodoxin oxidoreductase encoded by the *nifJ* gene. A pyruvate:ferredoxin (flavodoxin) oxidoreductase (POR) has also been purified from *R. capsulatus* (Yakunin and Hallenbeck, 1998a). POR can couple the oxidation of pyruvate to the reduction of nitrogenase in a coupled system with either NifF, FdxN or FdII. POR is constitutively synthesized, with synthesis augmented under N_2-fixing conditions, indicating that, in contrast to *K. pneumoniae* NifJ, *R. capsulatus* POR does not play a role unique to nitrogen fixation.

2.1.4. nif-*Associated Open Reading Frames (*orfs*)*

R. capsulatus contains several *nif*-regulated *orfs* not present in the *K. pneumoniae nif* gene cluster. These *orfs* are interspersed among genes of *nif* region A (Figure 1A; Table 1A). Highly conserved homologues of these *R. capsulatus orfs* are present in *A. vinelandii* and several other diazotrophic bacteria. Despite the high degree of sequence conservation among these *orfs*, none of them is essential for nitrogen fixation (at least under standard laboratory conditions). In some cases, the homologues are similarly organized in both *R. capsulatus* and *A. vinelandii*, *e.g.*, the *orf6* homologues belong to the *nifUSVW* operon and the *orf4* homologues are cotranscribed with the *nifENX* genes. On the other hand, *R. capsulatus orf1,* which encodes a leucine-rich periodic protein, is located downstream of *nifB1*, whereas its *A. vinelandii* homologue, *orf2*, is part of the *nifHDKTY-orf1-orf2* operon.

2.1.5. The Regulatory nifA1 *and* nifA2 *Genes*

As in all proteobacterial diazotrophs analyzed so far, expression of *R. capsulatus nif* genes is activated by NifA under nitrogen-limiting phototrophic conditions (*i.e.*, anaerobic growth in the light). In contrast to other N_2-fixing bacteria, *R. capsulatus* contains two copies of *nifA* (*nifA1*, *nifA2*) that are almost identical to each other (Figure 1; Hübner *et al.*, 1993; Masepohl *et al.*, 1988; Paschen *et al.*, 2001; see below). Mutants defective for either *nifA1* or *nifA2* are still able to grow diazotrophically, via molybdenum nitrogenase, whereas *nifA1*/*nifA2* double mutants exhibit a Nif^- phenotype. These results demonstrate that NifA1 and NifA2 can functionally substitute for each other (Table 1A, 1B; Masepohl *et al.*, 1988). Both NifA1 and NifA2 (as well as AnfA) activate expression of their target genes in concert with RNA polymerase (RNAP) containing the alternative sigma factor RpoN (σ^{54}, NtrA, NifR4; Cullen *et al.*, 1994; Hübner *et al.*, 1993; Jones and Haselkorn, 1989; Preker *et al.*, 1992; Schüddekopf *et al.*, 1993). Therefore, RpoN is not only essential for diazotrophic growth via molybdenum nitrogenase but also via the alternative nitrogenase (Table 1B; Schüddekopf *et al.*, 1993). The *R. capsulatus rpoN* gene is located in *nif* region B (Figure 1B) and is part of an autoactivated *nifU2-rpoN* superoperon (Cullen *et al.*, 1994). A primary promoter, which is directly upstream of the *rpoN* coding region, is expressed constitutively with respect to fixed-nitrogen status and seems to be responsible for the initial expression of *rpoN*. A secondary promoter, which is upstream of *nifU2*, is autoactivated by NifA and RpoN to increase expression of *rpoN* under fixed-nitrogen-limiting conditions.

*2.2. Genes Encoding the Alternative Nitrogenase (*anf *genes)*

In the absence of molybdenum, *R. capsulatus* can synthesize an alternative nitrogenase, which is similar to the iron-only nitrogenase of *A. vinelandii* (Schneider *et al.*, 1991b). In contrast to the situation in *A. vinelandii*, vanadium does not inhibit synthesis of the iron-only nitrogenase in *R. capsulatus* (Gollan *et al.*, 1993; Schneider *et al.*, 1991b; Wang *et al.*, 1993).

The structural genes of the alternative nitrogenase of *R. capsulatus*, *anfHDGK*, are located in the *anf* gene region shown in Figure 1D. As typically found for alternative nitrogenase systems, *R. capsulatus* contains an *anfG*-like gene, which codes for the small δ subunit of alternative dinitrogenases (FeFe proteins), situated between the *anfD* and *anfK* genes. Three genes, called *anf1*, *anf2*, and *anf3*, are cotranscribed with the *anfHDGK* genes. *R. capsulatus* strains carrying mutations in any of the seven genes of the *anfHDGK-1-2-3* operon are unable to grow diazotrophically under conditions in which the iron-only nitrogenase operates. However, mutants defective for any one of *anf1*, *anf2* or *anf3* (as well as double and triple mutants) are still able to reduce acetylene via the alternative nitrogenase (W. Klipp and coworkers, unpublished results). Although this phenotype resembles that of *K. pneumoniae nifV* mutants, which synthesize a modified form of FeMoco (Liang *et al.*, 1990), the molecular mechanisms leading to these changes in catalytic features in *anf1-2-3* mutants seem to be different. The *R. capsulatus anf1* and *anf2* genes are homologous to *A. vinelandii anfO* and *anfR*, respectively, which belong to the *anfHDGK-anfO-anfR* transcriptional unit (Mylona *et al.*, 1996). *A. vinelandii* mutants carrying in-frame deletions in *anfO* and *anfR* exhibit a phenotype similar to that of *R. capsulatus anf1-2-3* mutants and accumulate the subunits of the iron-only nitrogenase. In contrast to *R. capsulatus*, *A. vinelandii* does not seem to have a counterpart of *anf3*.

Expression of the *anfHDGK-1-2-3* operon is activated by the product of the *anfA* gene, which is located upstream of *anfH* (Figure 1D; Schüddekopf *et al.*, 1993). *R. capsulatus* strains lacking AnfA do not synthesize the alternative nitrogenase, suggesting that AnfA cannot be substituted by NifA, NtrC or any other transcriptional activator. Mutational analysis indicates that AnfA (like NifA) requires the alternative sigma factor, RpoN, to activate transcription from the *anfH* promoter. Expression of *anfA* itself occurs only under conditions of concomitant depletion of both fixed nitrogen and molybdenum (see below; Kutsche *et al.*, 1996).

In addition to *anfA*, *anfHDGK-1-2-3,* and *rpoN*, several *nif* and *rnf* genes are essential for the function of the alternative nitrogenase in *R. capsulatus* (Table 1A, 1B; Schüddekopf *et al.*, 1993). One of these genes is *nifB* (involved in synthesis of NifB-co, which is the iron- and sulfur-containing precursor of FeMoco) suggesting that NifB-co is the FeS precursor of the cofactor of the iron-only nitrogenase (FeFeco). The requirement of the *nifV* gene product (homocitrate synthase) for activity of the alternative nitrogenase indicates that homocitrate is not only a structural component of FeMoco but is also present in FeFeco. In addition to these genetic data, there are several lines of evidence, including spectroscopic results, which suggest that FeFeco is almost identical to FeMoco with the only fundamental structural difference being that the Mo atom of FeMoco is replaced by Fe in FeFeco (Krahn *et al.*, 2002).

NifEN is considered to form a molecular scaffold on which FeMoco is preassembled (Brigle *et al.*, 1987; Roll *et al.*, 1995). *A. vinelandii* strains carrying mutations in *nifEN* are still able to grow diazotrophically, via their alternative nitrogenase systems, because VnfEN can substitute for NifEN (Wolfinger and Bishop, 1991). Similarly, the *nifEN* genes are dispensable for activity of the iron-only nitrogenase of *R. capsulatus* (Schüddekopf *et al.*, 1993). However, *R. capsulatus* does not seem to harbor either *vnfEN* or any other *nifEN*-like genes, leaving the open question of how FeFeco is assembled and/or inserted into the *anfDGK*-encoded apo-dinitrogenase.

In contrast to what occurs in *A. vinelandii*, *R. capsulatus nifS* is not essential for activity of the alternative nitrogenase (Schüddekopf *et al.*, 1993). As mentioned above, *R. capsulatus* harbors three additional copies of *nifS*-like genes and one or more of these may substitute for *nifS* under conditions in which the alternative nitrogenase system operates.

The products of *rnfA*, *rnfC*, *rnfD*, *rnfG*, and *rnfF* are involved in electron transport to the *anfH*-encoded dinitrogenase reductase (Fe protein-3; Schüddekopf *et al.*, 1993). Interestingly, RnfB, which is essential for activity of molybdenum nitrogenase, is not required for the alternative system. In an *fdxN* mutant, the activity of both nitrogenase systems is dramatically decreased.

Because a *nifA1*/*nifA2* double deletion strain can still grow diazotrophically via the alternative nitrogenase, it seems likely that *R. capsulatus* AnfA is responsible for transcriptional activation of the above-mentioned *nif* and *rnf* genes. Indeed, the *nifB* promoter of *A. vinelandii* was shown to be activated not only by NifA but also by the transcriptional activators of both alternative nitrogenase systems, VnfA and AnfA (Drummond *et al.*, 1996).

*2.3. Genes Encoding a High-affinity Molybdate Transport System and Mo-dependent Repressor Proteins (*mod *and* mop *Genes)*

As in *A. vinelandii*, synthesis of the alternative nitrogenase in *R. capsulatus* is repressed by traces of molybdenum (Kutsche *et al.*, 1996). On the other hand, the molybdenum nitrogenase of *R. capsulatus* is maximally derepressed in media containing about 0.25 μM molybdate even though concentrations up to 1 mM do not inhibit nitrogenase activity (Gollan *et al.*, 1993). Below 0.25 μM MoO_4^{2-}, nitrogenase activity decreases and, at 0.09 μM MoO_4^{2-}, a concentration at which the alternative Mo-independent nitrogenase is still completely repressed, the specific activity is half-maximal. About 50% of the maximum activity of the alternative nitrogenase is observed at 0.006 μM MoO_4^{2-} and activity increases below this molybdate concentration. Even if no molybdenum is added to the growth medium of *R. capsulatus*, the alternative nitrogenase is almost completely repressed (Schneider *et al.*, 1991b). Selective removal of traces of molybdenum from the medium can be achieved with activated carbon (Schneider *et al.*, 1991a).

In both eubacteria and archea, molybdate is transported by an ABC-type high-affinity transporter comprising three proteins; the periplasmic molybdate-binding protein (ModA), the membrane protein (ModB), and the ATPase (ModC) (Self *et al.*, 2001). In *R. capsulatus*, the *modABC* genes belong to the *mopA-modABCD* transcriptional unit located in *nif* region B (Figure 1B; Table 1B). The exact role of MopD in molybdenum transport remains unknown (Wang *et al.* 1993), but ModD exhibits some similarity to NadC, which is known to be involved in the synthesis of NAD. A second *mopA*-like gene, *mopB*, forms a monocistronic operon reading in the opposite direction relative to *mopA-modABCD* (Figure 1B; Table 1B). Both MopA and MopB exhibit clear similarity to *E. coli* ModE, which acts in a molybdate-dependent manner as a repressor of the *modABCD* operon (Grunden *et al.*, 1996). Similarly, *R. capsulatus* MopA and MopB mediate Mo-dependent repression of *mopA-modABCD*. In addition, in the presence of molybdenum, MopA and MopB repress transcription of *anfA*, thereby inhibiting expression of the alternative nitrogenase in the presence of molybdenum (Figure 2; Kutsche *et al.*, 1996). ModE-like repressor proteins bind in a sequence-specific manner to DNA elements that show dyad symmetry and overlap the transcription start site (Pau *et al.*, 1997). A consensus sequence of CGNTATAT–$N_{7/8}$–TATATANCG has been suggested. Deletions within this sequence lead to transcription of *R. capsulatus anfA* in the presence of molybdenum, confirming the role of this element as the target of MopA/MopB-mediated repression (Kutsche *et al.*, 1996; Self *et al.*, 2001).

Mo-dependent repression is completely released in an *R. capsulatus mopA*/*mopB* double deletion mutant leading to synthesis of active iron-only nitrogenase even in the presence of 1 mM MoO_4^{2-} (Wang *et al.*, 1993). In contrast, Mo-dependent repression is retained in mutants lacking either MopA or MopB, indicating that both repressor proteins can substitute for each other in *anfA* repression. However, MopB (but not MopA) is essential for Mo-dependent expression of the *R. capsulatus dorCDA* operon (coding for dimethylsulfoxide reductase) demonstrating clear functional differences between MopA and MopB (Solomon *et al.*, 2000).

Molybdenum transport is not affected in an *R. capsulatus mopA*/*mopB* double mutant, whereas strains carrying lesions in the *modABC* genes are impaired in molybdenum uptake (Wang *et al.*, 1993). In contrast to the *mopA*/*mopB* double mutant, Mo-dependent repression of the alternative nitrogenase is only partially released in mutant strains lacking the ABC-type transport system, leading to synthesis of the alternative nitrogenase at concentrations up to about 10 μM MoO_4^{2-} (Wang *et al.*, 1993). This observation indicates that there are alternative low-affinity transport routes by which molybdenum can enter the cell.

It should be mentioned that *R. capsulatus* contains a Mop-like protein showing similarity to the *Clostridium pasteurianum* Mo-containing Mop products encoded by a multigene family (Hinton and Merritt, 1986; Hinton *et al.*, 1987). The *R. capsulatus mop* gene is located in *nif* region A (Figure 1A; Table 1A). The function of the Mop proteins is unknown, but they have been implicated in molybdate storage and homeostasis (Schüttelkopf *et al.*, 2002).

2.4. Structural Features of NifA1, NifA2, AnfA and RpoN, and cis-*Regulatory Elements*

2.4.1. Domain Structure of NifA1, NifA2 and AnfA

The NifA protein is a ubiquitous transcriptional activator in diazotrophic proteobacteria and has the typical modular domain structure common to RpoN-dependent activators (Dixon, 1998). NifA proteins consist of at least three distinct domains, namely a regulatory N-terminal domain, a central ATP-binding activator domain, and a DNA-binding C-terminal domain plus, in the case of the rhizobial type of NifA protein, an interdomain linker, which connects the central and the C-terminal domains and is believed to confer O_2 sensitivity on these types of NifA proteins (Dixon, 1998; Fischer, 1994). A pathway for transcriptional activation by NifA has been proposed (Dixon, 1998) to work as follows. First, NifA binds to the upstream activator sequences (UAS) and then makes contact with RNA polymerase containing RpoN (RNAP-σ^{54}) via a long-distance interaction. This interaction is facilitated by the bending of the intervening DNA by integration host factor (IHF). Subsequently, NifA catalyses the isomerization of closed promoter complexes to open promoter complexes.

As mentioned above, *R. capsulatus* contains two functional copies of *nifA*, called *nifA1* and *nifA2* (Masepohl *et al.*, 1988). The NifA1 and NifA2 proteins consist, respectively, of 579 and 582 amino acid residues (Paschen *et al.*, 2001). These NifA proteins differ only in their N-termini (either 19 or 22 amino acid residues, respectively), whereas the remainders of both proteins (560 amino acid residues) are identical to each other. The *R. capsulatus* NifA1 and NifA2 proteins contain the interdomain-linker and, consequently, NifA-mediated activation of *nifH* expression is sensitive to O_2 (Kern *et al.*, 1998). Activity of both NifA proteins is strongly inhibited by ammonium, but the NifA2 protein seems to be slighly more tolerant than NifA1 towards NH_4^+ inhibition (Paschen *et al.*, 2001). Because both NifA proteins differ only in their extreme N-terminal sequences, it seems likely that these sequences are responsible for the slight differences in NH_4^+ sensitivity between NifA1 and NifA2. Characterization of ammonium-tolerant NifA1 mutants demonstrates that the N-terminal domain is involved in posttranslational regulation of *R. capsulatus* NifA activity.

The involvement of the N-terminal domain in control of NifA activity by ammonium seems to be widespread among proteobacteria lacking NifL-like proteins. For example, in *Herbaspirillum seropedicae* and *Azospirillum brasilense*, the activity of NifA proteins lacking the N-terminal domain is no longer inhibited by ammonium (Arsene *et al.*, 1996; Monteiro *et al.*, 1999). Furthermore, the N-terminal domain of *H. seropedicae* NifA *in trans* restores ammonium control of the truncated NifA protein, suggesting that in this respect the N-terminal domain behaves in a manner similar to that of NifL in the γ-Proteobacteria. Analysis of an ammonium-tolerant *R. capsulatus* NifA1 mutant, which carries amino acid substitutions in both the N- and C-terminal domains, indicates that an interaction between these two domains might be involved in ammonium regulation of NifA activity (Paschen *et al.*, 2001). *R. capsulatus* NifA1 is highly stable in the presence of ammonium, which excludes the possibility that proteolytic degradation is involved in ammonium-dependent inactivation of NifA.

Like NifA, both *R. capsulatus* and *A. vinelandii* AnfA have the three-domain structure characteristic for the family of activators acting in concert with RNAP-σ^{54} (Fischer, 1994). In contrast to the central and the C-terminal domains, the N-terminal domains (believed to act as environmental sensors) are not highly homologous, but the cysteine motif Cys–X_2–Gly–X–Cys is conserved in the AnfA proteins of both organisms. In *A. vinelandii* AnfA, this motif seems to be involved in FeS cluster-binding, suggesting an involvement of the N-terminal domain in regulation of AnfA activity in a redox-sensitive manner (Jepson *et al.*, 2001). Interestingly, a similar cysteine motif is present in the interdomain linker of *R. capsulatus* NifA.

*2.4.2.*cis-*Regulatory Elements and the Alternative Sigma Factor, RpoN*

Transcriptional activation of *nif* gene expression requires binding of NifA to specific sites called upstream activator-binding sites (UAS). The consensus sequence for the NifA-specific UAS is TGT–N_{10}–ACA (Morett and Buck, 1988). Perfect UAS elements are located upstream of *R. capsulatus nifH*, *nifB1*, *nifB2*, and *rnfA* (Masepohl *et al.*, 1988; Schmehl *et al.*, 1993; Willison *et al.*, 1993). In contrast, only imperfect elements have been identified upstream of *R. capsulatus nifE*, *nifU2* and *orf6* (Masepohl *et al.*, 1993a; Moreno-Vivian *et al.*, 1989b; Preker *et al.*, 1992).

NifA proteins activate transcription in concert with RNAP containing σ^{54} (RpoN), which binds to a characteristic sequence motif, (C/T)TGG–N_8–TTGC that is typically located at position –24/–12 upstream of the transcription start site (Morett and Buck, 1989). Perfect –24/–12 sequences are present upstream of *rnfA*, *fprA*, *nifE*, *orf6*, *nifB1*, *nifB2*, *nifU2*, and *nifH* (Schmehl *et al.*, 1993). In contrast to RpoN-dependent promoters found in other diazotrophs, the *R. capsulatus nif* promoters contain not only the canonical –24/–12 motif but also additional invariant nucleotides between position –12 and the transcriptional start. The role of these conserved nucleotides in *nif*-gene regulation remains unknown because site-directed mutations in this part of the *nifB1* promoter do not lead to obvious changes in the transcriptional pattern of *nifB1* (W. Klipp and coworkers, unpublished results).

RpoN proteins (σ^{54}) consist of three functional regions; these include an N-terminal glutamine-rich region required for response to the activator, an acidic region involved in open complex formation, and a C-terminal domain characterized by a helix-turn-helix motif and the so-called RpoN box (Hoover, 2000). The C-terminal domain represents the DNA-binding domain of RpoN. *R. capsulatus* RpoN exhibits clear similarity to other σ^{54}-like proteins with the highest level of identity in the DNA-binding domain, which is consistent with the high degree of conservation of its target *nif* promoters (Cannon *et al.*, 1996; Cullen *et al.*, 1994). However, the *R. capsulatus* RpoN is the smallest σ^{54}-like protein identified so far and it contains a natural deletion of the acidic region. This virtual absence of the acidic region could contribute to the observed differences in DNA melting mediated by RpoN with either *R. capsulatus* or *K. pneumoniae*. As first described for the *K. pneumoniae* RpoN protein, promoter recognition by *R. capsulatus* RpoN is detectable in the absence of core RNA polymerase (Cannon *et al.*, 1996; Cullen *et al.*, 1994).

The interaction between NifA and RNAP-σ^{54}can be stimulated by IHF (Hoover *et al.*, 1990; Santero *et al.*, 1992; Wassem *et al.*, 2000). *R. capsulatus* IHF is an $\alpha\beta$ heterodimer (encoded by the *himA* and *hip* genes) similar to IHF of *E. coli* (Toussaint *et al.*, 1993). IHF is thought to bind to specific sequences located between the binding sites for NifA (UAS) and RpoN (–24/–12), thereby inducing a sharp bend in the DNA. *E. coli* IHF binds to the promoter regions of *nifH* genes from several bacteria including *R. capsulatus* (Hoover *et al.*, 1990). Furthermore, gel-shift assays and/or DNase I footprinting studies confirm binding of *R. capsulatus* IHF to the promoter regions of *hupSL* (structural genes of hydrogenase; Toussaint *et al.*, 1991) and to the *puf* and *puc* operons (coding for proteins of the photosynthetic light-harvesting complex; Kirndörfer *et al.*, 1998). The IHF-binding sites identified in the promoter regions of *nifH* and *hupSL* show good similarity to the consensus sequence ATCAA–N_4–TTG (Craig and Nash 1984), suggesting that the *R. capsulatus* IHF resembles other prokaryotic IHF proteins with regard to target selection. A putative IHF-binding site is also present in the *R. capsulatus anfH* promotor upstream region, implying that IHF may also modulate activation of this promoter by AnfA.

3. THE NITROGEN-FIXATION REGULON OF *R. CAPSULATUS*

3.1. R. capsulatus *NtrC Activates Transcription of its Target Genes in Concert with RNA Polymerase containing the Housekeeping Sigma Factor*

The *R. capsulatus* NtrB and NtrC proteins (RcNtrB and RcNtrC) are members of a two-component regulatory system, in which NtrB acts as a sensor kinase and NtrC functions as the response regulator (see below). Genetic evidence demonstrates that RcNtrC is essential for transcriptional activation of both *nif* and *anf* genes *in vivo* (Kranz and Foster-Hartnett, 1990; Schüddekopf *et al.*, 1993).

The enteric NtrC protein is the paradigm for a family of bacterial enhancer-binding proteins (EBPs) that activate transcription of RNA polymerase containing the alternative sigma factor σ^{54} (Xu and Hoover, 2001). NifA (see section 2.4.1.) is one member of this class of EBPs and, like NifA, NtrC is modular in structure, consisting of three functional domains. The N-terminal domains of NtrC and NifA do not, however, share homology. Phosphorylation of an aspartate residue, which is conserved within the N-terminal domains of NtrC proteins, results in their activation. The central domains of NtrC and NifA are highly conserved among activators of RNAP-σ^{54} and are responsible for transcriptional activation and ATP hydrolysis. The C-terminal domains of NtrC and NifA contain helix-turn-helix motifs that are responsible for recognition and binding of the specific upstream activator sequences of their respective target genes.

In contrast to almost all other known EBPs that mediate expression of σ^{54}-dependent genes, *R. capsulatus* NtrC activates transcription of its target genes in concert with RNAP containing the housekeeping σ^{70}-like sigma factor (Bowman and Kranz, 1998; Pasternak *et al.*, 1996). This fact reflects a natural deletion in the central domain of RcNtrC, encompassing part of the so-called C3 region that is thought to define the contact point with σ^{54} (Foster-Hartnett *et al.*, 1994; Jones and Haselkorn, 1989; Xu and Hoover, 2001). As with the σ^{54}-dependent activators, binding of RcNtrC at distant tandem UAS sites is required for transcriptional activation (Bowman and Kranz, 1998; Foster-Hartnett *et al.*, 1994). In contrast to σ^{54}-dependent activators, however, ATP binding but not hydrolysis is required for transcriptional activation by RcNtrC. Another example of an EBP acting in concert with RNAP- σ^{70} is *E. coli* TyrR (Pittard and Davidson, 1991). In contrast to RcNtrC, however, TyrR is not part of a two-component regulatory system and it lacks the N-terminal domain that is present on NtrC.

RNA polymerase purified from *R. capsulatus* (Rc–RNAP) shows the same subunit composition (α, β, β', and ω) as RNAP from *E. coli* and other bacteria (Cullen *et al.*, 1997). Rc–RNAP recognizes –35 elements (with the consensus sequence TTGAC) similar to those recognized by *E. coli* RNAP. In contrast to the well-conserved –35 element, the *R. capsulatus* –10 element is less well-defined as an AT-rich sequence. However, *in vitro* studies of the binding of linear DNA templates suggest that some differences do exist in promoter recognition and/or transcription initiation between photosynthetic and enteric RNAP holoenzymes (Cullen *et al.*, 1997). One possible explanation for such differences may be that bacteria like *R. capsulatus*, with a genomic composition of 65 % GC, have evolved some recognition/melting differences from organisms that, like *E. coli*, have a considerably lower GC content. In line with this hypothesis, the housekeeping sigma factor of *R. capsulatus* (76 kDa) has been shown to be significantly larger than its *E. coli* σ^{70} counterpart (Pasternak *et al.*, 1996).

The *R. capsulatus ntrC* gene is part of the *nifR3-ntrB-ntrC* transcriptional unit (Cullen *et al.*, 1998). The *nifR3-ntrB-ntrC* operon is constitutively transcribed from a single promoter upstream of *nifR3*. The levels of transcript are equivalent in both wild-type and *ntrC* mutants and also under either nitrogen-limited or nitrogen-sufficient conditions. Although the concentration of NtrC protein remains constant, synthesis of NifR3 is somehow translationally controlled by the NtrC protein in a nitrogen-dependent manner.

3.2. Targets of R. capsulatus *NtrC*

Originally, the *ntrC* mutant (strain J61) of *R. capsulatus* was identified by its inability to accumulate nitrogenase polypeptides (Wall and Braddock, 1984; Wall *et al.*, 1984). The inability of *R. capsulatus ntrC* mutants to grow diazotrophically is due to the lack of transcriptional activation of *nifA1*, *nifA2*, and *anfA*, which encode the transcriptional activators of all the other *nif* and *anf* genes (Foster-Hartnett and Kranz, 1992; Foster-Hartnett *et al.*, 1993; Hübner *et al.*, 1991; Kutsche *et al.*, 1996). In contrast to enteric *ntrC* mutants, strain J61 does not exhibit the "classical" Ntr (nitrogen regulation) phenotype, as defined by the inability to grow on different amino acids (including glutamine, proline or arginine) as a nitrogen source. Although mutant J61 also lacks the ability to transport methylammonium, which indicates that it is involved in regulation of at least one other aspect of nitrogen metabolism, the gene complementing mutant J61 was initially called *nifR1*, suggesting a (mainly) *nif*-specific function (Avtges *et al.*, 1985; Jones and Haselkorn, 1989; Kranz and Haselkorn, 1985).

In addition to the *nifA1*, *nifA2*, and *anfA* genes, several other *R. capsulatus* genes with roles in nitrogen metabolism require RcNtrC for transcriptional activation. Among these are: *mopA-modABCD* (coding for a molybdate-dependent repressor and a high-affinity molybdenum-transport system; Kutsche *et al.*, 1996); *glnB-glnA* (coding for a PII-like signal transduction protein and glutamine synthetase; Kranz *et al.*, 1990); *glnK-amtB* (coding for a GlnB paralogue and a methyl-ammonium transporter; Drepper *et al.*, 2003; Masepohl *et al.*, 2002; Yakunin and Hallenbeck, 2000b); *amtY* (coding for a putative ammonium transporter; Masepohl *et al.*, 2002; Yakunin and Hallenbeck, 2000b); and *ureDABCEFG* (required for synthesis and activity of urease; Masepohl *et al.*, 2001).

Both the binding of RcNtrC to its target promoters and the influence of the phosphorylation status of RcNtrC on binding have been analyzed by *in vitro* studies with purified RcNtrC and RcNtrB proteins (Cullen *et al.*, 1996). RcNtrB autophosphorylates *in vitro* to give RcNtrB-P, which in turn phosphorylates the RcNtrC protein. RcNtrC forms a dimer in solution, and RcNtrC-P binds to the upstream tandem binding sites as demonstrated for the promoters of *nifA1*, *nifA2*, *glnB*, *anfA*, *mopA*, and *ureD* (Bowman and Kranz, 1998; Masepohl *et al.*, 2001). A consensus NtrC-binding site, similar to the enteric UAS, has been defined for *R. capsulatus* as CGCC–N_2–A(T/A)(T/A)–N–(T/A)T–N_2–GC (Foster Hartnett and Kranz, 1994).

As mentioned above, mutations in *R. capsulatus ntrC* do not affect the use of proline as sole source of nitrogen (Keuntje *et al.*, 1995). The expression of *putA* (coding for proline dehydrogenase) is induced by proline and is not affected by either ammonium or other amino acids. Similarly, synthesis of arginase is induced by arginine regardless of the presence of ammonium and other nitrogen sources (Moreno-Vivian *et al.*, 1992).

3.3. Members of the Ntr System of R. capsulatus

In enteric bacteria, a general nitrogen regulation (Ntr) system that senses the cellular nitrogen status is responsible for controlling the expression of many genes concerned with nitrogen metabolism (for a review, see Ninfa and Atkinson, 2000). The enteric Ntr system comprises five gene products. These are: a bifunctional uridylyltransferase/uridylyl-removing enzyme, encoded by *glnD*; two trimeric PII signal-transduction proteins, encoded by *glnB* and *glnK*; and the two-component regulatory system, encoded by *ntrB-ntrC*. In addition, a bifunctional adenylyl-transferase/adenylyl-removing enzyme, which is encoded by *glnE*, regulates activity of glutamine synthetase (GS), encoded by *glnA*. GlnE catalyzes adenylylation of GS in concert with the unmodified form of PII, whereas the deadenylylation reaction requires teh modified PII-UMP (Jiang *et al.*, 1998).

In response to the cellular glutamine/2-ketoglutarate ratio, PII is regulated by reversible GlnD-mediated uridylylation. When cells are N-limited, PII is uridylylated and the resulting PII-UMP does not interact with the sensor kinase NtrB. Under these conditions, NtrB autophosphorylates and promotes the phosphorylation of the response regulator NtrC. In turn, NtrC-P acts as a transcriptional activator of its target genes. Under N-sufficient conditions, PII is in its unmodified form. PII now can interact with NtrB, which then acts as a phosphatase, mediating dephosphorylation (and thereby inactivation) of NtrC.

R. capsulatus contains genes homologous to *glnD*, *glnB*, *glnK*, *ntrB*, *ntrC*, *glnE*, and *glnA* and, therefore, regulatory mechanisms that are similar to those of the enteric Ntr system have been proposed (Hübner *et al.*, 1991; Kranz and Foster-Hartnett, 1990; Masepohl and Klipp, 1996; Masepohl *et al.*, 2002; Overbeek *et al.*, 2000). GlnD-mediated uridylylation of enteric PII proteins occurs at a conserved tyrosine residue at position 51. Because both *R. capsulatus* GlnB and GlnK exhibit high similarity to their *E. coli* counterparts, including the Tyr_{51} residue, it seems likely that both *R. capsulatus* signal-transduction proteins are also regulated via (reversible) uridylylation. It is worth noting that uridylylation of the GlnB protein has been demonstrated for another phototrophic bacterium, *R. rubrum* (Johansson and Nordlund, 1997, 1999). As for *A. vinelandii* (Colnaghi *et al.*, 2001), *Corynebacterium glutamicum* (Jakoby *et al.*, 1999), and *Sinorhizobium meliloti* (Rudnick *et al.*, 2001), the *glnD* gene seems to be essential for viability in *R.*

capsulatus (W. Klipp and coworkers, unpublished results). However, the lethality of *glnD* null mutations in *A. vinelandii* can be suppressed by preventing GS adenylylation. Similarly, *R. capsulatus glnD* null mutations can be obtained in a *glnE* mutant background (W. Klipp and coworkers, unpublished results). It seems likely that the observed lethality of *R. capsulatus glnD* mutants is due to the absence of PII-UMP, leading to accumulation of the adenylylated (inactive) form of GS.

4. AMMONIUM CONTROL OF SYNTHESIS AND ACTIVITY OF BOTH NITROGENASES

4.1. Transcriptional Regulation of nifA1*,* nifA2*, and* anfA *by the Ntr System*

The current model of ammonium regulation of nitrogen fixation in *R. capsulatus* is shown in Figure 2. Synthesis and activity of both the molybdenum nitrogenase and the alternative nitrogenase is controlled on at least three levels by ammonium and other environmental factors (see below; Masepohl *et al.*, 2002). At the first level, transcription of the *nifA1*, *nifA2*, and *anfA* genes is controlled by the Ntr system. At the second level, activity of the NifA1, NifA2, and AnfA proteins is inhibited by ammonium in an NtrC-independent manner. At the third level, activity of both nitrogenase reductases (Fe proteins), NifH and AnfH, is regulated by DraT/DraG-mediated reversible ADP-ribosylation.

As described above, the cellular nitrogen status in *R. capsulatus* is measured by an Ntr system thought to be similar to that of enteric bacteria. In response to ammonium availability, RcNtrB acts either as a kinase (phosphorylating and, thereby, activating RcNtrC) or as a phosphatase (mediating dephosphorylation and, thereby, inactivation of RcNtrC). Activity of *E. coli* NtrB as either a kinase or a phosphatase is controlled by either of the two PII signal-transduction proteins, GlnB and GlnK (Ninfa and Atkinson, 2000). In contrast to the situation in *E. coli*, and despite the high degree of similarity between GlnB and GlnK, activity of *R. capsulatus* NtrB seems to be mainly regulated through GlnB, whereas GlnK has only a limited effect (Masepohl *et al.*, 2002). Mutations in the *R. capsulatus glnB* gene lead to high-level expression of *nifA1*, *nifA2*, and *anfA* both in the presence and in the absence of ammonium. In contrast, *R. capsulatus glnK* mutants show only weak transcription of both *nifA* and *anfA* in the presence of ammonium, whereas wild-type levels of expression are found under N-limiting conditions.

4.2. Regulation of NifA and AnfA Activity

As mentioned above, both *R. capsulatus* NifA1 and NifA2 are functional and can substitute for each other in transcriptional activation of all of the other *nif* genes (Masepohl *et al.*, 1988). A *glnB* mutation leads to accumulation of (at least) the NifA1 protein in the presence of ammonium, but NifA-mediated *nifH* transcription is still inhibited by ammonium (Paschen *et al.*, 2001). This inhibition of NifA activity is independent of NtrC as demonstrated by analysis of *R. capsulatus* mutant strains overexpressing either *nifA1* or *nifA2* from a constitutive promoter (Drepper *et al.*, 2000, 2003; Hübner *et al.*, 1993; Kern *et al.*, 1998). In contrast to lesions in *glnB*, a *glnK* mutation leads to significant expression of *nifH* in the presence of ammonium (in *R. capsulatus* strains constitutively expressing *nifA*), indicating that NifA activity partially escapes ammonium inhibition in this mutant background.

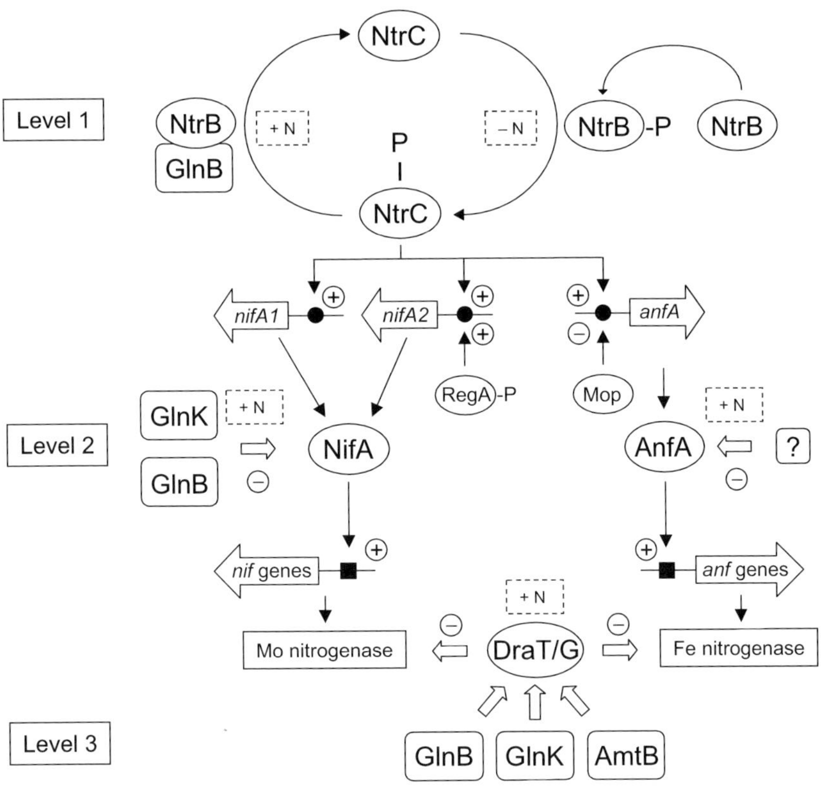

Figure 2. Model of the nitrogen-fixation regulon in R. capsulatus. *The model describes the regulatory cascade controlling expression of nitrogen-fixation genes and activity of the molybdenum-containing nitrogenase (*nif*-encoded) and the heterometal-free nitrogenase (*anf*-encoded) in response to ammonium availability. Ammonium control occurs on three levels (for details, see text). Nitrogenase-repressing conditions (presence of ammonium) or derepressing conditions (absence of ammonium) are symbolized by* [+N] *and* [–N], *respectively. Promoters indicated by solid circles are σ^{70}-dependent and activated by NtrC-P; promoters indicated by solid squares are σ^{54}-dependent and activated by NifA or AnfA.*

Modulation of NifA activity in response to ammonium is mediated by PII signal-transduction proteins as shown for many diazotrophic bacteria, including *Azorhizobium caulinodans* (Michel-Reydellet and Kaminski, 1999), *A. brasilense* (Arsene *et al.*, 1996, 1999), *A. vinelandii* (Little *et al.*, 2000), *H. seropedicae* (Souza *et al.*, 1999), *K. pneumoniae* (Arcondeguy *et al.*, 1999; He *et al.*, 1998; Jack *et al.*, 1999), *R. capsulatus* (Drepper *et al.*, 2003; Masepohl *et al.*, 2002), and *R. rubrum* (Zhang *et al.*, 2000). As shown by yeast two-hybrid studies, the two *R. capsulatus* PII proteins, GlnB and GlnK, interact with NifA1 and NifA2, suggesting that ammonium inhibition of activity of both NifA1 and NifA2 is directly mediated by these signal-transduction proteins (Masepohl *et al.*, 2002). It should be emphasized that three different mechanisms, which link PII proteins to ammonium control of NifA activity, are present in diazotrophic bacteria. First, in *A. vinelandii* and *K. pneumoniae*, GlnK is required for regulation of NifL-mediated inhibition of NifA activity. Second, in *A. brasilense* and *R. rubrum*, NifA activity (under nitrogenase-derepressing conditions) depends on GlnB's presence because NifA is inactive in a *glnB* mutant. Third, in *R. capsulatus* and *A. caulinodans*, neither GlnB nor GlnK is required for NifA activity because NifA is active in a *glnB*/*glnK* double mutant, but both GlnB and GlnK can mediate ammonium inhibition of NifA activity.

An *R. capsulatus glnB*/*glnK* double mutant synthesizes the alternative nitrogenase under nitrogen- and molybdenum-limiting conditions (Drepper *et al.*, 2003; Masepohl *et al.*, 2002), demonstrating that neither of the two PII proteins is required for AnfA activity. In contrast to NifA activity, AnfA activity is still inhibited by ammonium in *glnB*/*glnK* double mutants of *R. capsulatus* strains that constitutively express *anfA*. This inhibition is also independent of NtrC. Therefore, ammonium control of AnfA activity involves an as yet unidentified GlnB/GlnK-independent regulatory mechanism.

4.3. Post-translational Control of Nitrogenase Activity

Post-translational regulation of nitrogenase has been described for a few diazotrophic bacteria, including *R. capsulatus*, *R. rubrum*, *A. brasilense*, *Azospirillum lipoferum* and *Chromatium vinosum*. *R. capsulatus* is capable of switching-off nitrogenase activity in response to either addition of ammonium or changes in light intensity (exposure to darkness), when dinitrogenase reductase (Fe protein) is subjected to reversible ADP-ribosylation (Hallenbeck, 1992; Jouanneau *et al.*, 1989; Pierrard *et al.*, 1993). In contrast to *R. rubrum*, where ADP-ribosylation seems to be the sole mechanism responsible for switch-off of nitrogenase activity, *R. capsulatus* modulates its *in vivo* nitrogenase activity in response to the addition of ammonium not only by ADP-ribosylation of nitrogenase reductase but also with a switch-off response, which is independent of ADP-ribosylation, and with a so-called magnitude response (Förster *et al.*, 1999; Pierrard *et al.*, 1993; Yakunin and Hallenbeck, 1998b, 2000a; Yakunin *et al.*, 1999).

The *draT* gene product (dinitrogenase reductase ADP-ribosyltransferase) catalyzes ADP-ribosylation of dinitrogenase reductase at a highly conserved arginine residue (Arg_{102} in *R. capsulatus* NifH), whereas restoration of nitrogenase activity is performed by the DraG protein (dinitrogenase reductase-activating glycohydrolase). The *draTG* gene region of *R. capsulatus* is required for ADP-ribosylation of both dinitrogenase reductases, NifH and AnfH (Masepohl *et al.*, 1993b). In an *R. capsulatus glnB*/*glnK* double mutant, the ammonium control mechanisms of both nitrogenase systems are relieved (Drepper *et al.*, 2003; Klipp *et al.*, 2000; Masepohl *et al.*, 2002). In contrast, ADP-ribosylation of NifH and AnfH in response to darkness is not affected in a *glnB*/*glnK* mutant.

Most remarkably, in an *R. capsulatus glnB*/*glnK* double mutant, ammonium control of the molybdenum nitrogenase is completely relieved, leading to synthesis of active nitrogenase in the presence of high concentrations of ammonium (Drepper *et al.*, 2003; Masepohl *et al.*, 2002). Both in the presence and absence of ammonium, significantly higher amounts of the molybdenum nitrogenase accumulate in the *glnB*/*glnK* double mutant, resulting in higher nitrogenase activity than in the wild-type grown under nitrogen-limiting conditions. In contrast, the *glnB*/*glnK* double mutant does not synthesize the alternative nitrogenase in the presence of ammonium, most likely due to ammonium inhibition of AnfA activity (see above; Drepper *et al.*, 2003; Masepohl *et al.*, 2002).

It is worth noting that the *draTG* and *rnf* genes also play a role in nitrophenol reductase-mediated reduction of 2,4-dinitrophenol (DNP) by *R. capsulatus* (Saez *et al.*, 2001). Beause nitrophenol reductase itself is not the target of ADP-ribosylation, it has been speculated that either the electron-transfer system (encoded by *rnf*) or the DNP-uptake system may be controlled by DraT/DraG.

4.4. Role of (Methyl)-Ammonium Transporters in Regulation of Nitrogen Fixation

R. capsulatus contains two genes, *amtB* and *amtY*, coding for (methyl)-ammonium transporters (Masepohl *et al.*, 2002; Yakunin and Hallenbeck, 2000b). In bacteria and archaea, the *amtB* genes are invariably linked to *glnK* (Thomas *et al.*, 2000). GlnK of *E. coli* and *A. vinelandii* have been shown to bind to the membrane in an AmtB-dependent manner with GlnK acting as a negative regulator of the transport activity of AmtB (Coutts *et al.*, 2002). It is worth mentioning that *Azoarcus* GlnK is still detectable in association with membranes in an *amtB* mutant strain (Martin and Reinhold-Hurek, 2002). Membrane binding in *E. coli* and *A. vinelandii* is dependent on the uridylyation state of GlnK, such that it is maximal in nitrogen-sufficient conditions. An *R. capsulatus amtB* mutant (but not an *amtY* mutant), in addition to being ineffective in methyl-ammonium uptake, is also completely defective in carrying out ADP-ribosylation of NifH in response to ammonium addition (Yakunin and Hallenbeck, 2000b). As shown for the *glnB*/*glnK* double mutant, ADP-ribosylation in response to darkness is also unaltered in the *amtB* mutant. However, the *amtY* mutant shows higher levels of ADP-ribosylation than the wild-type strain in response to addition of the same amount of ammonium.

5. ENVIRONMENTAL FACTORS CONTROLLING NITROGEN FIXATION

Bcause nitrogen fixation is a highly energy-demanding process, different environmental factors control synthesis and activity of the two nitrogenases in *R. capsulatus*. Among these are (beside ammonium) the environmental O_2 partial pressure, light intensity, the C/N ratio, and the availability of molybdenum and iron. Nitrogenase activity and diazotrophic growth of *R. capsulatus* is optimal under photosynthetic (anaerobic) growth conditions. The formation of the photosynthetic apparatus of *R. capsulatus* is primarily regulated by O_2 tension (for a review, see Gregor and Klug, 1999). If the O_2 tension in the environment drops below a threshold value, the formation of the photosynthetic apparatus is induced. As mentioned above, the central activator of all the other *R. capsulatus nif* genes, NifA, is thought to be O_2-sensitive, a feature allowing *nif* gene expression only under conditions favourable for activity of the O_2-sensitive nitrogenase.

In anaerobic *R. capsulatus* cultures, the amount of photosynthetic complexes is influenced by light intensity, being greater under low *versus* high light intensities. This low-light induction is mediated by the *hvrA* gene product (Buggy *et al.*, 1994), which also influences *nif*-gene expression, thereby acting as a regulatory link between photosynthesis and nitrogen fixation (Kern *et al.*, 1998; see below). As described above, exposure to darkness results in switch-off of nitrogenase activity. In addition, light intensity controls the C/N ratio of *R. capsulatus* cells, which in turn controls synthesis and activity of nitrogenase (Maner and Oelze, 1999).

Availability of heterometals influences synthesis and activity of both the molybdenum-containing and alternative nitrogenases in different bacteria to different extents. In *R. capsulatus*, expression of the *anfA* gene is repressed by traces of molybdenum via the molybdate-dependent repressor proteins, MopA and MopB, whereas added vanadium has no effect on diazotrophic growth via the alternative nitrogenase (Gollan *et al.*, 1993; Kutsche *et al.*, 1996; Schneider *et al.*, 1991b; Wang *et al.*, 1993). In addition, molybdenum controls activity of the AnfA protein in a MopA/MopB-independent manner (Drepper *et al.*, 2000). Like *R. capsulatus*, *R. rubrum* harbors a molybdenum nitrogenase and an alternative (iron-only) nitrogenase (Davis *et al.*, 1996; Lehman and Roberts, 1991). In contrast to *R. capsulatus*, the alternative nitrogenase of *R. rubrum* does not appear to be metal-regulated (Lehman and Roberts, 1991). The alternative nitrogenase of *R. rubrum* is not repressed by molybdenum, but is expressed whenever a strain lacks an active molybdenum nitrogenase because of either physiological or genetic inactivation.

In *R. capsulatus*, iron availability influences electron transport to nitrogenase. Under iron-replete conditions, electron transport to nitrogenase is, mainly, mediated by the *rnf*-gene products and the two-[4Fe-4S]-comtaining ferredoxin, FdxN (Schmehl *et al.*, 1993). Under iron-limiting conditions, the flavodoxin, NifF, may be the only electron donor to nitrogenase (Genaro *et al.*, 1996). Accordingly, the *nifF* gene is highly expressed under iron limitation. Furthermore, iron limitation results in a decrease both in the cellular content of the iron-sulfur protein, RnfB, and in the level of transcription of the *rnfABCDGEH* operon (Jouanneau *et al.*, 1998).

6. LINKAGE OF NITROGEN FIXATION, PHOTOSYNTHESIS, AND CARBON DIOXIDE ASSIMILATION

R. capsulatus RegB/RegA form a classic two-component regulatory system, with RegB as a membrane-spanning histidine kinase capable of autophosphorylation and RegA as the cytosolic response regulator (Bird *et al.*, 1999; Sganga and Bauer, 1992). RegB/RegA functions as a global regulatory system that activates photosynthesis, carbon-assimilation, and nitrogen-fixation genes, and represses hydrogenase structural genes and its own expression (Elsen *et al.*, 2000). RegA indirectly activates synthesis of the molybdenum nitrogenase by binding to and activating the expression of *nifA2* but not *nifA1* (Figure 2). Thus, although RcNtrC is absolutely required for *nifA2* transcription, RegA acts as a coactivator of *nifA2*.

RegB/RegA-like systems are widespread among α-Proteobacteria, including the purple photosynthetic bacteria, *R. capsulatus*, *R. sphaeroides*, *Rhodovulum sulfidophilum*, and *Roseobacter denitrificans*, as well as the symbiotic bacteria, *Bradyrhizobium japonicum* and *S. meliloti* (Emmerich *et al.*, 2000; Joshi and Tabita, 1996; Qian *et al.*, 1996). RegA and its homologues exhibit an unprecedented high degree of conservation, especially in the C-terminal DNA-binding helix-turn-helix structure. In line with this finding, *R. capsulatus* RegA, *B. japonicum* RegR, and *S. meliloti* ActR were shown to be functionally similar.

The *R. capsulatus regA* gene is cotranscribed with *hvrA*, which encodes a histone-like protein exhibiting significant sequence similarity to *E. coli* H-NS. Both HvrA and H-NS are functionally similar because HvrA complements various phenotypes of an *E. coli hns* mutant (Bertin *et al.*, 1999). As mentioned above, HvrA is involved in low-light activation of the photosynthetic apparatus (Buggy *et al.*, 1994). A mutation in *hvrA* releases ammonium control of *nifH* expression in an *R. capsulatus* strain constitutively expressing *nifA1* (Kern *et al.*, 1998). This regulatory effect of the HvrA protein is mediated by the binding of HvrA to the *nifH* promoter as demonstrated by competitive gel-retardation studies (Raabe *et al.*, 2002).

7. NITROGEN FIXATION IN OTHER PHOTOSYNTHETIC PURPLE BACTERIA

7.1. Rates of Nitrogen Fixation in Purple Non-sulfur Bacteria

The capacity to fix molecular nitrogen (N_2) is virtually universal among members of the purple nonsulfur bacteria but the efficacy of the process varies considerably among species (for a review, see Madigan, 1995). Among 26 species placed in eight genera (*Rhodobacter*, *Rhodopseudomonas*, *Rhodospirillum*, *Rhodocyclus*, *Rhodopila*, *Rhodomicrobium*, *Rhodoferax*, and *Rubrivivax*), 24 species have been documented as fixing N_2. Species of *Rhodobacter*, such as *R. capsulatus* and *R. sphaeroides*, express high levels of nitrogenase *in vivo* and grow most rapidly on N_2, whereas species, like *Rhodopseudomonas palustris* and *R. Rubrum*, express relatively low nitrogenase levels and grow only very slow on N_2. *In vivo*, specific nitrogenase activities correlate to a great extent with rates of diazotrophic growth among purple nonsulfur bacteria, with strains of *R. capsulatus* consistently showing the highest specific nitrogenase activities and shortest generation times for photoheterotrophic growth on N_2 (Madigan *et al.*, 1984). Indeed, enrichment cultures for purple nonsulfur bacteria, using the ability to fix N_2 as selective agent, commonly yield *Rhodobacter* species, especially *R. capsulatus*.

Molecular analyses of the nitrogen-fixation process have been carried out in some detail in only three species of purple nonsulfur bacteria, *R. capsulatus*, *R. sphaeroides*, and *R. rubrum*, with most studies concerning N_2 fixation being performed in *R. capsulatus*. Two of these species, *R. capsulatus* and *R. rubrum*, have been described as containing an alternative (iron-only) nitrogenase (Davis *et al.*, 1996; Lehman and Roberts, 1991; Schneider *et al.*, 1991b, 1997). However, at present, it remains unknown if heterometal-free nitrogenases are widespread among purple nonsulfur bacteria and if any species of this family harbor a vanadium-containing nitrogenase.

7.2. Regulation of Nitrogen Fixation in Rhodobacter sphaeroides

As described above for *R. capsulatus*, *R. sphaeroides* contains a RegB/RegA-like global two-component regulatory system (PrrB/PrrA) that integrates the control of nitrogen fixation, photosynthesis, and carbon-dioxide assimilation (Joshi and Tabita, 1996). Although the *regA* and *regB* genes of both bacteria are organized similarly, and the RegB/RegA system exhibits similar functions in regulating photosynthesis gene expression, there appear to be differences relative to the severity of a mutation in *regA* (*prrA*) in the two organisms (Qian and Tabita, 1996).

In a ribulose 1,5-bisphosphate carboxylase/oxygenase (RubisCO)-deficient mutant of *R. sphaeroides* (strain 16PHC), nitrogenase activity is derepressed in the presence of ammonium under photoheterotrophic growth conditions (Qian and Tabita, 1998). In this case, the reduction of protons by the molybdenum-containing nitrogenase enzyme complex is thought to serve as an alternative redox-balancing mechanism. Reintroduction of a functional RubisCO suppresses the deregulation of nitrogenase synthesis in strain 16PHC. Like many other bacteria, *R. sphaeroides* contains two PII-encoding genes, *glnB* and *glnK* (Quian and Tabita, 1998). In contrast to the wild-type, where *glnK* transcription is under tight nitrogen control, expression of *glnK* is derepressed in strain 16PHC under photoheterotrophic conditions in the presence of ammonium. However, the mechanism underlying deregulation of *glnK* expression and nitrogenase synthesis remains to be elucidated. It should be mentioned at this point, that *R. sphaeroides ntrC* mutants have been isolated and these also show derepression of nitrogenase synthesis in the presence of ammonium (Zinchenko *et al.*, 1997).

In *R. capsulatus*, the *rpoN* gene is under nitrogen control, and the *rpoN* gene product is absolutely essential for expression of nitrogen-fixation genes. The *rpoN* gene from *R. sphaeroides* has been isolated via complementation of an *R. capsulatus rpoN* mutant, thereby demonstrating its functionality (Meijer and Tabita, 1992). The *R. sphaeroides rpoN* gene is part of the *nifUSVW-rpoN* transcriptional unit. In contrast to the situation in *R. capsulatus*, mutations in the *R. sphaeroides rpoN* gene do not impair the ability to grow diazotrophically and it has been discussed that a second, not yet identified, sigma factor might substitute for RpoN.

As described above, the *draT*/*draG* gene products are involved in post-translational regulation of nitrogenase activity in *R. capsulatus* and *R. rubrum*. In contrast to these bacteria, *R. sphaeroides* appears to have no *draTG* genes and no evidence of ADP-ribosylation of dinitrogenase reductase (Fe protein) was found in this bacterium (Yakunin *et al.*, 2001). However, transfer of *R. capsulatus draTG* is sufficient to confer on *R. sphaeroides* the ability to modify reversibly the NifH protein in response to either ammonium addition or darkness, suggesting that *R. sphaeroides* possesses all the elements necessary for the transduction of signals generated by either ammonium or darkness to the DraT/DraG proteins.

7.3. Regulation of Nitrogen Fixation in Rhodospirillum rubrum

Like *R. capsulatus*, *R. rubrum* harbors a molybdenum-containing and an alternative (iron-only) nitrogenase, and both bacteria contain *draTG* genes mediating reversible ADP-ribosylation of dinitrogenase reductase (see above). In contrast to *R. capsulatus*, the *R. rubrum* NifA protein requires GlnB for activation (Zhang *et al.*, 2000). Similar to the situation in *R. capsulatus*, the *R. rubrum glnB* gene is cotranscribed with *glnA*, which encodes glutamine synthetase (Johansson and Nordlund, 1996). As mentioned above, the *R. rubrum* GlnB protein is modified by reversible uridylylation in response to nitrogen status (Johansson and Nordlund, 1997; 1999). In contrast to many other bacteria, which harbor two PII-like proteins, *R. rubrum* contains three PII-like proteins, called GlnB, GlnK, and GlnJ (Zhang *et al.*, 2001). Unlike GlnB, both GlnK and GlnJ have no significant effect on *nif* gene expression. Although no effect on growth is associated with a *glnK* mutation, GlnK is synthesized in *R. rubrum* and supports normal modification of glutamine synthetase in response to ammonium. Either GlnB or GlnJ can serve as a critical element in regulating the reversible ADP-ribosylation of NifH. Both GlnB and GlnJ appear to be responsive not only to changes in nitrogen status but also to changes in energy status.

REFERENCES

Arcondeguy, T., van Heeswijk, W. C., and Merrick, M. (1999). Studies on the roles of GlnK and GlnB in regulating *Klebsiella pneumoniae* NifL-dependent nitrogen control. *FEMS Microbiol. Lett., 180*, 263-270.

Armengaud, J., Meyer, C., and Jouanneau, Y. (1994). Recombinant expression of the *fdxD* gene of *Rhodobacter capsulatus* and characterization of its product, a [2Fe-2S] ferredoxin. *Biochem. J., 300*, 413-418.

Armengaud, J., Meyer, C., and Jouanneau, Y. (1997). A [2Fe-2S] ferredoxin (FdVI) is essential for growth of the photosynthetic bacterium *Rhodobacter capsulatus*. *J. Bacteriol., 179*, 3304-3309.

Arsene, F., Kaminski, P. A., and Elmerich, C. (1996). Modulation of NifA activity by P_{II} in *Azospirillum brasilense*: evidence for a regulatory role of the NifA N-terminal domain. *J. Bacteriol., 178*, 4830-4838.

Arsene, F., Kaminski, P. A., and Elmerich, C. (1999). Control of *Azospirillum brasilense* NifA activity by P_{II}: effect of replacing Tyr residues of the NifA N-terminal domain on NifA activity. *FEMS Microbiol. Lett., 179*, 339-343.

Avtges, P., Kranz, R. G., and Haselkorn, R. (1985). Isolation and organization of genes for nitrogen fixation in *Rhodopseudomonas capsulata*. *Mol. Gen. Genet., 201*, 363-369.

Bertin, P., Benhabiles, N., Krin, E., Laurent-Winter, C., Tendeng, C., Turlin, E., *et al.* (1999). The structural and functional organization of H-NS-like proteins is evolutionary conserved in Gram-negative bacteria. *Mol. Microbiol., 31*, 319-329.

Bird, T. H., Du, S., and Bauer, C. E. (1999). Autophosphorylation, phosphotransfer, and DNA-binding properties of the RegB/RegA two-component regulatory system in *Rhodobacter capsulatus*. *J. Biol. Chem., 274*, 16343-16348.

Bishop, P. E., and Premakumar, R. (1992). Alternative nitrogen fixation systems. In G. Stacey, R. H. Burris, and H. J. Evans (Eds.), *Biological nitrogen fixation* (pp. 736-762). New York: Chapman and Hall.

Bowman, W. C., and Kranz, R. G. (1998). A bacterial ATP-dependent enhancer binding protein that activates the housekeeping RNA polymerase. *Genes Development,* 12, 1884-1893.

Buggy, J. J., Sganga, M. W., and Bauer, C. E. (1994). Characterization of a light-responding *trans*-activator responsible for differentially controlling reaction center and light-harvesting-I gene expression in *Rhodobacter capsulatus*. *J. Bacteriol., 176*, 6936-6943.

Burke, D. H., Alberti, M., and Hearst, J. E. (1993a). The *Rhodobacter capsulatus* chlorin reductase-encoding locus, *bchA*, consists of three genes, *bchX*, *bchY*, and *bchZ*. *J. Bacteriol., 175*, 2407-2413.

Burke, D. H., Hearst, J. E., and Sidow, A. (1993b). Early evolution of photosynthesis: Clues from nitrogenase and chlorophyll iron proteins. *Proc. Natl. Acad. Sci. USA*, *90*, 7134-7138.

Cannon, W., Missailidis, S., Austin, S., Moore, M., Drake, A., and Buck, M. (1996). Purification and activities of the *Rhodobacter capsulatus* RpoN (σ^N) protein. *Mol. Microbiol., 21*, 233-245.

Colnaghi, R., Rudnick, P., He, L., Green, A., Yan, D., Larson, E., and Kennedy, C. (2001). Lethality of *glnD* null mutations in *Azotobacter vinelandii* is suppressible by prevention of glutamine synthetase adenylylation. *Microbiol., 147*, 1267-1276.

Coutts, G., Thomas, G., Blakey, D., and Merrick, M. (2002). Membrane sequestration of the signal transduction protein GlnK by the ammonium transporter AmtB. *EMBO J., 21*, 536-545.

Craig, N. L., and Nash, H. A. (1984). *E. coli* integration host factor binds to specific sites in DNA. *Cell*, *39*, 707-716.

Cullen, P. J., Bowman, W. C., Foster-Hartnett, D., Reilly, S. C., and Kranz, R. G. (1998). Translational activation by an NtrC enhancer-binding protein. *J. Mol. Biol., 278*, 903-914.

Cullen, P. J., Bowman, W. C., and Kranz, R. G. (1996). *In vitro* reconstitution and characterization of the *Rhodobacter capsulatus* NtrB and NtrC two-component system. *J. Biol. Chem., 271*, 6530-6536.

Cullen, P. J., Foster-Hartnett, D., Gabbert, K. K., and Kranz, R. G. (1994). Structure and expression of the alternative sigma factor, RpoN, in *Rhodobacter capsulatus*; Physiological relevance of an autoactivated *nifU2-rpoN* superoperon. *Mol. Microbiol., 11*, 51-65.

Cullen, P. J., Kaufman, C. K., Bowman, W. C., and Kranz, R. G. (1997). Characterization of the Rhodobacter capsulatus housekeeping RNA polymerase. *J. Biol. Chem., 272*, 27266-27273.

Davis, R., Lehman, L., Petrovich, R., Shah, V. K., Roberts, G. P., and Ludden, P. W. (1996). Purification and characterization of the alternative nitrogenase from the photosynthetic bacterium *Rhodospirillum rubrum*. *J. Bacteriol., 178*, 1445-1450.

Dixon, R. (1998). The oxygen-responsive NIFL-NIFA complex: A novel two-component regulatory system controlling nitrogenase synthesis in γ-proteobacteria. *Arch. Microbiol., 169*, 371-380.

Drepper, T., Groß, S., Masepohl, B., and Klipp, W. (2000). Ammonium and molybdenum regulation of the molybdenum nitrogenase and the alternative nitrogenase in *Rhodobacter capsulatus*. In F. O. Pedrosa, M. Hungria, G. Yates and W. E. Newton (Eds.), *Nitrogen fixation: From molecules to crop productivity* (p. 129). Dordrecht: Kluwer Academic Publishers.

Drepper, T., Groß, S., Yakunin, A. F., Hallenbeck, P. C., Masepohl, B., and Klipp, W. (2003). Role of GlnB and GlnK in ammonium control of both nitrogenase systems in the phototrophic bacterium *Rhodobacter capsulatus*. *Microbiol., 149*, 2203-2212.

Drummond, M., Walmsley, J., and Kennedy, C. (1996). Expression from the *nifB* promoter of *Azotobacter vinelandii* can be activated by NifA, VnfA, or AnfA transcriptional activators. *J. Bacteriol., 178*, 788-792.

Duport, C., Jouanneau, Y., and Vignais, P. M. (1990). Nucleotide sequence of *fdxA* encoding a 7Fe ferredoxin of *Rhodobacter capsulatus*. *Nucl. Acids Res., 18*, 4618.

Duport, C., Jouanneau, Y., and Vignais, P. M. (1992). Transcriptional analysis and promoter mapping of the *fdxA* gene which encodes the 7Fe ferredoxin (FdII) of *Rhodobacter capsulatus*. *Mol. Gen. Genet., 231*, 323-328.

Elsen, S., Dischert, W., Colbeau, A., and Bauer, C. E. (2000). Expression of uptake hydrogenase and molybdenum nitrogenase in *Rhodobacter capsulatus* is coregulated by the RegB-RegA two-component regulatory system. *J. Bacteriol., 182*, 2831-2837.

Emmerich, R., Hennecke, H., and Fischer, H.-M. (2000). Evidence for a functional similarity between the two-component regulatory systems RegSR, ActSR, and RegBA (PrrBA) in α-proteobacteria. *Arch. Microbiol., 174*, 307-313.

Fani, R., Gallo, R., and Lio, P. (2000). Molecular evolution of nitrogen fixation: The evolutionary history of the *nifD*, *nifK*, *nifE*, and *nifN* genes. *J. Mol. Evol., 51*, 1-11.

Fischer, H.-M. (1994). Genetic regulation of nitrogen fixation in Rhizobia. *Microbiol. Rev., 58*, 352-386.

Fonstein, M., Koshy, E. G., Nikolskaya, T., Mourachov, P., and Haselkorn, R. (1995). Refinement of the high-resolution physical and genetic map of *Rhodobacter capsulatus* and genome surveys using blots of the cosmid encyclopedia. *EMBO J., 14*, 1827-1841.

Fonstein, M., Nikolskaya, T., Kogan, Y., and Haselkorn, R. (1998). Genome encyclopedias and their use for comparative analysis of *Rhodobacter capsulatus* strains. *Electrophoresis, 19*, 469-477.

Fonstein, M., Zheng, S., and Haselkorn, R. (1992). Physical map of the genome of *Rhodobacter capsulatus* SB 1003. *J. Bacteriol., 174*, 4070-4077.

Förster, B., Maner, K., Fassbinder, F., and Oelze, J. (1999). Reversible inactivation of nitrogenase in *Rhodobacter capsulatus* strain W107I deleted in the *draTG* gene region. *FEMS Microbiol. Lett., 170*, 167-171.

Foster-Hartnett, D., Cullen, P. J., Gabbert, K. K., and Kranz, R. G. (1993). Sequence, genetic, and *lacZ* fusion analyses of a *nifR3-ntrB-ntrC* operon in *Rhodobacter capsulatus*. *Mol. Microbiol., 8*, 903-914.

Foster-Hartnett, D., Cullen, P. J., Monika, E. M., and Kranz, R. G. (1994). A new type of NtrC transcriptional activator. *J. Bacteriol., 176*, 6175-6187.

Foster-Hartnett, D., and Kranz, R. G. (1992). Analysis of the promoters and upstream sequences of *nifA1* and *nifA2* in *Rhodobacter capsulatus*; activation requires *ntrC* but not *rpoN*. *Mol. Microbiol., 6*, 1049-1060.

Foster-Hartnett, D., and Kranz, R. G. (1994). The *Rhodobacter capsulatus glnB* gene is regulated by NtrC at tandem *rpoN*-independent promoters. *J. Bacteriol., 176*, 5171-5176.

Gennaro, G., Hübner, P., Sandmeier, U., Yakunin, A. F., and Hallenbeck, P. C. (1996). Cloning, characterization, and regulation of *nifF* from *Rhodobacter capsulatus*. *J. Bacteriol., 178*, 3949-3952.

Gollan, U., Schneider, K., Müller, A., Schüddekopf, K., and Klipp, W. (1993). Detection of the *in vivo* incorporation of a metal cluster into a protein. The FeMo cofactor is inserted into the FeFe protein of the alternative nitrogenase in *Rhodobacter capsulatus*. *Eur. J. Biochem., 215*, 25-35.

Gosink, M. M., Franklin, N. M., and Roberts, G. P. (1990). The product of the *Klebsiella pneumoniae nifX* gene is a negative regulator of the nitrogen fixation (*nif*) regulon. *J. Bacteriol., 172*, 1441-1447.

Grabau, C., Schatt, E., Jouanneau, Y., and Vignais, P. M. (1991). A new [2Fe-2S] ferredoxin from *Rhodobacter capsulatus*. *J. Biol. Chem., 266*, 3294-3299.

Gregor, J., and Klug, G. (1999). Regulation of bacterial photosynthesis genes by oxygen and light. *FEMS Microbiol. Lett., 179*, 1-9.

Grunden, A. M., Ray, R. M., Rosentel, J. K., Healy, F. G., and Shanmugam, K. T. (1996). Repression of the *Escherichia coli modABCD* (molybdate transport) operon by ModE. *J. Bacteriol., 178*, 735-744.

Hinton, S. M., and Merritt, B. (1986). Purification and characterization of a molybdenum-pterin-binding protein (Mop) in *Clostridium pasteurianum* W5. *J. Bacteriol., 168*, 688-693.

Hinton, S. M., Slaughter, C., Eisner, W., and Fisher, T. (1987). The molybdenum-pterin binding protein is encoded by a multigene family in *Clostridium pasteurianum*. *Gene, 54*, 211-219.

Hoover, T. R. (2000). Control of nitrogen fixation genes in *Klebsiella pneumoniae*. In E. W. Triplett (Ed.), *Prokaryotic nitrogen fixation* (pp. 131-147). Wymondham, UK: Horizon Scientific Press.

Hoover, T. R., Santero, E., Porter, S., and Kustu, S. (1990). The integration host factor stimulates interaction of RNA polymerase with NIFA, the transcriptional activator for nitrogen fixation operons. *Cell, 63*, 11-22.

Hübner, P., Masepohl, B., Klipp, W., and Bickle, T. A. (1993). *nif* gene expression studies in *Rhodobacter capsulatus*: *ntrC*-independent repression by high ammonium concentrations. *Mol. Microbiol., 10*, 123-132.

Hübner, P., Willison, J. C., Vignais, P. M., and Bickle, T. A. (1991). Expression of regulatory *nif* genes in *Rhodobacter capsulatus*. *J. Bacteriol., 173*, 2993-2999.

Imhoff, J. F., Trüper, H. G., and Pfennig, N. (1984). Rearrangement of the species and genera of the phototrophic "purple nonsulfur bacteria". *Int. J. Syst. Bacteriol., 34*, 340-343.

Jack, R., de Zamaroczy, M., and Merrick, M. (1999). The signal transduction protein GlnK is required for NifL-dependent nitrogen control of *nif* gene expression in *Klebsiella pneumoniae*. *J. Bacteriol., 181*, 1156-1162.

Jacobson, M. R., Brigle, K. E., Bennett, L. T., Setterquist, R. A., Wilson, M. S., Cash, V. L., *et al.* (1989). Physical and genetic map of the major *nif* gene cluster from *Azotobacter vinelandii*. *J. Bacteriol., 171*, 1017-1027.

Jakoby, M., Krämer, R., and Burkovski, A. (1999). Nitrogen regulation in *Corynebacterium glutamicum*: isolation of genes involved and biochemical characterization of corresponding proteins. *FEMS Microbiol. Lett., 173*, 303-310.

Jeong, H.-S., and Jouanneau, Y. (2000). Enhanced nitrogenase activity in strains of *Rhodobacter capsulatus* that overexpress the *rnf* genes. *J. Bacteriol., 182*, 1208-1214.

Jepson, B., Money, T., and Austin, S. (2002). *In vitro* activity of the transcriptional activator AnfA from *Azotobacter vinelandii*. In T. M. Finan, M. R. O´Brien, D. B. Layzell, J. K. Vessey and W. E. Newton (Eds.), *Nitrogen fixation: Global perspectives* (p. 398). Wallingford, UK: CABI Publishing.

Jiang, P., Peliska, J. A., and Ninfa, A. J. (1998). The regulation of *Escherichia coli* glutamine synthetase revisited: role of 2-ketoglutarate in the regulation of glutamine synthetase adenylylation state. *Biochem., 37*, 12802-12810.

Johansson, M., and Nordlund, S. (1996). Transcription of the *glnB* and *glnA* genes in the photosynthetic bacterium *Rhodospirillum rubrum*. *Microbiol., 142*, 1265-1272.

Johansson, M., and Nordlund, S. (1997). Uridylylation of the P_{II} protein in the photosynthetic bacterium *Rhodospirillum rubrum*. *J. Bacteriol., 179*, 4190-4194.

Johansson, M., and Nordlund, S. (1999). Purification of P_{II} and P_{II}-UMP and in vitro studies of regulation of glutamine synthetase in *Rhodospirillum rubrum*. *J. Bacteriol., 181*, 6524-6529.

Jones, R., and Haselkorn, R. (1989). The DNA sequence of the *Rhodobacter capsulatus ntrA*, *ntrB* and *ntrC* gene analogues required for nitrogen fixation. *Mol. Gen. Genet., 215*, 507-516.

Joshi, H. M., and Tabita, F. R. (1996). A global two component signal transduction system that integrates the control of photosynthesis, carbon dioxide assimilation, and nitrogen fixation. *Proc. Natl. Acad. Sci. USA, 93*, 14515-14520.

Jouanneau, Y., Duport, C., Meyer, C., and Gaillard, J. (1992). Expression in *Escherichia coli* and characterization of a recombinant 7Fe ferredoxin of *Rhodobacter capsulatus*. *Biochem. J., 286*, 269-273.

Jouanneau, Y., Jeong, H.-S., Hugo, N., Meyer, C., and Willison, J. C. (1998). Overexpression in *Escherichia coli* of the *rnf* genes from *Rhodobacter capsulatus*. Characterization of two membrane-bound iron-sulfur proteins. *Eur. J. Biochem., 251*, 54-64.

Jouanneau, Y., Meyer, C., Asso, M., Guigliarelli, B., and Willison, J. C. (2000). Characterization of a nif-regulated flavoprotein (FprA) from *Rhodobacter capsulatus*. Redox properties and molecular interaction with a [2Fe-2S] ferredoxin. *Eur. J. Biochem., 267*, 780-787.

Jouanneau, Y., Meyer, C., Gaillard, J., Forest, E., and Gagnon, J. (1993). Purification and characterization of a novel dimeric ferredoxin (FdIII) from *Rhodobacter capsulatus*. *J. Biol. Chem., 268*, 10636-10644.

Jouanneau, Y., Meyer, C., Gaillard, J., and Vignais, P. M. (1990). Purification and characterization of a 7Fe-ferredoxin from *Rhodobacter capsulatus*. *Biochem. Biophys. Res. Comm., 171*, 273-279.

Jouanneau, Y., Roby, C., Meyer, C. M., and Vignais, P. M. (1989). ADP-ribosylation of dinitrogenase reductase in *Rhodobacter capsulatus*. *Biochem., 28*, 6524-6530.

Kern, M., Kamp, P.-B., Paschen, A., Masepohl, B., and Klipp, W. (1998). Evidence for a regulatory link of nitrogen fixation and photosynthesis in *Rhodobacter capsulatus* via HvrA. *J. Bacteriol., 180*, 1965-1969.

Keuntje, B., Masepohl, B., and Klipp, W. (1995). Expression of the *putA* gene encoding proline dehydrogenase from *Rhodobacter capsulatus* is independent of NtrC regulation but requires an Lrp-like activator protein. *J. Bacteriol., 177*, 6432-6439.

Kirndörfer, M., Jäger, A., and Klug, G. (1998). Integration host factor affects the oxygen-regulated expression of photosynthesis genes in *Rhodobacter capsulatus*. *Mol. Gen. Genet., 258*, 297-305.

Klipp, W., Drepper, T., Groß, S., Masepohl, B., Raabe, K., Riedel, K.-U., *et al.* (2000). Genetics of nitrogen fixation in *Rhodobacter capsulatus*: Ammonium and molybdenum control of both nitrogenase systems. In F. O. Pedrosa, M. Hungria, G. Yates and W. E. Newton (Eds.), *Nitrogen fixation: From molecules to crop productivity* (pp. 141-142). Dordrecht: Kluwer Academic Publishers.

Klipp, W., Masepohl, B., and Pühler, A. (1988). Identification and mapping of nitrogen fixation genes of *Rhodobacter capsulatus*: Duplication of a *nifA-nifB* region. *J. Bacteriol., 170*, 693-699.

Krahn, E., Weiss, R., Krockel, M., Groppe, J., Henkel, G., Cramer, P., *et al.* (2002). The Fe-only nitrogenase from *Rhodobacter capsulatus*: Identification of the cofactor, an unusual, high-nuclearity iron-sulfur cluster, by Fe K-edge EXAFS and ^{57}Fe Mössbauer spectroscopy. *J. Biol. Inorg. Chem., 7*, 37-45.

Kranz, R. G., and Foster-Hartnett, D. (1990). Transcriptional regulatory cascade of nitrogen-fixation genes in anoxygenic photosynthetic bacteria: Oxygen- and nitrogen-responsive factors. *Mol. Microbiol., 4*, 1793-1800.

Kranz, R. G., and Haselkorn, R. (1985). Characterization of *nif* regulatory genes in *Rhodopseudomonas capsulata* using *lac* gene fusions. *Gene, 40*, 203-215.

Kranz, R. G., Pace, V. M., and Caldicott, I. M. (1990). Inactivation, sequence, and *lacZ* fusion analysis of a regulatory locus required for repression of nitrogen fixation genes in *Rhodobacter capsulatus*. *J. Bacteriol., 172*, 53-62.

Kumagai, H., Fujiwara, T., Matsubara, H., and Saeki, K. (1997). Membrane localization, topology, and mutual stabilization of the *rnfABC* gene products in *Rhodobacter capsulatus* and implications for a new family of energy-coupling NADH oxidoreductases. *Biochem., 36*, 5509-5521.

Kutsche, M., Leimkühler, S., Angermüller, S., and Klipp, W. (1996). Promoters controlling expression of the alternative nitrogenase and the molybdenum uptake system in *Rhodobacter capsulatus* are activated by NtrC, independent of σ^{54}, and repressed by molybdenum. *J. Bacteriol., 178*, 2010-2017.

Lehman, L. J., and Roberts, G. P. (1991). Identification of an alternative nitrogenase system in *Rhodospirillum rubrum*. *J. Bacteriol., 173*, 5705-5711.

Liang, J., Madden, M., Shah, V. K., and Burris, R. H. (1990). Citrate substitutes for homocitrate in nitrogenase of a *nifV* mutant of *Klebsiella pneumoniae*. *Biochem., 29*, 8577-8581.

Little, R., Reyes-Ramirez, F., Zhang, Y., van Heeswijk, W. C., and Dixon, R. (2000). Signal transduction to the *Azotobacter vinelandii* NIFL-NIFA regulatory system is influenced directly by interaction with 2-oxoglutarate and the PII regulatory protein. *EMBO J., 19*, 6041-6050.

Madigan, M. T. (1995). Microbiology of nitrogen fixation by anoxygenic photosynthetic bacteria. In R. E. Blankenship, M. T. Madigan and C. E. Bauer (Eds.), *Anoxygenic photosynthetic bacteria* (pp. 915-928). Dordrecht: Kluwer Academic Publishers.

Madigan, M., Cox, S. S., and Stegeman, R. A. (1984). Nitrogen fixation and nitrogenase activities in members of the family *Rhodospirillaceae*. *J. Bacteriol., 157*, 73-78.

Maner, K., and Oelze, J. (1999). Control by light of whole-cell nitrogenase activity and of nitrogenase and bacteriochlorophyll *a* formation in *Rhodobacter capsulatus* strain B10S. *Arch. Microbiol., 171*, 243-248.

Marrs, B. (1974). Genetic recombination in *Rhodopseudomonas capsulata*. *Proc. Natl. Acad. Sci. USA, 71*, 971-973.

Martin, D. E., and Reinhold-Hurek, B. (2002). Distinct roles of P_{II}-like signal transmitter proteins and *amtB* in regulation of *nif* gene expression, nitrogenase activity, and posttranslational modification of NifH in *Azoarcus* sp. strain BH72. *J. Bacteriol., 184*, 2251-2259.

Masepohl, B., Angermüller, S., Hennecke, S., Hübner, P., Moreno-Vivian, C., and Klipp, W. (1993a). Nucleotide sequence and genetic analysis of the *Rhodobacter capsulatus* ORF6-*nifU$_I$-SVW* gene region: Possible role of NifW in homocitrate processing. *Mol. Gen. Genet., 238*, 369-382.

Masepohl, B., Drepper, T., Paschen, A., Groß, S., Pawlowski, A., Raabe, K., *et al.* (2002). Regulation of nitrogen fixation in the phototrophic purple bacterium *Rhodobacter capsulatus*. *J. Mol. Microbiol. Biotechnol., 4*, 243-248.

Masepohl, B., Kaiser, B., Isakovic, N., Richard, C. L., Kranz, R. G., and Klipp, W. (2001). Urea utilization in the phototrophic bacterium *Rhodobacter capsulatus* is regulated by the transcriptional activator NtrC. *J. Bacteriol., 183*, 637-643.

Masepohl, B., and Klipp, W. (1996). Organization and regulation of genes encoding the molybdenum nitrogenase and the alternative nitrogenase in *Rhodobacter capsulatus*. *Arch. Microbiol., 165*, 80-90.

Masepohl, B., Klipp, W., and Pühler, A. (1988). Genetic characterization and sequence analysis of the duplicated *nifA/nifB* gene region of *Rhodobacter capsulatus*. *Mol. Gen. Genet., 212*, 27-37.

Masepohl, B., Krey, R., and Klipp, W. (1993b). The *draTG* gene region of *Rhodobacter capsulatus* is required for post-translational regulation of both the molybdenum and the alternative nitrogenase. *J. Gen. Microbiol., 139*, 2667-2675.

Meijer, W. G., and Tabita, F. R. (1992). Isolation and characterization of the *nifUSVW-rpoN* gene cluster from *Rhodobacter sphaeroides*. *J. Bacteriol., 174*, 3855-3866.

Michel-Reydellet, N., and Kaminski, P. A. (1999). *Azorhizobium caulinodans* P_{II} and GlnK proteins control nitrogen fixation and ammonia assimilation. *J. Bacteriol., 181*, 2655-2658.

Monteiro, R. A., Souza, E. M., Yates, M. G., Pedrosa, F. O., and Chubatsu, L. S. (1999). In-*trans* regulation of the N-truncated-NIFA protein of *Herbaspirillum seropedicae* by the N-terminal domain. *FEMS Microbiol. Lett., 180*, 157-161.

Moreno-Vivian, C., Hennecke, S., Pühler, A., and Klipp, W. (1989a). Open reading frame 5 (ORF5), encoding a ferredoxinlike protein, and *nifQ* are cotranscribed with *nifE, nifN, nifX*, and ORF4 in *Rhodobacter capsulatus*. *J. Bacteriol., 171*, 2591-2598.

Moreno-Vivian, C., Schmehl, M., Masepohl, B., Arnold, W., and Klipp, W. (1989b). DNA sequence and genetic analysis of the *Rhodobacter capsulatus nifENX* gene region: Homology between NifX and NifB suggests involvement of NifX in processing of the iron-molybdenum cofactor. *Mol. Gen. Genet., 216*, 353-363.

Moreno-Vivian, C., Soler, G., and Castillo, F. (1992). Arginine catabolism in the phototrophic bacterium *Rhodobacter capsulatus* E1F1. Purification and properties of arginase. *Eur. J. Biochem., 204*, 531-537.

Morett, E., and Buck, M. (1988). NifA-dependent in vivo protection demonstrates that the upstream activator sequence of *nif* promoters is a protein binding site. *Proc. Natl. Acad. Sci. USA, 85*, 9401-9405.

Morett, E., and Buck, M. (1989). *In vivo* studies on the interaction of RNA polymerase-σ^{54} with the *Klebsiella pneumoniae* and *Rhizobium meliloti nifH* promoters. The role of NifA in the formation of an open promoter complex. *J. Mol. Biol., 210*, 65-77.

Mylona, P. V., Premakumar, R., Pau, R. N., and Bishop, P. E. (1996). Characteristics of *orf1* and *orf2* in the *anfHDGK* genomic region encoding nitrogenase 3 of *Azotobacter vinelandii*. *J. Bacteriol., 178*, 204-208.

Naud, I., Vincon, M., Garin, J., Gaillard, J., Forest, E., and Jouanneau, Y. (1994). Purification of a sixth ferredoxin from *Rhodobacter capsulatus*. Primary structure and biochemical properties. *Eur. J. Biochem., 222*, 933-939.

Ninfa, A. J., and Atkinson, M. R. (2000). PII signal transduction proteins. *Trends in Microbiology, 8*, 172-179.

Overbeek, R., Larsen, N., Pusch, G. D., D´Souza, M., Selkov, E. Jr., Kyrpides, N., Fonstein, M., Maltsev, N., and Selkov, E. (2000). WIT: integrated system for high-throughput genome sequence analysis and metabolic reconstruction. *Nucl. Acids Res., 28*, 123-125.

Paschen, A., Drepper, T., Masepohl, B., and Klipp, W. (2001). *Rhodobacter capsulatus nifA* mutants mediating *nif* gene expression in the presence of ammonium. *FEMS Microbiol. Lett., 200*, 207-213.

Pasternak, C., Chen, W., Heck, C., and Klug, G. (1996). Cloning, nucleotide sequence and characterization of the *rpoD* gene encoding the primary sigma factor of *Rhodobacter capsulatus*. *Gene, 176*, 177-184.

Pau, R. N., Klipp, W., and Leimkühler, S. (1997). Molybdenum transport, processing and gene regulation. In G. Winkelmann and C. J. Carrano (Eds.), *Transition metals in microbial metabolism* (pp. 217-234). Amsterdam: Harwood Academic Publishers.

Pierrard, J., Ludden, P. W., and Roberts, G. P. (1993). Posttranslational regulation of nitrogenase in *Rhodobacter capsulatus*: Existence of two independent regulatory effects of ammonium. *J. Bacteriol., 175*, 1358-1366.

Pittard, A. J., and Davidson, B. E. (1991). TyrR protein of *Escherichia coli* and its role as repressor and activator. *Mol. Microbiol., 5*, 1585-1592.

Preker, P., Hübner, P., Schmehl, M., Klipp, W., and Bickle, T. A. (1992). Mapping and characterization of the promoter elements of the regulatory *nif* genes *rpoN, nifA1* and *nifA2* in *Rhodobacter capsulatus*. *Mol. Microbiol., 6*, 1035-1047.

Qian, Y., and Tabita, F. R. (1996). A global signal transduction system regulates aerobic and anaerobic CO_2 fixation in *Rhodobacter sphaeroides*. *J. Bacteriol., 178*, 12-18.

Qian, Y., and Tabita, F. R. (1998). Expression of *glnB* and a *glnB*-like gene (*glnK*) in a ribulose bisphosphate carboxylase/oxygenase-deficient mutant of *Rhodobacter sphaeroides*. *J. Bacteriol., 180*, 4644-4649.

Raabe, K., Drepper, T., Riedel, K.-U., Masepohl, B., and Klipp, W. (2002). The H-NS-like protein HvrA modulates expression of nitrogen fixation genes in the phototrophic purple bacterium *Rhodobacter capsulatus* by binding to selected *nif* promoters. *FEMS Microbiol. Lett.*, *216*, 151-158.

Rangaraj, P., Rüttimann-Johnson, C., Shah, V. K., and Ludden, P. W. (2000). Biosynthesis of the iron-molybdenum and iron-vanadium cofactors of the *nif*- and *vnf*-encoded nitrogenases. In E. W. Triplett (Ed.), *Prokaryotic nitrogen fixation* (pp. 55-79). Wymondham, UK: Horizon Scientific Press.

Rangaraj, P., Rüttimann-Johnson, C., Shah, V. K., and Ludden, P. W. (2001). Accumulation of ^{55}Fe-labeled precursors of the iron-molybdenum cofactor of nitrogenase on NifH and NifX of *Azotobacter vinelandii*. *J. Biol. Chem.*, *276*, 15968-15974.

Rich, P. R., Meunier, B., and Ward, F. B. (1995). Predicted structure and possible ionmotive mechanism of the sodium-linked NADH-ubiquinone oxidoreductase of *Vibrio alginolyticus*. *FEBS Lett.*, *375*, 5-10.

Rudnick, P. A., Arcondeguy, T., Kennedy, C. K., and Kahn, D. (2001). *glnD* and *mviN* are genes of an essential operon in *Sinorhizobium meliloti*. *J. Bacteriol.*, *183*, 2682-2685.

Saeki, K., and Kumagai, H. (1998). The *rnf* gene products in *Rhodobacter capsulatus* play an essential role in nitrogen fixation during anaerobic DMSO-dependent growth in the dark. *Arch. Microbiol.*, *169*, 464-467.

Saeki, K., Miyatake, Y., Young, D. A., Marrs, B. L., and Matsubara, H. (1990). A plant-ferredoxin-like gene is located upstream of ferredoxin I gene (*fdxN*) of *Rhodobacter capsulatus*. *Nucl. Acids Res.*, *18*, 1060.

Saeki, K., Suetsugu, Y., Tokuda, K.-i., Miyatake, Y., Young, D. A., Marrs, B. L., and Matsubara, H. (1991). Genetic analysis of functional differences among distinct ferredoxing in *Rhodobacter capsulatus*. *J. Biol. Chem.*, *266*, 12889-12895.

Saeki, K., Tokuda, K.-i., Fujiwara, T., and Matsubara, H. (1993). Nucleotide sequence and genetic analysis of the region essential for functional expression of the gene for ferredoxin I, *fdxN*, in *Rhodobacter capsulatus*: Sharing of one upstream activator sequence in opposite directions by two operons related to nitrogen fixation. *Plant Cell Physiol.*, *34*, 185-199.

Saez, L. P., Garcia, P., Martinez-Luque, M., Klipp, W., Blasco, R., and Castillo, F. (2001). Role for *draTG* and *rnf* genes in reduction of 2,4-dinitrophenol by *Rhodobacter capsulatus*. *J. Bacteriol.*, *183*, 1780-1783.

Santero, E., Hoover, T. R., North, A. K., Berger, D. K., Porter, S. C., and Kustu, S. (1992). Role of integration host factor in stimulating transcription from the σ^{54}-dependent *nifH* promoter. *J. Mol. Biol.*, *227*, 602-620.

Schmehl, M., Jahn, A., Meyer zu Vilsendorf, A., Hennecke, S., Masepohl, B., Schuppler, M., *et al.* (1993). Identification of a new class of nitrogen fixation genes in *Rhodobacter capsulatus*: A putative membrane complex involved in electron transport to nitrogenase. *Mol. Gen. Genet.*, *241*, 602-615.

Schneider, K., Gollan, U., Dröttboom, M., Selsemeier-Voigt, S., and Müller, A. (1997). Comparative biochemical characterization of the iron-only nitrogenase and the molybdenum nitrogenase from *Rhodobacter capsulatus*. *Eur. J. Biochem.*, *244*, 789-800.

Schneider, K., Müller, A., Johannes, K.-U., Diemann, E., and Kottmann, J. (1991a). Selective removal of molybdenum traces from growth media of N_2-fixing bacteria. *Analyt. Biochem.*, *193*, 292-298.

Schneider, K., Müller, A., Schramm, U., and Klipp, W. (1991b). Demonstration of a molybdenum- and vanadium-independent nitrogenase in a *nifHDK*-deletion mutant of *Rhodobacter capsulatus*. *Eur. J. Biochem.*, *195*, 653-661.

Schüddekopf, K., Hennecke, S., Liese, U., Kutsche, M., and Klipp, W. (1993). Characterization of *anf* genes specific for the alternative nitrogenase and identification of *nif* genes required for both nitrogenases in *Rhodobacter capsulatus*. *Mol. Microbiol.*, *8*, 673-684.

Schüttelkopf, A. W., Harrison, J. A., Boxer, D. H., and Hunter, W. N. (2002). Passive acquisition of ligand by the MopII molbindin from *Clostridium pasteurianum*; Structures of apo and oxyanion-bound forms. *J. Biol. Chem.*, *277*, 15013-15020.

Self, W. T., Grunden, A. M., Hasona, A., Shanmugam, K. T. (2001). Molybdate transport. *Res. Microbiol.*, *152*, 311-321.

Sganga, M. W., and Bauer, C. E. (1992). Regulatory factors controlling photosynthetic reaction center and light-harvesting gene expression in *Rhodobacter capsulatus*. *Cell*, *68*, 945-954.

Shah, V. K., Rangaraj, P., Chatterjee, R., Allen, R. M., Roll, J. T., Roberts, G. P., and Ludden, P. W. (1999). Requirement of NifX and other *nif* proteins for in vitro biosynthesis of the iron-molybdenum cofactor of nitrogenase. *J. Bacteriol., 181*, 2797-2801.

Siemann, S., Schneider, K., Behrens, K., Knöchel, A., Klipp, W., and Müller, A. (2001). FeMo cofactor biosynthesis in a *nifE*$^-$ mutant of *Rhodobacter capsulatus*. *Eur. J. Biochem., 268*, 1940-1952.

Solomon, P. S., Shaw, A. L., Young, M. D., Leimkühler, S., Hanson, G. R., Klipp, W., and McEwan, A. G. (2000). Molybdate-dependent expression of dimethylsulfoxide reductase in *Rhodobacter capsulatus*. *FEMS Microbiol. Lett., 190*, 203-208.

Souza, E. M., Pedrosa, F. O., Drummond, M., Rigo, L. U., and Yates, M. G. (1999). Control of *Herbaspirillum seropedicae* NifA activity by ammonium ions and oxygen. *J. Bacteriol., 181*, 681-684.

Suetsugu, Y., Saeki, K., and Matsubara, H. (1991). Transcriptional analysis of two *Rhodobacter capsulatus* ferredoxins by translational fusion to *Escherichia coli lacZ*. *FEBS Lett., 292*, 13-16.

Thomas, G., Coutts, G., and Merrick, M. (2000). The *glnKamtB* operon. A conserved gene pair in prokaryotes. *Trends Genet., 16*, 11-14.

Toussaint, B., Bosc, C., Richaud, P., Colbeau, A., and Vignais, P. M. (1991). A mutation in a *Rhodobacter capsulatus* gene encoding an integration host factor-like protein impairs *in vivo* hydrogenase expression. *Proc. Natl. Acad. Sci. USA, 88*, 10749-10753.

Toussaint, B., Delic-Attree, I., De Sury d`Aspremont, R., David, L., Vincon, M., and Vignais, P. M. (1993). Purification of the integration host factor homolog of *Rhodobacter capsulatus*: Cloning and sequencing of the *hip* gene, which encodes the b subunit. *J. Bacteriol., 175*, 6499-6504.

Wall, J. D., and Braddock, K. (1984). Mapping of *Rhodopseudomonas capsulata nif* genes. *J. Bacteriol., 158*, 404-410.

Wall, J. D., Love, J., and Quinn, S. P. (1984). Spontaneous Nif$^-$ mutants of *Rhodopseudomonas capsulata*. *J. Bacteriol., 159*, 652-657.

Wang, G., Angermüller, S., and Klipp, W. (1993). Characterization of *Rhodobacter capsulatus* genes encoding a molybdenum transport system and putative molybdenum-pterin-binding proteins. *J. Bacteriol., 175*, 3031-3042.

Wassem, R., Maltempi de Souza, E., Yates, M. G., de Oliveira Pedrosa, F., and Buck, M. (2000). Two roles for integration host factor at an enhancer-dependent *nifA* promoter. *Mol. Microbiol., 35*, 756-764.

Weaver, P. F., Wall, J. D., and Gest, H. (1975). Characterization of *Rhodopseudomonas capsulata*. *Arch. Microbiol., 105*, 207-216.

Williams, J. C., and Taguchi, A. K. W. (1995). Genetic manipulation of purple photosynthetic bacteria. In R. E. Blankenship, M. T. Madigan and C. E. Bauer (Eds.), *Anoxygenic photosynthetic bacteria* (pp. 1029-1065). Dordrecht: Kluwer Academic Publishers.

Willison, J. C., Pierrard, J., and Hübner, P. (1993). Sequence and transcript analysis of the nitrogenase structural gene operon (*nifHDK*) of *Rhodobacter capsulatus*: Evidence for intramolecular processing of *nifHDK* mRNA. *Gene, 133*, 39-46.

Wolfinger, E. D., and Bishop, P. E. (1991). Nucleotide sequence and mutational analysis of the *vnfENX* region of *Azotobacter vinelandii*. *J. Bacteriol., 173*, 7565-7572.

Xu, H., and Hoover, T. R. (2001). Transcriptional regulation at a distance in bacteria. *Curr. Opinion Microbiol., 4*, 138-144.

Yakunin, A. F., Fedorov, A. S., Laurinavichene, T. V., Glaser, V. M., Egorov, N. S., Tsygankov, A. A., *et al.* (2001). Regulation of nitrogenase in the photosynthetic bacterium *Rhodobacter sphaeroides* containing *draTG* and *nifHDK* genes from *Rhodobacter capsulatus*. *Can. J. Microbiol., 47*, 206-212.

Yakunin, A. F., Gennaro, G., and Hallenbeck, P. C. (1993). Purification and properties of a *nif*-specific flavodoxin from the photosynthetic bacterium *Rhodobacter capsulatus*. *J. Bacteriol., 175*, 6775-6780.

Yakunin, A. F., and Hallenbeck, P. C. (1998a). Purification and characterization of pyruvate oxidoreductase from the photosynthetic bacterium *Rhodobacter capsulatus*. *Biochim. Biophys. Acta, 1409*, 39-49.

Yakunin, A. F., and Hallenbeck, P. C. (1998b). Short-term regulation of nitrogenase activity by NH_4^+ in *Rhodobacter capsulatus*: multiple in vivo nitrogenase responses to NH_4^+ addition. *J. Bacteriol., 180*, 6392-6395.

Yakunin, A. F., and Hallenbeck, P. C. (2000a). Regulation of nitrogenase activity in *Rhodobacter capsulatus* under dark microoxic conditions. *Arch. Microbiol., 173*, 366-372.

Yakunin, A. F., and Hallenbeck, P. C. (2000b). AmtB is necessary for NH_4^+ induced nitrogenase switch-off and ADP-ribosylation in *Rhodobacter capsulatus*. In F. O. Pedrosa, M. Hungria, G. Yates and W. E. Newton (Eds.), *Nitrogen fixation: From molecules to crop productivity* (pp. 95-96). Dordrecht: Kluwer Academic Publishers.

Yakunin, A. F., Laurinavichene, T. V., Tsygankov, A. A., and Hallenbeck, P. C. (1999). The presence of ADP-ribosylated Fe protein of nitrogenase in *Rhodobacter capsulatus* is correlated with cellular nitrogen status. *J. Bacteriol., 181*, 1994-2000.

Yuvaniyama, P., Agar, J. N., Cash, V. L., Johnson, M. K., and Dean, D. R. (2000). NifS-directed assembly of a transient [2Fe-2S] cluster within the NifU protein. *Proc. Natl. Acad. Sci. USA*, *97*, 599-604.

Zhang, Y., Pohlmann, E. L., Ludden, P. W., and Roberts, G. P. (2000). Mutagenesis and functional characterization of the *glnB*, *glnA*, and *nifA* genes from the photosynthetic bacterium *Rhodospirillum rubrum*. *J. Bacteriol., 182*, 983-992.

Zhang, Y., Pohlmann, E. L., Ludden, P. W., and Roberts, G. P. (2001). Functional characterization of three GlnB homologs in the photosynthetic bacterium *Rhodospirillum rubrum*: Roles in sensing ammonium and energy status. *J. Bacteriol., 183*, 6159-6168.

Zinchenko, V., Babykin, M., Glaser, V., Mekhedov, S., and Shestakov, S. (1997). Mutation in *ntrC* gene leading to the derepression of nitrogenase synthesis in *Rhodobacter sphaeroides*. *FEMS Microbiol., Lett. 147*, 57-61.

Chapter 8

POST-TRANSLATIONAL REGULATION OF NITROGENASE IN PHOTOSYNTHETIC BACTERIA

S. NORDLUND[a] AND P. W. LUDDEN[b]
[a]*Department of Biochemistry and Biophysics, Stockholm University, SE-106 91 Stockholm, Sweden*
[b]*Department of Plant and Microbial Biology, 111 Koshland Hall, University of California, Berkeley, CA 94720-3102, USA*

1. INTRODUCTION

The photosynthetic bacteria are remarkable for their range of metabolic capabilities and, as a result, the ability to survive under a wide range of environmental conditions, including nitrogen limitation. The photosynthetic bacteria are members of the alpha subdivision of the proteobacteria (Imhoff *et al.*, 1984; 1998; Woese *et al.*, 1984) and, with a single known exception, the wild-type strains of all purple sulfur bacteria and all purple non-sulfur bacteria are capable of N_2 fixation (Madigan *et al.*, 1984). Most of the work that will be described in this chapter has been performed on a few species of purple, non-sulfur bacteria, including *Rhodospirillum rubrum* and *Rhodobacter capsulatus*. This chapter will focus on the biochemistry and physiology of N_2 fixation by photosynthetic bacteria and the regulation of nitrogenase activity by reversible ADP-ribosylation of the dinitrogenase reductase (Fe protein). ADP-ribosylation has also been found in *Azospirillum* and other genera and, where relevant, that work will be included. Other chapters in this series will cover the genes of the *nif*, *anf* and *rnf* regulons of the photosynthetic bacteria and the regulation of the expression of those genes. Although this chapter will discuss aspects of the nitrogenase enzyme, the mechanistic and structural details of nitrogenase will be covered in more detail in Volume 1 of this series.

W. Klipp, B. Masepohl, J. R. Gallon and W. E. Newton (eds.), Genetics and Regulation of Nitrogen Fixation in Free-Living Bacteria, 175-196.

2. THE DISCOVERY OF NITROGEN FIXATION BY PHOTOSYNTHETIC BACTERIA

The non-oxygenic photosynthetic bacteria were the subjects of extensive investigations following van Niel's classic studies on these organisms and the development of enrichment and culturing conditions. *R. rubrum* (van Niel's strain S-1) was a favorite organism for studies because of its ease of culture. Howard Gest and Martin Kamen employed van Niel's strain of *R. rubrum* in their investigations of light-dependent gas exchange under "inert" atmospheres. To their surprise, an atmosphere of N_2 was not "inert" in the presence of *R. rubrum* and this lead to the demonstration by isotope ratio mass spectrometry that *R. rubrum* could fix N_2. Several interesting accounts of these experiments have been published in which the personal side of science is highlighted (Gest, 1999; Kamen, 1986).

During these studies, Gest and Kamen observed copious amounts of H_2 produced by illuminated cultures of *R. rubrum* grown with glutamate as the N source (Gest and Kamen, 1949). The realization that nitrogenase itself was capable of reduction of protons to H_2 would await later experiments by Burns and Bulen (1965; 1966), but Gest and Kamen speculated about the linkage between H_2 evolution and nitrogen fixation. Their observation that N_2 inhibited the evolution of H_2 by resting cultures of glutamate-grown cells was a strong argument for the metabolic linkage of the two activities.

Gest and Kamen noted that cultures grown on NH_4^+ as the N source failed to evolve H_2, suggesting that NH_4^+ prevented the synthesis of the agent responsible for this activity and went on to observe that the product of N_2 fixation, NH_4^+, would inhibit both N_2 uptake and H_2 evolution. This observation would lead to later experiments in which it was demonstrated that the nitrogenase enzyme was subject to reversible inactivation by ADP-ribosylation.

Following the report of Gest and Kamen, a number of purple bacteria were tested for the ability to fix N_2 and all tested positive. In fact, of all purple bacteria tested to date, the great majority of species has been found to have the ability to grow with N_2 as the N source (Madigan, 1995). Schick also furthered the initial studies on gas exchange by *R. rubrum* and published a set manuscripts in which the light requirement for H_2 evolution and N_2 fixation was described as was the inhibitory effect of NH_4^+ (Schick, 1971). In these studies, the length of time of inhibition of N_2 uptake was found to correlate with the concentration of NH_4^+ added to the culture. Like Gest and Kamen, Schick used manometric techniques because these experiments predated the introduction of the acetylene-reduction assay.

It is important to note the context in which the early results were interpreted. Gest and Kamen's results predated the articulation of the operon model for gene expression in bacteria, the concept of allosteric regulation of enzymatic activity, and any knowledge of regulation of enzymatic activity by reversible covalent modification.

These early experiments were subsequently confirmed using the acetylene-reduction technique by Neilson and Nordlund (1975), who extended the studies to demonstrate the effects of glutamine and asparagine on *in vivo* nitrogenase activity. They further showed that none of the effectors that lead to "switch-off" of *in vivo* nitrogenase activity had significant effect on *in vitro* nitrogenase activity in extracts prepared from cells that had been sensitive to ammonium, glutamine or asparagine before lysis. Thus, they concluded, the effect of these compounds was not directly on nitrogenase. They speculated that the effectors might act on the availability of either reductant or ATP to nitrogenase *in vivo*. It was also found by Weare and Shanmugam (1976) that the glutamine synthetase inhibitor, methionine sulfoximine, both derepressed nitrogenase synthesis in the presence of ammonium and eliminated the "switch-off" of *in vivo* activity in response to ammonium in *R. rubrum*. The term "switch-off" is used by some to indicate any loss of *in vivo* nitrogenase acitivity but, for our purposes, we will use the term "switch-off" to refer specifically to the loss of *in vivo* nitrogenase activity resulting from the ADP-ribosylation of dinitrogenase reductase as described below (Zumft and Castillo, 1978).

3. *IN VITRO* STUDIES OF NITROGENASE IN PHOTOSYNTHETIC BACTERIA

Immediately after the breakthrough success of Mortenson and his co-workers in preparing consistently active crude extracts from *Clostridium pasteurianum* (Carnahan *et al.*, 1960), Schneider and co-workers adapted their techniques to prepare crude extracts of a number of N_2-fixing organisms, including *R. rubrum* (Schneider *et al.*, 1960). Although the results with *R. rubrum* were mixed and inconsistent, they achieved $^{15}N_2$ reduction to $^{15}NH_4^+$ *in vitro*. In their early studies on the enzymology of nitrogenase, Burns and Bulen employed the non-phototroph, *Azotobacter vinelandii*, as well as *R. rubrum* (Burns and Bulen, 1965). Although they had significant success with *A. vinelandii*, activities with extracts from *R. rubrum* were much lower and less consistent. They did achieve some success by removing the membrane fraction from their extracts by centrifugation and by adding a 2.5-fold excess of Mg^{2+} over MgATP in their assays. Although the latter result undoubtedly can be explained as allowing any Mn^{2+} in the extract to remain uncomplexed by ATP and thus available for DRAG to activate the dinitrogenase reductase (see below), the inhibitory effect of the chromatophore membranes in Burns and Bulen's extracts remains unexplained. Extracts of *R. rubrum*, along with those of *A. vinelandii*, were used in at least two seminal experiments during the early days of nitrogenase enzymology. The first of these experiments demonstrated that sodium dithionite could be used as the reductant for nitrogenase (Bulen *et al.*, 1965); the second showed that nitrogenase itself was capable of the ATP-dependent reduction of protons to H_2 (Burns and Bulen, 1966).

A significant advance in the handling of *R. rubrum* extracts came with the finding by Munson and Burris that consistently active preparations could be obtained with extracts prepared from cells grown with limiting N (NH_4^+) in continuous culture (Munson and Burris, 1969). Munson and Burris employed high Mg^{2+} concentrations in their assays and noted a lag in the acetylene-reduction activity; this lag was longer and much more pronounced in cells grown on either glutamate or N_2 as the N source. The results of Munson and Burris were also interesting because they grew cells under an H_2 atmosphere in the absence of N_2 and obtained excellent expression of nitrogenase. From this result, they concluded that N_2 is not required for the induction of genes for nitrogen fixation. Attempts by Munson to purify further the nitrogenase from *R. rubrum* were not successful.

The results of these workers are explained by the current model for regulation of nitrogenase in *R. rubrum* and related bacteria. As will be described in succeeding sections, the effects of ammonium, glutamine, and asparagine on N_2 uptake and nitrogenase-catalyzed H_2 evolution are explained by the response of a signal system that leads to inactivation of dinitrogenase reductase (Fe protein) in response to fixed nitrogen. The stimulatory effect of Mg^{2+} on crude extract activity results from a sparing effect, wherein high Mg^{2+} concentrations ensure that all of the ATP in the assay is bound to Mg^{2+} leaving any Mn^{2+} available for the "activating enzyme" (DRAG) required to activate dinitrogenase reductase. The lag phase observed by Munson and Burris reflects the slow activation of dinitrogenase reductase by DRAG during the assay.

It is worth noting at this point that the acetylene-reduction assay, introduced in 1967, was a tremendous aide to all N_2-fixation research and perhaps no research project was assisted more than the *R. rubrum* nitrogenase project. Acetylene is an excellent substrate for nitrogenase and can be used to measure accurately nitrogenase activity *in vivo* because acetylene is a very soluble and readily crosses the cell membrane. The product of acetylene reduction, ethylene, is a much less soluble gas, which also readily crosses the cell membrane. It, therefore, accumulates in the gas phase above the cell suspension, allowing it to be easily sampled and quantified by flame ionization gas chromatography. Although manometric techniques had allowed Gest, Schick and others to observe N_2 uptake by whole cells, these methods are very slow by comparison to acetylene-reduction assays and much less sensitive. The acetylene-reduction technique allowed workers to correlate the *in vivo* activity with other biochemical measurements in a way not possible for other systems of enzyme regulation.

4. THE PROTEIN ERA

The major breakthrough in understanding why N_2 fixation in *R. rubrum* is inhibited by the addition of ammonium ions and why extracts from actively fixing cells showed no or little nitrogenase activity was the demonstration that nitrogenase in extracts can be activated in the presence of a protein associated with the chromatophore membrane (Ludden and Burris, 1976; Nordlund *et al.*, 1977). This protein, the activating enzyme or the membrane component, was shown to catalyze activation of inactive nitrogenase in a reaction requiring ATP, Mg^{2+} and Mn^{2+}. It was also shown that dinitrogenase reductase (Fe protein) was the nitrogenase component that interacted with the activating enzyme from the chromatophore membrane. The activating enzyme can be released from the chromatophore membrane by washing with 0.5 M NaCl and requires either Mn^{2+} or Fe^{2+} for the activating process (Ludden and Burris, 1976; Nordlund *et al.*, 1977; 1978). These results explained the unusual behavior of nitrogenase in extracts of *R. rubrum* compared to other diazotrophs but led to a number of major questions that needed be answered in order to understand the system further. For example, what are the molecular features of the inactive form of dinitrogenase reductase? What are the characteristics of both the activating enzyme and the enzyme that catalyzes conversion to the inactive form? How are these enzymes regulated?

4.1. Composition of the Modifying Group and its Location

Purified nitrogenase from *R. rubrum* was purified (Ludden and Burris, 1978; Nordlund *et al.*, 1978) was shown to have adenine, ribose and phosphate present in the inactive form and all three were released upon activation of dinitrogenase reductase (Ludden and Burris, 1978; 1979). The inactive form of dinitrogenase reductase (Fe protein) was shown to interact poorly with dinitrogenase (MoFe protein) leading to essentially no nitrogenase activity and no ATP hydrolysis. Furthermore, there was no competition between the inactive and active forms when both were present in the nitrogenase reaction mixture (Ludden *et al.*, 1982). In spite of the inability of the inactive form to bind to dinitrogenase, it showed the same affinity for ATP and the same susceptibility to iron-chelating agents as the active, non-modified form (Guth and Burris, 1983). A very useful feature of the inactive form of dinitrogenase reductase is the appearance of two bands on SDS–PAGE (Dowling *et al.*, 1982; Gotto and Yoch, 1982; Ludden and Burris, 1978; Preston and Ludden, 1982). This characteristic, which is due to the modification of only one of the two subunits in the dimer, has been used as a diagnostic tool for identifying modification in other diazotrophs possessing this regulatory system.

Further studies leading to the final identification of the modifying group showed: that adenine is incorporated into the inactive form of dinitrogenase reductase (Nordlund and Ludden, 1983); that the modification could be released by heat treatment in a temperature-, pH-, and enzyme concentration-dependent manner (Dowling *et al.*, 1982); and that borate inhibits the activation by the activating enzyme (Ludden *et al.*, 1982). The final identification of the modifying group as ADP-ribose was the result of a number of different analytical methods, including NMR and mass spectrometry (Pope *et al.*, 1985a; 1985b). In addition to identifying the modifying group (which was released by heat treatment), a hexapeptide to which it was bound (which was obtained by proteolytic cleavage) was also identified. This latter information showed that the residue, which became modified, is an arginine (arg-101 in *R. rubrum* dinitrogenase reductase) located within an amino-acid sequence that is highly conserved in all dinitrogenase reductases.

When the 3-D structures of both nitrogenase components, plus that of the 2:1 complex between them, were solved (Schindelin *et al.*, 1997), the inhibitory effect of the modification of the dinitrogenase reductase was explained. The modified arginine residue is located close to the 4Fe4S cluster of dinitrogenase reductase and is within the area that participates in the docking to dinitrogenase during turnover. The linkage of the ADP-ribose moiety to an arginine residue had hitherto been demonstrated only in eukaryotic systems in which G-proteins are affected by ADP-ribosylation catalyzed by bacterial toxins, *e.g.*, pertussis and cholera (Moss and Vaughan, 1990). Furthermore, mono-ADP-ribosylation had also been shown in turkey erythrocytes (West and Moss, 1986). The identification of the modifying group in *R. rubrum* opened the possibility that ADP-ribosylation may be a more common way of regulating enzyme activity than previously thought and that the donor of the ADP-ribose moiety in *R. rubrum* is NAD^+ as in the eukaryotic systems.

Modification of the dinitrogenase reductase in *R. rubrum* is not only initiated by the addition of ammonium ions but it also occurs on shifting the cells to darkness (Kanemoto and Ludden, 1984). This observation helped explain the fact that, even if no ammonium was added before harvest, extracts of *R. rubrum* exhibited either no or little nitrogenase activity although anaerobic conditions were applied. When cells are harvested by centrifugation, they are subjected to darkness, which leads to conversion of dinitrogenase reductase to its inactive form. Moreover, addition of the uncoupler, CCCP, and O_2 had the same effect, although conversion was slower.

An important contribution to the analysis of the modification state was the application of a low cross-linker SDS-PAGE system that produced good separation of the modified and unmodified subunits of dinitrogenase reductase. The initial studies of the inhibition of nitrogenase were performed with N_2-fixing cultures grown with either glutamate or N_2 as nitrogen source. In contrast, cells that had been starved for fixed-nitrogen, either by growth on limiting concentrations of NH_4^+ or after growth on N_2 and then transfer to argon, showed no response to added NH_4^+ (Carithers *et al.*, 1979; Sweet and Burris, 1981; 1982; Yoch and Cantu, 1980). Extracts from cells, which were grown under fixed-nitrogen-limited conditions, show high nitrogenase activity and very little of the modified form, indicating that N-starved cells also are insensitive to darkness. The inhibition induced by addition of NH_4^+ was also demonstrated in *R. rubrum* grown under dark, fermentative conditions (Schultz *et al.*, 1985).

4.2. Metal Ions and the Activation Process

The initial observation that activation by the activating enzyme required the presence of Mn^{2+} lead to investigations aimed at establishing its role in the reaction (Nordlund and Noren, 1984; Yoch, 1979). The results showed that Mn^{2+} and Fe^{2+} are essentially equally efficient in supporting activation of inactive dinitrogenase reductase by the activating enzyme. These metal ions function both as free cation and in the form of Me^{2+}-ATP (Me = Mn or Fe) complexes that also are required for activation. However, Mg^{2+} could not fulfill the requirement for a free cation at physiological concentrations (Nordlund and Noren, 1984; Yoch, 1979) and Ba^{2+} was a potent inhibitor of activation by acting as a competitive inhibitor of Mn^{2+} (Nordlund and Noren, 1984). Evidence that Fe^{2+} could serve as the free cation in the activation was provided when *R. rubrum* grew in a medium without added Mn^{2+} and showed the same reversible inhibition of nitrogenase activity as in a medium that included Mn^{2+} (Cadez and Nordlund, 1991).

In spite of its regulatory function in N_2 fixation, the activating enzyme is expressed under all growth conditions tested, including those where no nitrogenase was synthesized, *i.e.*, in fixed nitrogen-sufficient medium (Triplett *et al.*, 1982). It was, therefore, suggested that the activating enzyme may have substrates other than the dinitrogenase reductase, although none has so far been identified in *R. rubrum*. However, DRAG (and DRAT) have been suggested to be involved in the regulation of 2,4-dinitrophenol reduction in *Rb. capsulatus* (Saez *et al.*, 2001).

4.3. Purification of the Activating (DRAG) and Modifying (DRAT) Enzymes

Because Mn^{2+} was established as a required cation for the activating enzyme, Mn^{2+} was included in the buffers used in the first partial purification of this enzyme (Zumft and Nordlund, 1981), which was present in only minute amounts in the cell. The final purification and characterization of the activating enzyme was achieved by Ludden and co-workers (Pope *et al.*, 1986; 1987; Saari *et al.*, 1984; 1986). The activating enzyme was purified more than 10,000-fold, emphasizing the low amount present in the cell. Its molecular mass was estimated as about 32 kDa by SDS-PAGE and about 21.5 kDa by gel filtration. The larger value agrees with the mass estimated from its amino-acid composition. The smaller mass value (Zumft and Nordlund, 1981) is probably due to some unknown structure-related properties leading to abnormal elution on gel filtration. These studies also showed that high ionic strength inhibits the activation reaction.

Using purified enzyme, the requirement for free Mn^{2+} was confirmed and the optimal Mn^{2+}/ATP ratio was determined as 2:1. Interestingly, denatured modified dinitrogenase reductase was as good a substrate as native modified dinitrogenase reductase. The whole ADP-ribose group is cleaved off in the activation reaction. ADP-ribose, ADP, or pyrophosphate inhibits the reaction, whereas NAD^+, nicotinamide, or glutamine had little effect. Using the hexapeptide containing the ADP-ribose (obtained by proteolysis of inactive dinitrogenase reductase) and a series of synthetic analogs, the reaction mechanism for the removal of the ADP-ribose moiety was developed from which it was proposed that the activating enzyme should be named **D**initrogenase **R**eductase **A**ctivating **G**lycohydrolase, DRAG. A spin-off from these studies is a fluorometric assay for studying ADP-ribosyl-arginine cleavage enzymes (Pope *et al.*, 1987). In all these studies, DRAG was believed to be O_2 sensitive. It was, therefore, surprising when it was later shown that DRAG is stable in the absence of dithionite, which was included in buffers to maintain anaerobic conditions, (Nielsen *et al.*, 1994). The inactivation of DRAG, which was shown to be increased by Mn^{2+}, was suggested to be due to radical formation from dithionite taking place at the active site. More recent studies hve shown that DRAG contains a binuclear Mn(II) center (Antharavally *et al.*, 1998). Both histidine-158 and asparagines-243 have been suggested to coordinate Mn^{2+} in the active site (Antharavally *et al.*, 2001). DRAG has also been purified from *Az. brasilense* (Ljungström *et al.*, 1989).

The purification and the characterization of the enzyme catalyzing the modification of the dinitrogenase reductase has also been achieved (Lowery *et al.*, 1986; 1989; Lowery and Ludden, 1988; 1989). Initially, it was shown that NAD^+ serves as the donor of the ADP-moiety, that the enzyme is not sensitive to O_2, and that dithionite inhibited the modification reaction. The complete purification and characterization of this enzyme showed that, like DRAG, it is present in exceedingly small amounts in the cell. A 20,000-fold purification was required to obtain a homogenous preparation. Its molecular mass was about 30 kDa. NAD^+ was confirmed as the ADP-ribose donor with a K_m of 2 mM. This enzyme was named **D**initrogenase **R**eductase **A**DP-ribosyl **T**ransferase, DRAT, and, in contrast to DRAG, it seems specific for dinitrogenase reductase as substrate. No other acceptor for the ADP-ribose group has been identified at the molecular level. However, etheno-NAD, nicotinamide hypoxanthine dinucleotide, or nicotinamide guanine dinucleotide will substitute for NAD^+ as donor of the ADP-ribose group. In addition, dinitrogenase reductase from *A. vinelandii*, *K. pneumoniae,* or *C. pasteurianum* will function as acceptors of the ADP-ribose group *in vitro*.

The reaction with dinitrogenase reductase from *R. rubrum* is absolutely dependent on the presence of MgADP in the activation mixture and addition of MgATP leads to inhibition. The effect of MgADP is exerted by its binding to dinitrogenase reductase, probably resulting in a conformational change of the dinitrogenase reductase that is required for ADP-ribosylation. In contrast to the *R. rubrum* dinitrogenase reductase, dinitrogenase reductase from *A. vinelandii* and *K. pneumoniae* could be ADP-ribosylated in the absence of MgADP. However, the K_m values were increased about 30-fold, from 22 μM to 0.7 mM for *A. vinelandii* and from 71 μM to 2.5 mM for *K. pneumoniae*.

An interesting observation is that ADP-ribosylated dinitrogenase reductase from *K. pneumoniae*, although incapable of acetylene reduction, would catalyze ATP hydrolysis. Similar results were obtained with a mutant in which the arginine-101 was substituted by histidine, indicating that the interaction between dinitrogenase reductase and dinitrogenase is different during inter-molecular electron transfer compared to the interaction during reductant-independent ATP-hydrolysis.

The reactions catalyzed by DRAT and DRAG together form a regulatory cycle in which dinitrogenase reductase is reversibly converted from an active unmodified form to the inactive ADP-ribosylated form (Figure 1). DRAT requires MgADP for activity, whereas DRAG is dependent on MgATP and a divalent cation, Mn^{2+} being the physiologically important one. This regulatory cycle is the molecular basis for the reversible inhibition caused either by addition of NH_4^+, glutamine, or asparagine to a culture of N_2-fixing *R. rubrum* or by subjecting the cells to darkness; the "switch-off" effect. However, although this cycle explains the conversion of dinitrogenase reductase, it does not explain the regulation of DRAT and DRAG at the molecular level, *i.e.*, the signal transduction from the "switch-off" effector to the enzymes. This interesting issue will be discussed in a later section.

5. EVIDENCE FOR THE DRAT/DRAG SYSTEM IN OTHER ORGANISMS

Although the majority of the work on the DRAT/DRAG system of ADP-ribosylation of dinitrogenase reductase has been done in *R. rubrum*, with significant contributions from *Az. brasilense* (Zhang *et al*., 1997) and *Rb. capsulatus*, evidence for this system exists in a wide range of organisms, indicating that it is not restricted to the purple bacteria and their close relatives. Evidence of loss of *in vivo* nitrogenase activity (acetylene reduction) in response to added ammonium is a first indication of the DRAT/DRAG system, but it is not definitive. Many organisms that lack DRAT/DRAG, including the most extensively studied model organisms, *K. pneumoniae*, *A. vinelandii* and *C. pasteurianum*, also show no "switch-off" of nitrogenase activity when NH_4^+ is added to actively fixing cultures. On the other hand, some organisms that lack DRAT/DRAG do show response to NH_4^+. *Rhodopseudomonas sphaeroides* was the first described to have an *in vivo* response but evidence for DRAT/DRAG in this organism is lacking (Yoch *et al.*, 1988). *Rb. capsulatus* clearly contains DRAT/DRAG, but it also contains at least one other system of ammonium response because *draT* mutants still show NH_4^+-dependent loss of *in vivo* activity. Hallenbeck has also characterized the "magnitude response" in this organism that is independent of DRAT/DRAG (Yakunin and Hallenbeck, 1998; Yakunin *et al.*, 1999). Below, the molecular characteristics of the DRAT/DRAG system as it is known in *R. rubrum* are discussed. Information about the details of the control(s) operating in *Rb. capsulatus* can be found in Hallenbeck (1992), Hallenbeck *et al.* (1982), Jouanneau *et al.* (1989), and Yakunin *et al.* (2001).

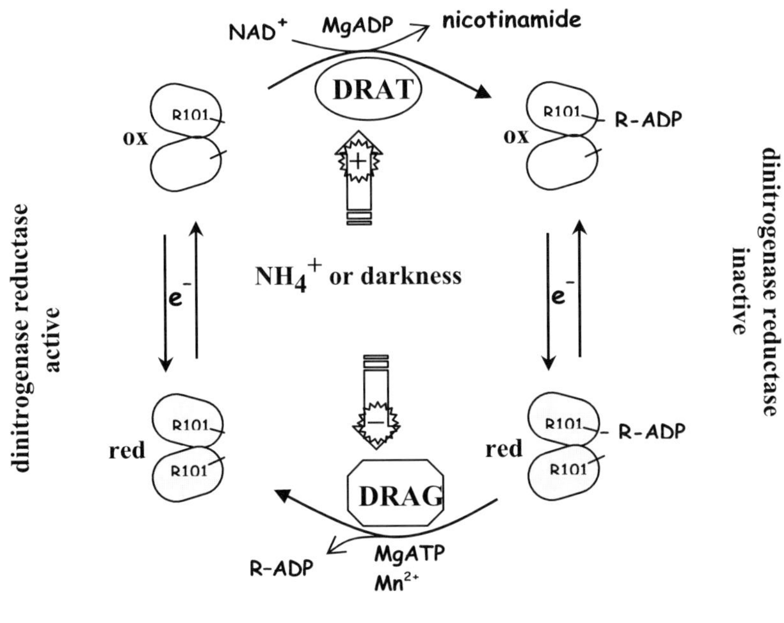

Figure 1. Model of the regulation of dinitrogenase reductase activity by reversible ADP-ribosylation. Darkness and NH_4^+ are shown as the "switch-off" effectors, probably being physiologically the most important.

6. OTHER ADP-RIBOSYLATIONS

Although the regulation of dinitrogenase reductase by ADP-ribosylation is the best characterized reversible mono-ADP-ribosylation system, numerous other ADP-ribosylations are found in nature. The requirement for NAD for action led to the discovery that diphtheria toxin catalyzes the NAD-dependent ADP-ribosylation of the elongation factor involved in eukaryotic protein synthesis (Honjo *et al.*, 1968; Kandel *et al.*, 1974). Other bacterial toxins, including cholera toxin and pertussis toxin, catalyze the ADP-ribosylations of the α subunits of eukaryotic G proteins, leading to cell death (Moss and Vaughan, 1977; West *et al.*, 1985). In these cases, the modification appears to be irreversible. *Clostridium botulinum* produces a number of toxins and among these are ADP-ribosylating toxins (Mauss *et al.*, 1990; Mohr *et al.*, 1992). Other examples of endogenous ADP-ribosylation in prokaryotes are relatively rare, but mycobacteria produce an ADP-ribosyl transferase activity capable of modifying several, as yet unidentified, proteins (Serres and Ensign, 1996). ADP-ribosylations of endogenous proteins in *Streptomyces griseus* have also been reported (Penyige *et al.*, 1990).

An early result in the studies of ADP-ribosylation was that a T4 phage-encoded ADP-ribosyl transferase modified a subunit of RNA polymerase (Goff, 1974). Subsequent work has shown that phage T4 encodes two ADP-ribosyl transferases, the *alt* gene product and the *mod* gene product (Mailhammer *et al.*, 1975). The *alt* gene product has been more extensively studied. It modifies the β, β' and σ^{70} subunits of RNA polymerase and that modification may direct the polymerase to specific early genes involved in phage replication.

In addition to mono-ADP-ribosylations, eukaryotic cells perform poly-ADP-ribosylation of histones and proteins involved in DNA repair (D'Amours *et al.*, 1999). The poly-ADP-ribosylation system is distinct from mono-ADP-ribosylation and will not be further discussed here.

The recognition that essential proteins in eukaryotic cells could be targets of ADP-ribosylation by bacterially produced toxins led Moss, Vaughan and others to ask if there were endogenous ADP-ribosylations. A number of ADP-ribosylating activities have been identified in eukaryotic cells (Koch-Nolte *et al.*, 1996; Moss *et al.*, 1999; West and Moss, 1986; Zolkiewska and Moss, 1993; Zolkiewska *et al.*, 1992; 1994), although to date none has been characterized as a reversible system of activity regulation. The eukaryotic ADP-ribosyl transferases have been purified and characterized and, in some cases, proteins capable of ADP-ribosylation *in vitro* have been identified, although these proteins may not be so modified *in vivo* (Klebl *et al.*, 1997; Koch-Nolte *et al.*, 1996; Zolkiewska and Moss, 1993).

A different approach was taken by Jacobson and co-workers, who sought proteins that were ADP-ribosylated *in vivo* by looking for the modification (Cervantes-Laurean *et al.*, 1993; 1995; 1996). They observed numerous proteins that were ADP-ribosylated. Furthermore, the Jacobson group sought to identify the amino-acid side chains that were modified. By determining the chemical stability of various ADP-ribosyl linkages, they were able to establish the presence of arginyl, lysyl-, cysteinyl-, aspartyl-, and seryl-linkages to ADP-ribose in eukaryotic cells (Jacobson *et al.*, 1994). There is significant sequence conservation among the ADP-ribosyl transferases, including DRAT (Bazan and Koch-Nolte, 1997). All transferases contain an XYE signature, where X is one of a small number of amino-acid side chains that determines what residue on the target protein will be ADP-ribosylated, *e.g.*, EXE is found in DRAT and designates an arginyl target.

The high resolution structures of three ADP-ribosylating toxins have been reported, including that of diphtheria toxin (Choe *et al.*, 1992). The diphtheria toxin structure was determined with NAD bound, revealing the site of NAD binding and the conformation of NAD. Interestingly, the NAD molecule has a significantly open configuration, which is unlike that observed with NAD binding in the Rossman fold of many dehydrogenases. The crystal structure also confirms the proximity of glutamate-148 of the diphtheria toxin to the nicotinamide of NAD. Collier *et al.* (1985) had previously shown that glutamate-148 becomes labeled following UV irradiation of the toxin to which (^{3}H-nicotinamide)-NAD was bound.

The fact that DRAT and DRAG are present in NH_4^+-grown (*nif*-repressed) cells suggests that these proteins might have some other role(s) in the cell. However, in *R. rubrum* to date, no other role(s) for either DRAT or DRAG has been established and *draTG* mutants have no observable phenotype other than the expected effects on the regulation of nitrogenase. Over-expression of DRAT and DRAG, either individually or in concert, also has no effect other than on nitrogenase regulation. If there is an additional role for either DRAT or DRAG in the cell, the effects are either very subtle or they have not been tested under appropriate conditions. Of potential relevance, the CooC protein, which is involved in Ni insertion into carbon monoxide dehydrogenase in *R. rubrum*, has significant sequence similarity to dinitrogenase reductase, including the gly-arg-gly site of ADP-ribosylation by DRAT. However, there is no evidence that this protein is a target for DRAT.

7. GENETICS OF THE DRAT/DRAG SYSTEM

The two genes encoding DRAT and DRAG are the *draT* and the *draG* gene, respectively, which in *R. rubrum* probably are co-transcribed together with an ORF downstream of *draG* encoding a protein of 15 kDa (Fitzmaurice *et al.*, 1989; Liang *et al.*, 1991). The ORF has been named *draB* but the role of the gene product has not been elucidated (Zhang *et al.*, 2001a). The *draTGB* region is localized 402 bp from the *nifHDK* genes, but is transcribed in the opposite direction. The region upstream of *draT* does not contain any typical *nif* promoter, suggesting that transcription of the *nif* and the *dra* genes are not co-regulated. This suggestion is supported by DRAG being detected under all growth conditions studied (Triplett *et al.*, 1982) and by the demonstration that *R. rubrum*, when grown fermentatively in the dark, responds to added NH_4^+ and so has an active DRAG (Schultz *et al.*, 1985).

As in *R. rubrum*, the *dra* genes in *Az. brasilense* and *Az. lipoferum* are located in the same region as *nifHDK* and also transcribed in the opposite direction (Zhang *et al.*, 1997). An ORF similar to *draB* was also identified. In *Rb. capsulatus*, however, the *draTG* genes were found widely separated from the *nifHDK* genes and no sequence corresponding to *draB* could be identified (Masepohl *et al.*, 1993).

8. SIGNAL TRANSDUCTION TO DRAT AND DRAG

Although the DRAT and DRAG proteins were purified and characterized more than 10 years ago, the signal(s) regulating their activities during "switch-off" conditions has not been identified. A number of models have been proposed but none has so far been shown to explain all features of the regulatory cascade. One of the most intriguing aspects is the response of the system to "switch-off" effectors that either affect the nitrogen status of the cell or the energy/redox status. Obviously, any model has to account for the signaling in both of these situations. In this context, a central issue is where the signal-transduction pathways from these different kinds of effectors merge. It might be either at DRAT/DRAG or before. In addition, it has not been shown whether the same signal(s) affects both DRAT and DRAG.

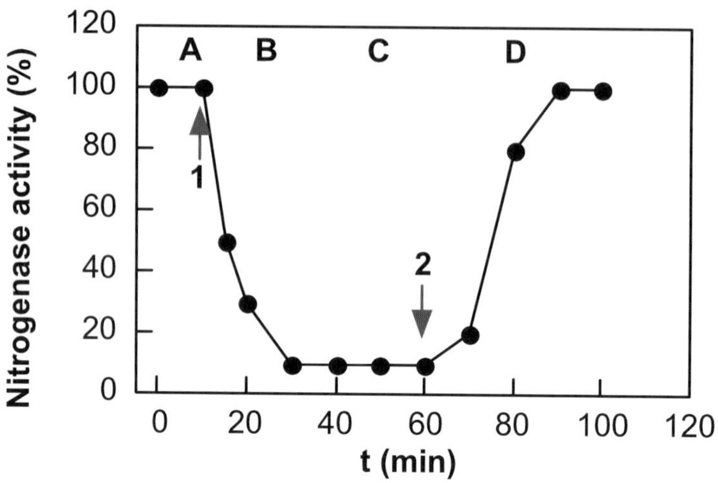

Figure 2. Idealized time course of "switch-off". Either the "switch-off" effector is added at 1 or light is turned off at 1 and light is turned on at 2.

Figure 2 shows an idealized time course of "switch-off" and the periods (A-D) during which the activities of DRAT and DRAG change. During period (A), prior to the addition of a "switch-off" effector, DRAT is inactive and DRAG is active. Support for this situation comes from studies on mutants in either *draT* or *draG* (Liang *et al.*, 1991). At (1), when either the effector is added or the light is turned off, and during period (B), DRAT is rapidly activated and DRAG inactivated, although probably more slowly. During period (C), DRAT is only transiently active, based on the fact that, in *R. rubrum*, nitrogenase activity only decreases to about 10%, indicating that fully active DRAT and inactive DRAG are not present at the same time. At (2), when either the added effector has been metabolized or the light turned on again, DRAG is activated and stays active through period (D).

One of the complications in determining what affects DRAT/DRAG is that both are fully active in cell extracts. This situation suggests that, whatever is the mechanism by which the enzymes are inactivated, it is dependent on the cell remaining intact and is lost upon cell breakage. Further, it has been assumed that at least some of the signals involved are general metabolites. This assumption is supported by studies in which the *draTG* genes from either *R. rubrum* or *Az. brasilense* were transferred to *K. pneumoniae*, which does not have an endogenous "switch-off" system (Fu *et al.*, 1990). The resulting *K. pneumoniae* strains showed typical "switch-off" when NH_4^+ was added. Although *Az. brasilense* and *R. rubrum* are sensing different energy signals, anaerobicity and darkness, respectively, the signal(s) must be functionally similar and this similarity was shown by transferring the *draTG* genes from *Az. brasilense* to a mutant of *R. rubrum*, which lacked DRAT and/or DRAG, and *vice versa* (Zhang *et al.*, 1995a). In both cases, the response to "switch-off" effectors was restored.

8.1. What Signals Are Involved in Activating DRAG and DRAT?

In efforts to establish the identity of the signal(s), a number of compounds that either reflect the nitrogen status, *e.g.*, glutamine, or indicate the energy/redox status, *e.g.*, the ATP/ADP ratio (Kanemoto and Ludden, 1987; Li *et al.*, 1987; Nordlund and Höglund, 1986; Paul and Ludden, 1984), have been suggested. However, none has been shown either to vary in concentration significantly during the initial phase of "switch-off" or to affect DRAT/DRAG activities *in vitro*. Another candidate for a direct role in the signal transduction is glutamine synthetase, but inhibitor studies have shown that glutamine synthetase is involved only when NH_4^+ is used as the "switch-off" effector (Falk *et al.*, 1982; Kanemoto and Ludden, 1984, 1987; Sweet and Burris, 1982). Under these conditions (or when either glutamine or asparagine is used as effector), a high flux through the reactions catalyzed by either glutamine synthetase or glutamate synthase is required for the "switch-off", *i.e.*, ADP-ribosylation, to occur (Norén *et al.*, 1997).

Although low molecular-mass molecules most likely participate in the signal transduction at some stage, a key issue is how the proteins, DRAT and DRAG, are affected and in what way their activities are controlled. Ludden, Roberts and co-workers suggested that complex formation between DRAT and DRAG is central to the regulation of DRAG activity and that inactive DRAT binds dinitrogenase reductase (Fe protein). Such binding has, however, essentially no effect on nitrogenase activity because there is a great excess of dinitrogenase reductase compared to DRAT in the cell. This model was largely based on studies of *R. rubrum* strains in which DRAT and DRAG were either over-expressed or absent in different combinations (Grunwald *et al.*, 1995). Furthermore, cross-linking techniques showed that DRAT can only be cross-linked to dinitrogenase reductase when NAD^+ is present, and that ATP is inhibitory to this interaction, in agreement with the requirements of the DRAT reaction (Grunwald and Ludden, 1997; Grunwald *et al.*, 2000). Additional support for this model was obtained in studies using mutants of DRAT and dinitrogenase reductase. Substituting arginine-101 of dinitrogenase reductase affected the reaction with NAD^+, but did not preclude the formation of a complex of the two proteins, suggesting that this residue is not absolutely required for either complex formation or NAD^+ binding, but is needed for NAD^+ cleavage (Ma and Ludden, 2001). Using DRAT mutants identified glutamate-112 in dinitrogenase reductase, along with lysine-103 and asparagine-248 in DRAT, as important in the regulation and activity of DRAT (Kim *et al.*, 1999; Zhang *et al.*, 2001a). The major unknown component in this model is the molecule that is postulated to affect DRAT, presumably as an inhibitor.

8.2. The Redox State of Dinitrogenase Reductase

A new insight, using *R. rubrum* and *K. pneumoniae*, into the regulation of DRAT and DRAG came from the effect of changes in the redox state of dinitrogenase reductase (Halbleib *et al.*, 2000a; 2000b). DRAT clearly shown to catalyze modification of only the oxidized form of dinitrogenase reductase, whereas the reduced form is the substrate for DRAG. These results agree with the observation that NAD^+ can act as a "switch-off" effector with whole cells, and that the NAD(P) pool is oxidized upon addition of "switch-off" effectors to *R. rubrum* cells (Norén and Nordlund, 1994; Norén *et al.*, 1997; Soliman and Nordlund, 1992). Further support for the importance of the redox state was derived with an *ntrBC* mutant of *R. rubrum*, which, when grown with glutamate as nitrogen source, showed a lower and slower response to "switch-off" effectors (Zhang *et al.*, 1995b). However, when grown photoheterotrophically with N_2 as nitrogen source, conditions under which the redox state of the NAD(P) pool is believed to be more oxidized, the "switch-off" effect was more like that in the wild type (Cheng and Nordlund, unpublished data). Taken together, these studies have led Nordlund and co-workers to propose that the regulation of DRAT is mainly through the availability of NAD^+ (for which DRAT has a surprisingly high K_m in *R. rubrum*) and that an important factor in the regulation of DRAG activity is its reversible association with the chromatophore membrane (Nordlund, 2000; Norén and Nordlund, 1997).

8.3. Association with the Chromatophore Membranes

It has been known since the mid-seventies that DRAG is associated with the chromatophores in cell extracts and can be washed off with 0.5 M NaCl, which is the first step in its purification. Whether the association is physiologically significant is a matter of discussion, but it should be noted that harvesting cells involves subjecting them to "switch-off" conditions, *i.e.*, darkness. If the association has a regulatory function, it should be specific and this aspect has been questioned (Halbleib and Ludden, 1999). On the other hand, DRAG forms two specific complexes when treating chromatophores with cross-linking agents (Norén and Nordlund, unpublished data). Furthermore, GDP/GTP can solubilize DRAG, whereas ADP/ATP has no effect (Halbleib and Ludden, 1999; Norén and Nordlund, 1997). These observations indicate that the association of DRAG with the membranes could be significant.

8.4. The Role of the P_{II} Protein

In most diazotrophs, the signal protein P_{II} plays a central role in the regulation of nitrogen metabolism. The regulatory effects of this trimeric protein are exerted through the two forms of the protein, the uridylylated form and the non-modified form. The bifunctional enzyme, uridylyl transferase/uridylyl-removing enzyme, catalyzes the reversible uridylylation in response to the nitrogen status, which is sensed as the concentration of glutamine, so that, under nitrogen-limiting conditions, P_{II} is uridylylated (Arcondéguy *et al.*, 2001). In a number of bacteria, two homologs, GlnB and GlnK, exist, where GlnB is the one originally identified. At the amino-acid level, these homologs typically show around 70% similarity. Although P_{II} has been proposed to have a role in the "switch-off" effect, there were no results in support of this hypothesis until recently.

In three recent papers, the role of P_{II} homologs in the "switch-off" in *R. rubrum* was clearly demonstrated. *R. rubrum* has three P_{II} proteins, encoded by *glnB*, *glnK,* and *glnJ*, respectively (Zhang *et al.*, 2001c). Although all three share a high amino-acid sequence similarity, they have different but, to some extent, overlapping roles. The GlnB protein is essential for *nif* expression and is believed to be required for activation of NifA (Zhang *et al.*, 2000; 2001c). Either GlnB or GlnJ is necessary for "switch-off"; little effect was seen on nitrogenase activity in the *glnB-glnJ* double mutant and the *glnB- glnJ- glnK* triple mutant strains after addition of NH_4^+ as "switch-off" effector (Zhang *et al.*, 2001c).

Interestingly and surprisingly, "switch-off" by darkness was also affected in these strains, indicating that the role of GlnB/GlnJ is downstream of the point in the signal-transduction pathway where signals reflecting the nitrogen status and the energy/redox status merge. In this context, it is important to note that, although the mutations discussed above may have polar effects on the two *amtB* genes located downstream of *glnK* and *glnJ*, respectively, this situation does not affect the "switch-off" because the *glnK- glnJ* double mutant strain shows wild-type "switch-off" (Zhang *et al.*, 2001c). This result contrasts with a *Rb. capsulatus* mutant that lacks AmtB, which does not respond to NH_4^+ as "switch-off" effector (Yakunin and Hallenbeck, 2002).

GlnK does not have any role in the "switch-off" in *R. rubrum* (Zhang *et al.*, 2001c), but different effects were seen in *glnB* and *glnK* mutants of *K. pneumoniae* harboring and expressing the *draTG* genes from *R. rubrum* (Zhang *et al.*, 2001b). In this case, in a *glnB* mutant, DRAT escapes regulation and stays active even under nitrogen-limiting conditions. In this mutant, inactivation of DRAG was slower. In a *glnK* mutant, no effect was seen on DRAT, but the regulation of DRAG was completely lost and DRAG remained active under "switch-off" conditions. In this context, it is worth noting that yeast two-hybrid studies indicate a direct interaction between GlnB and DRAT in *Rb. capsulatus* (Masepohl *et al.*, 2002a, 2002b). Taken together, these studies clearly demonstrate that P_{II} proteins play an important role in the signal transduction from a "switch-off" effector to DRAT and DRAG.

9. CONCLUSIONS

In conclusion, the last 25 years of research has identified the basis for the "switch-off" effect at the molecular level, including the enzymes catalyzing the reversible ADP-ribosylation of dinitrogenase reductase in *R. rubrum*. However, the components of the signal transduction to DRAT and DRAG have not been identified. The recent demonstration of the role of P_{II} proteins in this pathway is, however, most likely an important contribution to future studies. It is not surprising that P_{II} proteins are involved in the signaling of the nitrogen status, but it will be crucial to show whether or not there is a direct effect of these proteins in "switch-off" by darkness. The regulatory system, which controls nitrogenase activity in *R. rubrum* and other diazotrophs, has already shown a number of new features concerning enzyme control in response to metabolic changes that are important for understanding metabolic regulation in general. Some of the possible components that have been suggested, like the role of GTP and GTP hydrolysis, may, if confirmed, add further complexity to the control cascade.

ACKNOWLEDGEMENT

This chapter is dedicated to the memory of our friend and colleague, Werner Klipp. His life was like his seminars – passing too quickly for us to fully understand all of his brilliance. The many conversations with him at meetings and conferences are now treasured memories. Our small corner of the study of nitrogen fixation is irretrievably diminished with his loss.

REFERENCES

Antharavally, B. S., Poyner, R. R. and Ludden, P. W. (1998). EPR Spectral evidence for a binuclear Mn(II) center in dinitrogenase reductase-activating glycohydrolase from *Rhodospirillum rubrum*. *J. Am. Chem. Soc., 120*, 8897-8898.

Antharavally, B. S., Poyner, R. R., Zhang, Y., Roberts, G. P. and Ludden, P. W. (2001). Effects of specific amino acid substitutions on activities of dinitrogenase reductase-activating glycohydrolase from *Rhodospirillum rubrum*. *J. Bacteriol., 183*, 5743-5746.

Arcondéguy, T., Jack, R. and Merrick, M. (2001). P_{II} signal transduction proteins, pivotal players in microbial nitrogen control. *Microbiol. Mol. Biol. Rev., 65*, 80-105.

Bazan, J. F. and Koch-Nolte, F. (1997). Sequence and structural links between distant ADP-ribosyl-transferase families. *Adv. Exp. Med. Biol., 419*, 99-107.

Bulen, W. A., Burns, R. C. and LeComte, J. R. (1965). Nitrogen fixation: Hydrosulfite as electron donor with cell-free preparations of *Azotobacter vinelandii* and *Rhodospirillum rubrum*. *Proc. Natl. Acad. Sci. U.S.A., 53*, 532-539.

Burns, R. C. and Bulen, W. A. (1965). ATP-dependent hydrogen evolution by cell-free preparations of *Azotobacter vinelandii*. *Biochim. Biophys. Acta, 105*, 437-445.

Burns, R. C. and Bulen, W. A. (1966). A procedure for the preparation of extracts from *Rhodospirillum rubrum* catalyzing N_2 reduction and ATP-dependent H_2 evolution. *Arch. Biochem. Biophys., 113*, 461-463.

Cadez, P. and Nordlund, S. (1991). The requirement for Mn^{2+} and Ca^{2+} in nitrogen fixation by the photosynthetic bacterium *Rhodospirillum rubrum*. *FEMS Microbiol. Lett., 81*, 279-282.

Carithers, R. P., Yoch, D. C. and Arnon, D. I. (1979). Two forms of nitrogenase from the photosynthetic bacterium *Rhodospirillum rubrum*. *J. Bacteriol., 137*, 779-789.

Carnahan, J. E., Mortenson, L. E., Mower, H. F. and Castle, J. E. (1960). Nitrogen fixation in cell-free extracts of *Clostridium pasteurianum*. *Biochim. Biophys. Acta, 44*, 520-535.

Carroll, S. F., McCloskey, J. A., Crain, P. F., Oppenheimer, N. J., Marschner, T. M. and Collier, R. J. (1985). Photoaffinity labeling of diphtheria toxin fragment A with NAD: Structure of the photoproduct at position 148. *Proc. Natl. Acad. Sci. U.S.A., 82*, 7237-7241.

Cervantes-Laurean, D., Minter, D. E., Jacobson, E. L. and Jacobson, M. K. (1993). Protein glycation by ADP-ribose: Studies of model conjugates. *Biochemistry, 32*, 1528-1534.

Cervantes-Laurean, D., Loflin, P. T., Minter, D. E., Jacobson, E. L. and Jacobson, M. K. (1995). Protein modification by ADP-ribose via acid-labile linkages. *J. Biol. Chem., 270*, 7929-7936.

Cervantes-Laurean, D., Jacobson, E. L. and Jacobson, M. K. (1996). Glycation and glycoxidation of histones by ADP-ribose. *J. Biol. Chem., 271*, 10461-10469.

Choe, S., Bennett, M. J., Fujii, G., Curmi, P. M., Kantardjieff, K. A., Collier, R. J., *et al.* (1992). The crystal structure of diphtheria toxin. *Nature, 357*, 216-222.

D'Amours, D., Desnoyers, S., D'Silva, I. and Poirier, G. G. (1999). Poly(ADP-ribosyl)ation reactions in the regulation of nuclear functions. *Biochem. J., 342* (Pt 2), 249-268.

Dowling, T. E., Preston, G. G. and Ludden, P. W. (1982). Heat activation of the Fe protein of nitrogenase from *Rhodospirillum rubrum*. *J. Biol. Chem., 257*, 13987-13992.

Falk, G., Johansson, B. C. and Nordlund, S. (1982). The role of glutamine synthetase in the regulation of nitrogenase activity (switch-off effect) in *Rhodospirillum rubrum*. *Arch. Microbiol., 132*, 251-253.

Fitzmaurice, W. P., Saari, L. L., Lowery, R. G., Ludden, P. W. and Roberts, G. P. (1989). Genes coding for the reversible ADP-ribosylation system of dinitrogenase reductase from *Rhodospirillum rubrum*. *Mol. Gen. Genet., 218*, 340-347.

Fu, H., Burris, R. H. and Roberts, G. P. (1990). Reversible ADP-ribosylation is demonstrated to be a regulatory mechanism in prokaryotes by heterologous expression. *Proc. Natl. Acad. Sci. U.S.A., 87*, 1720-1724.

Gest, H. and Kamen, M. D. (1949). Photoproduction of molecular hydrogen by *Rhodospirillum rubrum*. *Science, 109*, 558-559.

Gest, H. (1999). Memoir of a 1949 railway journey with photosynthetic bacteria. *Photosynthesis Res., 61*, 91-96.

Goff, C. G. (1974). Chemical structure of a modification of the *Escherichia coli* ribonucleic acid polymerase alpha polypeptides induced by bacteriophage T4 infection. *J. Biol. Chem., 249*, 6181-6190.

Gotto, J. W. and Yoch, D. C. (1982). Regulation of *Rhodospirillum rubrum* nitrogenase activity. *J. Biol. Chem., 257*, 2868-2873.

Grunwald, S. K., Lies, D. P., Roberts, G. P. and Ludden, P. W. (1995). Posttranslational regulation of nitrogenase in *Rhodospirillum rubrum* strains overexpressing the regulatory enzymes dinitrogenase reductase ADP-ribosyltransferase and dinitrogenase reductase activating glycohydrolase. *J. Bacteriol., 177*, 628-635.

Grunwald, S. K. and Ludden, P. W. (1997). NAD-dependent cross-linking of dinitrogenase reductase and dinitrogenase reductase ADP-ribosyltransferase from *Rhodospirillum rubrum*. *J. Bacteriol., 179*, 3277-3283.

Grunwald, S. K., Ryle, M. J., Lanzilotta, W. N. and Ludden, P. W. (2000). ADP-ribosylation of variants of *Azotobacter vinelandii* dinitrogenase reductase by *Rhodospirillum rubrum* dinitrogenase reductase ADP-ribosyltransferase. *J. Bacteriol., 182*, 2597-2603.

Guth, J. H. and Burris, R. H. (1983). Comparative study of the active and inactive forms of dinitrogenase reductase from *Rhodospirillum rubrum*. *Biochim. Biophys. Acta, 749*, 91-100.

Halbleib, C. M. and Ludden, P. W. (1999). Characterization of the interaction of dinitrogenase reductase-activating glycohydrolase from *Rhodospirillum rubrum* with bacterial membranes. *Arch. Microbiol., 172*, 51-58.

Halbleib, C. M., Zhang, Y. and Ludden, P. W. (2000a). Regulation of dinitrogenase reductase ADP-ribosyltransferase and dinitrogenase reductase-activating glycohydrolase by a redox-dependent conformational change of nitrogenase Fe protein. *J. Biol. Chem., 275*, 3493-3500.

Halbleib, C. M., Zhang, Y., Roberts, G. P. and Ludden, P. W. (2000b). Effects of perturbations of the nitrogenase electron transfer chain on reversible ADP-ribosylation of nitrogenase Fe protein in *Klebsiella pneumoniae* strains bearing the *Rhodospirillum rubrum dra* operon. *J. Bacteriol., 182*, 3681-3687.

Hallenbeck, P. C., Meyer, C. M. and Vignais, P. M. (1982). Nitrogenase from the photosynthetic bacterium *Rhodopseudomonas capsulata*: Purification and molecular properties. *J. Bacteriol., 149*, 708-717.

Hallenbeck, P. C. (1992). Mutations affecting nitrogenase switch-off in *Rhodobacter capsulatus*. *Biochim. Biophys. Acta, 1118*, 161-168.

Honjo, T., Nishizuka, Y. and Hayaishi, O. (1968). Diphtheria toxin-dependent adenosine diphosphate ribosylation of aminoacyl transferase II and inhibition of protein synthesis. *J. Biol. Chem., 243*, 3553-3555.

Imhoff, J. F., Truper, H. G. and Pfennig, N. (1984). Rearrangement of the species and genera of the phototrophic "purple nonsulfur bacteria". *Int. J. Syst. Bacteriol., 34*, 340-343.

Imhoff, J. F., Petri, R. and Süling, J. (1998). Reclassification of species of the spiral-shaped phototrophic purple non-sulfur bacteria of the a-*Proteobacteria*: Description of the new genera *Phaeospirillum* gen. nov., *Rhodovibrio* gen. nov., *Rhodothalassium* gen. nov. and *Roseospira* gen. nov. as well as transfer of *Rhodospirillum fulvum* to *Phaeospirillum fulvum* comb. nov., of *Rhodospirillum molischianum* to *Phaeospirillum molischianum* comb. nov., of *Rhodospirillum salinarum* to *Rhodovibrio salinarum* comb. nov., of *Rhodospirillum sodomense* to *Rhodovibrio sodomensis* comb. nov., of *Rhodospirillum salexigens* to *Rhodothalassium salexigens* comb. nov. and of *Rhodospirillum mediosalinum* to *Roseospira mediosalina* comb. nov. *Int. J. Syst. Bacteriol., 48*, 793-798.

Jacobson, E. L., Cervantes-Laurean, D. and Jacobson, M. K. (1994). Glycation of proteins by ADP-ribose. *Mol. Cell. Biochem., 138*, 207-212.

Jouanneau, Y., Roby, C., Meyer, C. M. and Vignais, P. M. (1989). ADP-ribosylation of dinitrogenase reductase in *Rhodobacter capsulatus*. *Biochemistry, 28*, 6524-6530.

Kamen, M. D. (1986). Radiant Science, Dark Politics. Berkeley, CA: University of California Press.

Kandel, J., Collier, R. J. and Chung, D. W. (1974). Interaction of fragment A from diphtheria toxin with nicotinamide adenine dinucleotide. *J. Biol. Chem., 249*, 2088-2097.

Kanemoto, R. H. and Ludden, P. W. (1984). Effect of ammonia, darkness, and phenazine methosulfate on whole-cell nitrogenase activity and Fe protein modification in *Rhodospirillum rubrum*. *J. Bacteriol., 158*, 713-720.

Kanemoto, R. H. and Ludden, P. W. (1987). Amino acid concentrations in *Rhodospirillum rubrum* during expression and switch-off of nitrogenase activity. *J. Bacteriol., 169*, 3035-3043.

Kim, K., Zhang, Y. and Roberts, G. P. (1999). Correlation of activity regulation and substrate recognition of the ADP-ribosyltransferase that regulates nitrogenase activity in *Rhodospirillum rubrum*. *J. Bacteriol., 181*, 1698-1702.

Klebl, B. M., Gopel, S. O. and Pette, D. (1997). Specificity and target proteins of arginine-specific mono-ADP-ribosylation in T-tubules of rabbit skeletal muscle. *Arch. Biochem. Biophys., 347*, 155-162.

Koch-Nolte, F., Petersen, D., Balasubramanian, S., Haag, F., Kahlke, D., Willer, T., *et al.* (1996). Mouse T cell membrane proteins Rt6-1 and Rt6-2 are arginine/protein mono(ADPribosyl)transferases and share secondary structure motifs with ADP-ribosylating bacterial toxins. *J. Biol. Chem., 271*, 7686-7693.

Li, J., Hu, C.-Z. and Yoch, D. C. (1987). Changes in amino acid and nucleotide pools of *Rhodospirillum rubrum* during switch-off of nitrogenase activity initated by NH_4^+ or darkness. *J. Bacteriol., 169*, 231-237.

Liang, J., Nielsen, G. M., Lies, D. P., Burris, R. H., Roberts, G. P. and Ludden, P. W. (1991). Mutations in the *draT* and *draG* genes of *Rhodospirillum rubrum* result in loss of regulation of nitrogenase by reversible ADP-ribosylation. *J. Bacteriol., 173*, 6903-6909.

Ljungström, E., Yates, M. G. and Nordlund, S. (1989). Purification of the activating enyzme for the Fe-protein of nitrogenase from *Azospirillum brasilense*. *Biochim. Biophys. Acta, 994*, 210-214.

Lowery, R. G., Saari, L. L. and Ludden, P. W. (1986). Reversible regulation of the nitrogenase iron protein from *Rhodospirillum rubrum* by ADP-ribosylation *in vitro*. *J. Bacteriol., 166*, 513-518.

Lowery, R. G. and Ludden, P. W. (1988). Purification and properties of dinitrogenase reductase inactivating ADP-ribosyltransferase from the photosynthetic bacterium *Rhodospirillum rubrum*. *J. Biol. Chem., 263*, 16714-16719.

Lowery, R. G., Chang, C. L., Davis, L. C., McKenna, M. C., Stephens, P. J. and Ludden, P. W. (1989). Substitution of histidine for arginine-101 of dinitrogenase reductase disrupts electron transfer to dinitrogenase. *Biochemistry, 28*, 1206-1212.

Lowery, R. G. and Ludden, P. W. (1989). Effect of nucleotides on the activity of dinitrogenase reductase ADP-ribosyltransferase from *Rhodospirillum rubrum*. *Biochemistry, 28*, 4956-4961.

Ludden, P. W. and Burris, R. H. (1976). Activating factor for the iron protein of nitrogenase from *Rhodospirillum rubrum*. *Science, 194*, 424-426.

Ludden, P. W. and Burris, R. H. (1978). Purification and properties of nitrogenase from *Rhodospirillum rubrum* and evidence for phosphate, ribose and an adenine-like unit covalently bound to the iron protein. *Biochem. J., 175*, 251-259.

Ludden, P. W. and Burris, R. H. (1979). Removal of an adenine-like molecule during activation of dinitrogenase reductase from *Rhodospirillum rubrum*. *Proc. Natl. Acad. Sci. U.S.A., 76*, 6201-6205.

Ludden, P. W., Preston, G. G. and Dowling, T. E. (1982). Comparison of active and inactive forms of iron protein from *Rhodospirillum rubrum*. *Biochem. J., 203*, 663-668.

Ma, Y. and Ludden, P. W. (2001). Role of the dinitrogenase reductase arginine 101 residue in dinitrogenase reductase ADP-ribosyltransferase binding, NAD binding, and cleavage. *J. Bacteriol., 183*, 250-256.

Madigan, M., Cox, S. S. and Stegeman, R. A. (1984). Nitrogen fixation and nitrogenase activities in members of the family *Rhodospirillaceae*. *J. Bacteriol., 157*, 73-78.

Madigan, M. T. (1995). Microbiology of nitrogen fixation by anoxygenic photosynthetic bacteria. In R. E. Blankenship, M. T. Madigan and C. E. Bauer (Eds.), *Anoxygenic Photosynthetic Bacteria* (pp. 915-928). Dordrecht: Kluwer Academic Publishers.

Mailhammer, R., Yang, H. L., Reiness, G. and Zubay, G. (1975). Effects of bacteriophage T4-induced modification of *Escherichia coli* RNA polymerase on gene expression *in vitro*. *Proc. Natl. Acad. Sci. U.S.A., 72*, 4928-4932.

Masepohl, B., Krey, R. and Klipp, W. (1993). The *draTG* gene region of *Rhodobacter capsulatus* is required for post-translational regulation of both the molybdenum and the alternative nitrogenase. *J. Gen. Microbiol., 139*, 2667-2675.

Masepohl, B., Dreiskemper, P., Drepper, T., Gross, S., Isakovic, N., Pawlowski, A., *et al.* (2002a). Protein-protein interactions of regulatory and other gene products involved in nitrogen fixation in *Rhodobacter capsulatus*: Identification of new *anf* genes. In T. Finan, M. O'Brian, D. Layzell, K. Vessey and W. E. Newton (Eds.), *Nitrogen fixation: Global perspectives* (pp. 424). Wallingford, UK: CABI Publishing.

Masepohl, B., Drepper, T., Paschen, A., Gross, S., Pawlowski, A., Raabe, K., *et al.* (2002b). Regulation of nitrogen fixation in the phototrophic purple bacterium *Rhodobacter capsulatus*. *J. Mol. Microbiol. Biotechnol., 4*, 243-248.

Mauss, S., Chaponnier, C., Just, I., Aktories, K. and Gabbiani, G. (1990). ADP-ribosylation of actin isoforms by *Clostridium botulinum* C2 toxin and *Clostridium perfringens* iota toxin. *Eur. J. Biochem., 194*, 237-241.

Mohr, C., Koch, G., Just, I. and Aktories, K. (1992). ADP-ribosylation by *Clostridium botulinum* C3 exoenzyme increases steady-state GTPase activities of recombinant RhoA and RhoB proteins. *FEBS Lett., 297*, 95-99.

Moss, J. and Vaughan, M. (1977). Mechanism of action of choleragen. Evidence for ADP-ribosyltransferase activity with arginine as an acceptor. *J. Biol. Chem., 252*, 2455-2457.

Moss, J. and Vaughan, M. (1990). *ADP-ribosylating toxins and G proteins. Insights into signal transduction*. Washington D. C.: American Society for Microbiology Press.

Moss, J., Balducci, E., Cavanaugh, E., Kim, H. J., Konczalik, P., Lesma, E. A., *et al.* (1999). Characterization of NAD:arginine ADP-ribosyltransferases. *Mol. Cell. Biochem., 193*, 109-113.

Munson, T. O. and Burris, R. H. (1969). Nitrogen fixation by *Rhodospirillum rubrum* grown in nitrogen limited continuous culture. *J. Bacteriol., 97*, 1093-1098.

Neilson, A. H. and Nordlund, S. (1975). Regulation of nitrogenase synthesis in intact cells of *Rhodospirillum rubrum*: inactivation of nitrogen fixation by ammonia, L-glutamine and L-asparagine. *J. Gen. Microbiol., 91*, 53-62.

Nielsen, G. M., Bao, Y., Roberts, G. P. and Ludden, P. L. (1994). Purification and characterization of an oxygen-stable form of dinitrogenase reductase-activating glycohydrolase from *Rhodospirillum rubrum*. *Biochem. J., 302*, 801-806.

Nordlund, S. (2000). Regulation of nitrogenase activity in phototrophic bacteria by reversible covalent modification. In E. W. Triplett (Ed.), *Prokaryotic nitrogen fixation: A model system for analysis of a biological process* (pp. 149-163). Wymondham, UK: Horizon Scientific Press.

Nordlund, S., Eriksson, U. and Baltscheffsky, H. (1977). Necessity of a membrane component for nitrogenase activity in *Rhodospirillum rubrum*. *Biochim. Biophys. Acta, 462*, 187-195.

Nordlund, S., Eriksson, U. and Baltscheffsky, H. (1978). Properties of the nitrogenase system from a photosynthetic bacterium, *Rhodospirillum rubrum*. *Biochim. Biophys. Acta, 504*, 248-254.

Nordlund, S. and Höglund, L. (1986). Studies of the adenylate and pyridine nucleotide pools during nitrogenase 'switch-off' in *Rhodospirillum rubrum*. *Plant Soil, 90*, 203-209.

Nordlund, S. and Ludden, P. W. (1983). Incorporation of adenine into the modifying group of inactive iron protein of nitrogenase from *Rhodospirillum rubrum*. *Biochem. J., 209*, 881-884.

Nordlund, S. and Noren, A. (1984). Dependence on divalent cations of the activation of inactive Fe-protein of nitrogenase from *Rhodospirillum rubrum*. *Biochim. Biophys. Acta, 791*, 21-27.

Norén, A. and Nordlund, S. (1994). Changes in the NAD(P)H concentration caused by addition of nitrogenase 'switch-off' effectors in *Rhodospirillum rubrum* G-9, as measured by fluorescence. *FEBS Lett., 356*, 43-45.

Norén, A. and Nordlund, S. (1997). Dinitrogenase reductase activating glycohydrolase can be released from chromatophores of *Rhodospirillum rubrum* by treatment with MgGDP. *J. Bacteriol., 179*, 7872-7874.

Norén, A., Soliman, A. and Nordlund, S. (1997). The role of NAD^+ as a signal during nitrogenase switch-off in *Rhodospirillum rubrum*. *Biochem. J., 322*, 829-832.

Paul, T. D. and Ludden, P. W. (1984). Adenine nucleotide levels in *Rhodospirillum rubrum* during switch-off of whole-cell nitrogenase activity. *Biochem J., 224*, 961-969.

Penyige, A., Barabas, G., Szabo, I. and Ensign, J. C. (1990). ADP-ribosylation of membrane proteins of *Streptomyces griseus* strain 52-1. *FEMS Microbiol. Lett., 57*, 293-297.

Pope, M. R., Murrell, S. A. and Ludden, P. W. (1985a). Purification and properties of the heat released nucleotide modifying group from the inactive Fe protein of nitrogenase from *Rhodospirillum rubrum*. *Biochemistry, 24*, 2374-2380.

Pope, M. R., Murrell, S. A. and Ludden, P. W. (1985b). Covalent modification of the iron protein of nitrogenase from *Rhodospirillum rubrum* by adenosine diphosphoribosylation of a specific arginyl residue. *Proc. Natl. Acad. Sci. U.S.A., 82*, 3173-3177.

Pope, M. R., Saari, L. L. and Ludden, P. W. (1986). *N*-glycohydrolysis of adenosine diphosphoribosyl arginine linkages by dinitrogenase reductase activating glycohydrolase (activating enzyme) from *Rhodospirillum rubrum*. *J. Biol. Chem., 261*, 10104-10111.

Pope, M. R., Saari, L. L. and Ludden, P. W. (1987). Fluorometric assay for ADP-ribosylarginine cleavage enzymes. *Anal. Biochem., 160*, 68-77.

Preston, G. G. and Ludden, P. W. (1982). Change in subunit composition of the iron protein of nitrogenase from *Rhodospirillum rubrum* during activation and inactivation of iron protein. *Biochem. J., 205*, 489-494.

Saari, L. L., Pope, M. R., Murrell, S. A. and Ludden, P. W. (1986). Studies on the activating enzyme for iron protein of nitrogenase from *Rhodospirillum rubrum*. *J. Biol. Chem., 261*, 4973-4977.

Saari, L. L., Triplett, E. W. and Ludden, P. W. (1984). Purification and properties of the activating enzyme for iron protein of nitrogenase from the photosynthetic bacterium *Rhodospirillum rubrum*. *J. Biol. Chem., 259*, 15502-15508.

Saez, L. P., Garcia, P., Martinez-Luque, M., Klipp, W., Blasco, R. and Castillo, F. (2001). Role for *draTG* and *rnf* genes in reduction of 2,4-dinitrophenol by *Rhodobacter capsulatus*. *J. Bacteriol., 183*, 1780-1783.

Schick, H.-J. (1971). Substrate and light dependent fixation of molecular nitrogen in *Rhodospirillum rubrum*. *Arch. Mikrobiol., 75*, 89-101.

Schindelin, H., Kisker, C., Schlessman, J. L., Howard, J. B. and Rees, D. C. (1997). Structure of ADP AlF_4-stabilized nitrogenase complex and its implications for signal transduction. *Nature, 387*, 370-376.

Schneider, K. C., Bradbeer, C., Singh, R. N., Wang, L. C., Wilson, P. W. and Burris, R. H. (1960). Nitrogen fixation by cell-free preparations from microorganisms. *Proc. Natl. Acad. Sci. U.S.A., 46*, 726-733.

Schultz, J. E., Gotto, J. W., Weaver, P. F. and Yoch, D. C. (1985). Regulation of nitrogen fixation in *Rhodospirillum rubrum* grown under dark, fermentative conditions. *J. Bacteriol., 162*, 1322-1324.

Serres, M. H. and Ensign, J. C. (1996). Endogenous ADP-ribosylation of proteins in *Mycobacterium smegmatis*. *J. Bacteriol., 178*, 6074-6077.

Soliman, A. and Nordlund, S. (1992). Studies on the effect of NAD(H) on nitrogenase activity in *Rhodospirillum rubrum*. *Arch. Microbiol., 157*, 431-435.

Sweet, W. J. and Burris, R. H. (1981). Inhibition of nitrogenase activity by NH_4^+ in *Rhodospirillum rubrum*. *J. Bacteriol., 145*, 824-831.

Sweet, W. J. and Burris, R. H. (1982). Effects of *in vivo* treatments on the activity of nitrogenase isolated from *Rhodospirillum rubrum*. *Biochim. Biophys. Acta, 680*, 17-21.

Triplett, E. W., Wall, J. D. and Ludden, P. W. (1982). Expression of the activating enzyme and Fe protein of nitrogenase from *Rhodospirillum rubrum*. *J. Bacteriol., 152*, 786-791.

Weare, N. M. and Shanmugam, K. T. (1976). Photoproduction of ammonium ion from N_2 in *Rhodospirillum rubrum*. *Arch. Microbiol., 110*, 207-213.

West, R. E., Jr., Moss, J., Vaughan, M., Liu, T. and Liu, T. Y. (1985). Pertussis toxin-catalyzed ADP-ribosylation of transducin. Cysteine 347 is the ADP-ribose acceptor site. *J. Biol. Chem.*, *260*, 14428-14430.

West, R. E., Jr. and Moss, J. (1986). Amino acid specific ADP-ribosylation: specific NAD:arginine mono-ADP-ribosyltransferases associated with turkey erythrocyte nuclei and plasma membranes. Biochemistry, 25, 8057-8062.

Woese, C. R., Stackebrandt, E., Weisburg, W. G., Paster, B. J., Madigan, M. T., Folwer, V. J., *et al.* (1984). The phylogeny of purple bacteria: The alpha subdivision. System. *Appl. Microbiol.*, *5*, 315-326.

Yakunin, A. F., Fedorov, A. S., Laurinavichene, T. V., Glaser, V. M., Egorov, N. S., Tsygankov, A. A., *et al.* (2001). Regulation of nitrogenase in the photosynthetic bacterium *Rhodobacter sphaeroides* containing *draTG* and *nifHDK* genes from *Rhodobacter capsulatus*. *Can. J. Microbiol.*, *47*, 206-212.

Yakunin, A. F. and Hallenbeck, P. C. (1998). Short-term regulation of nitrogenase activity by NH_4^+ in *Rhodobacter capsulatus*: multiple *in vivo* nitrogenase responses to NH_4^+ addition. *J. Bacteriol.*, *180*, 6392-6395.

Yakunin, A. F. and Hallenbeck, P. C. (2002). AmtB is necessary for NH_4^+-induced nitrogenase switch-off and ADP-ribosylation in *Rhodobacter capsulatus*. *J. Bacteriol.*, *184*, 4081-4088.

Yakunin, A. F., Laurinavichene, T. V., Tsygankov, A. A. and Hallenbeck, P. C. (1999). The presence of ADP-ribosylated Fe protein of nitrogenase in *Rhodobacter capsulatus* is correlated with cellular nitrogen status. *J. Bacteriol.*, *181*, 1994-2000.

Yoch, D. C. (1979). Manganese, an essential trace element for N_2 fixation by *Rhodospirillum rubrum* and *Rhodopseudomonas capsulata*: role in nitrogenase regulation. *J. Bacteriol.*, *140*, 987-995.

Yoch, D. C. and Cantu, M. (1980). Changes in the regulatory form of *Rhodospirillum rubrum* nitrogenase as influenced by nutritional and environmental factors. *J. Bacteriol.*, *142*, 899-907.

Yoch, D. C., Li, J., Hu, C.-Z. and Scholin, C. (1988). Ammonia switch-off of nitrogenase from *Rhodobacter sphaeroides* and *Methylosinus trichosporium*: no evidence for Fe protein modification. *Arch. Microbiol.*, *150*, 1-5.

Zhang, Y., Burris, R. H., Ludden, P. W. and Roberts, G. P. (1995a). Comparison studies of dinitrogenase reductase ADP-ribosyl transferase/dinitrogenase reductase activating glycohydrolase regulatory systems in *Rhodospirillum rubrum* and *Azospirillum brasilense*. *J. Bacteriol.*, *177*, 2354-2359.

Zhang, Y. P., Burris, R. H., Ludden, P. W. and Roberts, G. P. (1997). Regulation of nitrogen fixation in *Azospirillum brasilense*. *FEMS Microbiol. Lett.*, *152*, 195-204.

Zhang, Y. P., Cummings, A. D., Burris, R. H., Ludden, P. W. and Roberts, G. P. (1995b). Effect of an *ntrBC* mutation on the posttranslational regulation of nitrogenase acitvity in *Rhodospirillum rubrum*. *J. Bacteriol.*, *177*, 5322-5326.

Zhang, Y., Kim, K., Ludden, P. W. and Roberts, G. R. (2001a). Isolation and characterization of *draT* mutants that have altered regulatory properties of dinitrogenase reductase ADP-ribosyltransferase in *Rhodospirillum rubrum*. *Microbiol.*, *147*, 193-202.

Zhang, Y., Pohlmann, E. L., Halbleib, C. M., Ludden, P. W. and Roberts, G. P. (2001b). Effect of P_{II} and its homolog GlnK on reversible ADP-ribosylation of dinitrogenase reductase by heterologous expression of the *Rhodospirillum rubrum* dinitrogenase reductase ADP-ribosyl transferase-dinitrogenase reductase-activating glycohydrolase regulatory system in *Klebsiella pneumoniae*. *J. Bacteriol.*, *183*, 1610-1620.

Zhang, Y., Pohlmann, E. L., Ludden, P. W. and Roberts, G. P. (2000). Mutagenesis and functional characterization of the *glnB*, *glnA*, and *nifA* genes from the photosynthetic bacterium *Rhodospirillum rubrum*. *J. Bacteriol.*, *182*, 983-992.

Zhang, Y., Pohlmann, E. L., Ludden, P. L. and Roberts, G. P. (2001c). Functional characterization of three GlnB homologs in the photosynthetic bacterium *Rhodospirillum rubrum*: Roles in sensing ammonium and energy status. *J. Bacteriol.*, *183*, 6159-6168.

Zolkiewska, A. and Moss, J. (1993). Integrin alpha 7 as substrate for a glycosylphosphatidylinositol-anchored ADP-ribosyltransferase on the surface of skeletal muscle cells. *J. Biol. Chem.*, *268*, 25273-25276.

Zolkiewska, A., Nightingale, M. S. and Moss, J. (1992). Molecular characterization of NAD:arginine ADP-ribosyltransferase from rabbit skeletal muscle. *Proc. Natl. Acad. Sci. U.S.A.*, *89*, 11352-11356.

Zolkiewska, A., Okazaki, I. J. and Moss, J. (1994). Vertebrate mono-ADP-ribosyltransferases. *Mol. Cell. Biochem.*, *138*, 107-112.

Zumft, W. G. and Castillo, F. (1978). Regulatory properties of the nitrogenase from *Rhodopseudomonas palustris*. *Arch. Microbiol., 117*, 53-60.
Zumft, W. G. and Nordlund, S. (1981). Stabilization and partial characterization of the activating enzyme for dinitrogenase reductase (Fe-protein) from *Rhodospirillum rubrum*. *FEBS Lett., 127*, 79-82.

Chapter 9

REGULATION OF NITROGEN FIXATION IN FREE-LIVING DIAZOTROPHS

M. J. MERRICK
Department of Molecular Microbiology, John Innes Centre, Norwich NR4 7UH, UK

1. INTRODUCTION

Although the process of nitrogen fixation is found distributed amongst a taxonomically very diverse range of microorganisms, research over the last thirty years has, perhaps not surprisingly, revealed that the physiological constraints that these organisms face are often very similar. The ways in which free-living diazotrophs deal with these problems have been examined by physiologists, biochemists, and geneticists and this research is now providing a unified picture of the exquisitely elaborate mechanisms that have evolved. Furthermore, whereas each group of organisms has evolved its own particular genetic and biochemical strategies to maximise the efficiency of nitrogen fixation, some common themes are now becoming clear.

The physiological issues to be addressed by a free-living diazotroph are well defined; namely, to optimise nitrogen fixation while coping with the O_2 sensitivity of nitrogenase, to satisfy the substantial energy demands of the fixation process, to supply metals for a range of metalloenzymes, and to utilise other sources of fixed nitrogen before fixing atmospheric N_2. It is now apparent that all of these aspects of regulation are not only inter-related but also have to be integrated with the rest of the cell's metabolism.

This chapter focuses on the mechanisms by which free-living diazotrophs regulate nitrogen fixation in response to changes in the cellular nitrogen status. In all organisms, this regulation operates at the level of transcription of the nitrogen

W. Klipp, B. Masepohl, J. R. Gallon and W. E. Newton (eds.), Genetics and Regulation of Nitrogen Fixation in Free-Living Bacteria, 197-223.

fixation (*nif*) genes. Furthermore, such regulation is usually effected at two levels; a general nitrogen control system that co-ordinates cellular nitrogen metabolism, and a *nif*-specific mechanism that facilitates regulation in response to particular signals. In addition to this transcriptional control, a number of organisms have evolved special mechanisms that allow very rapid short-term regulation of the activity of the nitrogenase enzyme in response to fluctuations in availability of fixed nitrogen.

2. GENERAL NITROGEN CONTROL SYSTEMS

2.1. Proteobacteria

The general nitrogen regulation (Ntr) system was discovered, and is most fully described, in the γ Proteobacteria but is now known to be present also in the α and β Proteobacteria (Merrick and Edwards, 1995). The Ntr system comprises four proteins: GlnB, which is a member of the P_{II} family of signal transduction proteins; GlnD, which is a uridylyltransferase; and a two-component regulatory pair, the sensor histidine kinase, NtrB, and the response regulator, NtrC (Figure 1). Among the free-living diazotrophs, the Ntr system has been characterised, in full or in part, in *Klebsiella pneumoniae, Azotobacter vinelandii, Azospirillum brasilense, Rhodobacter capsulatus, Rhodospirillum rubrum, Gluconacetobacter diazotrophicus, Herbaspirillum seropedicae*, and *Azoarcus*. There have been no studies of nitrogen regulation in diazotrophic members of the δ-Proteobacteria but genome sequence analysis in *Desulfovibrio vulgaris* and *Geobacter sulfurreducens* suggests that these organisms may have at least *glnB* and *glnD* genes.

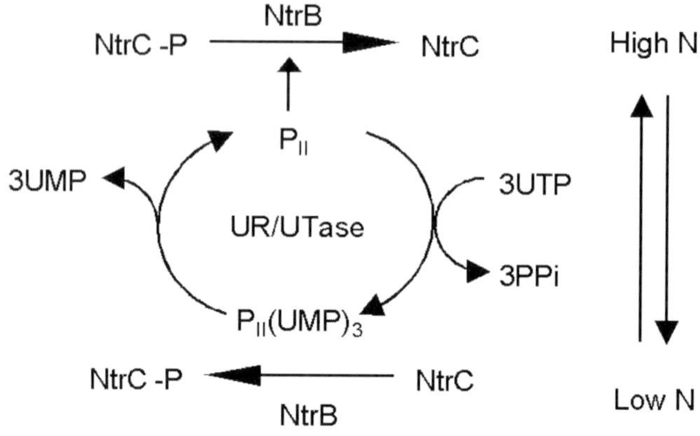

Figure 1. The nitrogen regulation (Ntr) system of enteric bacteria. The activity of the response regulator, NtrC, is regulated in response to the intracellular nitrogen status. UTase (glnD *product) catalayses the uridylylation and deuridylylation of P_{II}* (glnB *product). P_{II} in turn regulates the activity of the sensor histidine kinase NtrB, which catalyses the phosphorylation and dephosphorylation of NtrC.*

The GlnB protein, like all P_{II} proteins, is a small trimeric protein that constitutes an intracellular sensor of nitrogen status (Arcondéguy *et al.*, 2001). The protein forms a squat barrel 30 Å high with each 12kDa monomer arranged such that a rather unstructured loop of around 18 residues, called the T-loop, extends some 13 Å above the surface of the barrel (Carr *et al.*, 1996; Xu *et al.*, 1998). A principal characteristic of the P_{II} proteins in the Proteobacteria is their ability to be switched between two forms by covalent modification of tyrosine-51, which lies at the apex of the T-loop.

Both uridylylation and deuridylylation of GlnB are effected by the GlnD protein, which responds to the intracellular glutamine concentration in such a way that glutamine specifically inhibits the uridylylation reaction (Jiang *et al.*, 1998). Consequently, GlnB is uridylylated under nitrogen-limiting (low intracellular glutamine) conditions and de-uridylylated in nitrogen-sufficient (high intracellular glutamine) conditions. The modification state of GlnB reflects the intracellular nitrogen status and also modulates the ability of GlnB to interact with other proteins in the cell and thereby to regulate their activity.

One of these GlnB targets is the NtrB/NtrC two-component system. NtrC is a typical σ^N (σ^{54})-dependent transcriptional activator protein that comprises a DNA-binding carboxy terminal domain, a highly conserved central domain, which is required for the activation of transcription, and an N-terminal domain, which is characteristic of two-component response regulator proteins (Stock *et al.*, 1989). These proteins have a highly conserved tertiary structure in the N-terminal region, containing an aspartic acid residue at position 54 which, in the case of NtrC, is phosphorylated in response to low nitrogen, thereby activating the protein (Kern *et al.*, 1999; Volkman *et al.*, 1995). NtrC is a dimer in its non-phosphorylated form but for the activation of transcription to take place, oligomerisation to either a tetramer or higher order oligomer must occur. Phosphorylation induces both DNA binding and the oligomerisation of NtrC and many NtrC-dependent promoters contain more than one binding site, facilitating oligomerisation (Mettke *et al.*, 1995; Weiss *et al.*, 1991, 1992; Wyman *et al.*, 1997).

Control of NtrC activity in response to nitrogen status is mediated by NtrB, which acts as a phosphate donor to NtrC. NtrB consists of three domains: an N-terminal domain involved in signal transduction, a central phosphotransferase/phosphatase/dimerisation domain, and a C-terminal kinase domain. The NtrB protein is a dimer that is autophosphorylated on a conserved histidine (His-139 in *K. pneumoniae*) at the amino end of the C-terminal domain (Ninfa and Bennett, 1991). ATP binds to one subunit and phosphorylation occurs on the conserved histidine residue of the second subunit. It is NtrB that is a target for regulation by GlnB. NtrB is stimulated to dephosphorylate NtrC in the presence of GlnB and ATP - this is termed regulated phosphatase activity (Kamberov *et al.*, 1994, 1995; Keener and Kustu, 1988; Liu and Magasanik, 1995). Binding of GlnB to the kinase domain of NtrB inhibits kinase activity and appears to result in an altered conformation that is transmitted to the other two domains, thereby causing the central domain to assume a conformation with potent phosphatase activity (Pioszak *et al.*, 2000).

In nitrogen excess, unmodified GlnB stimulates dephosphorylation of NtrC by NtrB and conversely, in nitrogen starvation when GlnB is modified, GlnB-UMP no longer interacts with NtrB and the kinase activity predominates so that NtrC is phosphorylated and transcriptionally active. Hence, the modification of GlnB in response to the cellular nitrogen status provides the intracellular switch that in turn regulates the phosphatase and kinase activities of NtrB and so the transcriptional activity of NtrC.

Expression of the *nif* genes is dependent on NtrC in *K. pneumoniae* (Espin *et al.*, 1982; Merrick, 1983), *A. brasilense* (Liang *et al.*, 1993; Pedrosa and Yates, 1984), *R. capsulatus* (Kranz *et al.*, 1990; Kranz and Foster-Hartnett, 1990; Kranz and Haselkorn, 1985), and *H. seropedicae* (Persuhn *et al.*, 2000). In each of these organisms, NtrC regulates expression of a *nif*-specific regulatory system (described below). The exceptions to this pattern are in *A. vinelandii*, *R. rubrum,* and *G. diazotrophicus*, where mutations in *ntrC* do not affect *nif* expression (Toukdarian *et al.*, 1990; Toukdarian and Kennedy, 1986; Zhang *et al.*, 1995). However, NtrC is required for expression in *A. vinelandii* of *vnfH,* the structural gene for the vanadium nitrogenase Fe protein (Bali *et al.*, 1988).

Members of the α–, β–, γ– (and probably the δ–) Proteobacteria encode two or more P_{II}-like proteins. Typically, these are a GlnB protein and one or more homologues designated GlnK, the structural genes for which are almost invariably linked to a second gene (*amtB*) that encodes a high-affinity ammonium transporter (Arcondéguy *et al.*, 2001; Thomas *et al.*, 2000). Expression of *glnK* is regulated by NtrC such that GlnK levels are markedly elevated under nitrogen limitation (Atkinson and Ninfa, 1998). Recent studies indicate that, in *Escherichia coli,* GlnK is necessary to control the levels of NtrC-P during nitrogen starvation when the fixed levels of GlnB are insufficient to provide this control (Atkinson *et al.*, 2002a, 2002b; Blauwkamp and Ninfa, 2002a; 2002b). GlnK also regulates the activity of AmtB and may, therefore, play a part in controlling the intracellular ammonium pool (Coutts *et al.*, 2002). The roles of the P_{II} proteins extend to a wide variety of aspects of nitrogen metabolism (Arcondéguy *et al.*, 2001) and (as discussed later) they play a critical role in *nif*-specific gene regulation in a number of organisms.

Nitrogen and carbon metabolism must invariably be kept in balance and, in *R. capsulatus* and in *Rhodobacter sphaeroides*, they are co-ordinated through the actions of the RegB-RegA two-component system. This control system was originally discovered through its role in anaerobic activation of the *puf, puc* and *puh* photosynthetic gene operons in *R. capsulatus* (Sganga and Bauer, 1992). In *R. sphaeroides*, the RegB-RegA system is required for positive regulation of the *cbbI* and *cbbII* genes, which encode the enzymes of the Calvin-cycle CO_2-fixation pathway (Qian and Tabita, 1996). The precise signal to which RegB responds is unknown but the current model favours a response to redox state rather than to molecular O_2, and studies in *R. sphaeroides* suggest that the signal originates from the *cbb*$_3$-type oxidase (O'Gara *et al.*, 1998).

That this system is also involved in regulation of nitrogen fixation became apparent when it was found that, in a *cbb* mutant of *R. sphaeroides*, nitrogenase synthesis is derepressed in the presence of ammonium and that this derepression requires RegB (Qian and Tabita, 1996). In *R. capsulatus,* RegA acts as a coactivator, together with NtrC, of *nifA2* expression but the precise mechanism of this co-activation is not known (Elsen *et al.*, 2000). The derepression of nitrogenase in a *cbb* mutant appears to serve as a means of controlling redox balance, because the hydrogenase activity of nitrogenase allows removal of excess reducing equivalents in the absence of the ability to use CO_2 as an electron acceptor. In *R. sphaeroides,* the regulatory circuits controlled by NtrBC and RegBA are tightly integrated. Both *glnB* and *glnK* are normally regulated in response to nitrogen source; *glnB* expression being three-fold higher in glutamate than ammonium, whereas *glnK* is induced more than seventy-fold. However, a mutant that is devoid of a functional reductive pentose pathway fails to express *glnB* and shows constitutive expression of *glnK* (Qian and Tabita, 1998). The factor(s) responsible for this control are presently unknown.

RegBA homologues are present in a number of α-Proteobacteria, including *Bradyrhizobium japonicum* where the RegA homologue, RegR, activates transcription of the *fixR-nifA* operon (Bauer *et al.*, 1998). *R. capsulatus regA* can complement a *B. japonicum regR* mutant confirming that the two genes are functionally similar (Emmerich *et al.*, 2000).

2.2. *Gram-positive Bacteria*

There is very little information concerning nitrogen control among the diazotrophic Gram-positive organisms. Global regulation of nitrogen metabolism in *Bacillus subtilis* is mediated by the transcription factor TnrA (Fisher, 1999; Wray *et al.*, 2000) but whether this system is present in the diazotrophic *Bacillus* species is unknown. A number of *nif* genes have been cloned from *Clostridium pasteurianum*, including six copies of *nifH* homologues (Wang *et al.*, 1988). Under nitrogen-fixing conditions, transcripts of most of these genes are present, suggesting that they are functional. Sequences either identical to or very similar to the consensus *Escherichia coli* promoter are found in the –35 and –10 regions and a common upstream sequence (ATCAATAT–$N_{6\text{-}10}$–ATGGATTC) is present at around position –100, but the role of this sequence is not known (Wang *et al.*, 1988).

Homologues of the standard *ntr* genes found in the Proteobacteria are not present in the genera, *Bacillus, Clostridium* and *Streptomyces.* However where genome information is available, all members of these genera encode an *amtB-glnK* operon, which in the Actinobacteria is an *amtB-glnK-glnD* operon. Whether any of these genes influences nitrogen fixation in the diazotrophic members of these genera is unknown.

2.3. *Cyanobacteria*

Nitrogen-fixation genes have been identified in a number of cyanobacteria but have been characterised in most detail in *Anabaena* PCC7120, *Anabaena variabilis* ATCC 29413, and *Synechococcus* strain RF-1. Nitrogen-fixing cyanobacteria can be divided into those species in which nitrogen fixation occurs in differentiated cells called heterocysts, *e.g.*, *Anabaena* PCC7120 or *Nostoc punctiforme*, and non-heterocystous organisms, such as *Gloeothece*. When species such as *Anabaena* are deprived of a source of fixed nitrogen, about every tenth vegetative cell in the cyanobacterial filament differentiates into a morphologically and physiologically distinct heterocyst. During heterocyst differentiation, an 11kb DNA element is excised from within the *nifD* gene to form a contiguous *nifHDK* operon (Golden *et al.*, 1985). Although this process is triggered by nitrogen deprivation, the signal transduction cascade that brings it about is not yet completely described.

Although species, such as *N. punctiforme,* contain a single major *nif*-gene cluster, *Anabaena variabilis* (ATCC29413) contains two *nif* clusters (*nif1* and *nif2*) and a *vnf*-gene cluster, which encodes a vanadium nitrogenase (Thiel *et al.*, 2002). Whereas the *nif1*- and *vnf*-encoded nitrogenases function in the heterocyst, the *nif2*-encoded enzyme functions in all cells but only under anaerobic conditions (Thiel *et al.*, 1997). Consequently, whereas expression of *nif1* and *vnf* is linked to the differentiation process, the *nif2* genes appear to be similar to the *nif* genes of non-heterocystous cyanobacteria and to be regulated directly by environmental factors, such as availability of fixed nitrogen and O_2 tension.

The cyanobacteria have a distinct general nitrogen control system in which the central transcriptional regulator is NtcA, a member of the CRP (cAMP receptor protein) family (Herrero *et al.*, 2001). NtcA is widely distributed among cyanobacteria and has been identified in a number of nitrogen-fixing species, including *Anabaena* PCC7120, *N. punctiforme*, *Cyanothece, Synechococcus*, and *Trichodesmium*. The NtcA amino-acid sequence is highly conserved and shows three strongly conserved regions. Region I covers around 70 residues and has many characteristics of cyclic-nucleotide-binding proteins; this structural feature may represent a metabolite-binding site. The binding site for NtcA on DNA has been extensively studied and is characterised by a palindromic sequence, GTA–N_8–TAC (Jiang *et al.*, 2000; Luque *et al.*, 1994). *In vitro,* NtcA shows sequence-specific binding to this motif in the *glnA* promoter but fails to activate transcription from either the *glnA* or *ntcA* promoters even at saturating concentrations (Tanigawa *et al.*, 2002). However, DNA binding by NtcA is significantly stimulated in the presence of 2-oxoglutarate and is then competent to activate transcription from both promoters (Tanigawa *et al.*, 2002; Vazquez-Bermudez *et al.*, 2002). These data strongly suggest that 2-oxoglutarate is the major regulator of NtcA-dependent transcription and that, unlike the Proteobacteria, 2-oxoglutarate does not act through the P_{II} protein. However, recent data from *Synechocystis* indicate that P_{II} may be required for NtcA-dependent transcription activation under certain physiological conditions (Aldehni *et al.*, 2003).

NtcA mutants of *Anabaena* PCC7120, *Anabaena variabilis* ATCC 29413 and *N. punctiforme* are impaired in heterocyst development and are unable to fix N_2 (Frias *et al.*, 1994; Herrero *et al.*, 2001; Wei *et al.*, 1994). As heterocyst development occurs in response to combined nitrogen deprivation, the involvement of NtcA is understandable and two groups of genes, *hetC* and *devBCA*, which are required for heterocyst development, appear to be directly regulated by NtcA (Herrero *et al.*, 2001; Muro-Pastor *et al.*, 1999).

An *Anabaena* PCC7120 *ntcA* mutant does not express the *nifHDK* genes (Frias *et al.*, 1994) and the NtcA protein has been reported to bind weakly *in vitro* to the *nifH* upstream region (Ramasubramanian *et al.*, 1994). However, sequences in the *nifH* promoter, although showing some similarity to the NtcA consensus, do not match precisely (Herrero *et al.*, 2001). This situation could mean either that the *nif* genes are regulated by an as yet unidentified protein or that the lack of a consensus NtcA promoter reflects a requirement for a higher intracellular NtcA concentration.

2.4. *Archaea*

Within the Archaea, nitrogen fixation has been found only in methanogenic species but not all methanogens fix N_2 (Leigh, 2000). Studies of the regulation of *nif* genes in these organisms have been hampered by the problems of genetic analysis but genetic systems have now been developed in *Methanococcus maripaludis*, which is also relatively fast growing and provides a good model system.

Evidence for a nitrogen regulon, *i.e.*, a group of genes that are subject to nitrogen control by a common transcriptional regulator, comes from analysis of potential common sequences within promoters of genes that are either expected or known to be nitrogen regulated. The upstream sequences of the *nifH* genes in *Methanobacterium thermoautotrophicum* and *Methanobacterium ivanovii* and of the *glnA* (glutamine synthetase) gene in *Methanococcus voltae* contain a common palindromic sequence, GGAA–N_6–TTCC (Possot *et al.*, 1989; Sibold *et al.*, 1991; Souillard and Sibold, 1989). Subsequent searches identified an identical sequence upstream of these same genes in *M. maripaludis*, of the *amtB-glnK* operons of *M. thermoautotrophicum* and *M. jannaschii*, and of the *M. jannaschii nadE* gene (Kessler and Leigh, 1999). The role of this regulatory sequence was demonstrated in the case of both the *nifH* and *glnA* genes of *M. maripuladis*, where mutations that disrupt the palindrome lead to failure of ammonium to repress transcription (Cohen-Kupiec *et al.*, 1997; 1999). These data are consistent with a mechanism of negative control in which a repressor protein inhibits transcription of these N-regulated genes in the presence of ammonium. The proposed protein has yet to be identified, although components of cell extracts have been shown to bind the *nifH* promoter (Cohen-Kupiec *et al.*, 1997).

The mechanism described above is apparently not universal among methanogenic diazotrophs. The proposed operator sequence is not present in the *nif* promoter of *M. thermoautotrophicum* strain ΔH but is found in strain Marburg of the same species. The sequence is also not present in the *nifH* promoters of *Methanosarcina barkeri* and *Methanosarcina mazei* so, in these species, a different regulation mechanism may be present (Chien *et al.*, 1998; Ehlers *et al.*, 2002). In *M. barkeri*, extracts from nitrogen-limited cells bind to the *nifH* promoter and a substance present in ammonium-grown cells inhibits DNA binding by a transcription-associated protein(s), suggesting that *nif* expression is positively controlled (Chien *et al.*, 1998).

3. *NIF*-SPECIFIC NITROGEN CONTROL

Transcriptional regulation of nitrogen-fixation genes in response to availability of fixed nitrogen can minimally occur at just one level. In such cases, the *nif* structural genes, *e.g.*, *nifHDK,* are part of the general nitrogen regulon and are under the direct control of the global transcriptional regulator, which may be either an activator or a repressor. This is apparently the situation in at least some of the Archaea, such as *M. maripaludis*, where the *nif* genes are subject to the same regulation as other genes, such as either *amtB* or *nadE*. In the Gram-positive bacteria and the cyanobacteria, as discussed above, the regulatory circuits are not well characterised, although it would appear that the cyanobacterial *nif* genes are not directly controlled. However, in the very well-documented Gram-negative systems, a second level of regulation is almost always present. In these systems, the *nif* structural genes are controlled by a *nif*-specific regulator and it is expression of this protein that is subject to global nitrogen control.

3.1. The NifA Protein

The *nif*-specific regulator NifA, first identified in *K. pneumoniae*, is found in all diazotrophic members of the α–, β– and γ–Proteobacteria. In the γ–Proteobacteria *K. pneumoniae* and *A. vinelandii* and in the β–Proteobacterium *Azoarcus, nifA* is part of a two-gene operon and is located downstream of the *nifL* gene (Bali *et al.*, 1992; Egener *et al.*, 2002; Martin and Reinhold-Hurek, 2002; Merrick *et al.*, 1982). In all other cases, it constitutes a single cistron. The presence of NifA in the δ–Proteobacteria has not been investigated but, in *Desulfovibrio gigas*, the *nifH* gene does not have a characteristic upstream NifA-binding site (Kent *et al.*, 1989).

Either the *nifA* or *nifLA* promoter provides the interface between the global and the *nif*-specific regulatory circuits and it is this promoter that is activated by the NtrC protein under nitrogen-limiting conditions. The exception to this is *A. vinelandii nifA*, which is not NtrC-dependent (Blanco *et al.*, 1993).

The NifA protein, like NtrC, is a σ^N-dependent transcriptional activator and consequently all *nifA*-dependent promoters are characterised by the recognition site for σ^N RNA polymerase, namely a –24/–12 type promoter with a consensus sequence TGGCAC–N_5–TTGCA in which the GG at –24/–25 is invariant and the GC at –12/–13 is very highly conserved (Barrios *et al.*, 1999; Merrick, 1993). These promoters also contain binding sites for NifA (otherwise known as upstream activator sequences, UAS) that are typically located 80 to 100 bp upstream of the –24/–12 sequence and that conform to a consensus TGT–N_{10}–ACA (Buck *et al.*, 1986). Because σ^N RNA polymerase absolutely requires an activator protein in order to initiate transcription, the *nif* genes are only expressed in the presence of an active form of NifA. The protein binds to the UAS and contacts the holoenzyme by a DNA-looping mechanism (Buck *et al.*, 1987) in which the looping event is often facilitated by the binding of Integration Host Factor (IHF) to the region between the UAS and the promoter (Hoover *et al.*, 1990). Formation of the open promoter complex and subsequent transcription initiation requires ATP hydrolysis, which is catalysed by NifA (Austin *et al.*, 1990; Lee *et al.*, 1993).

The NifA proteins have three distinct domains: an N-terminal domain of between 170 and 250 residues that is a member of the GAF-domain family (Ho *et al.*, 2000); a central domain of around 240 residues that is characteristic of all σ^{54}-dependent activators or Enhancer Binding Proteins (EBP) and that belongs to the AAA+ domain family (Neuwald *et al.*, 1999); and a C-terminal DNA-binding domain of around 50 residues. Although the NifA protein shares many features with NtrC, it is not a classical response-regulator protein. The N-terminal domain does not have the typical features of those proteins, including the conserved aspartate residue (typically Asp-54) that is the site of phosphorylation in proteins such as NtrC and indeed there is no evidence that NifA is phosphorylated under any conditions. Nevertheless, this domain is of particular interest as it constitutes the regulatory domain of the protein and plays a major role in determining whether NifA is active.

The central domain is responsible for interaction with σ^N RNA polymerase and for ATP hydrolysis. The central domain of EBP proteins has been modelled and shown to contain seven highly conserved motifs of which the first (C1) is typical of the Walker A motif found in a wide range of proteins that bind and hydrolyse ATP (Osuna *et al.*, 1997). The roles of the other six motifs are unknown but studies of positive control mutants in the NifA protein of *Bradyrhizobium japonicum* identifed region C3 and a highly conserved sequence (GAFTGA) as a candidate for interaction with σ^N RNA polymerase (Gonzalez *et al.*, 1998).

Between the central domain and the C-terminal domain is a variable region that characteristically divides the NifA proteins into two sub-families (Figure 2). One group includes the NifA proteins from the γ–Proteobacteria, *i.e.*, *Klebsiella, Enterobacter*, and *Azotobacter.* These proteins contain a single conserved cysteine residue very near the end of the central domain. The second group includes the NifA proteins in the α– and β–Proteobacteria, *i.e.*, all the symbiotic rhizobia and members of the genera *Azospirillum, Rhodobacter, Rhodospirillum*, and *Herbaspirillum*. These proteins contain two conserved cysteine residues at the end of the central domain. The second of these is equivalent to the cysteine in the γ–Proteobacteria. They also have an additional two conserved cysteine residues, in a CXXXXC motif, in a region between the central domain and the DNA-binding domain. The NifA proteins, which carry the CXXXXC motif, are also distinguished by the fact that the activity of these proteins is O_2 sensitive, whereas those without the motif are O_2 resistant (Fischer *et al.*, 1988; Kullik *et al.*, 1989; Souza *et al.*, 1999).

The O_2 sensitivity of NifA has been studied in most detail in *Bradyrhizobium japonicum*, where it was found that mutations in the cysteines of the CXXXXC motif resulted in an inactive protein. Mutation of the residues that separate the cysteines have no effect on activity, whereas a change in the spacing significantly decreases activity (Fischer *et al.*, 1989). All these data suggest that the motif may be involved in binding a metal ion but to date confirmation of this by purification of a NifA protein from this group has not been achieved. O_2-tolerant variants of NifA were successfully isolated in *Rhizobium meliloti* and all the alleles sequenced contained the same mutation, causing a change from methionine to isoleucine in residue 217 near to the putative ATP-binding site (Krey *et al.*, 1992). This result led to a model, which proposes that, at high O_2 concentrations, the loss of NifA activity is due to a conformational change in the ATP-binding site, thereby abolishing ATP binding or hydrolysis.

```
AV    --IRLLMSHRWPGNVRELENCLERSAIMSEDGTI|TRDVVSLTGVDNESPPLAAP--
KP    --IRLLMEYSWPGNVRELENCLERSAVLSESGLI|DRDVILFNHRDNPPKALASS--
EA    --VRLLMTYSWPGNVRELENCLERASVMTDEGLI|DRDVILFNHHESPALSVKPG--

BJ    --IDVLMSCKFPGNVRELENCIERTATLSAGTSI|VRSDFACSQGQCLSTTLWKS--
RE    --IEVMSQCYFPGNVRELENCVRRTATLARSSSI|VSSDFACKNSQCLSSLLWKT--
RL    --IDILSKCAFPGNVRELDNCVQRTATLASSNTI|TSSDFACQQDQCSSALLRKA--
RM    --LDHLSKCKFPGNVRELENCVRRTATLARSKTI|TSSDFACQTDQCFSSRLWKG--
AC    --IDVLRRCYFPGNVRELENCIRRTATLAHDAVI|TPHDFACDSGQCLSAMLWKG--

HS    --MKVMMNCYWPGNVRELENCVERTATMMRGDLI|TEVHFSCQQNKCLTKVLHEP--
AB    --LEVLNRCTWPGNVRELENCIERAATQSRDGII|RTESLSCSLNLCNSSVLFQY--
RC    --FDQICRCQFPGNERELENCVNRAAALSDGAIV|LAEELACRQGACLSAELFRL--
RR    --LTAMGGCNFPGNVRELENCVCRAATLAQDEVI|QELGLSCHNDKCLSASLWQR--
```

Figure 2. Location of conserved cysteine residues in the C-terminal region of NifA proteins. AV – A. vinelandii , KP – K. pneumoniae, EA – E. agglomerans, BJ – B. japonicum, RE – R. etli, RL – R. leguminosarum, RM – R. meliloti, AC – A. caulinodans, HS – H. seropedicae, AB – A. brasilense, RC – R. capsulatus, RR – R. rubrum. *The vertical line indicates the boundary between the central and C-terminal domains.*

3.2. The VnfA and AnfA Proteins

Those diazotrophs that can synthesise alternative nitrogenases, *i.e.*, either vanadium (*vnf*) or iron (*anf*) or both nitrogenases, regulate expression of the structural genes for these enzymes with specific NifA homologues designated as VnfA and AnfA (Joerger *et al.*, 1989; Masepohl and Klipp, 1996). These proteins differ primarily from NifA in their N-terminal domains, which are particularly characterised by the presence of two cysteine residues in a CXXXXC motif. Studies of amino-acid substitutions in this region of *A. vinelandiii* AnfA demonstrated that both cysteines (Cys-21 and Cys-26) are required for AnfA activity (Jepson and Austin, 2002; Premakumar *et al.*, 1994). The data suggest the involvement of a metal cluster and purification of the isolated N-terminal domain of *Av*AnfA shows it to have a red/brown colour and spectral characteristics consistent with a 2Fe-2S cluster (Jepson and Austin, 2002).

Expression of both *vnf* and *anf* structural genes is under nitrogen control. In both *A. vinelandii* and *R. capsulatus, anfA* expression is regulated by ammonium but, whereas in *R. capsulatus* this control is mediated by NtrC, this is not the case in *A. vinelandii* (Kennedy *et al.*, 1991; Kutsche *et al.*, 1996; Premakumar *et al.*, 1998; Toukdarian and Kennedy, 1986). Expression of the *A. vinelandii vnfA* gene is not subject to nitrogen control so, in this case, all nitrogen regulation occurs at the level of VnfA activity (Premakumar *et al.*, 1998).

3.3. Regulation of NifA Activity

In NifA-dependent systems, the regulation of *nif*-gene expression is achieved through regulation of *nifA* expression and by regulation of NifA activity. Although *nifA* expression is almost invariably Ntr regulated, the control of NifA activity occurs in a variety of ways that vary from organism to organism. In those organisms that have an O_2-resistant NifA protein, namely *Klebsiella, Azotobacter,* and *Enterobacter,* nitrogen control of NifA activity is regulated by its partner protein NifL in concert with a P_{II} protein. By comparison, in those organisms where NifA is O_2 sensitive, nitrogen control of NifA activity is mediated directly by a P_{II} protein.

3.3.1. Klebsiella pneumoniae

In *K. pneumoniae,* NifA activity is regulated in response to the cellular nitrogen status by its partner protein NifL (Merrick *et al.*, 1982). Expression of the *nifL* and *nifA* genes is controlled by NtrC and is translationally coupled so that NifL and NifA are expressed stoichiometrically, consistent with the proposal that they form a protein complex (Govantes *et al.*, 1996; Henderson *et al.*, 1989). Either inactivation of *nifL* or overexpression of *nifA* leads to constitutive activation of the *nif* genes in the presence of either fixed nitrogen or O_2 (Arnott *et al.*, 1989).

The NifL protein is predicted to comprise two domains. The N-terminal domain of around 290 residues contains a PAS motif, a widely distributed motif that is found in many sensor proteins and particularly associated with sensors of O_2, redox status, and light (Taylor and Zhulin, 1999; Zhulin *et al.*, 1997). The *K. pneumoniae* NifL is a flavoprotein with FAD bound to the N-terminal domain and most probably associated with the PAS motif (Schmitz, 1998). The C-terminal domain of around 200 residues shows significant homology to the histidine protein kinase transmitter domain but does not contain the highly conserved histidine residue that characterises these proteins (Drummond and Wootton, 1987). Studies of mutations in *nifL* suggested that both the O_2- and fixed nitrogen-sensing properties of the protein were separable (Sidoti *et al.*, 1993) and led to the search for a factor that could modulate the nitrogen response of NifL.

Studies of the classical components of the Ntr system, namely GlnB and GlnD, excluded both these proteins (Edwards and Merrick, 1995; Holtel and Merrick, 1989) and it was the recognition that the γ–Proteobacteria, such as *E. coli* and *K. pneumoniae,* encode a second P_{II} protein, which was designated GlnK (van Heeswijk *et al.*, 1996), that finally suggested another candidate. Mutations in *glnK* were then indeed found to impair nitrogen control by *Klebsiella* NifL, leading to a model in which GlnK interacted with NifL under nitrogen-limiting conditions, thereby allowing NifA to be active (He *et al.*, 1997, 1998; Jack *et al.*, 1999). The GlnK and GlnB proteins are 68% identical and hence, as both can be present in the cell at the same time, the question arises as to how they are distinguished with respect to their interaction with NifL (Arcondéguy *et al.*, 1999). Studies in *E. coli* have suggested that the T-loop of the P_{II} proteins plays a major role in interactions with P_{II} target proteins (Jiang *et al.*, 1997; Martinez-Argudo and Contreras, 2002; Pioszak *et al.*, 2000). The T-loops of both *E. coli* and *K. pneumoniae* GlnB and GlnK differ at just three residues and changing a single residue (Asp-54) in GlnB to the Asn-54 found in GlnK dramatically increases the ability of GlnB to behave like GlnK with regard to regulation of NifL activity. Two changes in the T-loop (giving a Asp54Asn/Thr43Ala GlnB protein) essentially confers full GlnK activity on GlnB (Arcondéguy *et al.*, 2000).

Regulation of NifL activity by GlnK is surprisingly not dependent upon uridylylation of GlnK, raising the question of how NifL could respond to a rapid change in the intracellular nitrogen status (Edwards and Merrick, 1995; He *et al.*, 1998). However, recent studies suggest that changes in the cellular localisation of both GlnK and NifL in response to nitrogen status may play a major role in this regulatory system. In *K. pneumoniae*, as in nearly all bacteria, the *glnK* gene is linked to *amtB*, which encodes a high affinity ammonium transporter. The *glnK-amtB* operon is regulated by NtrC and is only expressed under nitrogen limitation (Jack *et al.*, 1999). In *E. coli,* GlnK associates specifically with AmtB in the membrane and this association is markedly increased in response to a rapid increase in N status, *e.g.*, after an ammonium shock (Coutts *et al.*, 2002). This rapid association of GlnK with AmtB serves to regulate AmtB activity but will also quickly deplete the intracellular pool of GlnK. If, as expected, a similar mechanism operates in *K. pneumoniae,* then an increase in N status could serve to provide the rapid depletion of the GlnK pool with consequent liberation of NifL, which would then inhibit NifA activity. This model is further enhanced by the recognition that some 55% of the NifL protein is membrane-associated in nitrogen-limited conditions, whereas less than 10% of NifL is found in the membrane fraction of cells growing in nitrogen sufficiency (Klopprogge *et al.*, 2002).

3.3.2. Azotobacter vinelandii

The NifLA system of *A. vinelandii,* although sharing many properties with that of *K. pneumoniae*, has a number of distinct features. Firstly, the *nifLA* operon is not regulated by the Ntr system but is constitutively expressed (Blanco *et al.*, 1993), consequently all *nif* regulation in *Azotobacter* is achieved by controlling the activity of NifA. The NifL and NifA proteins of *Azotobacter* have proved to be much more amenable to purification and analysis *in vitro* than those of *Klebsiella* and hence much more is known of the detailed molecular nature of the proteins and their mode of action (Austin *et al.*, 1994; Dixon, 1998).

The *A. vinelandii* NifL is highly homologous to *K. pneumoniae* NifL. The protein is also a flavoprotein, containing FAD in the N-terminal domain (Hill *et al.*, 1996). The C-terminal domain shows a higher level of homology to the histidine protein kinases and binds both ATP and ADP (Söderbäck *et al.*, 1998). This domain is competent to inhibit NifA activity in response to either ADP *in vitro* or the level of fixed nitrogen *in vivo* (Söderbäck *et al.*, 1998). Furthermore, the domain includes a histidine within the motif that typically contains the phosphorylated histidine (Blanco *et al.*, 1993). However, mutational analysis showed that many amino-acid substitutions at this position do not impair NifL activity and there is no evidence that *Azotobacter* NifL is ever subject to phosphorylation (Woodley and Drummond, 1994).

A mutation in the *A. vinelandii glnD* gene (formerly called *nfrX*) prevented *nif*-gene expression but could be supressed by a secondary mutation in *nifL* (Contreras *et al.*, 1991). This result suggested that uridylylation of a regulatory component may be necessary to prevent inhibition of NifA activity by NifL. *A. vinelandii* is atypical amongst the γ-Proteobacteria in having only a single P_{II} protein, which is encoded in a *glnK-amtB* operon (Meletzus *et al.*, 1998). Studies both *in vivo* and *in vitro* show that the inhibitory activity of the NifL protein is stimulated by interaction with non-uridylylated P_{II} (either the native *Azotobacter* GlnK or *E. coli* GlnB) (Little *et al.*, 2000, 2002; Reyes-Ramirez *et al.*, 2001; Rudnick *et al.*, 2002). Consequently, the system is quite distinct from that in *Klebsiella* because, in *Azotobacter*, GlnK interacts with NifL under conditions of nitrogen excess and stimulates it to inhibit transcriptional activation by NifA. The inhibitory activity of NifL is then relieved under nitrogen-limiting conditions because the elevated levels of 2-oxoglutarate modulate the interaction of NifL with NifA (Little *et al.*, 2000). The interaction of GlnK is with the C-terminal histidine protein kinase-like domain of NifL and is abolished either by a single amino-acid substitution (Glu44Cys) in the T-loop of GlnK or by uridylylation of GlnK (Little *et al.*, 2002).

3.3.3. Azoarcus

The nitrogen-control system in *Azoarcus* is not yet fully described. However, *Azoarcus* has been shown to synthesise three P_{II}-like proteins, GlnB, GlnK, and GlnY, of which the last two are encoded in operons (*glnK-amtB; glnY-amtY*) along with AmtB-like proteins. Transcription of *nif* genes is repressed by either ammonium or nitrate in both wild-type and *glnB* or *glnK* mutants but, in a *glnB/glnK* double mutant, some transcription was detectable in the presence of ammonium and almost full expression was found in nitrate-grown cells (Martin and Reinhold-Hurek, 2002). These data suggest that either GlnB or GlnK can mediate *nif* repression (possibly through NifLA) in *Azoarcus*.

3.3.4. Azospirillum brasilense

The *A. brasilense* NifA belongs to the O_2-sensitive group of NifA proteins and does not have a partner NifL protein. The first indication that a P_{II}-like protein might be involved in the nitrogen regulation of NifA activity came from the observation that, in *A. brasilense*, a *glnB* mutant was Nif^- (Liang *et al.*, 1992), whereas a *glnK* mutant had no affect on nitrogen fixation (de Zamaroczy, 1998). As the *glnB* mutation did not affect *nifA* expression, it appeared that GlnB is required to maintain the active form of NifA. Deletions within the N-terminal domain of NifA restore *nif*-gene expression, suggesting that GlnB is required to activate NifA by preventing the inhibitory effect of its N-terminal domain (Arsene *et al.*, 1996).

Mutation of residue Tyr-18 to Phe in the N-terminal domain of NifA results in an active NifA that does not require GlnB. However, whether GlnB either interacts directly with NifA or modulates the activity of another protein that, in turn, regulates NifA activity remains unsolved (Arsene *et al.*, 1999). Both a *glnB* Tyr51Phe mutant and a *glnD* mutant exhibit a Nif⁻ phenotype (Arsene *et al.*, 1999; van Dommelen *et al.*, 2002), which is consistent with the fact that, during N_2 fixation, *A. brasilense* GlnB is uridylylated (de Zamaroczy, 1998) and suggests that it is GlnB-UMP, which is required for NifA activation.

3.3.5. Herbaspirillum seropedicae

As for *A. brasilense,* in *H. seropedicae*, a *glnB* mutant is Nif⁻, although *nifA* expression, which is NtrC-dependent, would be expected to be constitutive in this background (Benelli *et al.*, 1997). Studies of the *H. seropedicae* NifA protein *in vivo* show that the full-length protein, when expressed in *A. brasilense*, is active only under low O_2 and in the absence of ammonium, but NifA is not active when expressed in either *E. coli* or *K. pneumoniae* (Souza *et al.*, 1999). In contrast, an amino terminally truncated NifA is still active in the presence of ammonium in *A. brasilense*, *E. coli*, or *K. pneumoniae*, indicating that the N-terminal domain is involved in nitrogen control. Furthermore, when expressed in *trans*, this domain can inhibit the activity of the truncated NifA (Monteiro *et al.*, 1999a, 1999b; Souza *et al.*, 1999). If GlnB were to interact with the N-terminal domain, then the inactivity of *H. seropedicae* NifA in *E. coli* could be due to the absence of the cognate P_{II}.

3.3.6. Rhodobacter capsulatus

R. capsulatus is unique in having two *nifA* genes, which encode virtually identical proteins that can substitute for each other and which differ only in their 19 N-terminal residues (Masepohl *et al.*, 1988; Paschen *et al.*, 2001). Expression of both genes is regulated by NtrC (Foster-Hartnett and Kranz, 1992), but strains expressing either *nifA*$_1$ or *nifA*$_2$ from a constitutive promoter in an *ntrC* mutant still show inhibition of *nifH* transcription in ammonium, again suggesting post-translational control of NifA activity (Hübner *et al.*, 1993). Mutations in the N-terminal domain of NifA1 result in ammonium-tolerant NifA proteins, suggesting that, as in other NifA proteins, this domain is involved in regulation of activity (Paschen *et al.*, 2001).

R. capsulatus encodes two P_{II} proteins, GlnB and GlnK. Mutations in *glnB* do not affect NifA activity, whereas in a *glnK* mutant, NifA partially escapes ammonium inhibition and, in a *glnB/glnK* double mutant, ammonium control is completely abolished (Masepohl *et al.*, 2002). Yeast two-hybrid studies indicate that both GlnB and GlnK can interact with NifA, suggesting that these proteins directly mediate nitrogen control of NifA activity, although GlnB apparently only partially substitutes for GlnK (Masepohl *et al.*, 2002).

In contrast to *nif* regulation, *anfH* is still inhibited by ammonium in a *glnB/glnK* double mutant and this inhibition is also independent of NtrC. Hence, the ammonium control of AnfA activity must occur by a completely different mechanism (Masepohl *et al.*, 2002).

3.3.7. Rhodospirillum rubrum

R. rubrum can synthesise three P_{II} -like proteins, GlnB, GlnJ and GlnK, of which the last two are encoded in operons that also encode ammonium transport-like proteins, *glnJ-amtB*$_1$ and *glnK-amtB*$_2$ (Johansson and Nordlund, 1997, 1999; Zhang *et al.*, 2000, 2001). A Δ*glnB* mutant has no nitrogenase activity, whereas a *glnB* Tyr51Phe mutant shows around 10% of wild-type nitrogenase activity. Expression of *R. rubrum nifA* from a multicopy plasmid does not restore nitrogenase activity in a Δ*glnB* mutant, whereas a *glnB* Tyr51Phe mutant is complemented (Zhang *et al.*, 2000). The effects of the *glnB* mutations are not on *nifA* expression and, therefore, GlnB appears to be essential for NifA activity. There is no specific evidence for a GlnB-NifA interaction but the data do suggest that GlnB-UMP promotes activation of NifA (Zhang *et al.*, 2001).

3.4. Regulation of VnfA and AnfA Activity

The activities of *A. vinelandii* VnfA and AnfA and of *R. capsulatus* AnfA are all subject to nitrogen control. In *R. capsulatus,* this control appears to occur by a novel mechanism that is independent of NtrC, GlnB, and GlnK (Masepohl *et al.*, 2002). In *A. vinelandii,* the activities of both VnfA and AnfA are nitrogen regulated but ammonium repression of both proteins is relieved when their N-terminal domain is deleted (Drummond *et al.*, 1995). The role of the *A. vinelandii* GlnK in this regulation has not been investigated.

4. NITROGEN CONTROL OF NITROGENASE ACTIVITY

4.1. ADP-Ribosylation

The final level at which nitrogen fixation can be subject to control in response to the availability of fixed nitrogen is by regulation of the activity of the nitrogenase enzyme itself. In 1950, studies of nitrogen fixation in *Rhodospirillum rubrum* first demonstrated that the addition of ammonium to nitrogen-fixing cultures led to a rapid decrease in nitrogenase activity (Gest *et al.*, 1950). This property was not exhibited by other diazotrophs that were available to study at the time. Subsequent studies showed the inhibitory effect to be reversible and the length of the inhibitory period to be correlated with the concentration of ammonium added (Schick, 1971). Extensive biochemical studies finally led to the recognition that inactivation of nitrogenase was due to ADP-ribosylation of residue Arg-101 on one of the two subunits of NifH, the Fe protein of nitrogenase (also called dinitrogenase reductase) (Pope *et al.*, 1985). The process requires the concerted action of two enzymes; dinitrogenase reductase ADP-ribosyl transferase (DRAT), which transfers the ADP-ribose moiety of NAD to Arg-101 of NifH (Lowery *et al.*, 1986), and dinitrogenase reductase activating glycohydrolase (DRAG), which reactivates NifH by removing ADP-ribose (Saari *et al.*, 1986). The structural genes for both DRAG (*draG*) and DRAT (*draT*) have been cloned from *R. rubrum* (Fitzmaurice *et al.*, 1989).

ADP-ribosylation of nitrogenase has been reported in a variety of other diazotrophs and indeed, the *draT* and *draG* genes have also been cloned from *R. capsulatus* (Masepohl *et al.*, 1993), *A. brasilense* (Zhang *et al.*, 1992), and *A. lipoferum* (Inoue *et al.*, 1996). In all cases, they appear to constitute a *draT-draG* operon and, in all but *R. capsulatus,* they are located adjacent to the *nifHDK* operon. There is also evidence for this system in *Chromatium vinosum, Rhodopseudomonas palustris,* and *Rhodopseudomonas viridis* and physiological responses consistent with ADP-ribosylation have been reported in *Azoarcus, Azotobacter chroococcum, Azorhizobium sesbaniae*, and *R. sphaeroides*. However, *R. sphaeroides* does not contain *draTG* genes and does not apparently show ADP-ribosylation of nitrogenase (Yakunin *et al.*, 2001).

As DRAG and DRAT are encoded in an operon, they are expressed together and their activities are post-translationally regulated *in vivo* as evidenced by studies in *R. rubrum, A. brasilense,* and *R. capsulatus* of mutants lacking either of the enzymes (Liang *et al.*, 1991; Masepohl *et al.*, 1993; Zhang *et al.*, 1992, 1993). In nitrogen-fixing conditions, DRAG is active and DRAT is completely inactive but, following addition of a good nitrogen source, *e.g.*, ammonium, DRAG becomes inactive within a few minutes and DRAT becomes active, leading to inactivation of NifH. Exhaustion of the nitrogen source leads to a reversal of this process. In *R. capsulatus*, ADP-ribosylation occurs not only on a subunit of the molybdenum-nitrogenase NifH, but also on a subunit of the iron-nitrogenase AnfH (Masepohl *et al.*, 1993).

The introduction of the *R. rubrum* and *A. brasilense draTG* genes into *K. pneumoniae* proved to be a critical experiment because it demonstrated that DRAT and DRAG could regulate the activity of a heterologous nitrogenase in response to added fixed nitrogen, indicating that the signal transduction system responsible for regulating the system was present in *K. pneumoniae* (Fu *et al.*, 1990). These data focussed attention on the Ntr system and in particular on the P_{II} proteins. Subsequent studies of P_{II} mutants in the heterologous *K. pneumoniae* system, in *R. Rubrum*, and in *R. capsulatus* have confirmed that the P_{II} proteins do indeed play a role in regulating ADP-ribosylation.

R. capsulatus synthesizes two P_{II} proteins, GlnB and GlnK, and in a *glnB/glnK* double mutant, nitrogenase is not subject to post-translational ammonium control, indicating that either one or both P_{II} proteins are necessary for DRAT activity (Masepohl *et al.*, 2002). The situation in *R. rubrum* is more complex in having three P_{II} proteins. A *glnK*, a *glnJ* or a *glnK/glnJ* mutant is unaffected in ADP-ribosylation, whereas either a *glnB* or a *glnB/glnK* mutant shows some impairment in the response to ammonium, and a *glnB/glnJ* mutant shows almost no ammonium response (Zhang *et al.*, 2001). Hence, it would appear that either GlnB or GlnJ can influence the activities of the DRAG/DRAT proteins. The precise mechanism of this control is not yet understood but, given the mode of action of P_{II} in other systems, it seems likely that the P_{II} proteins directly affect the DRAG and/or DRAT activities. ADP-ribosylation in *R. rubrum* also occurs in response to an energy shift as effected by moving cells into the dark, but a *glnB/glnJ* double mutant is also impaired in this response, indicating that these proteins also mediate a signal of energy limitation (Zhang *et al.*, 2001). This situation is distinct from that in *R. capsulatus*, where a *glnB/glnK* double mutant responds normally to a light-dark shift (Masepohl *et al.*, 2002).

Although the P_{II} proteins have clearly been implicated in the signal-transduction pathway controlling ADP-ribosylation, the precise signals are not yet identified. However, in *R. capsulatus*, mutations in ammonium-transport genes also affect this process. *R. capsulatus* has two *amt* genes; *amtB*, which is linked to *glnK* in a *glnK-amtB* operon, and a second gene, *amtY*, which is unlinked. An *amtB* mutant is completely defective in ADP-ribosylation, whereas an *amtY* mutant shows elevated levels of nitrogenase modification compared to wild-type in response to the same amount of ammonium (Masepohl *et al.*, 2002; Yakunin and Hallenbeck, 2000; 2002). The role of the Amt proteins in the signal-transduction pathway is still unclear and, given the potential for Amt proteins to sequester P_{II} proteins in response to ammonium shock, these phenotypes could reflect indirect effects on the free intracellular pools of P_{II} protein (Coutts *et al.*, 2002).

NAD has been implicated as a potential signal molecule in some aspects of the control of ADP-ribosylation in *R. rubrum* (Noren *et al.*, 1997; Noren and Nordlund, 1994; Soliman and Nordlund, 1992) and it is, therefore, of potential interest that P_{II} may also play a role in regulation of NAD synthetase. In both *H. seropedicae* and *Azoarcus*, the structural gene for NAD synthetase (*nadE*) is genetically linked to a *glnB*-like gene, giving rise to the possibility that this linkage reflects the potential for an interaction between the two proteins (Arcondéguy *et al.*, 2001; Benelli *et al*, 1997; Martin *et al.*, 2000).

4.2. Ammonium Switch-off in Methanogens

The phenomenon of ammonia switch-off is also found in the diazotrophic methanogens, *Methanosarcina barkeri* and *Methanococcus maripaludis* (Kessler and Leigh, 1999; Lobo and Zinder, 1990). However, there is no evidence for ADP-ribosylation in these organisms (Kessler *et al.*, 2001) and *draG* and *draT* homologues are not encoded in the completed genome sequences of the related diazotrophs, *M. thermoautotrophicum* and *Methanosarcina mazei* Gö1. Hence, these organisms may use an alternative method of post-translational modification.

The *nif* structural-gene clusters of all the diazotrophic methanogens are characterised by the presence of two *glnB*-like genes, now designated *nifI*$_1$ and *nifI*$_2$, which are located between the nitrogenase structural genes *nifH* and *nifDK* (Chien and Zinder, 1996; Ehlers *et al.*, 2002; Kessler *et al.*, 1998; 2001; Kessler and Leigh, 1999; Sibold *et al.*, 1991; Souillard *et al.*, 1988). The P_{II} proteins encoded by these genes are from a quite distinct sub-group. The *nifH*-proximal gene (*nifI*$_1$) encodes a polypeptide of around 105 residues, whereas the *nifH*-distal gene (*nifI*$_2$) encodes a polypeptide of 120-130 residues. Both polypeptides are highly homologous to proteins such as GlnB and GlnK and they are distinguished by the regions that correspond to the T-loop. Whereas the T-loop in either the GlnB or GlnK proteins is almost invariably 18 residues long, sequence alignments suggest that, in $NifI_1$, the equivalent region is only 13 residues long and, in $NifI_2$, it is 27-32 residues long. There is no conserved site for either uridylylation or phosphorylation within the predicted T-loops, suggesting that the proteins are either subject to a novel form of modification or, perhaps, are not modified at all.

The NifI proteins are not restricted to the diazotrophic methanogens but are also found in the Firmibacteria (in *Clostridium acetobutylicum* and *Clostridium cellobioparum*) and in the δ-Proteobacteria (in *Desulfovibrio gigas*). In all cases, they are located downstream of *nifH*, suggesting that they have a conserved function (Arcondéguy *et al.*, 2001).

The roles of the NifI proteins have been analysed in detail in *M. maripaludis* where the construction of in-frame deletions coupled with complementation analysis showed that both proteins are required for switch-off. The process of switch-off was reversible and did not affect *nif*-gene transcription, *nifH* mRNA stability, or NifH protein stability (Kessler *et al.*, 2001). The process mediated by the NifI proteins, therefore, appears to be quite novel and a number of possible mechanisms have been suggested. These suggestions include non-covalent association of nitrogenase with another protein factor (which could be NifI itself), reversible covalent modification of a Nif protein other than NifH, or a reversible covalent modification that is not resolved on SDS PAGE (Kessler *et al.*, 2001).

The structure of the *nifH-nifI₁-nifI₂-nifD-nifK* operon implies a stoichiometric relationship between these five polypeptides and raises interesting questions about the likely structures of the $NifI_1$ and $NifI_2$ proteins. By comparison with all other P_{II} proteins, they would be expected to form trimers, but the 1:1 ratio of $NifI_1$ to $NifI_2$ could allow the formation of a hexameric protein. The occurrence in *Aquifex aeolicus* of a novel P_{II} gene, which appears to constitute a "duplicated" *glnB*-like gene where two similar coding sequences are fused in-frame, suggests that such a hexameric structure may be possible (Arcondéguy *et al.*, 2001).

4.3. Other Mechanisms of Ammonia Switch-off

Mechanisms of ammonium-induced switch-off of nitrogenase, which are independent of ADP-ribosylation, have been described. In the unicellular cyanobacterium *Gleothece*, a novel modification of NifH by palmitoylation has been reported (Gallon *et al.*, 2000). In *R. capsulatus* and *A. brasilense,* a second mechanism for post-translational regulation of nitrogenase activity, in addition to ADP-ribosylation, has also been reported but the mechanism underlying this control is unknown (Pierrard *et al.*, 1993; Yakunin and Hallenbeck, 1998; Zhang *et al.*, 1996).

5. CONCLUSIONS

Our understanding of the regulation of the nitrogen-fixation process in free-living diazotrophs has advanced considerably in the last decade with information coming from a wide range of model systems. These advances have led to a much more global view of the mechanisms, which facilitate the very stringent control that is necessary to maximise the physiological benefits from diazotrophy. With regard to nitrogen control, members of the P_{II} protein family, *i.e.*, the GlnB, GlnK, NifI, etc. proteins, clearly play a pivotal role in nearly all organisms and P_{II} is now being recognised as the critical signal-transduction protein in a wide variety of aspects of bacterial nitrogen metabolism (Arcondéguy *et al.*, 2001).

At the molecular level, our understanding of the regulatory processes is advancing considerably in those organims that have a NifA-dependent mode of control, although the ability to purify active NifA proteins (particularly of the O_2-sensitive group) is still a major hurdle.

The heterocystous cyanobacteria are particularly complex owing to the integration of the regulation of both nitrogenase expression and activity with that of heterocyst development. Nevertheless, considerable information is now emerging and a broad outline of the major signal-transduction pathways may be achieved fairly soon. The groups of organisms where there is still much to be learnt are the Archaea and the Gram-positive diazotrophs. In the Archaea, the advent of genome sequences and good genetic systems in model organisms, *e.g.*, *Methanobacterium thermoautotrophicum* and *Methanosarcina mazei*, shows considerable promise for the future but similar opportunities are not yet apparent in the Gram-positive diazotrophs.

In summary, the challenge in the future is to begin to integrate our current knowledge into a whole-cell perspective of the genetic, biochemical and physiological processes that contribute to successful diazotrophy.

REFERENCES

Aldehni, M. F., Sauer, J., Spielhaupter, C., Schmid, R., and Forchhammer, K. (2003). Signal transduction protein P_{II} is required for NtcA-regulated gene expression during nitrogen deprivation in the cyanobacterium *Synechococcus elongatus* strain PCC 7942. *J. Bacteriol., 185,* 2582-2591.

Arcondéguy, T., Jack, R., and Merrick, M. (2001). P_{II} signal transduction proteins: Pivotal players in microbial nitrogen control. *Microbiol. Mol. Biol. Rev., 65,* 80-105.

Arcondéguy, T., Lawson, D., and Merrick, M. (2000). Two residues in the T-loop of GlnK determine NifL-dependent nitrogen control of *nif* gene expression. *J. Biol. Chem., 275,* 38452-38456.

Arcondéguy, T., van Heeswijk, W. C., and Merrick, M. (1999). Studies on the roles of GlnK and GlnB in regulating *Klebsiella pneumoniae* NifL-dependent nitrogen control. *FEMS Microbiol. Lett., 180,* 263-270.

Arnott, M., Sidoti, C., Hill, S., and Merrick, M. (1989). Deletion analysis of the nitrogen fixation regulatory gene *nifL* of *Klebsiella pneumoniae*. *Arch. Microbiol., 151,* 180-182.

Arsene, F., Kaminski, P. A., and Elmerich, C. (1996). Modulation of NifA activity by P_{II} in *Azospirillum brasilense*: Evidence for a regulatory role of the NifA N-terminal domain. *J. Bacteriol., 178,* 4830-4838.

Arsene, F., Kaminski, P. A., and Elmerich, C. (1999). Control of *Azospirillum brasilense* NifA activity by P_{II}: Effect of replacing Tyr residues of the NifA N-terminal domain on NifA activity. *FEMS Microbiol. Lett., 179,* 339-343.

Atkinson, M. and Ninfa, A. J. (1998). Role of the GlnK signal transduction protein in the regulation of nitrogen assimilation in *Escherichia coli*. *Mol. Microbiol., 29,* 431-447.

Atkinson, M. R., Blauwkamp, T. A., Bondarenko, V., Studitsky, V., and Ninfa, A. J. (2002a). Activation of the *glnA, glnK,* and *nac* promoters as *Escherichia coli* undergoes the transition from nitrogen excess growth to nitrogen starvation. *J. Bacteriol., 184,* 5358-5363.

Atkinson, M. R., Blauwkamp, T. A., and Ninfa, A. J. (2002b). Context-dependent functions of the PII and GlnK signal transduction proteins in *Escherichia coli. J. Bacteriol., 184,* 5364-5375.

Austin, S., Buck, M., Cannon, W., Eydmann, T., and Dixon, R. (1994). Purification and *in vitro* activities of the native nitrogen fixation control proteins NIFA and NIFL. *J. Bacteriol., 176,* 3460-3465.

Austin, S., Henderson, N., and Dixon, R. (1990). Characterisation of the *Klebsiella pneumoniae* nitrogen-fixation regulatory proteins NIFA and NIFL *in vitro. Eur. J. Biochem., 187,* 353-360.

Bali, A., Blanco, G., Hill, S., and Kennedy, C. (1992). Excretion of ammonium by a *nifL* mutant of *Azotobacter vinelandii* fixing nitrogen. *Appl. Environ. Microbiol., 58,* 1711-1718.

Bali, A., Hill, S., Santero, E., Toukdarian, A., Walmsley, J., and Kennedy, C. (1988). NifA and NtrC activate separate nitrogen fixation pathways in *Azotobacter.* In H. Bothe, F. J. DeBruijn, and W. E. Newton (Eds.), *Nitrogen fixation: Hundred years after* (p. 316). Stuttgart: Gustav Fischer.

Barrios, H., Valderrama, B., and Morett, E. (1999). Compilation and analysis of sigma^{54}-dependent promoter sequences. *Nucleic Acids Res., 27,* 4305-4313.

Bauer, E., Kaspar, T., Fischer, H.-M., and Hennecke, H. (1998). Expression of the *fixR-nifA* operon in *Bradyrhizobium japonicum* depends on a new response regulator, RegR. *J. Bacteriol., 180,* 3853-3863.

Benelli, E. M., Souza, E. M., Funayama, S., Rigo, I. U., and Pedrosa, F. O. (1997). Evidence for two possible *glnB*-type genes in *Herbaspirillum seropedicae. J. Bacteriol., 179,* 4623-4626.

Blanco, G., Drummond, M., Woodley, P., and Kennedy, C. (1993). Sequence and molecular analysis of the *nifL* gene of *Azotobacter vinelandii. Mol. Microbiol., 9,* 869-880.

Blauwkamp, T. A. and Ninfa, A. J. (2002a). Nac-mediated repression of the *serA* promoter of *Escherichia coli. Mol. Microbiol., 45,* 351-363.

Blauwkamp, T. A. and Ninfa, A. J. (2002b). Physiological role of the GlnK signal transduction protein of *Escherichia coli*: survival of nitrogen starvation. *Mol. Microbiol., 46,* 203-214.

Buck, M., Cannon, W., and Woodcock, J. (1987). Transcriptional activation of the *Klebsiella pneumoniae* nitrogenase promoter may involve DNA loop formation. *Mol. Microbiol., 1,* 243-249.

Buck, M., Miller, S., Drummond, M., and Dixon, R. (1986). Upstream activator sequences are present in the promoters of nitrogen fixation genes. *Nature, 320,* 374-378.

Carr, P. D., Cheah, E., Suffolk, P. M., Vasudevan, S. G., Dixon, N. E., and Ollis, D. L. (1996). X-ray structure of the signal transduction protein P_{II} from *Escherichia coli* at 1.9Å. *Acta Cryst., D52,* 93-104.

Chien, Y., Helmann, J. D., and Zinder, S. H. (1998). Interactions between the promoter regions of nitrogenase structural genes (*nifHDK*$_2$) and DNA-binding proteins from N_2- and ammonium-grown cells of the archaeon *Methanosarcina barkeri* 227. *J. Bacteriol., 180,* 2723-2728.

Chien, Y. T. and Zinder, S. H. (1996). Cloning, functional organization, transcript studies, and phylogenetic analysis of the complete nitrogenase structural genes (*nifHDK2*) and associated genes in the archaeon *Methanosarcina barkeri* 227. *J. Bacteriol., 178,* 143-148.

Cohen-Kupiec, R., Blank, C., and Leigh, J. A. (1997). Transcriptional regulation in Archaea: *In vivo* demonstration of a repressor binding site in a methanogen. *Proc. Natl. Acad. Sci. U.S.A., 94,* 1316-1320.

Cohen-Kupiec, R., Marx, C. J., and Leigh, J. A. (1999). Function and regulation of *glnA* in the methanogenic archaeon *Methanococcus maripaludis. J. Bacteriol., 181,* 256-261.

Contreras, A., Drummond, M., Bali, A., Blanco, G., Garcia, E., Bush, G. *et al.* (1991). The product of the nitrogen fixation regulatory gene *nfrX* of *Azotobacter vinelandii* is functionally and structurally homologous to the uridylyltransferase encoded by *glnD* in enteric bacteria. *J. Bacteriol., 173,* 7741-7749.

Coutts, G., Thomas, G., Blakey, D., and Merrick, M. (2002). Membrane sequestration of the signal transduction protein GlnK by the ammonium transporter AmtB. *EMBO J., 21,* 1-10.

de Zamaroczy, M. (1998). Structural homologues P_{II} and P_Z of *Azospirillum brasilense* provide intracellular signalling for selective regulation of various nitrogen-dependent functions. *Mol. Microbiol., 29,* 449-463.

Dixon, R. (1998). The oxygen-responsive NIFL-NIFA complex: a novel two-component regulatory system controlling nitrogenase synthesis in gamma-proteobacteria. *Arch. Microbiol., 169,* 371-380.

Drummond, M., Frise, E., Green, A., and Jacob, J. (1995). Nitrogen regulation of VnfA and AnfA function in *Azotobacter vinelandii* analysed by chimera formation. In I. A. Tikonovich, N. A. Provorov, V. I. Romanov, and W. E. Newton (Eds.), *Nitrogen fixation: Fundamentals and applications* (p. 214). Dordrecht: Kluwer Academic Publishers.

Drummond, M. H. and Wootton, J. C. (1987). Sequence of *nifL* from *Klebsiella pneumoniae*: mode of action and relationship to two families of regulatory proteins. *Mol. Microbiol., 1,* 37-44.

Edwards, R. and Merrick, M. (1995). The role of uridylyltransferase in the control of *Klebsiella pneumoniae nif* gene regulation. *Mol. Gen. Genet., 247,* 189-198.

Egener, T., Sarkar, A., Martin, D. E., and Reinhold-Hurek, B. Identification of a NifL-like protein in a diazotroph of the β-subgroup of the *Proteobacteria, Azoarcus* sp. strain BH72. *Microbiol., 148,* 3203-3210.

Ehlers, C., Grabbe, R., Veit, K., and Schmitz, R. A. (2002). Characterization of GlnK(1) from *Methanosarcina mazei* Strain Go1: complementation of an *Escherichia coli glnK* mutant strain by GlnK(1). *J. Bacteriol., 184,* 1028-1040.

Elsen, S., Dischert, W., Colbeau, A., and Bauer, C. E. (2000). Expression of uptake hydrogenase and molybdenum nitrogenase in *Rhodobacter capsulatus* is coregulated by the RegB-RegA two-component regulatory system. *J. Bacteriol., 182,* 2831-2837.

Emmerich, R., Hennecke, H., and Fischer, H. M. (2000). Evidence for a functional similarity between the two-component regulatory systems RegSR, ActSR, and RegBA (PrrBA) in α-Proteobacteria. *Arch. Microbiol., 174,* 307-313.

Espin, G., Alvarez-Morales, A., Cannon, F., Dixon, R., and Merrick, M. J. (1982). Cloning of the *glnA*, *ntrB* and *ntrC* genes of *Klebsiella pneumoniae* and studies of their role in regulation of the nitrogen fixation (*nif*) gene cluster. *Mol. Gen. Genet., 186,* 518-524.

Fischer, H.-M., Bruderer, T., and Hennecke, H. (1988). Essential and non-essential domains in the *Bradyrhizobium japonicum* NifA protein: Identification of indispensible cysteine residues potentially involved in redox activity and/or metal binding. *Nucleic Acids Res., 16,* 2207-2224.

Fischer, H.-M., Fritsche, S., Herzog, B., and Hennecke, H. (1989). Critical spacing between two essential cysteine residues in the interdomain linker of the *Bradyrhizobium japonicum* NifA protein. *FEBS Lett., 255,* 167-171.

Fisher, S. H. (1999). Regulation of nitrogen metabolism in *Bacillus subtilis*: Vive la difference! *Mol. Microbiol., 32,* 223-232.

Fitzmaurice, W. P., Saari, L. L., Lowery, R. G., Ludden, P. W., and Roberts, G. P. (1989). Genes coding for the reversible ADP-ribosylation system of dinitrogenase reductase from *Rhodospirillum rubrum*. *Mol. Gen. Genet., 218,* 340-347.

Foster-Hartnett, D. and Kranz, R. G. (1992). Analysis of the promoters and upstream sequences of *nifA1* and *nifA2* in *Rhodobacter capsulatus*; activation requires *ntrC* but not *rpoN*. *Mol. Microbiol., 6,* 1049-1060.

Frias, J. E., Flores, E., and Herrero, A. (1994). Requirement of the regulatory protein NtcA for the expression of nitrogen assimilation and heterocyst development genes in the cyanobacterium *Anabaena* sp. PCC 7120. *Mol. Microbiol., 14,* 823-832.

Fu, H., Burris, R. H., and Roberts, G. P. (1990). Reversible ADP-ribosylation is demonstrated to be a regulatory mechanism in prokaryotes by heterologous expression. *Proc. Natl. Acad. Sci. U.S.A., 87,* 1720-1724.

Gallon, J. R., Cheng, J., Dougherty, L. J., Gallon, V. A., Hilz, H., Pederson, D. M. *et al.* (2000). A novel covalent modification of nitrogenase in a cyanobacterium. *FEBS Lett., 468,* 231-233.

Gest, H., Kamen, M. D., and Bergoff, H. M. (1950). Studies on the metabolism of photsynthetic bacteria. V. Photoproduction of hydrogen and nitrogen fixation by *Rhodospirillum rubrum*. *J. Biol. Chem., 182,* 153-170.

Golden, J. W., Robinson, S. J., and Haselkorn, R. (1985). Rearrangement of nitrogen fixation genes during heterocyst differentiation in the cyanobacterium *Anabaena*. *Nature, 314,* 419-423.

Gonzalez, V., Olvera, L., Soberon, X., and Morett, E. (1998). *In vivo* studies on the positive control function of NifA: A conserved hydrophobic amino acid patch at the central domain involved in transcriptional activation. *Mol. Microbiol., 28,* 55-67.

Govantes, F., Molina-Lopez, J. A., and Santero, E. (1996). Mechanism of coordinated synthesis of the antagonistic regulatory proteins NifL and NifA of *Klebsiella pneumoniae*. *J. Bacteriol., 178,* 6817-6823.

He, L., Soupene, E., and Kustu, S. (1997). NtrC is required for control of *Klebsiella pneumoniae* NifL activity. *J. Bacteriol., 179,* 7446-7455.

He, L., Soupene, E., Ninfa, A. J., and Kustu, S. (1998). Physiological role for the GlnK protein of enteric bacteria: relief of NifL inhibition under nitrogen-limiting conditions. *J. Bacteriol., 180,* 6661-6667.

Henderson, N., Austin, S. A., and Dixon, R. A. (1989). Role of metal ions in negative regulation of nitrogen fixation by the *nifL* gene product from *Klebsiella pneumoniae. Mol. Gen. Genet., 216,* 484-491.

Herrero, A., Muro-Pastor, A. M., and Flores, E. (2001). Nitrogen control in cyanobacteria. *J. Bacteriol., 183,* 411-425.

Hill, S., Austin, S., Eydmann, T., Jones, T., and Dixon, R. (1996). *Azotobacter vinelandii* NIFL is a flavoprotein that modulates transcriptional activation of nitrogen-fixation genes via a redox-sensitive switch. *Proc. Natl. Acad. Sci. U.S.A., 93,* 2143-2148.

Ho, Y. S., Burden, L. M., and Hurley, J. H. (2000). Structure of the GAF domain, a ubiquitous signaling motif and a new class of cyclic GMP receptor. *EMBO J., 19,* 5288-5299.

Holtel, A. and Merrick, M. J. (1989). The *Klebsiella pneumoniae* P_{II} protein (*glnB* gene product) is not absolutely required for nitrogen regulation and is not involved in NifL-mediated *nif* gene regulation. *Mol. Gen. Genet., 217,* 474-480.

Hoover, T. R., Santero, E., Porter, S., and Kustu, S. (1990). The integration host factor stimulates interaction of RNA polymerase with NIFA, the transcriptional activator for nitrogen fixation operons. *Cell, 63,* 11-22.

Hübner, P., Masepohl, B., Klipp, W., and Bickle, T. A. (1993). *nif* gene expression studies in *Rhodobacter capsulatus*: *ntrC*-independent repression by high ammonium concentrations. *Mol. Microbiol., 10,* 123-132.

Inoue, A., Shigematsu, T., Hidaka, M., Masaki, H., and Uozumi, T. (1996). Cloning, sequencing and transcriptional regulation of the *draT* and *draG* genes of *Azospirillum lipoferum* FS. *Gene, 170,* 101-106.

Jack, R., de Zamaroczy, M., and Merrick, M. (1999). The signal transduction protein GlnK is required for NifL-dependent nitrogen control of *nif* expression in *Klebsiella pneumoniae. J. Bacteriol., 181,* 1156-1162.

Jepson, B. and Austin, S. (2002). A mutation in the conserved cysteine motif impairs the activity of AnfA. In T. Finan, M. R. O'Brian, D. B. Layzell, J. K. Vessey, and W. E. Newton (Eds.), *Nitrogen fixation: Global perspectives* (p. 398). Wallingford, UK: CABI Publishing.

Jiang, F., Wisen, S., Widersten, M., Bergman, B., and Mannervik, B. (2000). Examination of the transcription factor NtcA-binding motif by *in vitro* selection of DNA sequences from a random library. *J. Mol. Biol., 301,* 783-793.

Jiang, P., Peliska, J. A., and Ninfa, A. J. (1998). Enzymological characterization of the signal-transducing uridylyltransferase/uridylyl-removing enzyme (EC 2.7.7.59) of *Escherichia coli* and its interaction with the PII protein. *Biochemistry, 37,* 12782-12794.

Jiang, P., Zucker, P., Atkinson, M. R., Kamberov, E. S., Tirasophon, W., Chandran, P. *et al.* (1997). Structure/function analysis of the PII signal transduction protein of *Escherichia coli*: Genetic separation of interactions with protein receptors. *J. Bacteriol., 179,* 4342-4353.

Joerger, R. D., Jacobson, M. R., and Bishop, P. E. (1989). Two *nifA*-like genes required for expression of alternative nitrogenases by *Azotobacter vinelandii. J. Bacteriol., 171,* 3258-3267.

Johansson, M. and Nordlund, S. (1997). Uridylylation of the P_{II} protein of the photosynthetic bacterium *Rhodospirillum rubrum. J. Bacteriol., 179,* 4190-4194.

Johansson, M. and Nordlund, S. (1999). Purification of P_{II} and P_{II}-UMP and *in vitro* studies of regulation of glutamine synthetase in *Rhodospirillum rubrum. J. Bacteriol., 181,* 6524-6529.

Kamberov, E. S., Atkinson, M. R., Feng, J., Chandran, P., and Ninfa, A. J. (1994). Sensory components controlling bacterial nitrogen assimilation. *Cell. Mol. Biol. Res., 40,* 175-191.

Kamberov, E. S., Atkinson, M. R., and Ninfa, A. J. (1995). The *Escherichia coli* PII signal transduction protein is activated upon binding 2-ketoglutarate and ATP. *J. Biol. Chem., 270,* 17797-17807.

Keener, J. and Kustu, S. (1988). Protein kinase and phosphoprotein phosphatase activities of nitrogen regulatory proteins NTRB and NTRC of enteric bacteria: Roles of the conserved amino-terminal domain of NTRC. *Proc. Natl. Acad. Sci. U.S.A., 85,* 4976-4980.

Kennedy, C., Bali, A., Blanco, G., Contreras, A., Drummond, M., Merrick, M. *et al.* (1991). Regulation of expression of genes for three nitrogenases in *Azotobacter vinelandii.* In M.Polsinelli, R. Materassi, and M. Vincenzini (Eds.), *Nitrogen fixation. Proceedings of the fifth international symposium on nitrogen fixation with non-legumes* (pp. 13-23). Dordrecht: Kluwer Academic Publishers.

Kent, H. M., Buck, M., and Evans, D. J. (1989). Cloning and sequencing of the *nifH* gene of *Desulfovibrio gigas*. *FEMS Microbiol. Lett., 61,* 73-78.

Kern, D., Volkman, B. F., Luginbuhl, P., Nohaile, M. J., Kustu, S., and Wemmer, D. E. (1999). Structure of a transiently phosphorylated switch in bacterial signal transduction. *Nature, 402,* 894-898.

Kessler, P. S., Blank, C., and Leigh, J. A. (1998). The *nif* gene operon of the methanogenic archaeon *Methanococcus maripaludis*. *J. Bacteriol., 180,* 1504-1511.

Kessler, P. S., Daniel, C., and Leigh, J. A. (2001). Ammonia switch-off of nitrogen fixation in the methanogenic archaeon *Methanococcus maripaludis*: Mechanistic features and requirement for the novel GlnB homologues, $NifI_1$ and $NifI_2$. *J. Bacteriol., 183,* 882-889.

Kessler, P. S. and Leigh, J. A. (1999). Genetics of nitrogen regulation in *Methanococcus maripaludis*. *Genetics, 152,* 1343-1351.

Klopprogge, K., Grabbe, R., Hoppert, M., and Schmitz, R. A. (2002). Membrane association of *Klebsiella pneumoniae* NifL is affected by molecular oxygen and combined nitrogen. *Arch. Microbiol., 177,* 223-234.

Kranz, R. G. and Foster-Hartnett, D. (1990). Transcriptional regulatory cascade of nitrogen-fixation genes in anoxygenic photosynthetic bacteria: Oxygen- and nitrogen-responsive factors. *Mol. Microbiol., 4,* 1793-1800.

Kranz, R. G. and Haselkorn, R. (1985). Characterisation of the *nif* regulatory genes in *Rhodopseudomonas capsulata* using *lac* gene fusions. *Gene, 40,* 203-215.

Kranz, R. G., Pace, V. M., and Caldicott, I. M. (1990). Inactivation, sequence, and *lacZ* fusion analysis of a regulatory locus required for repression of nitrogen fixation genes in *Rhodobacter capsulatus*. *J. Bacteriol., 172,* 53-62.

Krey, R., Pühler, A., and Klipp, W. (1992). A defined amino acid exchange close to the putative nucleotide binding site is responsible for an oxygen-tolerant variant of the *Rhizobium meliloti* NifA protein. *Mol. Gen. Genet., 234,* 433-441.

Kullik, I., Hennecke, H., and Fischer, H.-M. (1989). Inhibition of *Bradyrhizobium japonicum nifA*-dependent *nif* gene activation by oxygen occurs at the NifA protein level and is irreversible. *Arch. Microbiol., 151,* 191-197.

Kutsche, M., Leimkühler, S., Angermüller, S., and Klipp, W. (1996). Promoters controlling expression of the alternative nitrogenase and the molybdenum uptake system in *Rhodobacter capsulatus* are activated by NtrC, independent of σ^{54}, and repressed by molybdenum. *J. Bacteriol., 178,* 2010-2017.

Lee, H. S., Berger, D. K., and Kustu, S. (1993). Activity of purified NIFA, a transcriptional activator of nitrogen fixation genes. *Proc. Natl. Acad. Sci. U.S.A., 90,* 2266-2270.

Leigh, J. A. (2000). Nitrogen fixation in methanogens - the archaeal perspective. In E. W. Triplett (Ed.), *Prokaryotic nitrogen fixation: A model system for analysis of a biological process* (pp. 657-669). Wymondham: Horizon Scientific Press.

Liang, J. H., Nielsen, G. M., Lies, D. P., Burris, R. H., Roberts, G. P., and Ludden, P. W. (1991). Mutations in the *draT* and *draG* genes of *Rhodospirillum rubrum* result in loss of regulation of nitrogenase by reversible ADP-ribosylation. *J. Bacteriol., 173,* 6903-6909.

Liang, Y. Y., Arsene, F., and Elmerich, C. (1993). Characterization of the *ntrBC* genes of *Azospirillum brasilense* Sp7: Their involvement in the regulation of nitrogenase synthesis and activity. *Mol. Gen. Genet., 240,* 188-196.

Liang, Y. Y., de Zamaroczy, M., Arsene, F., Paquelin, A., and Elmerich, C. (1992). Regulation of nitrogen fixation in *Azospirillum brasilense* Sp7: Involvement of *nifA, glnA* and *glnB* gene products. *FEMS Microbiol. Lett., 79,* 113-119.

Little, R., Colombo, V., Leech, A., and Dixon, R. (2002). Direct interaction of the NifL regulatory protein with the GlnK signal transducer enables the *Azotobacter vinelandii* NifL-NifA regulatory system to respond to conditions replete for nitrogen. *J. Biol. Chem., 277,* 15472-15481.

Little, R., Reyes-Ramirez, F., Zhang, Y., van Heeswijk, W. C., and Dixon, R. (2000). Signal transduction to the *Azotobacter vinelandii* NIFL-NIFA regulatory system is influenced directly by interaction with 2-oxoglutarate and the PII regulatory protein. *EMBO J., 19,* 6041-6050.

Liu, J. and Magasanik, B. (1995). Activation of the dephosphorylation of nitrogen regulator I-phosphate of *Escherichia coli*. *J. Bacteriol., 177,* 926-931.

Lobo, A. L. and Zinder, S. H. (1990). Nitrogenase in the archaebacterium *Methanosarcina barkeri* 227. *J. Bacteriol., 172,* 6789-6796.

Lowery, R. G., Saari, L. L., and Ludden, P. W. (1986). Reversible regulation of the nitrogenase iron protein from *Rhodospirillum rubrum* by ADP-ribosylation *in vitro*. *J. Bacteriol., 166,* 513-518.

Luque, I., Flores, E., and Herrero, A. (1994). Molecular mechanism for the operation of nitrogen control in cyanobacteria. *EMBO J., 13,* 2862-2869.

Martin, D. E., Hurek, T., and Reinhold-Hurek, B. (2000). Occurrence of three PII-like signal transmitter proteins in the diazotrophic proteobacterium *Azoarcus* sp. BH72. *Mol. Microbiol., 38,* 276-288.

Martin, D. E. and Reinhold-Hurek, B. (2002). Distinct roles of P_{II}-like signal transmitter proteins and *amtB* in regulation of *nif* gene expression, nitrogenase activity, and posttranslational modification of NifH in *Azoarcus* sp. strain BH72. *J. Bacteriol., 184,* 2251-2259.

Martinez-Argudo, I. and Contreras, A. (2002). PII T-Loop mutations affecting signal transduction to NtrB also abolish yeast two-hybrid interactions. *J. Bacteriol., 184,* 3746-3748.

Masepohl, B., Drepper, T., Paschen, A., Gross, S., Pawlowski, A., Raabe, K. *et al.* (2002). Regulation of nitrogen fixation in the phototrophic purple bacterium *Rhodobacter capsulatus*. *J. Mol. Microbiol. Biotechnol., 4,* 243-248.

Masepohl, B. and Klipp, W. (1996). Organisation and regulation of genes encoding the molybdenum nitrogenase and the alternative nitrogenase in *Rhodobacter capsulatus*. *Arch. Microbiol., 165,* 80-90.

Masepohl, B., Klipp, W., and Pühler, A. (1988). Genetic characterization and sequence analysis of the duplicated *nifA*/*nifB* gene region of *Rhodobacter capsulatus*. *Mol. Gen. Genet., 212,* 27-37.

Masepohl, B., Krey, R., and Klipp, W. (1993). The *draTG* gene region of *Rhodobacter capsulatus* is required for post-translational regulation of both the molybdenum and the alternative nitrogenase. *J. Gen. Microbiol., 139,* 2667-2675.

Meletzus, D., Rudnick, P., Doetsch, N., Green, A., and Kennedy, C. (1998). Characterization of the *glnK-amtB* operon of *Azotobacter vinelandii*. *J. Bacteriol., 180,* 3260-3264.

Merrick, M., Hill, S., Hennecke, H., Hahn, M., Dixon, R., and Kennedy, C. (1982). Repressor properties of the *nifL* gene product of *Klebsiella pneumoniae*. *Mol. Gen. Genet., 185,* 75-81.

Merrick, M. J. (1983). Nitrogen control of the *nif* regulon in *Klebsiella pneumoniae*: Involvement of the *ntrA* gene and analogies between *ntrC* and *nifA*. *EMBO J., 2,* 39-44.

Merrick, M. J. (1993). In a class of its own - the RNA polymerase sigma factor σ^{54} (σ^N). *Mol. Microbiol., 10,* 903-909.

Merrick, M. J. and Edwards, R. A. (1995). Nitrogen control in bacteria. *Microbiol.Revs., 59,* 604-622.

Mettke, I., Fiedler, U., and Weiss, V. (1995). Mechanism of activation of a response regulator: interaction of NtrC-P dimers induces ATPase activity. *J. Bacteriol., 177,* 5056-5061.

Monteiro, R. A., Souza, E. M., Funayama, S., Yates, M. G., Pedrosa, F. O., and Chubatsu, L. S. (1999a). Expression and functional analysis of an N-truncated NifA protein of *Herbaspirillum seropedicae*. *FEBS Lett., 447,* 283-286.

Monteiro, R. A., Souza, E. M., Yates, M. G., Pedrosa, F. O., and Chubatsu, L. S. (1999b). In-*trans* regulation of the N-truncated-NIFA protein of *Herbaspirillum seropedicae* by the N-terminal domain. *FEMS Microbiol. Lett., 180,* 157-161.

Muro-Pastor, A. M., Valladares, A., Flores, E., and Herrero, A. (1999). The *hetC* gene is a direct target of the NtcA transcriptional regulator in cyanobacterial heterocyst development. *J. Bacteriol., 181,* 6664-6669.

Neuwald, A. F., Aravind, L., Spouge, J. L., and Koonin, E. V. (1999). AAA+: A class of chaperone-like ATPases associated with the assembly, operation, and disassembly of protein complexes. *Genome Res., 9,* 27-43.

Ninfa, A. J. and Bennett, R. L. (1991). Identification of the site of autophosphorylation of the bacterial protein kinase/phosphatase NR_{II}. *J. Biol. Chem., 266,* 6888-6893.

Noren, A. and Nordlund, S. (1994). Changes in the NAD(P)H concentration caused by addition of nitrogenase 'switch-off' effectors in *Rhodospirillum rubrum* G-9, as measured by fluorescence. *FEBS Lett., 356,* 43-45.

Noren, A., Soliman, A., and Nordlund, S. (1997). The role of NAD^+ as a signal during nitrogenase switch-off in *Rhodospirillum rubrum*. *Biochem. J., 322,* 829-832.

O'Gara, J. P., Eraso, J. M., and Kaplan, S. (1998). A redox-responsive pathway for aerobic regulation of photosynthesis gene expression in *Rhodobacter sphaeroides* 2.4.1. *J. Bacteriol., 180,* 4044-4050.

Osuna, J., Soberon, X., and Morett, E. (1997). A proposed architecture for the central domain of the bacterial enhancer-binding proteins based on secondary structure prediction and fold recognition. *Protein Sci., 6,* 543-555.

Paschen, A., Drepper, T., Masepohl, B., and Klipp, W. (2001). *Rhodobacter capsulatus nifA* mutants mediating *nif* gene expression in the presence of ammonium. *FEMS Microbiol. Lett., 200,* 207-213.

Pedrosa, F. O. and Yates, M. G. (1984). Regulation of nitrogen fixation (*nif*) genes of *Azospirillum brasilense* by *nifA* and *ntr* (*gln*) type gene products. *FEMS Microbiol. Lett., 23,* 95-101.

Persuhn, D. C., Souza, E. M., Steffens, M. B., Pedrosa, F. O., Yates, M. G., and Rigo, L. U. (2000). The transcriptional activator NtrC controls the expression and activity of glutamine synthetase in *Herbaspirillum seropedicae*. *FEMS Microbiol. Lett., 192,* 217-221.

Pierrard, J., Ludden, P. W., and Roberts, G. P. (1993). Posttranslational regulation of nitrogenase in *Rhodobacter capsulatus*: Existence of two independent regulatory effects of ammonium. *J. Bacteriol., 175,* 1358-1366.

Pioszak, A. A., Jiang, P., and Ninfa, A. J. (2000). The *Escherichia coli* PII signal transduction protein regulates the activities of the two-component system transmitter protein NRII by direct interaction with the kinase domain of the transmitter module. *Biochemistry, 39,* 13450-13461.

Pope, M. R., Murrell, S. A., and Ludden, P. W. (1985). Covalent modification of the iron protein of nitrogenase from *Rhodospirillum rubrum* by adenosine diphosphorylation of a specific arginyl residue. *Proc. Natl. Acad. Sci. U.S.A., 82,* 3173-3177.

Possot, O., Sibold, L., and Aubert, J. P. (1989). Nucleotide sequence and expression of the glutamine synthetase structural gene, *glnA*, of the archaebacterium *Methanococcus voltae*. *Res. Microbiol., 140,* 355-371.

Premakumar, R., Loveless, T. M., and Bishop, P. E. (1994). Effect of amino acid substitutions in a potential metal-binding site of AnfA on expression from the *anfH* promoter in *Azotobacter vinelandii*. *J. Bacteriol., 176,* 6139-6142.

Premakumar, R., Pau, R. N., Mitchenall, L. A., Easo, M., and Bishop, P. E. (1998). Regulation of the transcriptional activators AnfA and VnfA by metals and ammonium in *Azotobacter vinelandii*. *FEMS Microbiol. Lett., 164,* 63-68.

Qian, Y. and Tabita, F. R. (1996). A global signal transduction system regulates aerobic and anaerobic CO_2 fixation in *Rhodobacter sphaeroides*. *J. Bacteriol., 178,* 12-18.

Qian, Y. and Tabita, F. R. (1998). Expression of *glnB* and a *glnB*-like gene (*glnK*) in a ribulose bisphosphate carboxylase/oxygenase-deficient mutant of *Rhodobacter sphaeroides*. *J. Bacteriol., 180,* 4644-4649.

Ramasubramanian, T. S., Wei, T. F., and Golden, J. W. (1994). Two *Anabaena* sp. strain PCC 7120 DNA-binding factors interact with vegetative cell- and heterocyst-specific genes. *J. Bacteriol., 176,* 1214-1223.

Reyes-Ramirez, F., Little, R., and Dixon, R. (2001). Role of *Escherichia coli* nitrogen regulatory genes in the nitrogen response of the *Azotobacter vinelandii* NifL-NifA complex. *J. Bacteriol., 183,* 3076-3082.

Rudnick, P., Kunz, C., Gunatilaka, M. K., Hines, E. R., and Kennedy, C. (2002). Role of GlnK in NifL-mediated regulation of NifA activity in *Azotobacter vinelandii*. *J. Bacteriol., 184,* 812-820.

Saari, L. L., Pope, M. R., Murrell, S. A., and Ludden, P. W. (1986). Studies on the activating enzyme for iron protein of nitrogenase from *Rhodospirillum rubrum*. *J. Biol. Chem., 261,* 4973-4977.

Schick, H. J. (1971). Substrate and light dependent fixation of molecular nitrogen in *Rhodospirillum rubrum*. *Arch. Mikrobiol., 75,* 89-101.

Schmitz, R. A. (1998). NifL of *Klebsiella pneumoniae* carries an N-terminally bound FAD cofactor, which is not directly required for the inhibitory function of NifL. *FEMS Microbiol. Lett., 157,* 313-318.

Sganga, M. W. and Bauer, C. E. (1992). Regulatory factors controlling photosynthetic reaction center and light-harvesting gene expression in *Rhodobacter capsulatus*. *Cell, 68,* 945-954.

Sibold, L., Henriquet, M., Possot, O., and Aubert, J.-P. (1991). Nucleotide sequence of *nifH* regions from *Methanobacterium ivanovii* and *Methanosarcina barkeri* 227 and characterisation of *glnB*-like genes. *Res. Microbiol., 142,* 5-12.

Sidoti, C., Harwood, G., Ackerman, R., Coppard, J., and Merrick, M. (1993). Characterisation of mutations in the *Klebsiella pneumoniae* nitrogen fixation regulatory gene *nifL* which impair oxygen regulation. *Arch. Microbiol., 159,* 276-281.

Söderbäck, E., Reyes-Ramirez, F., Eydmann, T., Austin, S., Hill, S., and Dixon, R. (1998). The redox- and fixed nitrogen-responsive regulatory protein NIFL from *Azotobacter vinelandii* comprises discrete flavin and nucleotide-binding domains. *Mol. Microbiol., 28,* 179-192.

Soliman, A. and Nordlund, S. (1992). Studies on the effect of NAD(H) on nitrogenase activity in *Rhodospirillum rubrum. Arch. Microbiol., 157,* 431-435.

Souillard, N., Magot, M., Possot, O., and Sibold, L. (1988). Nucleotide sequence of regions homologous to *nifH* (nitrogenase Fe protein) from the nitrogen-fixing Archaebacteria *Methanococcus_thermolithotropus* and *Methanobacterium ivanovii.* Evolutionary implications. *J. Mol. Evol., 27,* 65-76.

Souillard, N. and Sibold, L. (1989). Primary structure, functional organization and expression of nitrogenase structural genes of the thermophilic archaebacterium *Methanococcus thermolithotrophicus. Mol. Microbiol., 3,* 541-551.

Souza, E. M., Pedrosa, F. O., Drummond, M., Rigo, L. U., and Yates, M. G. (1999). Control of *Herbaspirillum seropedicae* NifA activity by ammonium ions and oxygen. *J. Bacteriol., 181,* 681-684.

Stock, J. B., Ninfa, A. J., and Stock, A. M. (1989). Protein phosphorylation and regulation of adaptive responses in bacteria. *Microbiol. Revs., 53,* 450-490.

Tanigawa, R., Shirokane, M., Maeda, S. S., Omata, T., Tanaka, K., and Takahashi, H. (2002). Transcriptional activation of NtcA-dependent promoters of *Synechococcus* sp. PCC 7942 by 2-oxoglutarate *in vitro. Proc. Natl. Acad. Sci. U.S.A., 99,* 4251-4255.

Taylor, B. L. and Zhulin, I. B. (1999). PAS domains: internal sensors of oxygen, redox potential, and light. *Microbiol. Mol. Biol. Rev., 63,* 479-506.

Thiel, T., Lyons, E. M., and Erker, J. C. (1997). Characterization of genes for a second Mo-dependent nitrogenase in the cyanobacterium *Anabaena variabilis. J. Bacteriol., 179,* 5222-5225.

Thiel, T., Meeks, J. C., Elhai, J., Potts, M., Larimer, F., Lamerdin, J. *et al.* (2002). Nitrogen fixation: Analysis of the genome of the cyanobacterium *Nostoc punctiforme.* In T. Finan, M. R. O'Brian, D. B. Layzell, J. K. Vessey, and W. E. Newton (Eds.), *Nitrogen fixation: Global perspectives.* (pp. 88-91). Wallingford, UK: CABI Publishing.

Thomas, G., Coutts, G., and Merrick, M. (2000). The *glnKamtB* operon: A conserved gene pair in prokaryotes. *Trends Genet., 16,* 11-14.

Toukdarian, A. and Kennedy, C. (1986). Regulation of nitrogen metabolism in *Azotobacter vinelandii*: isolation of *ntr* and *glnA* genes and construction of *ntr* mutants. *EMBO J., 5,* 399-407.

Toukdarian, A., Saunders, G., Selman-Sosa, G., Santero, E., Woodley, P., and Kennedy, C. (1990). Molecular analysis of the *Azotobacter vinelandii glnA* gene encoding glutamine synthetase. *J. Bacteriol., 172,* 6529-6539.

van Dommelen, A., Keijers, V., Somers, E., and Vanderleyden, J. (2002). Cloning and characterisation of the *Azospirillum brasilense glnD* gene and analysis of a *glnD* mutant. *Mol. Genet. Genomics, 266,* 813-820.

van Heeswijk, W. C., Hoving, S., Molenaar, D., Stegeman, B., Kahn, D., and Westerhoff, H. V. (1996). An alternative P_{II} protein in the regulation of glutamine synthetase in *Escherichia coli. Mol. Microbiol., 21,* 133-146.

Vazquez-Bermudez, M. F., Herrero, A., and Flores, E. (2002). 2-Oxoglutarate increases the binding affinity of the NtcA (nitrogen control) transcription factor for the *Synechococcus glnA* promoter. *FEBS Lett., 512,* 71-74.

Volkman, B. F., Nohaile, M. J., Amy, N. K., Kustu, S., and Wemmer, D. E. (1995). 3-dimensional solution stucture of the N-terminal receiver domain of NtrC. *Biochemistry, 34,* 1413-1424.

Wang, S.-Z., Chen, J.-S., and Johnson, J. L. (1988). The presence of five *nifH*-like sequences in *Clostridium pasteurianum*: Sequence divergence and transcription properties. *Nucleic Acids Res., 16,* 439-454.

Wei, T. F., Ramasubramanian, T. S., and Golden, J. W. (1994). *Anabaena* sp. strain PCC 7120 *ntcA* gene required for growth on nitrate and heterocyst development. *J. Bacteriol., 176,* 4473-4482.

Weiss, D. S., Batut, J., Klose, K. E., Keener, J., and Kustu, S. (1991). The phosphorylated form of the enhancer-binding protein NTRC has an ATPase activity that is essential for activation of transcription. *Cell, 67,* 155-167.

Weiss, V., Claverie-Martin, F., and Magasanik, B. (1992). Phosphorylation of nitrogen regulator I of *Escherichia coli* induces strong cooperative binding to DNA essential for activation of transcription. *Proc. Natl. Acad. Sci. U.S.A., 89,* 5088-5092.

Woodley, P. and Drummond, M. (1994). Redundancy of the conserved His residue in *Azotobacter vinelandii* NifL, a histidine autokinase homologue which regulates transcription of nitrogen fixation genes. *Mol. Microbiol., 13,* 619-626.

Wray, L. V., Zalieckas, J. M., and Fisher, S. H. (2000). Purification and *in vitro* activities of the *Bacillus subtilis* TnrA transcription factor. *J. Mol. Biol., 300,* 29-40.

Wyman, C., Rombel, I., North, A. K., Bustamante, C., and Kustu, S. (1997). Unusual oligomerisation required for activity of NtrC, a bacterial enhancer-binding protein. *Science, 275,* 1658-1661.

Xu, Y., Cheah, E., Carr, P. D., van Heeswijk, W. C., Westerhoff, H. V., Vasudevan, S. G. *et al.* (1998). GlnK, a PII-homologue: Structure reveals ATP binding site and indicates how the T-loops may be involved in molecular recognition. *J. Mol. Biol., 282,* 149-165.

Yakunin, A. F., Fedorov, A. S., Laurinavichene, T. V., Glaser, V. M., Egorov, N. S., Tsygankov, A. A. *et al.* (2001). Regulation of nitrogenase in the photosynthetic bacterium *Rhodobacter sphaeroides* containing *draTG* and *nifHDK* genes from *Rhodobacter capsulatus*. *Can. J. Microbiol., 47,* 206-212.

Yakunin, A. F. and Hallenbeck, P. C. (1998). Short-term regulation of nitrogenase activity by NH_4^+ in *Rhodobacter capsulatus:* Multiple *in vivo* nitrogenase response to NH_4^+ addition. *J. Bacteriol., 180,* 6392-6395.

Yakunin, A. F. and Hallenbeck, P. C. (2000). AmtB is necessary for NH_4^+ induced nitrogenase switch-off and ADP-ribosylation in *Rhodobacter capsulatus*. In F. O. Pedrosa, M. Hungria, M. G. Yates, and W. E. Newton (Eds.), *Nitrogen fixation: From molecules to crop productivity* (pp. 95-96). Dordrecht: Kluwer Academic Publishers.

Yakunin, A. F. and Hallenbeck, P. C. (2002). AmtB is necessary for NH_4^+-induced nitrogenase switch-off and ADP-ribosylation in *Rhodobacter capsulatus*. *J. Bacteriol., 184,* 4081-4088.

Zhang, Y., Burris, R. H., Ludden, P. W., and Roberts, G. P. (1993). Posttranslational regulation of nitrogenase activity by anaerobiosis and ammonium in *Azospirillum brasilense*. *J. Bacteriol., 175,* 6781-6788.

Zhang, Y., Burris, R. H., Ludden, P. W., and Roberts, G. P. (1996). Presence of a second mechanism for the posttranslational regulation of nitrogenase activity in *Azospirillum brasilense* in response to ammonium. *J. Bacteriol., 178,* 2948-2953.

Zhang, Y., Burris, R. H., and Roberts, G. P. (1992). Cloning, sequencing, mutagenesis, and functional characterization of *draT* and *draG* genes from *Azospirillum brasilense*. *J. Bacteriol., 174,* 3364-3369.

Zhang, Y., Cummings, A. D., Burris, R. H., Ludden, P. W., and Roberts, G. P. (1995). Effect of an *ntrBC* mutation on the posttranslational regulation of nitrogenase activity in *Rhodospirillum rubrum*. *J. Bacteriol., 177,* 5322-5326.

Zhang, Y., Pohlmann, E. L., Ludden, P. W., and Roberts, G. P. (2000). Mutagenesis and functional characterization of the *glnB*, *glnA*, and *nifA* genes from the photosynthetic bacterium *Rhodospirillum rubrum*. *J. Bacteriol., 182,* 983-992.

Zhang, Y., Pohlmann, E. L., Ludden, P. W., and Roberts, G. P. (2001). Functional characterization of three GlnB homologs in the photosynthetic bacterium *Rhodospirillum rubrum*: Roles in sensing ammonium and energy status. *J. Bacteriol., 183,* 6159-6168.

Zhulin, I. B., Taylor, B. L., and Dixon, R. (1997). PAS domain S-boxes in archaea, bacteria and sensors for oxygen and redox. *Trends Biochem. Sci., 22,* 331-333.

Chapter 10

MOLYBDENUM UPTAKE AND HOMEOSTASIS

R. N. PAU
Department of Biochemistry and Molecular Biology,
The University of Melbourne, Parkville 3010, Australia

1. MOLYBDENUM OUTSIDE CELLS

1.1. Bio-availability

The chemical properties of molybdenum are harnessed for life by incorporating it into two distinct enzyme catalytic centers, the unique nitrogenase cofactor and versatile molybdenum cofactor (Hille, 1996; Stiefel, 1993). This chapter discusses how cells take up molybdenum, and the proteins involved in molybdenum homeostasis before it enters the pathways for cofactor biosynthesis.

Molybdenum is not a very abundant metal on earth, ranking 53rd in abundance among elements, it comprises about 1.5 mg kg^{-1} in the earth's crust. In rocks, it is present in molybdenum minerals, such as molybdenite (MoS_2), powellite ($CaMoO_4$), and wulfenite ($PbMoO_4$). It is, however, readily available for living cells as the soluble tetraoxo anion, molybdate (MoO_4^{2-}), one of several oxo species produced by the hydrolysis of molybdenum at neutral pH.

$$MoO_4^{2-} \underset{K=10^{-6.5}}{\rightleftharpoons} MoO_3^{+} \underset{K=10^{-12.9}}{\rightleftharpoons} MoO_2^{2+}$$

Equilibria between the hydrolytic products of molybdenum shown above favour molybdate over the other species, so that the concentration of molybdate is greater than that of the cation MoO_2^{2+} by a factor of 10^{19} (Frausto da Silva and Williams, 2001; Pope *et al.*, 1980). Most metal oxides, hydroxides, and sulfides have low solubilities. In contrast, at a concentration of about 110 nM, molybdenum, as

W. Klipp, B. Masepohl, J. R. Gallon and W. E. Newton (eds.), Genetics and Regulation of Nitrogen Fixation in Free-Living Bacteria, 225-256.

molybdate, is the most abundant transition metal in the sea. Here, the concentration of sulfate (2.7 mM), a very similar oxo anion to molybdate, is very much higher. This situation suggested that the low productivity of the sea for nitrogen-fixing cyanobacteria is due to competition of sulfate with molybdate (Howarth and Cole, 1985). However, it is now clear that bacteria can evolve transport systems which have both high selectivity and affinity for molybdate. Thus, high concentrations of sulfate do not affect nitrogenase-dependent growth of the free-living aerobic nitrogen-fixing soil bacterium *Azotobacter vinelandii*, which has a highly selective high-affinity molybdate uptake system. In a broader environmental study, molybdenum limitation showed no effect on estuarine cyanobacterial growth (Paulsen *et al.*, 1991).

On land, the distribution of molybdenum is very uneven, ranging from 0.8-3.9 mg kg^{-1} in alkaline soils. Its concentration in fresh water varies from 2-100 nM. Many factors, such as binding to clays or iron oxides, affect its availability (Kletzin and Adams, 1996). Biological factors may deplete its concentration locally, as in microbial mats and plant rhizospheres. Furthermore, the formation of polyoxo-molybdates decreases the availability of molybdate in acidic soils, making soil pH a major factor affecting molybdenum availability. Because molybdate availability affects agricultural productivity, information about the molybdenum content of soils has proved important (Gupta, 1997). For example, remediation of molybdenum deficiency in Australia made it possible to develop extensive land previously neglected because of the failure of legume pastures (Brennan and Bruce, 1999).

1.2. Competing Metals and Organic Ligands in the Environment

1.2.1. Tungsten and Vanadium

The chemistry of molybdenum, tungsten, and vanadium are very similar (Frausto da Silva and Williams, 2001; Pope et al., 1980). The higher oxidation states of all three metals have roles in biology, although only molybdenum has a widespread role. In contrast to molybdenum, whose biological role was recognized since the 1930's, the role of tungsten has only recently been established in prokaryotes but, even so, it is more widespread than originally thought (Kletzin and Adams, 1996). In tungstoenzymes, tungsten replaces molybdenum bound to pterin cofactors. The widespread use of molybdenum in living organisms is favoured both by its availability and by the more positive reduction potentials of molybdenum compared to tungsten complexes (Frausto da Silva and Williams, 2001). Although molybdenum and tungsten are present in about equal concentrations in rocks, the concentration of the oxo anion tungstate in soil water is several orders of magnitude lower than that of molybdate, except in areas with tungsten-rich minerals and special environments, such as hot-spring waters.

Most microbes cannot distinguish tungstate from molybdate, and substitution of one for the other usually affects enzymic activity. When the concentration of tungstate is high, it inhibits nitrogenase activity and consequently the growth of diazotrophic bacteria (Lei *et al.*, 2000; Premakumar *et al.*, 1996; Siemann *et al.*,

2003). Tungstate similarly inactivates many enzymes with molybdopterin cofactors (Trautwein *et al.*, 1994). Conversely, substitution by molybdenum inactivates tungstoenzymes (Mukund and Adams, 1996). In a few cases, molybdenum replaces tungsten functionally and *vice versa*. For example, tungsten can replace molybdenum in the catalytic center of *Rhodobacter capsulatus* dimethylsufoxide reductase to give an active enzyme that accesses the same range of oxidation states as the molybdenum enzyme (Stewart *et al.*, 2000).

The concentration of tungstate in seawater is extremely low (10 pM), about four orders of magnitude lower than that of molybdate. However, tungstoenzymes predominantly occur in thermophilic and hyperthermophilic organisms in the specialised niche of oceanic hydrothermal vents (Kletzin and Adams, 1996; Zierenberg *et al.*, 2000). In this environment with high sulfide concentrations, the tungstate concentration of about 50 nM is more than two orders of magnitude higher than that of molybdate because of the relative insolubility of molybdenum sulfide. The selection of tungsten over molybdenum is related to its availability, and there is no need for selective uptake of tungstate against molybdate. However, molybdo- and tungsto-enzymes coexist in some bacteria, which must be able to discriminate between the two anions (Hochheimer *et al.*, 1998).

Although vanadium is more abundant than molybdenum on land, it is restricted to a few enzymes, the vanadium nitrogenase and haloperoxidases (Willsky, 1990). The oxo species of vanadium that are relevant to biological systems are $H_2VO_4^-$ and HVO_4^{2-}, which are similar to phosphate rather than to molybdate. This difference implies that vanadate does not compete with molybdate. It nevertheless affects many metabolic processes related to phosphate metabolism.

1.2.2. Iron

The availability of iron is severely limited under oxidising conditions at neutral pH when Fe(III) hydrolyses to highly insoluble ferric hydroxides and oxides. To obtain iron, many bacteria synthesise and secrete a variety of siderophores, which are highly efficient organic chelators that solubilize ferric hydroxides and oxides. A special feature of the chemistry of molybdate, which distinguishes it from sulfate and phosphate, is the ability of molybdenum oxo species to react with hydroxo ligands, such as hydroxycarboxylates and catecholates, to form stable complexes in which molybdenum has octahedral coordination geometry. The cationic species MoO_2^{2+} has four vacant coordination sites that are ideally suited to form complexes with catecholate siderophores. As the cation is in equilibrium with molybdate, addition of molybdate to catechols results in the formation highly stable soluble molybdenum complexes.

Siderophores have been extensively studied in the nitrogen-fixing bacterium *A. vinelandii*, which synthesizes three catecholamide siderophores (Corbin and Bulen, 1969; Cornish and Page, 1995; Page and von Tigerstrom, 1988). They are the bidentate mono(catecholamide) aminochelin (2,3-dihydroxybenzoyl putrescine), the tetradentate bis(catecholoamide) azotochelin (*N,N'*-bis-(2,3-dihyroxybenzoyl-L-lysine), and the hexadentate tris(catecholamide) protochelin (Figure 1). Protochelin is formally the condensation product of aminochelin and azotochelin.

Like most siderophores, protochelin is hexadentate, ideally suited to complexing the six-fold coordination sphere of iron. On the other hand, the tetradentate azotochelin matches the four vacant coordination sites of the cation MoO_2^{2+}. However, the azotochelin-molybdenum complex may be a dimer, similar to the molybdenum complex formed by the azotochelin analogue *N,N'*-bis(2,3-dihydroxybenzoyl)-1,5-diaminopentane in which two metal centers are bridged by two ligands, rather than a single molybdenum center coordinated to a single ligand (Duhme *et al.*, 1996). Because of its low denticity, the bidentate aminochelin is unusual for a siderophore, although it clearly functions as such (Page and von Tigerstrom, 1988).

Not surprisingly, the hexadentate protochelin is most effective at solubilizing iron, followed by the bidentate aminochelin, which forms a tris(catecholamide) iron complex. Azotochelin, the least efficient at solubilizing iron, first forms a tris(catecholamide) iron complex, followed by a bis(catecholamide)iron complex. The formation constants at neutral pH of the Fe(III) complexes of protochelin, azotochelin, and aminochelin are $10^{-44.6}$, $10^{-35.0}$, and $10^{-41.3}$ M^{-1}, respectively (Duhme *et al.*, 1997; 1998; Khodr *et al.*, 2002).

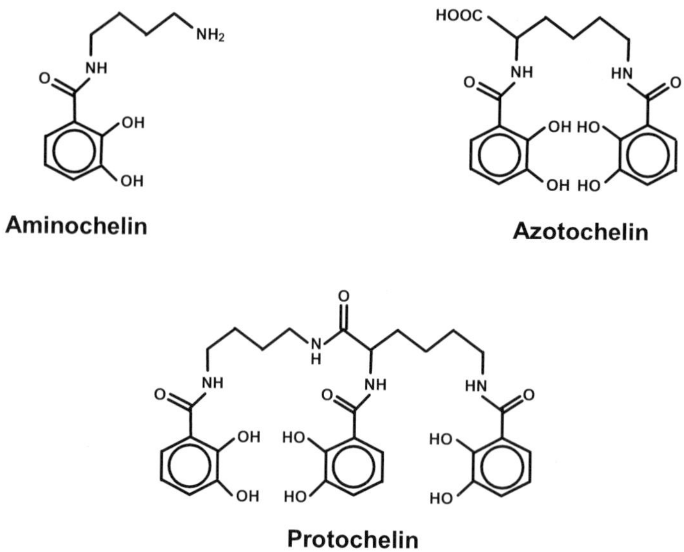

Figure 1. Catecholamide siderophores of Azotobacter vinelandii.

Molybdate hardly affects the solubilization of iron by protechelin, but it significantly delays the solubilization of iron by azotochelin and aminochelin (Khodr *et al.*, 2002). Again, this reactivity relates to the different denticities of the siderophores, protochelin being able to bind Fe(III) in a 1:1 stoichiometric fashion. In contrast, when azotochelin is added to an aqueous solution containing both iron and molybdate, the molybdenum-azotochelin complex forms instantly and very much faster than the thermodynamically more stable iron complex. Several weeks are required to reach equilibrium. Molybdate also delays the solubilization of iron by aminochelin, although equilbrium is reached in hours rather than days, making aminochelin a more efficient siderophore than azotochelin in the presence of molybdate.

Given the different efficiencies of the siderophores in solubilizing iron, it is of interest that molybdate concentration differentially regulates their synthesis in iron-limited medium. *A. vinelandii* secretes aminochelin and azotochelin at low molybdate concentrations and protochelin at high molybdate concentrations (Cornish and Page, 1995, 2000). In iron-limited cultures with concentrations of molybdate lower than 100 μM, *A. vinelandii* synthesizes about 100 μM azotochelin (Duhme *et al.*, 1998). All the molybdenum is, therefore, complexed under these conditions, while iron slowly solubilizes. In iron-limited cultures with greater than 100 μM molybdate, the synthesis of azotochelin and aminochelin ceases, and protochelin is produced instead. The molybdate-dependent synthesis of the hexadentate siderophore to the exclusion of other catecholamide siderophores is most likely an adaptation evolved to enable iron uptake at extremely high molybdate concentrations. It clearly shows the importance of competition between molybdate and iron for catecholate siderophores. The most efficient hexadentate siderophore, protochelin, is only produced when the concentration of molybdate precludes the formation of Fe(III) complexes with the lower denticity siderophores. The mechanism for this switch is not fully understood, but it may involve sensing the activity of cytoplasmic ferric iron reductase activity (Cornish and Page, 2000).

It is conceivable that *A. vinelandii* takes up the molybdenum-siderophore complexes. Specific porin channels transport siderophores through the outer membrane of Gram-negative cells. It is not known whether molybdenum-catecholamide complexes traverse the outer membrane of *A. vinelandii* through such channels, however, specific membrane proteins are repressed by iron and molybdenum in *A. vinelandii* (Page and von Tigerstrom, 1982). Iron and molybdenum siderophore complexes must release the metals. The dissociation constant for the molybdenum azotochelin complex is only 100 μM (Duhme *et al.*, 1998). Because the periplasmic binding protein binds molybdate in the nanomolar range and, therefore, has a greater affinity than the siderophore for molybdate, no special mechanism is necessary for it to acquire molybdate from the siderophore complex in the periplasm (Imperial *et al.*, 1998).

1.2.3. Homocitrate

Of the hydroxy carboxylate ligands that form molybdenum complexes, homocitrate is of greatest interest because it coordinates molybdenum in the molybdenum nitrogenase cofactor (Einsle *et al.*, 2002). Both *A. vinelandii* and *Klebsiella pneumoniae* secrete homocitrate when they express molybdenum-nitrogenase, and mutants that cannot synthesise homocitrate take it up from medium. However, it is not known whether a molybdenum-homocitrate complex is an important molybdenum species either outside the cell or in the cytoplasm (Hoover *et al.*, 1987; Madden *et al.*, 1991).

2. TRANSPORT

2.1. High-affinity Molybdate Transport

Many bacteria and Archaea actively take up molybdate when its concentration in the environment is low. *Escherichia coli* concentrates molybdenum more than 100-fold (Scott and Amy, 1989). The concentration of molybdate in the medium required to achieve the half-maximal rate of molybdenum uptake in *E. coli* and *K. pneumoniae* is 25-50 nM (Corcuera *et al.*, 1993; Imperial *et al.*, 1985). The cyanobacterium *Anabaena variabilis* accumulates molybdenum when its concentration is less than 1 mM (Thiel *et al.*, 2002). *A. vinelandii*, which is known for its high affinity for molybdate, was used to deplete traces of molybdate from minimal medium (Schneider *et al.*, 1991).

An ATP-binding cassette (ABC) transporter performs this high-affinity uptake. ABC transporters are arguably the most important molecular pumps that transport small solute molecules into and out of prokaryotic and eukaryotic cells. They consist of functionally distinct protein components that are usually separate proteins in prokaryotes. Specific substrate-binding proteins capture their cognate solute molecules and transfer them to a protein channel in the cell membrane. Transport through the membrane involves conformational changes, which are driven by ATP hydrolysis in an ATPase that is associated with the cytoplasmic side of a membrane-spanning protein. The names ModA, ModB and ModC are usually given, respectively, to the substrate-binding protein, the transmembrane protein, and the ATP-binding protein of the molybdate-transport system.

The genes encoding these three proteins of the molybdate transporter are usually found in the same operon, in the order *modA, modB, modC*. Some organisms show variations from this organisation. There are two substrate-binding protein orthologues in *Haemophilus influenzae*. In *Synechocystis* sp., *modB* and *modC* are fused. In *Clostridium pasteurianum*, the genes nifN-B and nifVΩ, which are required for nitrogenase function, are interspersed by putative *modA* and *modB* genes (Chen *et al.*, 2001). Genes similar to *modC* are present in the molybdenum-nitrogenase (*nif*) gene cluster of *A. vinelandii* and *C. pasteurianum* (Jacobson *et al.*, 1989; Wang *et al.*, 1990). The role of these additional *modC* orthologues is not known. In a few bacteria, such as *Bacillus subtilis*, *Caulobacter crescentus*, and *Deinococcus thermus*, the gene encoding the ATP-binding protein is not in the same operon as the substrate-binding and transmembrane proteins. *A. vinelandii*, with its high molybdenum-uptake capability, has duplicated the genes for the molybdate transporter. The molybdate-transport system has been characterized in *E. coli* (Maupin-Furlow *et al.*, 1995; Walkenhorst *et al.*, 1995), *A. vinelandii* (Mouncey *et al.*, 1995), *R. capsulatus* (Wang *et al.*, 1993), and *Staphylococcus carnosus* (Neubauer *et al.*, 1999).

An open reading frame, *modD*, is present immediately following the *modABC* genes in the molybdate-transport operons of *E. coli*, *H. influenzae*, and *R. capsulatus*. A mutation in *modD* had no effect on molybdenum transport in *E. coli* (Maupin-Furlow *et al.*, 1995). On the other hand, a strain of *R. capsulatus* with a mutation in *modD* required a four-fold higher molybdate concentration than the wild-type strain for the molybdate-dependent repression of the alternative iron-only nitrogenase (Wang *et al.*, 1993).

2.2. The First Protein Ligand for Molybdate

The protein ModA of the molybdate transporter is the first protein ligand for molybdate. It is present in the periplasm of Gram-negative bacteria. In Gram-positive bacteria, such as *Staphylococcus carnosa*, which have only a single membrane and no periplasmic region, ModA is attached to the outer side of the membrane by an N-terminal lipid anchor (Neubauer *et al.*, 1999). ModA enables the cell to select molybdate in the presence of competing anions. Only molybdate and tungstate bind to *E. coli* ModA (Rech *et al.*, 1996). The anions vanadate, sulfate, phosphate, arsenate, nitrate, chlorate, selenate, perchlorate, and permanganate do not bind. Tungstate, but neither vanadate nor sulfate, competes for radioactive molybdate transported into *A. vinelandii* and *An. variabilis* (Mouncey *et al.*, 1995; Thiel *et a*l., 2002). The dissociation constant of *E. coli* ModA for both molybdate and tungstate is 20 nM, in line with the dissociation constants of a wide range of periplasmic binding proteins (Imperial *et al.*, 1998).

2.3. The Structure of ModA

Substrate-binding proteins of ABC transporters range in size from 25 to 59 kDa. Despite showing similar tertiary structures, binding proteins for different substrates have low amino-acid sequence homology. At 25 kDa, ModA is one of the smallest periplasmic binding proteins. The molybdate-binding proteins from *A. vinelandii* and *E. coli* contain 232 and 233 residues, respectively. They consist of two similar globular domains separated by a deep cleft (Figure 2). Each domain consists of three layers ($\alpha/\beta/\alpha$). The mixed β-sheets each have five strands. Two strands connect the domains at the bottom of the cleft. Both the N and C-termini are located in domain I, and domain II arises as an insertion between strands 4 and 5 of domain I.

Molybdate and tungstate both bind in a completely dehydrated site at the interface between the two domains. Side-chain hydroxyls, a side-chain amide, and backbone amides donate hydrogen bonds to the anions. These arise from the turns at the N-termini of four α-helices, two from each domain, which converge on the anion. A predominance of backbone amide hydrogen bonds provides a rigid binding site. Charged residues are not involved in binding, and multiple local dipoles dissipate the charge on the anion. The five amino acids that donate seven hydrogen bonds to molybdate are not the same in *A. vinelandii* and *E. coli* (Figure 3). In *A. vinelandii*, backbone amides from Asn-10, Ser-37, Tyr-118, Val-147, an amide from the side-chain of Asn-10, and side-chain hydroxyls from Thr-9 and Ser-37 donate hydrogen bonds. In *E. coli*, five backbone amides, Ser-12, Ser-39, Ala-125, and Val-152, and side-chain hydroxyls from Ser-10, Ser-39, and Tyr-170 donate hydrogen bonds. An amino-acid sequence comparison shows that the binding proteins from *E. coli* and *A. vinelandii* belong to distinct phylogenetic clusters (Lawson *et al*., 1998).

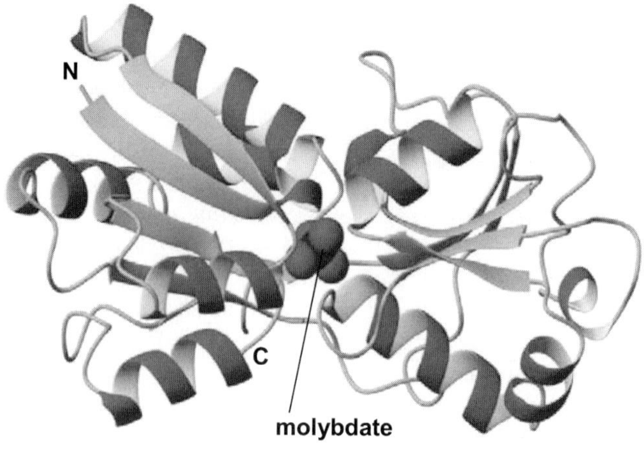

Figure 2. A ribbon representation of ModA from A. vinelandii *showing molybdate (or tungstate) as a CPK model bound in the cleft between the two domains. The Protein Data Bank entry is 1ATG. All the figures of protein models in this chapter were prepared with the program MolMol (Koradi et al., 1996).*

2.4. Binding Specificity

The affinity of the *E. coli* molybdate-binding protein for molybdate and tungstate is orders of magnitude greater than that for sulfate (Rech *et al*., 1996). Conversely, the sulfate-binding protein of *Salmonella typhimurium* has a much higher affinity for sulfate than molybdate (Pardee, 1966). Nevertheless, the structures of the molybdate- and sulfate-binding proteins are very similar (Pflugarth and Quiocho, 1985). The principal difference is that the molybdate-binding protein has about 50 fewer amino acids at its C-terminus. In the sulfate-binding protein, three instead of two strands form the hinge between the two domains, and the C-terminus is in domain II instead of domain I.

The binding sites of molybdate- and sulfate-binding proteins are very similar, although the residues involved in binding differ. In the sulfate-binding protein, seven hydrogen bonds donated by seven amino acids bind sulfate. Five are from the backbone amides of Asp-11, Ser-45, Gly-131, Ala-173, two from the backbone amide and the side-chain hydroxyl of Ser-130, and one from the imide from Trp-192. However, the volume of the binding site in the sulfate-binding protein (64 Å^3) is smaller than that of ModA (80 Å^3) (Hu *et al*., 1997; Lawson *et al*., 1998). Because the sulfur-oxygen bonds of sulfate are shorter (1.47 Å) than the molybdenum- and tungsten-oxygen bonds of molybdate and tungstate (1.77 Å and 1.78 Å, respectively), the differences in volume are significant and play an important role in anion selectivity. The volume of the smallest sphere that encloses sulfate is 99 $Å^3$ compared to 132 $Å^3$ for both molybdate and tungstate. These volumes are about 60% greater than the volumes of the binding pockets in their respective binding proteins, which indicates that there is a close hand-in-glove complementary fit in the shape and volume between the anions and their binding pockets.

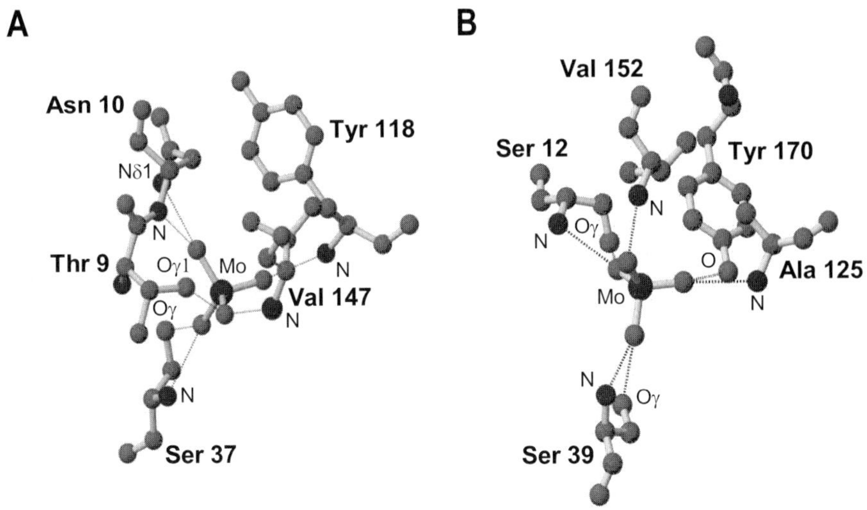

Figure 3. The molybdate/tungstate-binding sites in ModA from A. vinelandii *(A) and* E. coli *(B). The Protein Data Bank entries are 1ATG and 1WOD respectively. The nitrogen and oxygen atoms that donate hydrogen bonds to molybdate are labelled.*

Other factors, such as protein ligand type and electrostatic factors, may also affect the specificity (Hu *et al.*, 1997). More hydroxyl residues are involved in the molybdate- than in the sulfate-binding protein, which may relate to the affinity of molybdate and tungstate for polyhydroxyl compounds. The faces of the binding pocket in ModA are more hydrophobic than those of the sulfate-binding protein. Electrostatic considerations suggest that it is energetically more favourable to move a larger anion to an apolar environment than to a polar one (Hu *et al.*, 1997).

Vanadate resembles mono- and di-basic phosphate oxo anions in being protonated at neutral and alkaline pHs. In this respect, both vanadate and phosphate differ from molybdate, sulfate, and tungstate. The high specificity for phosphate of the periplasmic phosphate-binding protein depends on the presence of an aspartate residue in the binding site. This residue accepts a hydrogen bond from the proton of the phosphate anion and repels unprotonated sulfate anions (Luecke and Quiocho, 1990; Quiocho and Ledvinda, 1996).

The vanadium-nitrogenase of *A. vinelandii* is synthesised when vanadate is present and molybdate absent. Under these conditions, nanomolar concentrations of vanadate stimulate vanadium-nitrogenase synthesis, suggesting that there is an efficient transport system for vanadate (Robson *et al.*, 1986). A specific vanadate-transport system has not been identified, and vanadate may be transported by the high-affinity phosphate-transport system.

2.5. The Permease and ATP-binding Protein

In addition to the binding protein, ModA, the molybdate transporter consists of an integral membrane protein or permease, and its associated ATP-binding protein. Crystal structures for both the permease and ATP-binding protein of ABC transporters are available (Locher *et al.*, 2002). By homology with other ABC transporters, the permease and the ATP-binding proteins of the molybdate transporter are homodimers.

In *E. coli* and *A. vinelandii*, the monomer of the permease of the molybdate transporter, ModB, is a 24 kDa protein with five potential transmembrane helices. ModB from *R. capsulatus*, which has the same molecular mass, has six predicted transmembrane helices. The ATP-binding protein, ModC of *E. coli*, *A. vinelandii*, and *R. capsulatus*, ranges in size from 38-41.5 kDa. In common with the ATP-binding proteins of ABC transporters, ModC has a domain of about 200-240 amino acids that is highly conserved from bacteria to man. This domain belongs to the large AAA protein-domain family (ATPases Associated with diverse cellular Activities) that is involved in effecting molecular movements (Maurizi and Li, 2001). A domain C-terminal to the ATP-binding domain is the TOBE (Transport associated OB) domain. It is common to a subset of ABC transporters (see section 3.1.; Koonin *et al.*, 2000).

The gene *modF* is located adjacent to the operon that encodes the molybdate transporter of *E. coli* and several other bacteria (Grunden *et al.*, 1996; Walkenhorst *et al.*, 1995). In *E. coli* and *A. vinelandii*, modF is transcribed in the opposite direction from the operon encoding the molybdate transporter. It encodes a 54-kDa protein that has conserved ATPase domains that are similar to ModC. However, the domains are duplicated, presumably making ModF a monomer with duplicated domains as opposed to a homodimer. It also lacks the C-terminal TOBE domain that is present in ModC. The role of ModF is unknown because a mutant modF strain of *E. coli* showed no detectable phenotype, (Grunden and Shanmugam, 1997).

2.6. Specific High-affinity Tungstate Transport

As the high-affinity molybdate uptake system does not discriminate between molybdate and tungstate, uptake of these anions in organisms that have a high-affinity molybdate transporter must reflect their availability in the environment, and this situation usually favours molybdate. However, some organisms can distinguish between tungstate and molybdate. In the thermophilic methanogenic Archaea *Methanobacterium wolfei* and *M. thermoautotrophicum*, the enzyme formylmethanofuran dehydrogenase catalyzes the first step in methane formation from carbon dioxide. Both organisms synthesise distinct tungsten- and molybdenum-containing enzymes. Expression of the tungstoenzyme is constitutive. Molybdate induces expression of the molybdoenzyme, whereas tungstate does not (Hochheimer *et al.*, 1998).

Some organisms have a high-affinity ABC transport system that specifically transports tungstate in the presence of molybdate. In the Gram-positive anaerobe *Eubacterium acidaminophilum*, which contains at least two tungstoenzymes, an equimolar amount of molybdate does not affect the transport of tungstate. The genes *tupABC*, which encode a tungstate-specific ABC transporter, are located downstream from the genes for the tungsten-containing formate dehydrogenase (Makdessi *et al.*, 2001). The substrate-binding protein is a 30.9-kDa protein that belongs to a phylogenetic group distinct from the molybdate- and sulfate-binding proteins. A number of organisms have homologues of TupA. Tungstate binds to TupA with high affinity, but other anions, including sulfate and molybdate, do not. The apparent dissociation constant for tungstate binding determined by a gel-shift assay is ~0.5 μM. The basis for this extraordinary discrimination between tungstate and molybdate is not yet known, but it could depend on differences in pK between molybdate and tungstate, as the latter is more basic.

2.7. Low-affinity Transport

Early studies on molybdate transport showed that high concentrations of molybdate can overcome molybdenum deficiency in strains with mutations, which inactivate the high-affinity transport system (Glaser and DeMoss, 1971; Imperial *et al.*, 1985; Sperl and DeMoss, 1975). Consequently, molybdate must be able to enter cells by at least two pathways; a high-affinity system that functions at low molybdate concentrations, and a system or systems that operate at high concentrations. Wild-type cells use the low-affinity route, when high molybdate concentrations repress expression of the high-affinity transport system. Measurement of intracellular molybdate concentrations suggests that *E. coli* regulates both uptake systems. It concentrates molybdate when grown in low molybdate concentrations, and the intracellular concentration of molybdate is only 5-12 μM when grown in medium with greater than 100 μM molybdate (Scott and Amy, 1989).

The low-affinity uptake pathways have not been well characterised. One possible alternative route is by the high-affinity sulfate transporter. Despite its lower affinity for molybdate, the sulfate transporter is the most likely route for molybdate transport in organisms that do not have a high-affinity molybdate transporter. In contrast to *A. vinelandii*, sulfate is a competitive inhibitor of molybdate uptake in *C. pasteurianum* (Elliot and Mortenson, 1975). In *E. coli*, there is evidence that some molybdate may enter the cell *via* the sulfate transporter (Lee *et al.*, 1990; Rosentel *et al.*, 1995). In the presence of high concentrations of sulfate, which repress the sulfate-uptake system, higher concentrations of molybdate are required for activity of the molybdoenzyme, formate hydrogen lyase, in *modABC* mutants. In addition, a double mutant strain with defective sulfate and molybdate high-affinity uptake systems shows no activity in low sulfate medium, whereas a strain with a defective molybdate-uptake system and a functional sulfate-uptake system shows formate hydrogen lyase activity. It was estimated that the sulfate transporter is about ten-times less efficient at transporting molybdate than the molybdate transporter.

3. CYTOPLASMIC MOLYBDATE-BINDING PROTEINS

3.1. A Molybdate-binding Domain

Extensive computer analysis shows that cells use only about twenty distinct small domains (<100 amino acids) to bind small molecules (Anantharaman *et al.*, 2001). In some proteins, these binding regions occur either as stand-alone modules or as tandem repeats of the domains. In others, they join to unrelated protein domains. Proteins with small-molecule-binding domains are involved in transport, metabolism, and signal transduction. Early studies in *C. pasteurianum* showed that ^{99}Mo bound to a protein called Mop (Molybdenum Protein) before it was incorporated into formate dehydrogenase and nitrogenase (Elliot and Mortenson, 1975; Hinton and Mortenson, 1985a; 1985b; Hinton *et al.*, 1987). The monomer of this Mop protein has 68 amino acids. It subsequently proved to represent the small domain that binds molybdate (Lawson *et al.*, 1997). Because the initial characterisation of Mop suggested that it also bound pterin, current protein databases refer to Mop protein homologues as molybdopterin-binding proteins. All recent work shows that Mop domains bind molybdate and the early evidence for molybdopterin binding has not been substantiated. Mop domains have about 70 amino acids with highly conserved residues that are involved in both molybdate- and tungstate-binding (Figure 4).

Mop domains occur in three cytoplasmic proteins with distinct functions. Molbindins are proteins consisting solely of Mop domains. They are present as either 7-kDa domains or 14-kDa tandem repeats (biMop) (Hinton and Merritt, 1986; Mouncey *et al.*, 1995). BiMop domains are also found present in the molybdate-dependent regulatory protein, ModE, and a Mop domain occurs in the C-terminal domain of ModC, the ATP-binding protein of the molybdate transporter.

3.2. The Structure of Molbindins

The domains of molbindins and molybdate-binding domains of the regulatory protein ModE belong to a family of domains called the TOBE (transport-associated OB)-domain, which has a tandem repeat of OB-folds. OB (oligonucleotide/ oligosaccharide-binding)-folds are β-barrels of 5 strands arranged in a Greek-key motif (Koonin *et al.*, 2000; Murzin, 1993). The distinguishing feature of the TOBE-domains is that, unlike other OB-folds, two OB-folds interact so that the each

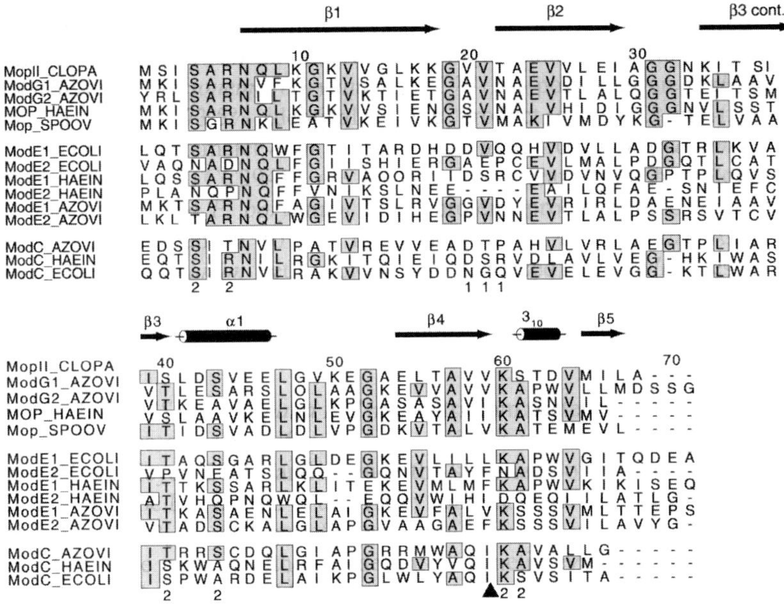

Figure 4. Alignment of the amino-acid sequences of the Mop domains from: the molbindins, Mop, MopII, and ModG; the regulatory protein, ModE; and the transporter ATPase, ModC. The bacteria are C. pasteurianum *(CLOPA),* A. vinelandii *(AZOVI),* Sporomusa ovata *(SPOVA),* E. coli *(ECOLI) and* H. influenzae *(HAEIN). The tandem domains in ModG and ModE are labelled ModG1 and ModG2, ModE1 and ModE2, respectively. The secondary structural elements of the first Mop domain of ModG are shown above the alignment. The solid triangle shows the position of the inter-domain boundary. The numbers 1 and 2 below the alignment show the positions of the amino acids in the type 1 and type 2 binding sites, respectively. All alignments in this chapter were formatted with ALSCRIPT (Barton, 1993).*

partner supplies the C-terminal strand of the other's β-barrel. Thus, in the biMop molbindin ModG, the C-terminal strand in each Mop domain exchanges, so that C-terminal strand of domain I becomes associated with the N-terminal strand of domain II and, reciprocally, the C-terminal strand of domain II is associated with domain I (Figure 5). The tandem Mop domains are joined by a short surface loop. The molbindins from *A. vinelandii*, *C. pasteurianum*, and *Sporomusa ovata* consist of three TOBE-domains (for detailed descriptions of the X-ray crystal structures see: Delarbre *et al.*, 2001; Schüttelkopf *et al.*, 2002; Wagner *et al.*, 2000). They are, therefore, either hexamers of Mop subunits or trimers of tandem biMop subunits. Most molbindins are hexameric. BiMop domains occur less frequently, as in *A. vinelandii* ModG and Mop from both *Streptomyces coelicolor* and *Chlorobium tepidum*. The β-barrels of the molybdate-binding proteins are capped by a short α-helix. They may have additional short strands or helices. For example, ModG has a short 3_{10} helix before strand 5, which is uncharacteristic of OB-folds. The equivalent of the 3_{10} helix in ModG is an α-helix in ModE.

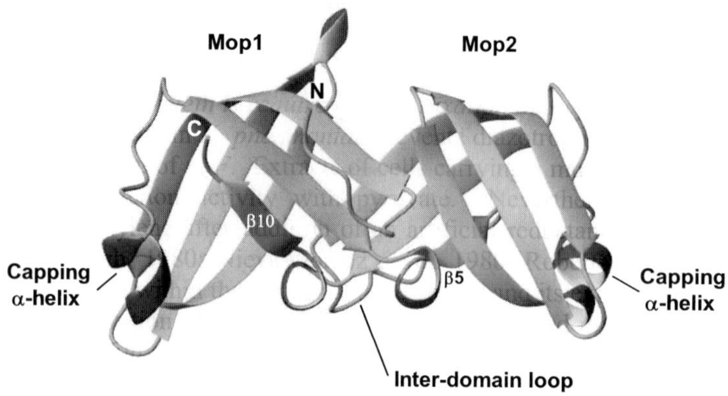

Figure 5. The structure of a single subunit of the molbindin ModG from A. vinelandii, *showing the interlocking Mop domains. The subdomains of the biMop TOBE domain are shown as Mop1 and Mop2. The strands that exchange between the two Mop domains are labelled β5 and β10.*

Molbindins are roughly spherical hollow molecules in which the Mop dimers or the biMop subunits are arranged with their long axes inclined around a 3-fold axis (Figure 6). The structures are extremely symmetrical, with Mop units that superimpose closely. The interaction between closely complementary inter-subunit interfaces results in the burial of about 40% of their accessible surfaces. They have eight potential molybdate-binding sites of two distinct types (Wagner *et al.*, 2000). Two sites, called type 1, are located on the 3-fold axis at the opposite poles of the spherical molecule. Six type 2 sites are disposed symmetrically at three inter-subunit interfaces between the equator and the poles, so there are three type 2 sites in each hemisphere.

In *A. vinelandii* ModG and *C. pasteurianum* MopII, either molybdate or tungstate is bound to the type 1 site by twelve hydrogen bonds, four from each subunit (Figure 7). The metal atom lies on the 3-fold axis and one of the anion oxygen atoms points inwards along the axis. Three hydrogen bonds, one from an asparagine side-chain amide (Asn-22) in each subunit in ModG, and from a threonine side-chain hydroxyl (Thr-22) in MopII, bind this axial oxygen atom. The three non-axial anion oxygen atoms point outward, each towards a different subunit. They each receive three hydrogen bonds from three consecutive amides in a turn between β-strands in a different subunit. Therefore, the three subunits donate nine hydrogen bonds to the three non-axial oxygen atoms. The residues involved are Ala-20, Val-21, and Asn-22 in ModG, and Val-20, Val-21 and Thr-22 in MopII.

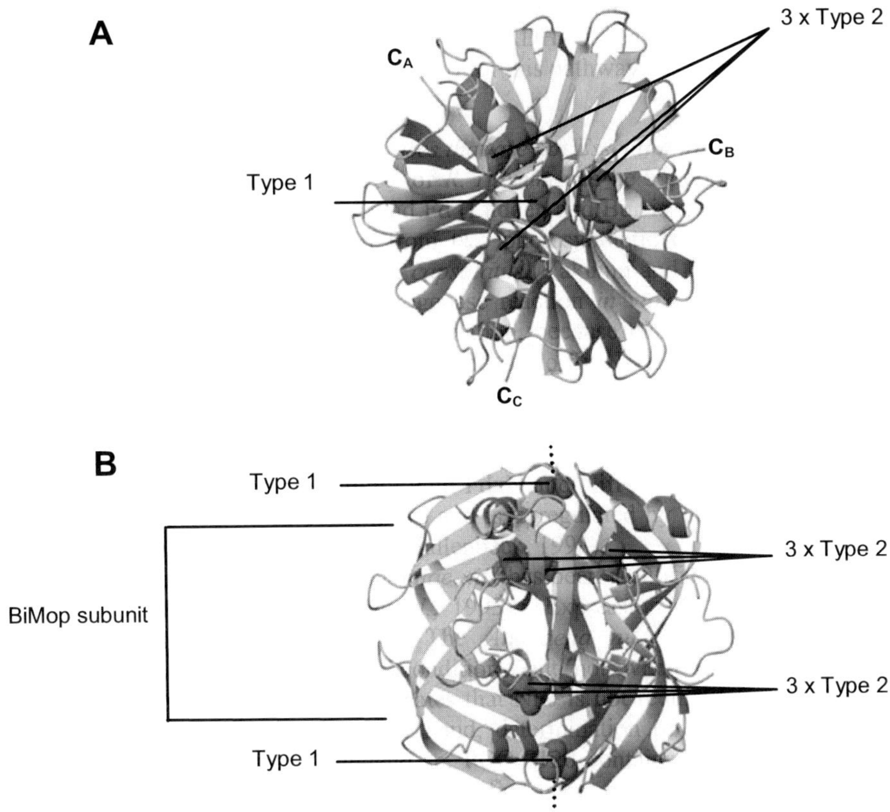

Figure 6. Fully loaded ModG protein from A. vinelandii *(PDB entry 1H9M) as viewed down the 3-fold axis (A) and at 90° to the 3-fold axis (B). The 3-fold axis is indicated by dotted lines in B. Molybdate in the type 1 and type 2 molybdate-binding sites are drawn as CPK models. The C-termini of the three subunits in A are labelled C_A, C_B and C_C.*

The type 1 site in *S. ovata* Mop is essentially the same as the MopII site, although in the published crystal form, it is only half occupied by a rotationally disordered anion. The axial oxygen atom is displaced from the 3-fold axis and points outwards instead of inwards. This orientation brings this oxygen atom within hydrogen-bonding distance of the main-chain amide of Thr-22 in one subunit. Backbone amides in three loops donate six more hydrogen bonds to three non-axial oxygen atoms. One crystal form of *C. pasteurianum* MopII also has a partially occupied type 1 site with the axial oxygen of the anion pointing outwards.

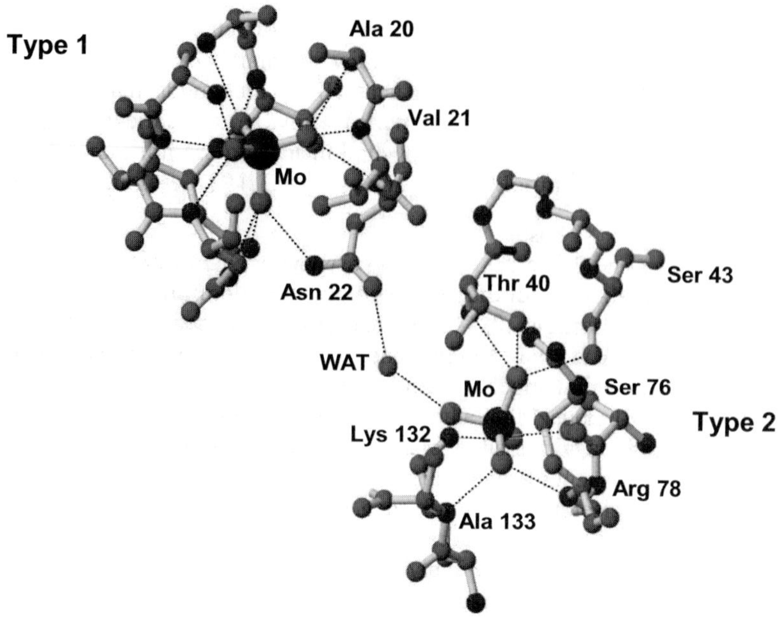

Figure 7. Type 1' and type 2' molybdate-binding sites in A. vinelandii *ModG with bound ligand. Site 1 is formed from equivalent loops from three subunits. The bonding amino acids of only one loop in the type 1 site are labelled. In the type 2 site, Ser-76, Arg-78, Lys-132 and Ala-133 are from one subunit and Thr-40 and Ser-43 from the other. A short hydrogen-bond network joins the anions bound in the two sites via a water molecule (WAT) and Asn-22.*

The type 2 sites bind either molybdate or tungstate asymmetrically between pairs of neighbouring subunits (Figure 7). The binding site opens to the interior of the molecule and is inaccessible to bulk solvent in the ligand-bound protein. Taking the type 2' binding site of ModG as representative (the apostrophes distinguish binding sites in one hemisphere from equivalent sites in the other), the protein donates eight hydrogen bonds to the anion. Two loops from one subunit donate five hydrogen bonds. The side-chain hydroxyl of Ser-76, backbone amides from Arg-78 and Ala-133, and the side-chain amine of Lys-132 donate hydrogen bonds from one subunit. Remarkably, the strictly conserved Arg-78 donates a hydrogen bond from its backbone amide rather than from its side-chain. Its charged side-chain may either play a role in the initial acquisition of the anion or have a role in the structural integrity of the site. Lys-132 is discretely disordered and interacts with two ligand oxygen atoms. Its charge partially compensates for the negative charge on the ligand. The adjacent subunit donates two hydrogen bonds from the side-chain hydroxyls of Thr-40 and Ser-43. An ordered water molecule donates a further hydrogen bond. In MopII, Ser-61 donates two hydrogen bonds from the side-chain hydroxyl and backbone amide. The gain in a hydrogen bond from Ser-61 compensates for the absence of the hydrogen bond donated by a water molecule as seen in ModG. Lys-60 donates a single hydrogen bond.

3.3. Specificity and Affinity

In the *H. influenzae* Mop protein, the type 1 site is more selective than the type 2 site. It binds only molybdate, in contrast to the type 2 site, which binds sulfate weakly as well as molybdate. The dissociation constants for sulfate and molybdate binding in site 2 are 1.6 mM and 117 nM, respectively (Masters, S., Howlett, G., Pau, R.N. unpublished data). ModG also shows this higher selectivity of the type 1 sites because, when molybdate-loaded ModG is crystallised from high concentrations of phosphate, diphosphate anions replace molybdate in one of the two distinct type 2 sites, but not in the type 1 site. As in the periplasmic binding protein ModA, the volume of the binding site plays an important role in selectivity. The type 1 site has roughly the minimal possible volume, 82-84 $Å^3$, which is close to 80 $Å^3$, the volume of the binding site of ModA. Further, hydrogen bonds from backbone amides predominate in the type 1 site, contributing to its inflexibility. No charged groups are involved. In contrast, binding site 2 has a relatively large volume 135 $Å^3$ and many side-chains are involved in ligand binding, including the charged amino group of Lys-132.

In *H. influenzae* Mop, the type I site has a lower affinity than the type 2 site for molybdate (the dissociation constants are 500 nM for site 1 and 117 nM for site 2). These values indicate that the type 2 site is occupied before the type 1 site. Clearly this protein binds molybdate tightly. Intriguingly, the relative affinities between the sites are reversed in *C. pasteurianum* MopII, and the affinities are lower, with dissociation constants of 1.6 μM and 48 μM for sites 1 and 2, respectively (Schüttelkopf *et al.*, 2002). These lower affinities suggest that the *C. pasteurianum* protein has a passive storage role, binding the ligand less strongly than proteins that use the ligand.

In fully loaded ModG, the ordered water molecule, which donates a hydrogen bond to an oxygen atom of molybdate bound in site 2, also donates a hydrogen bond to the carbonyl oxygen in the side-chain of Asn-22 (Figure 7). The latter, in turn, donates a hydrogen bond to the inwardly pointing axial oxygen atom of molybdate bound in the type 1 site. Thus, a short hydrogen-bond network (<10 Å) involving only three hydrogen bonds connects the anions bound in the two binding sites. Equivalent hydrogen-bond networks link the axial oxygen atom of molybdate bound in the type 1 site to molybdate molecules bound in the two other type 2 sites. These hydrogen-bond networks are absent from MopII because the Thr-22 side-chain can only donate a single hydrogen bond that binds molybdate in the type 1 site. Again, in *S. ovata*, the corresponding residue is modelled as Met, providing only a single hydrogen bond. The asparagine side-chain involved in the hydrogen-bond network is found in a minority of molbindins. Where present, the hydrogen-bond network may provide a means of cooperative binding. Binding of molybdate in site 2 would align the Asn-22 side-chain and alter the conformation of the site 1. *H. influenzae* Mop, which has an Asn residue and presumably forms a hydrogen-bond network, shows cooperative binding of molybdate (Masters, S., Howlett, G., and Pau, R.N., unpublished data). On the other hand, the kinetics of molybdate binding to *C. pasteurianum* MopII, which lacks the hydrogen-bond network, provides little evidence for cooperative binding (Schüttelkopf *et al.*, 2002).

The conformations of the bound and unbound forms of MopII are quite similar (Schüttelkopf *et al.*, 2002). Exceptions are the side-chain of Val-20 in site 1, which moves inwards, so limiting access to the site in the bound form. Similarly, in ligand-bound ModG, the C^{β}-atoms of the three Ala-20 residues interlock and close over the bound oxo anion. More noteworthy, the Lys-60 side-chain in the type 2 site of MopII adopts very different conformations in the apo and ligand-bound forms. In the apo form, it points away from the binding site and towards the centre of the cavity, an orientation that may affect the relative affinity of the sites.

Molbindins not only bind eight molybdate molecules, but also have a minimum of two sites with distinct binding affinities for molybdate. These different affinities allow binding over a broader range of concentrations than is possible in a protein with a single binding site. The concentration range of anion binding may be extended in other ways. There are three distinct Mop proteins, MopI, MopII, and MopIII, in *C. pasteurianum* (Hinton *et al.*, 1987). These proteins could potentially combine to produce hexameric Mop proteins with different constituent monomers and, thus, provide binding sites of different affinities. Again, molbindins, like ModG, have two distinct type 2 binding sites. The type 2 binding sites have different binding constants as shown by the crystal form with bound diphosphate (Delarbre *et al.*, 2001). The iron-storage protein, ferritin, provides a possible parallel to the molbindins. Its twenty-four subunits are a mixture of one-to-three similar subunits that confer a range of properties depending on their subunit composition (Boyd *et al.*, 1985).

The structures and binding kinetics of the molbindins are, therefore, consistent with a role as both a buffer and store for molybdate. ModG is repressed by high molybdate concentrations, which restricts its function to molybdate-limited conditions. The only observations on the phenotypes of strains with mutant molbindins comes from *A. vinelandii* (Mouncey *et al.*, 1995). A strain with a mutation in the *modG* gene had the wild-type phenotype for the activity of the molybdoenzyme, nitrate reductase, which is not surprising as we now know that *A. vinelandii* has two ModG homologues. A strain with mutations in both *modG* and *modE* showed nitrate-reductase activity at much lower concentrations than the wild-type, suggesting that binding of molybdate to ModE and ModG limits nitrate-reductase activity at low molybdate concentrations (Mouncey *et al.*, 1995). Further analysis is required because of the duplication of the *modG* and *modE* genes in *A. vinelandii.*

4. THE ATP-BINDING PROTEIN OF THE MOLYBDATE TRANSPORTER

A number of ABC transporter ATPases, including those for sugar, glycerol, iron, and molybdate transport, have a C-terminal TOBE-domain (Koonin *et al.*, 2000). However, although TOBE domains have two obligatory OB-folds, the ATPase of the molybdate transporter, ModC, only has a single Mop-like sequence (Pau *et al.*, 1997). The second predicted OB-fold, which is required to form the obligate dimer of the TOBE-domain, lacks the conserved molybdate-binding residues of the Mop sequence, and so is not evident without the use of position-specific score matrices and fold-recognition algorithms (Kelley *et al.*, 2000). Because molybdate-binding sites in molbindins are formed at domain interfaces, the single Mop domains of ModC may interact with each other in the ModC dimer (Koonin *et al.*, 2000).

In the absence of a structure for ModC, a homology model can be derived from the structural model of *E. coli* MalK, the ATP-binding protein of the sugar transporter. Each MalK monomer consists of a conserved ATP-binding domain and a C-terminal TOBE-domain (Bohm *et al.*, 2002). However, the TOBE-domains of the MalK dimer are on opposite sides of the molecule and do not interact with each other. If the homology model is correct, the TOBE domains of ModC do not interact with each other to form molybdate-binding sites as previously suggested. To form molybdate-binding sites, they must interact with another protein containing an appropriate TOBE domain. The most obvious protein for ModC to interact with is ModE, which has an open potentially interactive face in its C-terminal molybdate-binding domain that consists of two and not three TOBE domains as found in molbindins (see section 5.5). It is also conceivable that molbindin subunits may associate with both ModC and ModE.

The role of the TOBE domain with a Mop-like sequence at the C-terminus of ModC is yet not known but, again, MalK may provide an analogy. The C-terminal domain of MalK, called the regulatory domain, has an important regulatory function (Panagiotidis *et al.*, 1998). Regulation of maltose-dependent gene expression involves protein-protein interactions between MalK and the regulatory domain of another protein, MalT (Böhm *et al.*, 2002). As MalT also interacts with other regulatory proteins, it is at the center of a regulatory network. In this novel transport-controlled regulation of gene expression, the activity of the transporter rather than the intracellular ligand concentration plays a part in regulating gene expression. It is tempting to speculate that the Mop-like domain at the C-terminus of ModC may have a similar role.

5. THE REGULATORY PROTEIN, MODE

5.1. The modE *Genes and Molybdate-dependent DNA Binding*

Metals usually regulate the transcription of genes by feedback loops that involve metal-binding regulatory proteins. For molybdenum, the regulatory protein ModE either activates or represses transcription of genes involved in molybdate transport and metabolism in response to intracellular molybdate concentration. ModE is one of the few regulatory proteins found in both bacteria and Archaea (Figure 8). In Archaea, ModE is restricted to methanogenic organisms. In many organisms, such as *A. vinelandii*, the gene *modE* immediately precedes *modABC* and is the first gene in the molybdate-transport operon (Mouncey *et al.*, 1995). In *E. coli*, *modE* is located immediately upstream from the *modABCD* operon, and is transcribed in the opposite direction to *modABCD* (Grunden *et al.*, 1996; Walkenhorst *et al.*, 1995). Presumably, the expression of *modE* is autoregulated in *A. vinelandii*. In *E. coli*, ModE is expressed constitutively at a low level independent of molybdate concentration (Grunden *et al.*, 1996; Walkenhorst *et al.*, 1995).

A

```
BACTERIA                 GSISQGAKDAGISYKSAWDAINEMNQLS
E_coli                   GSISQGAKDAGISYKSAWDAINDMNQLS
S_typhimunium            GSINQAAKNAKVSYKSAWDHLEAMNKIS
H_infuenze               GSINRAAKVVPLSYKAAWDAIDTMNNLA
A_vinelandii             GSISAAGRAMDMSYRRAWLLVDALNHMF
A_tumelaciem             GSITRAAKAVGISYKGAWDAIDELNNLA
P_aeruginosa             GSIASAARKLGYSYKFIWTYIKKLEDVF
P_aerophilum             GTIAGAAREVGLSYKTAWDAVGTLNNIF
R_capsulatus

ARCHAEA
M_kandleri               GTVKGAAERLGWSYGYARRSIEALERAF
M_thermoautotrophicum    GSITVAARDAGIPYRSALAYIRNVEDIL
                             α3              α4
```

B

```
modA_ECOLI    TTCGTTATATTGTCGCCTACATAACGT
dmsA_ECOLI    TTCGATGTATACAAGCCTATATAGCGA
mosA_ECOLI    GACGCTATATACATGATTACATAGCGA
napF_ECOLI    ATCGCTATATAAATATATTTATAACGA
modA_HAEIN    TTCAATATATAATCAAGTAAATAACGT
modG_AZVIN    GACAATTTATAGTGCAAATGATATAGC
modE_AZVIN    TGCTTTATATAAACTGGATAAATAGAT
mopA_RCAPS    ATCGCTATTAGTCGGGTCTATATAACG
antA_RCAPS    GTCGTTATATGGAATCACTATATAACG
```

Figure 8. (A) Alignment of amino acids in the first DNA-recognition helices of the winged helix-turn-helix from ModE homologues in bacteria and Archaea. The locations of the helices α1 and α2 in the winged helix-turn-helix of E. coli *ModE are shown below the alignment. (B) Alignment of ModE-binding sites in the promoters of molybdate-regulated genes from* E. coli *(ECOLI),* H. influenzae *(HAEIN),* A. vinelandii *(AZOVIN) and* R. capsulatus *(RCAPS). The experimentally determined binding sites are from the promoters of modA, dmsA and napF of* E. coli. *The most frequent either amino acid or base at each position is shown with a grey background.*

A number of bacteria have more than one ModE homologue. In *R. capsulatus*, these are encoded on genes *mopA* and *mopB* (Wang *et al.*, 1993). *mopA* is the first gene in the molybdate-transport operon, *mopAmodABCD*, whereas *mopB* is located immediately upstream and transcribed divergently from *mopA*. The two ModE homologues appear to have different roles. MopA is primarily involved in the regulation of nitrogen-fixation gene expression in response to molybdate, whereas MopB has a role in both nitrogen fixation and dimethylsulfoxide respiration (Solomon *et al.*, 2000). In *A. vinelandii*, there are two ModE homologues. One is encoded on a gene that is associated with molybdate-transport genes in the operon *modEABC*. Although there is a second operon, *modA2B2C2*, with homologues of the molybdate-transporter genes, the second *modE* homologue, *modE2*, is located upstream of *vnfA2*, the gene encoding one of two homologues of the alternative nitrogenase regulatory protein, VnfA. Similarly, *H. influenzae* has only one of two *modE* homologues associated with the molybdate-transport operon. As in *R. capsulatus*, these *modE* homologues may have distinct functions.

Both molybdate and tungstate bind *E. coli* ModE with a dissociation constant of 0.8 μM (Anderson *et al.*, 1997; McNicholas *et al.*, 1997). Molybdate is required for high-affinity binding to a palindromic DNA-binding site (Figure 8) (Anderson *et al.*, 1997; Grunden *et al.*, 1996; McNicholas *et al.*, 1997). The apparent dissociation constant for the binding of ModE in the presence of molybdate to DNA in a number of promoters is approximately 25-35 nM (Anderson *et al.*, 1997; McNicholas *et al.*, 1997; 1998; McNicholas and Gunsalus, 2002).

5.2. Activation of Transcription

ModE activates the transcription of operons that contain genes encoding proteins involved in the biosynthesis of the molybdenum cofactor and molybdoenzymes. The regulation of molybdate-dependent expression of operons for the first steps in the biosynthesis of the molybdenum cofactor, and the molybdoproteins, dimethylsulfoxide reductase and respiratory nitrate reductase, have been studied in detail (Anderson *et al.*, 2000; McNicholas *et al.*, 1997; 1998; McNicholas and Gunsalus, 2002). The regulation of these genes is complex as other regulatory proteins are involved. ModE is involved to a lesser extent in the expression of formate hydrogen lyase (Rosentel *et al.*, 1995). A number of other genes may also be regulated by ModE in *E. coli* (McNicholas and Gunsalus, 2002).

The *E. coli moaABCDE* operon encodes proteins for the first committed steps in the biosynthesis of the demolybdo-cofactor (paradoxically called molybdopterin), the precursor of the molybdenum cofactor. The proteins, MoaA, B, and C, are involved in the conversion of a guanosine derivative into the molybdopterin precursor Z, then MoaD and E convert precursor Z to molybdopterin. Expression of the *moaABCDE* operon requires molybdate and anaerobic conditions. There are two promoters. The redox-dependent regulatory protein, Fnr, which controls genes in response to anoxia, regulates the distal promoter. Molybdate-dependent activation depends on ModE. There is a ModE-binding site adjacent to the –35 RNA polymerase-binding site in the proximal promoter. Unexpectedly, tungstate has a different phenotype from molybdate. Because molybdate-dependent regulation is only observed in strains with mutations in genes for molybdenum-cofactor biosynthesis, this situation can be interpreted as negative feedback that depends on an active molybdenum cofactor, with replacement of molybdenum in the cofactor by tungsten giving rise to an inactive cofactor.

Dimethylsulfoxide reductase is a molybdoenzyme of broad specificity that is required for anaerobic respiration with dimethylsulfoxide, nitrate, or trimethylamine N-oxide as terminal electron acceptors (Stewart, 1993). It is composed of three subunits, the reductase, which contains the molybdenum cofactor, an electron-transfer protein, and a membrane-anchor protein. They are encoded by *dmsABC*. Anaerobic conditions induce the expression of *dmsABC* and nitrate represses it. Both these responses require molybdate, and ModE is involved in molybdate-dependent regulation. Strains that are defective in molybdate uptake have impaired anaerobic induction and show no nitrate-dependent repression of the *dmsABC* operon. Like *moaA*, the *dmsABC* operon has two putative promoters, named P1 and P2 for the distal and proximal promoters, respectively. The P1 promoter has binding sites for the transcription factors Fnr and NarL, which regulate gene expression in response to anoxia and nitrate. The ModE-binding site is centred 52.5 bp downstream of the major *dmsA* transcript start site (P1) and overlaps the putative –10 consensus sequence of the P2 promoter. Molybdate-dependent expression from the P2 promoter depends on ModE. The expression from both promoters is not fully understood. There is a binding site for integration host factor (IHF), a specific DNA-binding protein, immediately upstream from the ModE-binding site. IHF mutant strains show impaired molybdate-dependent regulation. ModE may, therefore, regulate *dmsABC*-gene expression in concert with another transcription factor.

Component proteins of the periplasmic respiratory nitrate reductase of *E. coli* are encoded by the operon *napFDAGHC*. The reductase consists of the molybdoprotein, NapA, which is associated with the small di-heme subunit, NapB. The *napFDAGHC* operon also encodes the electron-transfer proteins, NapF, NapG, NapH, and NapC, and the chaperone, NapD, which is involved in the assembly of the NapAB enzyme. NapAB is responsible for anaerobic respiration in the presence of nitrate and, to a lesser extent, nitrite. Its transcription is activated by anaerobic conditions in the presence of nitrate, responses that depend on the regulatory proteins Fnr and NarP, respectively. The *napF* promoter has Fnr- and NarP-binding sites immediately upstream of the –35 consensus polymerase-binding site. ModE mediates a 10-fold induction of *napF* transcription by molybdate. The ModE-binding site is centred at –133.5 bp upstream from the transcription-start site, suggesting that ModE-dependent activation may require DNA bending. However, integration host factor does not appear to be involved and the details of how molybdate effects its regulation are not fully understood.

5.3. Repression of Transcription of Genes for Transport and Homeostasis

ModE regulates the internal concentration of molybdate by repressing transcription of the operon for high-affinity molybdate transport and, thus, limiting the amount of the transporter at high molybdate concentrations (Grunden *et al.*, 1996; McNicholas *et al.*, 1997; Mouncey *et al.*, 1996). In *A. vinelandii*, it also represses transcription of the gene for the molbindin, ModG (Mouncey *et al.*, 1996).

5.4. Metal-dependent Regulation of Nitrogenase Biosynthesis

The presence and absence of molybdate and vanadate regulates the expression of the genes for the distinct molybdenum-, vanadium-, and iron-only nitrogenases of *A. vinelandii*. The molybdenum-nitrogenase is only synthesised in the presence of molybdate, vanadium-nitrogenase synthesis requires molybdate to be absent and vanadate to be present, and the iron-only nitrogenase synthesis requires both molybdate and vanadate to be absent (Eady, 1995; Joerger *et al.*, 1989). Thus, molybdate represses the expression of both the alternative nitrogenases. *R. capsulatus* has only a single alternative nitrogenase, an iron-only nitrogenase, that is expressed in the absence of molybdate (Masepohl *et al.*, 2002).

The ModE homologues, MopA and MopB, regulate synthesis of the iron-only nitrogenase in *R. capsulatus* and a putative ModE-binding site is located in the promoter of *anfA*, which encodes the regulatory protein for this alternative nitrogenase (Kutsche *et al.*, 1996). The expression of both the vanadium- and iron-only alternative nitrogenases of *A. vinelandii* is repressed by molybdate. Deletion of ModE did not affect alternative nitrogenase expression (Mouncey *et al.*, 1996). However, the regulation of nitrogenase biosynthesis by ModE in *A. vinelandii* needs revaluation because its genome sequence shows that it contains two ModE homologues. There are putative ModE-binding sites upstream of the genes for VnfA2 and AnfA, which are a homologue of the regulatory protein associated with the vanadium-nitrogenase and the regulatory protein for the iron-only nitrogenase, respectively.

5.5. ModE Structure and Molybdate-dependent DNA Binding

ModE from *E. coli* is a 2 x 28.3-kDa homodimer with subunits of 262 amino acids. It consists of two functionally distinct and almost equally sized domains: an N-terminal DNA-binding domain and a C-terminal molybdate-binding domain (Hall *et al.*, 1999) (Figure 9).

The N-terminal domain consists of 121 amino acids. It is mainly α-helical and contains a DNA-binding winged helix-turn-helix motif in which there is a C-terminal β-hairpin, called a wing, in addition to the helix-turn-helix (Brennan, 1993). The fold of the N-terminal domain of ModE is classified as an HTH9 fold in the Pfam families database (Bateman *et al.*, 2002). This fold family consists largely of ModE homologues. The winged helix-turn-helix motif precedes a long helix (α4, residues 80-108) that forms the backbone of the domain (Figure 9). A model of the ModE-DNA interactions, based on the interactions of the bacterial catabolite activator protein and the regulatory protein DTR with DNA, identified putative residues involved in the interaction. These residues are highly conserved in ModE homologues throughout bacteria and Archaea (Hall *et al.*, 1999) (see Figure 8).

There are few interactions between the N- and C-terminal domains. A short linker region (residues 115-126) joins the two domains and introduces an angle between the two axes, resulting in asymmetry in the homodimer. The majority of the dimer interactions are in the N-terminal domain.

The C-terminal domains (residues 127-262) correspond to two of the three biMop TOBE domains of molbindins. The structures of biMop domains of ModE and ModG are very similar and superimpose with a rms value of 1.3 Å over 121 equivalent residues. The lack of a third subunit in ModE precludes the trimeric type 1 binding site seen in molbindins.

The two molybdate-binding sites in *E. coli* ModE are equivalent to the type 2 binding sites of molbindins. They are located in the N-terminal regions of the two Mop sub-domains. There are only two, instead of the four, molybdate-binding sites because the C-terminal sub-domain does not contain the conserved amino acids involved in binding molybdate (Figure 4). In the C-terminal Mop sub-domain, the residues of the highly conserved SARN motif of ModG are changed to NADN, and the equivalents of Thr-110 and Lys-132 are changed to Pro and Asn, respectively.

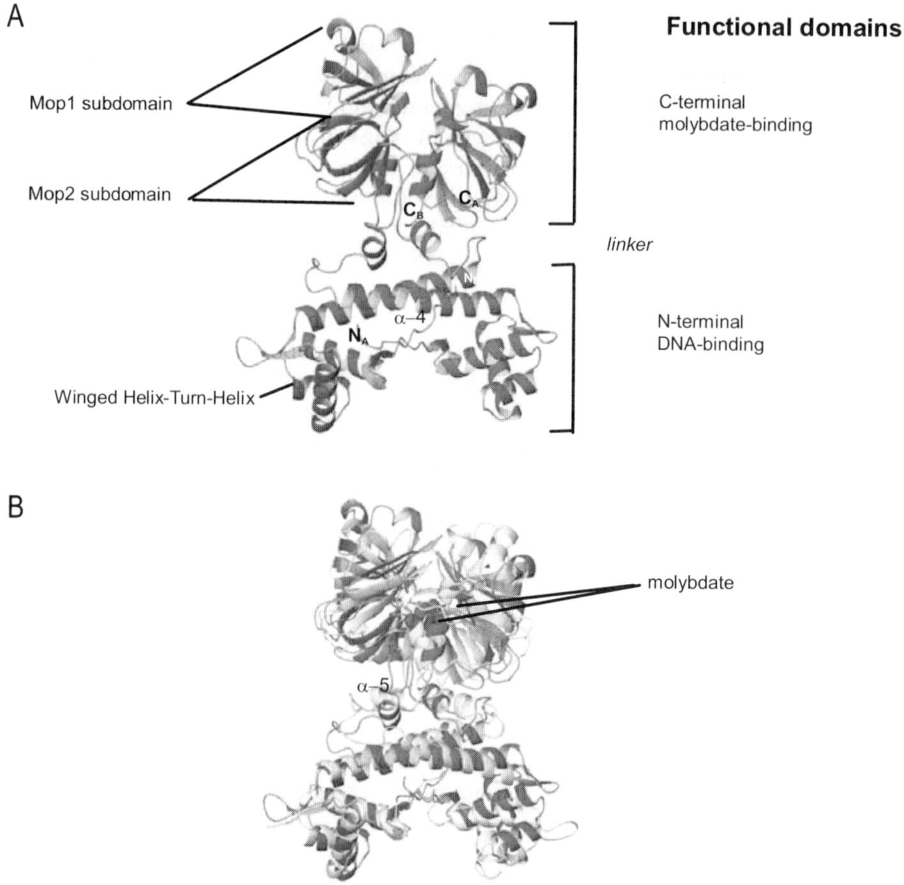

Figure 9. (A) E. coli *apo-ModE structure. The C-terminal molybdate-binding domain of each subunit consists of two sub-domains, labelled Mop1 and Mop2. The N-terminal DNA-binding domains have winged helix-turn-helix motifs. (B) Comparison of the apo and ligand-bound forms of* E. coli *ModE. Molybdate ions in the ligand-bound form are shown as CPK models. The ligand-bound form is coloured lighter grey. The Protein Data Bank entries of the apo and ligand-bound forms are 1B9M and 1O7L, respectively.*

Unlike molbindins, where the biMop domains are oriented head-to-head with respect to each other, in ModE they are arranged head-to-tail so that the molybdate-binding surfaces face each other (Delarbre *et al.* 2001). *H. influenzae* ModE also lacks the conserved molybdate-binding sites in the C-terminal sub-domain. The second Mop sub-domain is lost altogether in some ModE homologues, such those of *Methanosarcina acetivorans* and *M. mazei*. In contrast, both sub-domains of *A. vinelandii* ModE have the residues associated with molybdate binding.

The isolated C-terminal domain of *E. coli* ModE retains its ability to bind two moles of molybdate and tungstate per mole of dimer (Gourley *et al.*, 2001). Comparison of apo and molybdate-bound ModE structures and those of the isolated C-terminal domain show that the binding of molybdate leads to an extensive movement of the sub-domains against each other, trapping the ligands, but with little conformational change within each domain.

The crystal structure of ModE with bound molybdate shows that molybdate binding results in considerable movements of the secondary structure elements relative to the unbound protein (Schüttelkopf *et al.*, 2003; Figure 9). The asymmetry between the polypeptide chains seen in the apo form is decreased. The movements in the molybdate-binding domains are similar to those observed in the isolated domains, the domains effecting a closing motion. One of the helices, $\alpha5$ in the linker region, swivels by approximately 60°. There are major conformational changes in the DNA-binding domain. It moves up relative to the molybdate-binding domain, and the two recognition helices of the winged helix-turn-helix move slightly closer to each other, which positions them to insert into two consecutive major grooves of DNA.

6. CONCLUSIONS

Over thirty enzymes, which have significant roles in the nitrogen, carbon and sulfur cycles, use molybdenum in their catalytic centers (Stiefel, 1993). The simple organisation and small size of the high-affinity molybdate transporter may be an indication of its ancient origin, as is the distribution of ModE in Archaea as well as in the Eubacteria. Nevertheless, many bacteria and Archaea do not have a high-affinity molybdate transporter, molbindin, or a molybdenum-dependent regulatory protein, and there appear to be no such specialised systems for molybdate in eukaryotes.

The free intracellular concentration of metals, such as copper, is kept exceedingly low, less than one free atom per cell, because of its toxicity (O'Halloran and Culotta, 2000). Metal ions are tightly bound to chaperones that pass the metal from one protein to another. Even less toxic metals, such as zinc, have femtomolar concentrations of free metal ion despite having millimolar cellular concentrations (Outten and O'Halloran, 2001). In contrast, bacteria may tolerate significant intracellular concentrations of free molybdate. There is no molybdate-efflux system. The total cellular concentration of molybdate in *E. coli* is about 5 μM, and the dissociation constant for the ModE is 0.8 μM, which suggests that *E. coli* may have micromolar intracellular concentrations of free molybdate. The molbindin ModG of *A. vinelandii* is repressed at high molybdate concentrations, indicating that it functions only at low molybdate concentrations and is not required to sequester molybdate as its concentration increases.

Molybdate toxicity in ruminants is due to the decrease in the availability of copper as a consequence of the formation of thiomolybdate, which sequesters copper, in the gut. Clearly, this ion does not affect the bacteria that produce it (Suttle, 1991). Molybdate might potentially interfere with the first step of sulfur metabolism, involving the enzyme ATP sulfurylase, which catalyzes the formation of adenosine 5'-phosphosulfate (APS) from ATP and inorganic sulfate. However, there is no evidence for the formation of an analogous molybdenum species under chemical conditions (Hill *et al.*, 2002). The molybdonucleotides that do form have a pentamolybdenum core and, as they appear to be stable only at acid pH, it is unlikely that they have a biological role. *A. vinelandii* has evolved a special response to allow the uptake of ferric iron in the presence of extracellular molybdate concentrations greater than 100 μM, and may use the chemical reactivity of catechols to capture both molybdate and iron.

The differing physiological demands of organisms may, therefore, dictate the requirements for molybdate transport, storage, and regulation. *E. coli* lacks a molbindin, whereas the closely related *H. influenzae* has one, and *A. vinelandii* has duplicated the transporter, molbindin and regulatory protein. *A. vinelandii* has a high requirement for molybdate because of its high nitrogenase activity and the low turnover number of the enzyme. Up to 10% of cellular protein may be nitrogenase. The extreme adaptation is the specific tungstate-transport system evolved in some organisms that acquire tungstate to the exclusion of molybdate.

The cytoplasmic molybdate-binding proteins all have a common TOBE domain. Presumably, this evolved as a regulatory domain in ABC transporters and then gave rise to molbindins and ModE. There is, as yet, no evidence for interaction between the cytoplasmic molybdate-binding proteins, such as between ModE and ModC. Nor is there evidence for the interaction of molybdate-binding proteins with proteins in the pathways for the biosynthesis of molybdenum-nitrogenase or molybdenum-pterin cofactors. The structure of molbindins, where the bound ligands are shielded from bulk solvent, is different from metal chaperones, such as those for copper, where the bound ligand is exposed, ready for transfer to another site (Arnesano *et al.*, 2001a; 2001b). Although the cells may regulate their intracellular molybdate content, if they can tolerate free molybdate without affecting other metabolic processes, the transfer of molybdate from the cytoplasmic binding proteins to other proteins may not require protein-protein interactions.

ACKNOWLEDGEMENTS

I owe a debt of gratitude to D. M. Lawson and L. A. Mitchenall. I am also grateful to D. Studholme for analysis of ModE and its binding sites and to A.G. Wedd for advice and support. The genome of *A. vinelandii* is being sequenced at the US Department of Energy's Joint Genome Institute as part of the Microbial Genome Program (available at www.azotobacter.org). This work is supported by the Australian Research Council Discovery grant DP0344143.

REFERENCES

Anantharaman, V., Koonin, E. V., and Aravind, L. (2001) Regulatory potential, phyletic distribution and evolution of ancient, intracellular small-molecule-binding domains. *J. Mol. Biol., 307*, 1271-1292.

Anderson, L. A., McNairn, E., Leubke, T., Pau, R. N., and Boxer, D. H. (2000). ModE-dependent molybdate regulation of the molybdenum cofactor operon *moa* in *Escherichia coli. J. Bacteriol., 182,* 7035-7043.

Anderson, L. A., Palmer, T., Price, N. C., Bornemann, S., Boxer, D. H., and Pau, R. N. (1997). Characterisation of the molybdenum-responsive ModE regulatory protein and its binding to the promoter region of the *modABCD* (molybdenum transport) operon of *Escherichia coli. Eur. J. Biochem., 246*, 119-126.

Arnesano, F., Banci, L., Bertini, I., Cantini, F., Ciofi-Baffoni, S., Huffman, D. L., et al. (2001a). Characterization of the binding interface between the copper chaperone Atx1 and the first cytosolic domain of Ccc2 ATPase. *J. Biol. Chem., 276*, 41365-41376.

Arnesano, F., Banci, L., Bertini, I., Huffman, D. L., and O'Halloran, T. V. (2001b). Solution structure of the Cu(I) and apo forms of the yeast metallochaperone, Atx1. *Biochemistry, 40,* 1528-1539.

Barton, G. J. (1993). ALSCRIPT. A tool to format multiple sequence alignments. *Prot. Eng., 6*, 37-40.

Bateman, A., Birney, E., Cerruti, E., Durbin, R., Etwiller, L., Eddy, S. R., et al. (2002). The Pfam protein families database. *Nucleic Acids Res., 30*, 276-280.

Bohm, A., Diez, J., Diederichs, K., Welte, W., and Boos, W. (2002). Structural model of MalK, the ABC subunit of the maltose transporter of *Escherichia coli*: Implications for mal gene regulation, inducer exclusion, and subunit assembly. *J. Biol. Chem., 277*, 3708-3717.

Boyd, D., Vecoli, C., Belcher, D., Jain, S., and Drysdale, J. (1985). Structural and functional relationships of human ferritin H and L chains deduced from cDNA clones. *J. Biol. Chem., 260*, 11755-11761.

Brennan, R. (1993). The winged-helix DNA-binding motif: another helix-turn-helix takeoff. *Cell, 74*, 773-776.

Brennan, R., and Bruce, R. (1999). Molybdenum. In K. Peverill, L. Sparrow and D. Reuter (Eds.), *Soil analysis: An interpretation manual* (pp. 303-305). Collingwood, Vic., Australia: CSIRO Publishing.

Chen, J. S., Toth, J., and Kasap, M. (2001). Nitrogen-fixation genes and nitrogenase activity in *Clostridium acetobutylicum* and *Clostridium beijerinckii. J. Ind. Microbiol. Biotech., 27*, 281-286.

Corbin, J. L., and Bulen, W. A. (1969). The isolation and identification of 2,3-dihydroxybenzioc acid and 2-N,6-N-di(2,3-hydroxybenzoyl)-L-lysine formed by iron-deficient *Azotobacter vinelandii. Biochemistry, 27*, 2745-2752.

Corcuera, G. L., Bastidas, M., and Dubourdieu, M. (1993). Molybdenum uptake in *Escherichia coli* K12. *J. Gen. Microbiol., 139*, 1869-1875.

Cornish, A. S., and Page, W. J. (1995). Production of the triacetecholate siderophore protochelin by *Azotobacter vinelandii. BioMetals, 8*, 332-338.

Cornish, A. S., and Page, W. J. (2000). Role of molybdate and other transition metals in the accumulation of protochelin by *Azotobacter vinelandii. Appl. Environ. Microbiol., 66*, 1580-1586.

Delarbre, L., Stevenson, C. E. M., White, D. J., Mitchenall, L. A., Pau, R. N., and Lawson, D. M. (2001). Two crystal structures of the cytoplasmic molybdate-binding protein ModG suggest a novel cooperative binding mechanism and provide insights into ligand-binding specificity. *J. Mol. Biol., 308*, 1063-1079.

Duhme, A.-K., Dauter, Z., Hider, R. C., and Pohl, S. (1996). Complexation of molybdenum by siderophores: Synthesis and structure of the double-helical cis-dioxomolybdenum(VI) complex of a bis(catecholamide) siderophore analogue. *Inorg. Chem., 35*, 3059-3061.

Duhme, A. K., Hider, R. C., and Khodr, H. H. (1997). Synthesis and iron-binding properties of protochelin, the tris(catecholamide) siderophore of *Azotobacter vinelandii. Chem. Ber. 130,* 969-973.

Duhme, A.-K., Hider, R. C., Naldrett, M. J., and Pau, R. N. (1998). The stability of the molybdenum-azotochelin complex and its effect on siderophore production of *Azotobacter vinelandii. J. Biol. Inorg. Chem., 3*, 520-526.

Eady, R. R. (1995). Vanadium nitrogenases of *Azotobacter. Met. Ions Biol. Syst., 31*, 363-405.

Einsle, O., Tezcan, F., Andrade, S., Schmid, B., Yoshida, M., Howard, J., et al. (2002). Nitrogenase MoFe-protein at 1.16 Å resolution: A central ligand in the FeMo-cofactor. *Science, 297*, 1696-1700.

Elliot, B. B., and Mortenson, L. E. (1975). Transport of molybdate by *Clostridium pasteurianum. J. Bacteriol., 124*, 1295-1301.

Frausto da Silva, J. J. R., and Williams, R. J. P. (2001). *The biological chemistry of the elements - The inorganic chemistry of life*, 2 ed. Oxford, UK: Clarendon Press.

Glaser, J. H., and DeMoss, J. A. (1971). Phenotypic restoration by molybdate of nitrate reductase activity in chlD mutants of *Escherichia coli. J. Bacteriol., 108*, 854-860.

Gourley, D. G., Schüttelkopf, A. W., Anderson, L. A., Price, N. C., Boxer, D. H., and Hunter, W. N. (2001). Oxyanion binding alters conformation and quaternary structure of the c-terminal domain of the transcriptional regulator ModE. Implications for molybdate-dependent regulation, signaling, storage, and transport. *J. Biol. Chem., 276*, 20641-20647.

Grunden, A. M., Ray, R. M., Rosentel, J. K., Healy, F. G., and Shanmugam, K. T. (1996). Repression of the *Escherichia coli modABCD* (molybdate transport) operon by ModE. *J. Bacteriol., 178*, 735-744.

Grunden, A. M., and Shanmugam, K. T. (1997). Molybdate transport and regulation in bacteria. *Arch. Microbiol., 168*, 345-354.

Gupta, U. C. (Ed.). (1997). *Molybdenum in agriculture*. Cambridge, UK: Cambridge University Press.

Hall, D. R., Gourley, D. G., Leonard, G. A., Duke, E. M., Anderson, L. A., Boxer, D. H., et al. (1999). The high-resolution crystal structure of the molybdate-dependent transcriptional regulator (ModE) from *Escherichia coli*: A novel combination of domain folds. *EMBO J., 18*, 1435-1446.

Hill, L. M. R., George, G. N., Duhme-Klair, A.-K., and Young, C. G. (2002). Solution structural studies of molybdate-nucleotide polyanions. *J. Inorg. Biochem., 99*, 274-283.

Hille, R. (1996). The mononuclear molybdenum enzymes. *Chem. Rev., 96*, 2757-2816.

Hinton, S. M., and Merritt, B. (1986). Purification and characterization of a molybdenum-pterin-binding protein (Mop) in *Clostridium pasteurianium* W5. *J. Bacteriol., 168*, 688-693.

Hinton, S. M., and Mortenson, L. E. (1985a). Identification of molybdoproteins in *Clostridium pasteurianum. J. Bacteriol., 162*, 477-484.

Hinton, S. M., and Mortenson, L. E. (1985b). Regulation and order of involvement of molybdoproteins during synthesis of molybdoenzymes in *Clostridium pasteurianum*. *J. Bacteriol., 162*, 485-493.

Hinton, S. M., Slaughter, C., Eisner, W., and Fisher, T. (1987). The molybdenum-pterin binding protein is encoded by a multigene family in *Clostridium pasteurianum*. *Gene, 54*, 211-219.

Hochheimer, A., Hedderich, R., and Thauer, R. K. (1998). The formylmethanofuran dehydrogenase isoenzymes in *Methanobacterium wolfei* and *Methanobacterium thermoautotrophicum*: Induction of the molybdenum isoenzyme by molybdate and constitutive synthesis of the tungsten isoenzyme. *Arch. Microbiol., 170*, 389-393.

Hoover, T. R., Robertson, A. D., Cerny, R. L., Hayes, R. N., Imperial, V. K., Shah, V. K., et al. (1987). Identification of the V factor needed for synthesis of the iron-molybdenum cofactor of nitrogenase as homocitrate. *Nature, 329*, 855-857.

Howarth, R. W., and Cole, J. J. (1985). Molybdenum availability, nitrogen limitation, and phytoplankton growth in natural waters. *Science, 229*, 653-655.

Hu, Y., Rech, S., Gunsalus, R. P., and Rees, D. C. (1997). Crystal structure of the molybdate binding protein ModA. *Nature Struct. Biol., 4*, 703-707.

Imperial, J., Hadi, M., and Amy, N. K. (1998). Molybdate binding by ModA, the periplasmic component of the *Escherichia coli mod* molybdate transport system. *Biochim. Biophys. Acta, 1370*, 337-346.

Imperial, J., Ugalde, R. A., Shah, V. K., and Brill, W. J. (1985). Mol$^-$ mutants of *Klebsiella pneumoniae* requiring high levels of molybdate for nitrogenase activity. *J. Bacteriol., 163*, 1285-1287.

Jacobson, M. R., Brigle, K. E., Bennett, L. T., Setterquist, R. A., Wilson, M. S., Cash, V. L., et al. (1989). Physical and genetic map of the major nif gene cluster from *Azotobacter vinelandii*. *J. Bacteriol., 171*, 1017-1027.

Joerger, R. D., Jacobson, M. R., Premakumar, R., Wolfinger, E. D., and Bishop, P. E. (1989). Nucleotide sequence and mutational analysis of the structural genes (*anfHDGK*) for the second alternative nitrogenase from *Azotobacter vinelandii*. *J. Bacteriol., 171*, 1075-1086.

Kelley, L. A., MacCallum, R. M., and Sternberg, M. J. E. (2000). Enhanced genome annotation using structural profiles in the program 3D-PSSM. *J. Mol. Biol., 299*, 499-520.

Khodr, H. H., Hider, R. C., and Duhme-Klair, A. K. (2002). The iron-binding properties of aminochelin, the mono(catecholamide) siderophore of *Azotobacter vinelandii*. *J. Biol. Inorg. Chem., 7*, 891-896.

Kletzin, A., and Adams, M. W. W. (1996). Tungsten in biological systems. *FEMS Microbiol. Rev., 18*, 5-63.

Koonin, E. V., Wolf, Y. I., and Aravind, L. (2000). Protein fold recognition using sequence profiles and its application in structural genomics. In P. Bork (Ed.), *Advances protein chemistry, vol. 54* (pp. 246-275). San Diego, CA: Academic Press.

Koradi, R., Billeter, M., and Wüthrich, K. (1996). MOLMOL: A program for display and analysis of macromolecular structures. *J. Mol. Graphics, 14*, 51-55.

Kutsche, M., Leimkühler, S., and Klipp, W. (1996). Promoters controlling the expression of the alternative nitrogenase and the moybdenum uptake system in *Rhodobacter capsulatus* are activated by NtrC, independent of s54, and repressed by molybdenum. *J. Bacteriol., 178*, 2010-2017.

Lawson, D. M., Williams, C. E., Mitchenall, L. A., and Pau, R. N. (1998). Ligand size is a major determinant of specificity in periplasmic oxyanion-binding proteins: The 1.2 A resolution crystal structure of *Azotobacter vinelandii* ModA. *Structure, 6*, 1529-1539.

Lawson, D. M., Williams, C. E., White, D. J., Choay, A. P., Mitchenall, L. A., and Pau, R. N. (1997). Protein ligands for molybdate. Specificity and charge stabilisation at anion-binding sites of periplasmic and intracellular molybdate-binding proteins of *Azotobacter vinelandii*. *J. Chem. Soc., Dalton Trans.*, 3981-3984.

Lee, J. H., Wendt, J. C., and Shanmugam, K. T. (1990). Identification of a new gene, molR, essential for utilization of molybdate by *Escherichia coli*. *J. Bacteriol., 172*, 2079-2087.

Lei, S., Pulakat, L., and Gavini, N. (2000). Activation of vanadium nitrogenase expression in *Azotobacter vinelandii* DJ54 revertant in the presence of molybdenum. *FEBS Lett., 482*, 149-153.

Locher, K., Lee, A., and Rees, D. (2002). The *E. coli* BtuCD structure: A framework for ABC transporter architecture and mechanism. *Science, 296*, 1091-1098.

Luecke, H., and Quiocho, F. A. (1990). High specificity of a phosphate transport protein determined by hydrogen bonds. *Nature, 347*, 402-406.

Madden, M. S., Paustian, T. D., Ludden, P. W., and Shah, V. K. (1991). Effects of homocitrate, homocitrate lactone, and fluorohomocitrate on nitrogenase in *nifV*⁻ mutants of *Azotobacter vinelandii*. *J. Bacteriol., 173*, 5403-5405.

Makdessi, K., Andreesen, J. R., and Pich, A. (2001). Tungstate uptake by a highly specific ABC transporter in *Eubacterium acidaminophilum*. *J. Biol. Chem., 276*, 24557-24564.

Masepohl, B., Drepper, T., Paschen, A., Gross, S., Pawlowski, A., Raabe, K., et al. (2002). Regulation of nitrogen fixation in the phototrophic purple bacterium *Rhodobacter capsulatus*. *J. Mol. Microbiol. Biotech., 4*, 243-248.

Maupin-Furlow, J. A., Rosentel, J. K., Lee, J. H., Deppenmeier, U., Gunsalus, R. P., and Shanmugam, K. T. (1995) Genetic analysis of the *modABCD* (molybdate transport) operon of *Escherichia coli*. *J. Bacteriol., 177*, 4851-4856.

Maurizi, M. R., and Li, C. H. (2001) AAA proteins: In search of a common molecular basis. *EMBO Reports, 21*, 980-985.

McNicholas, P. M., Chiang, R. C., and Gunsalus, R. P. (1998). Anaerobic regulation of the *Escherichia coli dmsABC* operon requires the molybdate-responsive regulator ModE. *Mol. Microbiol., 27*, 197-208.

McNicholas, P. M., and Gunsalus, R. P. (2002). The molybdate-responsive *Escherichia coli* ModE transcriptional regulator coordinates periplasmic nitrate reductase (*napFDAGHBC*) operon expression with nitrate and molybdate availability. *J. Bacteriol., 184*, 3253-3259.

McNicholas, P. M., Rech, S. A., and Gunsalus, R. P. (1997). Characterization of the ModE DNA-binding sites in the control regions of *modABCD* and *moaABCDE* of *Escherichia coli*. *Mol. Microbiol., 23*, 515-524.

Mouncey, N. J., Mitchenall, L. A., and Pau, R. N. (1995). Mutational analysis of genes of the mod locus involved in molybdenum transport, homeostasis, and processing in *Azotobacter vinelandii*. *J. Bacteriol., 177*, 5294-5302.

Mouncey, N. J., Mitchenall, L. A., and Pau, R. N. (1996). The modE gene product mediates molybdenum-dependent expression of genes for the high-affinity molybdate transporter and *modG* in *Azotobacter vinelandii*. *Microbiol., 142*, 1997-2004.

Mukund, S., and Adams, M. W. W. (1996). Molybdenum and vanadium do not replace tungsten in the catalytically active forms of the three tungstoenzymes in the hyperthermophilic archaeon *Pyrococcus furiosus*. *J. Bacteriol., 178*, 163-167.

Murzin, A. G. (1993). OB(oligonucleotide oligosaccharide binding)-fold - common structural and functional solution for nonhomologous sequences. *EMBO J., 12*, 861-867.

Neubauer, H., Pantel, I., Lindgren, P. E., and Gotz, F. (1999). Characterization of the molybdate transport system ModABC of *Staphylococcus carnosus*. *Arch. Microbiol., 172*, 109-115.

O'Halloran, T. V., and Culotta, V. C. (2000). Metallochaperones, an intracellular shuttle service for metal ions. *J. Biol. Chem., 275*, 25057-25060.

Outten, C. E., and O'Halloran, T. V. (2001). Femtomolar sensitivity of metalloregulatory proteins controlling zinc homeostasis. *Science, 292*, 2488-2492.

Page, W. J., and von Tigerstrom, M. (1988). Aminochelin, a catecholamine siderophore produced by *Azotobacter vinelandii*. *J. Gen. Microbiol., 134*, 453-460.

Page, W. J., and von Tigerstrom, M. (1982). Iron- and molybdenum-repressible outer membrane proteins in competent *Azotobacter vinelandii*. *J. Bacteriol., 151*, 237-242.

Pardee, A. B. (1966) Purification and properties of a sulfate-binding protein from *Salmonella typhimurium*. *J. Biol. Chem., 241*, 5886-5892.

Pau, R. N., Klipp, W., and Leimkühler, S. (1997). Molybenum transport, processing and gene regulation. In G. Winkelmann and C. J. Carrano (Eds.), *Transition metals in microbial metabolism* (pp. 217-234). Amsterdam: Harwood Academic Publishers.

Paulsen, D. M., Paerl, H. W., and Bishop, P. E. (1991). Evidence that molybdenum-dependent nitrogen fixation is not limited by high sulfate concentrations in marine environments. *Limnol. Oceanog., 36*, 1325-1334.

Pflugarth, J. W., and Quiocho, F. A. (1985). Sulphate sequestered in sulphate-binding protein of *Salmonella typhimurium* is bound solely by hydrogen bonds. *Nature, 314*, 257-260.

Pope, M. T., Still, E. R., and Williams, J. P. (1980). A comparison between the chemistry and biochemistry of molybdenum and related elements. In M. Coughlan (Ed.), *Molybdenum and molybdenum-containing enzymes* (pp. 3-40). Oxford: Pergamon Press.

Premakumar, R., Jacobitz, S., Ricke, S. C., and Bishop, P. E. (1996). Phenotypic characterization of a tungstate-tolerant mutant of *Azotobacter vinelandii*. *J. Bacteriol., 178*, 691-696.

Quiocho, F. A., and Ledvinda, P. S. (1996). Atomic structure and specificity of bacterial periplasmic receptors for active transport and chemotaxis: variation of common themes. *Mol. Microbiol., 20*, 17-25.

Rech, S., Wolin, C., and Gunsalus, R. P. (1996). Properties of the periplasmic ModA molybdate-binding protein of *Escherichia coli*. *J. Biol. Chem., 271*, 2557-2562.

Robson, R. L., Eady, R. R., Richardson, T. H., Miller, R. W., Hawkins, M., and Postgate, J. R. (1986). The alternative nitrogenase of *Azotobacter chroococcum* is a vanadium enzyme. *Nature, 322*, 388-390.

Rosentel, J. K., Healy, F., Maupin-Furlow, J. A., Lee, J. H., and Shanmugam, K. T. (1995). Molybdate and regulation of *mod* (molybdate transport), *fdhF*, and *hyc* (formate hydrogen lyase) operons in *Escherichia coli*. *J. Bacteriol., 177*, 4857-4864.

Schneider, K., Müller, A., Johannes, K.-U., Diemann, E., and Kottmann, J. (1991). Selective removal of molybdenum traces from growth media of N_2-fixing bacteria. *Anal. Biochem., 193*, 292-298.

Schüttelkopf, A. W., Boxer, D. H., and Hunter, W. N. (2003). Crystal structure of activated ModE reveals conformational changes involving both oxyanion and DNA-binding domains. *J. Mol. Biol., 326*, 761-767.

Schüttelkopf, A. W., Harrison, J. A., Boxer, D. H., and Hunter, W. N. (2002). Passive acquisition of ligand by the MopII molbindin from *Clostridium pasteurianum*: Structures of apo and oxyanion-bound forms. *J. Biol. Chem., 277*, 15013-15020.

Scott, D., and Amy, N. K. (1989). Molybdenum accumulation in *chlD* mutants of *Escherichia coli. J. Bacteriol., 171*, 1284-1287.

Siemann, S., Schneider, K., Oley, M., and Müller, A. (2003). Characterization of a tungsten-substituted nitrogenase isolated from *Rhodobacter capsulatus. Biochemistry 42*, 3846-3857.

Solomon, P. S., Shaw, A. L., Young, M. D., Leimkühler, S., Hanson, G. R., Klipp, W. et al. (2000). Molybdate-dependent expression of dimethylsulfoxide reductase in *Rhodobacter capsulatus. FEMS Microbiol. Lett., 190*, 203-208.

Sperl, G. T., and DeMoss, J. A. (1975). *chlD* gene function in molybdate activation of nitrate reductase. *J. Bacteriol., 122*, 1230-1238.

Stewart, L. J., Bailey, S., Bennett, B., Charnock, J. M., Garner, C. D., and McAlpine, A. S. (2000). Dimethylsulfoxide reductase: An enzyme capable of catalysis with either molybdenum or tungsten at the active site. *J. Mol. Biol. 299,* 593-600.

Stewart, V. (1993). Nitrate regulation of anaerobic respiratory gene expression in *Escherichia coli. Mol. Microbiol., 9*, 425-434.

Stiefel, E. I. (1993). Molybdenum enzymes, cofactors and chemistry. In E. I. Stiefel, D. Coucouvanis and W. E. Newton (Eds.), *Molybdenum enzymes, cofactors and model systems* (pp. 1-18). Washington DC: American Chemical Society.

Suttle, N. F. (1991). The interactions between copper, molybdenum, and sulphur in ruminant nutrition. *Annu. Rev. Nutr.,11,* 121-140.

Thiel, T., Pratte, B., and Zahalak, M. (2002). Transport of molybdate in the cyanobacterium *Anabaena variabilis* ATCC 29413. *Arch. Microbiol., 179*, 50-56.

Trautwein, T., Krauss, F., Lottspeich, F., and Simon, H. (1994). The (2R)-hydroxy-carboxylate-viologen-oxyreductase from *Proteus vulagris* is a molybdenum-containing iron-sulfur protein. *Eur. J. Biochem., 222*, 1025-1032.

Wagner, U. G., Stupperich, E., and Kratky, C. (2000). Structure of the molybdate/tungstate binding protein Mop from *Sporomusa ovata. Structure Fold Des., 8*, 1127-1136.

Walkenhorst, H. M., Hemschemeier, S. K., and Eichenlaub, R. (1995). Molecular analysis of the molybdate uptake operon, *modABCD*, of *Escherichia coli* and *modR*, a regulatory gene. *Microbiol. Res., 150*, 347-361.

Wang, G., Angermüller, S., and Klipp, W. (1993). Characterization of *Rhodobacter capsulatus* genes encoding a molybdenum transport system and putative molybdenum-pterin-binding proteins. *J. Bacteriol., 175*, 3031-3042.

Wang, S. Z., Chen, J. S., and Johnson, J. L. (1990). A nitrogen-fixation gene (*nifC*) in *Clostridium pasteurianum* with sequence similarity to *chlJ* of *Escherichia coli. Biochem. Biophys. Res. Commun., 169*, 1122-1128.

Willsky, G. R. (1990). Vanadium in the biosphere. In N. D. Chasteen (Ed.), *Vanadium in biological systems* (pp. 1-24). Amsterdam: Kluwer Academic Publishers.

Zierenberg, R. A., Adams, M. W. W., and Arp, A. J. (2000). Life in extreme environments: Hydrothermal vents. *Proc. Natl. Acad. Sci. U.S.A., 97*, 12961-12962.

Chapter 11

ELECTRON TRANSPORT TO NITROGENASE: DIVERSE ROUTES FOR A COMMON DESTINATION

K. SAEKI
Department of Biology, Graduate School of Science, Osaka University, 1-1 Machikaneyama, Toyonaka, Osaka 560-0043, Japan

1. INTRODUCTION

Nitrogenase, which consists of dinitrogenase (*e.g.*, the MoFe protein) and dinitrogenase reductase (the Fe protein), catalyzes the biological reduction of molecular nitrogen (N_2) according to the following equation:

$$N_2 + (6+2n)\,H^+ + (6+2n)\,e^- + (12+4n)\,MgATP \rightarrow 2\,NH_3 + n\,H_2 + (12+4n)\,MgADP + (12+4n)\,Pi$$

(where $n \geq 1$ depending on the availability of both H^+ and e^-)

Electrons are first transferred to dinitrogenase reductase, which in turn donates electrons to dinitrogenase. This reaction is common to all three conventional nitrogenases; the molybdenum-containing nitrogenase as well as the alternative vanadium-containing and the iron-only nitrogenases. It does not apply to the unusual N_2-fixation system reported for the thermophile, *Streptomyces thermoautotrophicus*, which was isolated from a burning charcoal pile (Gadkari *et al.*, 1992; Ribbe *et al.*, 1997). As shown in the above equation, the availability of reducing power is as important as the availability of ATP for the throughput of the nitrogenase reaction. When the supply of reducing power is limited, the value of *n* increases (Haaker and Klugkist, 1987) and, as a consequence, net fixation of N_2 decreases, with more reducing equivalents diverted towards H_2 production.

W. Klipp, B. Masepohl, J. R. Gallon and W. E. Newton (eds.), Genetics and Regulation of Nitrogen Fixation in Free-Living Bacteria, 257-290.

Despite the importance of electron supply to nitrogenase, the electron transport pathway is often described along the lines: "the electron donor to nitrogenase is either reduced ferredoxin (Fd) or reduced flavodoxin (Fld), which is reduced in turn by the activity of either pyruvate-Fd(Fld) oxidoreductase or some alternative reductase(s)". One reason for this sketchy view is that the plausible electron carriers have been regarded simply either as "a cellular currency for low redox equivalents" or as general cellular electron carriers that function in diverse metabolic pathways. This view is supported by the fact that most of these small electron carriers, regardless of their source, can interact not only with the Fe protein but also with many other enzymes, such as hydrogenases, plant ferredoxin-$NADP^+$ reductases (FNRs), and with photosystem I of photosynthesis. Another reason for neglecting these electron carriers is that nitrogenase biochemistry can be studied without any knowledge of the physiological electron transport pathway because the Fe protein (dinitrogenase reductase) readily accepts electrons from non-physiological chemical reductants, such as dithionite, and efficiently transfers these electrons to the dinitrogenase (the MoFe protein).

This simplified view of a general electron-transfer system has been supported biochemically and genetically by the observation that a Fld and a pyruvate-Fld oxidoreductase constitute a specific, non-branched, electron-transport pathway to nitrogenase in *Klebsiella pneumoniae* (Nieva-Gomez *et al.*, 1980; Shah *et al.*, 1983). However, this simple system in *Klebsiella* must be considered as an exceptional case. In *K. pneumoniae*, both genes (*nifF* and *nifJ*), encoding a single electron carrier and its sole reductase, are dispensable under non-diazotrophic conditions yet their mutants display a clear Nif^- phenotype. This is not the case in other diazotrophs, in which the systems appear more complex. For example, many bacteria possess multiple Fds and Flds and these electron carriers display both functional specialization and crosstalk. In addition, some of the small iron-sulfur proteins can have functions in the synthesis and protection of other iron-sulfur proteins (Nakamura *et al.*, 1999; Zheng *et al.*, 1998) and also in the regulation of transcription (Beinert and Kiley, 1999; Hidalgo and Demple, 1994; Schwartz *et al.*, 2001). Moreover, a reactivity measured *in vitro* does not always guarantee a function *in vivo*, which depends on both affinity (determined by protein structure), and cellular concentration (based on gene expression). In some cases, disruption of a gene, which encodes an effective electron carrier for *in vitro* activity, does not affect activity *in vivo*. In other cases, a particular electron carrier has some other, essential (house-keeping) physiological role, so mutations in its gene are practically lethal. Because of these complexities, it is often difficult to determine conclusively whether a specific electron carrier donates electrons directly to the Fe protein in a bacterium of interest. Nonetheless, recent genetic, genomics, and biochemical studies have implicated some major electron carriers and the systems that reduce them in nitrogen fixation in several organisms, such as *Azotobacter vinelandii, Anabaena* sp. PCC 7120, and *Rhodobacter capsulatus*. This chapter deals firstly with the general properties of electron carriers and their significance in representative species, and then with the three major reductases that reduce these electron carriers, namely, pyruvate-Fld oxidoreductase (*nifJ*-gene product), FNR, and the presumed ion-translocating NADH-Fd oxidoreductase (*rnf*-gene products).

2. DIRECT ELECTRON DONORS TO THE FE PROTEIN OF NITROGENASE

2.1. General Properties of Electron-Donor Proteins

2.1.1. Flavodoxin (Fld)

Fld was first identified as a flavoprotein that can substitute for the function of ferredoxin (Fd) (Knight and Hardy, 1966; 1967). Flds are acidic proteins (pI = 3.5 to 4.0) of 140-180 amino acids with a single flavin mononucleotide (FMN) prosthetic group. They occur not only in many prokaryotes but also in some eukaryotic algae (Mayhew and Ludwig, 1975; Simondsen and Tollin, 1980). Flds display three redox forms; oxidized (Ox), semiquinone (SQ), and hydroquinone (HQ). Redox potentials range from –50 to –250 mV for the SQ/Ox couple and from –370 to –450 mV for the HQ/SQ couple. The redox potential of –172 mV for the HQ/SQ couple in free FMN is, therefore, negatively shifted by more than –200 mV as a consequence of non-covalent binding with the apoprotein. Because the redox potential of the Fe protein of nitrogenase is lower than –400 mV, the HQ form of Fld is the only functional electron donor (Duyvis *et al.*, 1998; Yates, 1972). The Ox form of Fld is stable under aerobic conditions. Crystal structures for seven Flds, including one from the diazotrophic *Anabaena* sp. PCC 7120 (Rao *et al.*, 1992), are now available and all show an open twisted α/β structure, consisting of five parallel β-sheets connected by α-helices that surround the sheets (Freigang *et al.*, 2002). Flds can be divided into a short-chain group with less than 150 amino acids and a long-chain group with an insertion of 15-to-30 amino acids interrupting the fifth β-sheet (Mayhew and Ludwig, 1975). Both classes are found in diazotrophs.

Flds from various sources, such as *K. pneumoniae, A. vinelandii* and *Azotobacter chroococcum*, can directly donate electrons to both their own Fe proteins *in vitro* and Fe proteins from other sources, at greater rates than dithionite (Shah *et al.*, 1983; Yates, 1972). Early studies had shown that *A. vinelandii* Fld increased the specific activity of *A. vinelandii* Fe protein by more than 50% relative to that in the presence of dithionite alone (Hageman and Burris, 1978a). Kinetic studies, using stopped-flow spectroscopy, confirmed the high affinity between Fld and Fe protein by showing that these proteins form a tight complex; the reported dissociation constants (K_D) for the pairs from *K. pneumoniae* and *R. capsulatus* are 13 μM and 0.44 μM, respectively (Hallenbeck and Gennaro, 1998; Thorneley and Deistung, 1988). Another study suggested that *A. vinelandii* Fld (HQ) might be an effective reductant for the Fe protein-MoFe protein complex (Duyvis *et al.*, 1998), although the physiological significance of this observation is yet to be revealed.

2.1.2. Ferredoxin (Fd)

Fd is a non-heme iron protein that was first isolated from *Clostridium pasteurianum* (Mortenson *et al.*, 1962). Soon afterward, another non-heme protein from spinach chloroplasts was shown to be functionally exchangeable with the *C. pasteurianum* protein (Tagawa and Arnon, 1962). Fds are acidic proteins (pI is from below 3 to 4), usually of 50-120 amino acids with either one or two acid-labile iron-sulfur clusters. However, some variants, such as poly-Fds, are exceptionally large, and others, like the zinc-containing Fd mostly found in archaea, contain hetero-metals (Iwasaki *et al.*, 1997; Steigerwald *et al.*, 1992). They can be divided into several groups depending on the number and types of iron-sulfur cluster(s), *e.g.*, [2Fe-2S], [4Fe-4S], [3Fe-4S], 2[4Fe-4S] and [3Fe-4S][4Fe-4S] Fds are known (Matsubara and Saeki, 1992; Otaka and Ooi, 1989). Among these, the significant electron donors to the Fe protein may be limited to the [2Fe-2S] and 2[4Fe-4S] groups (Fukuyama, 2001; Sieker and Adman, 2001; Stout, 2001; Zanetti *et al.*, 2001). The redox potentials of Fds are modulated by their polypeptide structure and occur in the ranges of –240 to –460 mV, –50 to –420 mV and 0 to –645 mV, respectively, for the couples $[2Fe\text{-}2S]^{2+/1+}$, $[3Fe\text{-}4S]^{1+/0}$ and $[4Fe\text{-}4S]^{2+/1+}$ (Cammack, 1992).

Three-dimensional structures of a number of Fds are available, including several from diazotrophs, such as the [2Fe-2S]-type from *Anabaena* sp. PCC 7120, the 2[4Fe-4S]-type from *C. pasteurianum*, and the [3Fe-4S][4Fe-4S]-type from *A. vinelandii*, along with their mutant forms (see references in Sieker and Adman, 2001; Stout, 2001; Zanetti *et al.*, 2001). The Fe-S clusters are covalently bound through cysteinyl (Cys) residues, which display characteristic spacing (Figure 1), the so-called "cluster-binding motifs" (Bruschi and Guerlesquin, 1988; Cammack, 1992; Matsubara and Saeki, 1992). In the [2Fe-2S]-type Fds from plants and cyanobacteria, a cluster is typically bound by Cys-X_4-Cys-X_2-Cys and a distal Cys residue. A cubane [4Fe-4S]-type cluster is often bound by a Cys-X_2-Cys-X_2-Cys motif and a distal Cys residue, whereas in the 2[4Fe-4S]-type Fds, two Cys-X_2-Cys-X_2-Cys-X_3-Cys motifs are used in which the last Cys residue corresponds to the distal Cys residue of another motif. Sometimes, the second motif is substituted by Cys-X_2-Cys-$X_{(7\text{-}9)}$-Cys-X_3-Cys. The motifs in [3Fe-4S][4Fe-4S]-type are similar to those in the 2[4Fe-4S]-type, except that, to accommodate the [3Fe-4S] cluster, the first motif is either Cys-X_7-Cys-X_3-Cys or Cys-X_2-Asp-X_2-Cys-X_3-Cys. Some 2[4Fe-4S]-type Fds are highly sensitive to O_2, whereas most of the [3Fe-4S][4Fe-4S]-type Fds are resistant. Rhizobia and *A. vinelandii* also contain Fd-like genes, designated *fixX*, that encode products with a Cys-X_2-Cys-X_2-Cys motif near the C-terminus and a distal Cys residue near the N-terminus but unlike the typical Fds, the predicted products possess isoelectric points around 6-to-8 (Dusha *et al.*, 1987; Earl *et al.*, 1987; Iismaa and Watson, 1989; Reyntjens *et al.*, 1997; Wientjens, 1993).

Fds from various sources can donate electrons, *in vitro*, to Fe proteins from various sources. For example, spinach [2Fe-2S]-type Fd can support nitrogenase activity in extracts of *Rhodospirillum rubrum* (Yoch and Arnon, 1975; Yoch and Carithers, 1979). However, the ability to donate electrons *in vitro* does not necessarily mean capability *in vivo*. For example, the 2[4Fe-4S]-type Fds (FdxN proteins) from *R. capsulatus* and *Sinorhizobium meliloti* both almost equally support *in vitro* acetylene reduction by *R. capsulatus* nitrogenase in the presence of spinach chloroplasts. In contrast, heterologous *in vivo* complementation was unsuccessful. Thus, nitrogen fixation by mutants of either *S. meliloti* or *R. capsulatus*, which are defective in *fdxN*, could not be supported by expressing *fdxN* genes from the other species (Masepohl *et al.*, 1992; Riedel *et al.*, 1995). This situation could be caused either by different specificities of reductases for Fd in the two organisms or by currently unidentified function(s) of Fd, such as in metal-cluster biosynthesis, but the precise reason remains an open question.

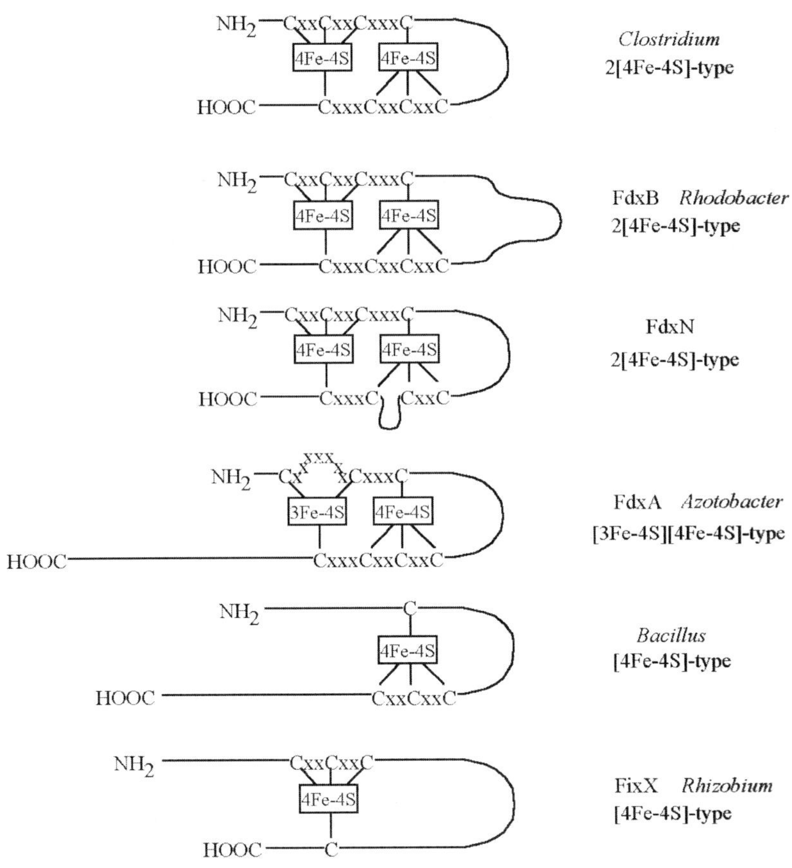

Figure 1. Schematic representation of various ferredoxins.
C and X indicate, respectively, cysteine residues involved in Fe-S cluster-binding and intermediate unspecified amino-acid residues.

In contrast to the situation with Flds, the formation of tight complexes between Fds and Fe proteins has been little studied. However, direct interaction between these proteins has been supported by the observation that FdxN from *R. capsulatus* could be cross-linked with the Fe protein of this organism in the presence of carbodiimide (Naud *et al.*, 1996).

2.2. Physiological Electron Carriers to the Fe Protein in Various Bacteria

Because *in vitro* biochemistry and *in vivo* molecular genetics are mutually complementary techniques that have been used to deduce physiological electron carriers to nitrogenase, this section will concentrate on five representative organisms that have been studied by both means. Other organisms, for which there is less genetic data, will be described in relation to these five organisms.

2.2.1. The NifF Protein is the Sole Electron Carrier in Klebsiella pneumoniae

Genetic manipulation of *K. pneumoniae* generated *nifF* and *nifJ* mutants that were impaired in the ability to fix N_2 *in vivo* but showed nitrogenase activity *in vitro* in the presence of dithionite (Roberts *et al.*, 1978; St. John *et al.*, 1975). Although cell extracts from *nifF* mutants displayed virtually no acetylene-reduction activity with pyruvate as reductant, addition of purified Fld from *A. chroococcum* restored activity to rates greater than those in wild-type extracts (Hill and Kavanagh, 1980). This *in vitro* complementation was confirmed using Fld purified from derepressed cells of either wild-type or *nifJ* mutants of *K. pneumoniae* (Nieva-Gomez *et al.*, 1980). These experiments demonstrated that the NifF product, Fld, is a sole electron carrier to the Fe protein. Later, nucleotide sequence analysis revealed that the *nifF* gene lies within the *nif* gene cluster (Arnold *et al.*, 1988).

More recently, a second Fld gene, *fldA,* was identified in *K. pneumoniae* (Achenbach and Genova, 1997). This gene is distinct from *nifF* and encodes a product more similar to *Escherichia coli* Fld than to the NifF protein. Its expression seems to be regulated by the Fur repressor and is, thus, repressed by the presence of high concentrations of iron. Expression of *fldA* is also derepressed by heat shock. However, it is not known if the FldA protein participates in nitrogen fixation under iron-starved conditions. Of note, *K. pneumoniae* strains carrying mutations in *nifF* have low, but significant, *in vivo* nitrogenase activity (Gennaro *et al.*, 1996), despite being unable to grow diazotrophically (Roberts *et al.*, 1978; St. John *et al.*, 1975).

2.2.2. The NifF Protein is the Major, but not Sole, Electron Carrier in Azotobacter vinelandii

A. vinelandii strain OP contains two major electron carriers, a [3Fe-4S][4Fe-4S]-type ferredoxin, Fd I (Yoch and Arnon, 1972), and an Fld (Benemann *et al.*, 1969; Hinkson and Bulen, 1967; Shethna *et al.*, 1966; van Lin and Bothe, 1972), both of which are effective electron donors to the Fe protein *in vitro* (Benemann *et al.*, 1971; Hageman and Burris, 1978b; Yoch and Arnon, 1972). Of the two, Fld is more important, as shown by independent gene-cloning and gene-disruption studies (Bennett *et al.*, 1988; Morgan *et al.*, 1988). Firstly, the *nifF* gene, which encodes Fld, is located within the major *nif*-gene cluster (Jacobson *et al.*, 1989), whereas the *fdxA* gene, which encodes Fd I, is remote from the major *nif*-gene cluster. Secondly, the *nifF* gene is associated with a characteristic *nif*-consensus promoter, which is preceded by a typical NifA-binding sequence, and its expression is up-regulated under N_2-fixing conditions, whereas the *fdxA* is constitutively expressed. Thirdly, disruption of the *nifF* gene caused a 30% decrease in acetylene-reduction activity by whole cells, although the disrupted mutant grew as fast under N_2-fixing conditions as the wild-type cells. In contrast, disruption of *fdxA* caused almost no difference in either acetylene-reduction activity or growth rate.

Interestingly, however, further genetic studies revealed that a *nifF* / *fdxA* double mutant still grew diazotrophically at a rate comparable to that of wild-type cells (Martin *et al.*, 1989). These experiments suggested that Fld is a major, but not the sole, electron carrier and that there must be one or more alternative electron carriers either to co-function with Fld or to substitute for Fld when it is not available. Candidates for the alternative electron carrier(s) include the products of several Fd-like genes (Gao-Sheridan *et al.*, 1998; Jacobson *et al.*, 1989; Raina *et al.*, 1988, 1993; Reyntjens *et al.*, 1997; Rodriguez-Quinones *et al.*, 1993; Wientjens, 1993; Yoch and Arnon, 1972) together with three iso-Flds that have been reported for *A. vinelandii* ATCC 478 (Klugkist *et al.*, 1986). The amino-acid sequences of three of these Fds possess one typical Cys-X_2-Cys-X_2-Cys-X_3-Cys motif and another more unusual motif of Cys-X_2-Cys-$X_{7\text{-}9}$-Cys-X_3-Cys. In this respect, they are similar to the FdxN proteins in *S. meliloti* and *R. capsulatus,* both of which have important role(s) in N_2 fixation (see below). Even so, the designation *fdxN* has been applied to the gene just downstream from *nifB* (Joerger and Bishop, 1988) and suggested designations for the other two Fd genes are *fixFd* and *vnfFd* (Reyntjens *et al.*, 1997).

The *A. vinelandii fdxN* and *fixFd* genes are each preceded by both a characteristic *nif*-consensus promoter and a NifA-binding site and may, therefore, have roles involving the conventional molybdenum-nitrogenase system. On the other hand, *vnfFd* is situated just downstream of the *vnfH* gene, which encodes the Fe protein of the vanadium-containing nitrogenase (Joerger *et al.*, 1990; Raina *et al.*, 1988). Disruption of the *nifB*-linked *fdxN* causes a decrease of about 60-to-65% in both whole-cell acetylene-reduction activity and *in vitro* Mo-nitrogenase activity (Rodriguez-Quinones *et al.*, 1993). Although the authors suggest that FdxN functions mainly in iron-molybdenum cofactor synthesis, cooperating with NifB protein, it is possible that the FdxN protein also donates electrons to the Fe protein. In this context, relevant information would accrue if strains doubly and/or triply disrupted in *fdxN, fixFd* and *nifF* were available and if the effects of purified FdxN protein in appropriate assays were defined.

In addition to Mo-nitrogenase, *A. vinelandii* possesses the alternative V-nitrogenase and Fe-only nitrogenase systems. Information concerning electron donors for the V- and Fe-only systems is very limited. Interestingly, expression of *fdxN* is approximately 50% down-regulated in the presence of vanadium (Rodriguez-Quinones *et al.*, 1993), whereas the *vnfFd* gene is expressed only in the absence of molybdenum and in the presence of vanadium (Raina *et al.*, 1993). Disruption of *vnfFd* results in an inability to grow when N_2 fixation depends on V-nitrogenase (Raina *et al.*, 1993). Thus, although the biochemical properties of VnfFd and the function of NifF under conditions that favor the alternative systems remain unclear, it is possible that VnfFd is a major electron donor to V-nitrogenase.

2.2.3. The 2[4Fe-4S]-type FdxN Protein is probably the Sole Electron Carrier in Sinorhizobium meliloti

Electron carriers in the alfalfa microsymbiont, *S. meliloti*, have been little studied at the protein level. However, the complete genome sequence of *S. meliloti* 1021 (Galibert et al., 2001; http://sequence.toulouse.inra.fr/meliloti.html) revealed that this bacterium contains as many as five ORFs that encode either Fd or Fd-like proteins, but no Fld gene. Two of these ORFs, designated *fdx* and *Y03875*, are located on the chromosome; the other three (*fdxN, fdxB* and *fixX*) are on the megaplasmid, pSymA. The *fdxN* gene is located close to *nifB* and encodes a product with one typical Cys-X_2-Cys-X_2-Cys-X_3-Cys sequence motif and another more unusual Cys-X_2-Cys-$X_{7\text{-}9}$-Cys-X_3-Cys sequence motif (Klipp *et al.*, 1989; Mulligan *et al.*, 1988). These features of *fdxN* in *S. meliloti* are similar to those of other genes designated *fdxN*. The *fdxN* gene is co-transcribed with upstream *nifA* and *nifB* genes, primarily under the control of a promoter in front of *nifA*, but also secondarily under the control of a characteristic *nif*-consensus promoter with two putative NifA-binding sites in front of *nifB* (Klipp *et al.*, 1989).

Disruption of *fdxN* in *S. meliloti* 2011 caused complete loss of the capacity for root nodules to fix N_2 (the Fix$^-$ phenotype), indicating an absolute requirement for FdxN for symbiotic N_2 fixation (Klipp *et al.*, 1989). The purified FdxN protein, expressed in *E. coli*, contains two [4Fe-4S] clusters and can donate electrons to *R. capsulatus* nitrogenase *in vitro* in the presence of illuminated chloroplasts (Riedel *et al.*, 1995). The resulting acetylene-reduction activity, promoted by *S. meliloti* FdxN, depends on the concentration of the FdxN protein and is comparable to the activity obtained using *R. capsulatus* FdxN. Unlike FdxN from *R. capsulatus*, which seems to stabilize the nitrogenase proteins (Jouanneau *et al.*, 1995), no similar function has been observed for FdxN from *S. meliloti* (Masepohl *et al.*, 1993). Taken together, these data indicate that the FdxN protein is probably the sole electron donor to the Fe protein in *S. meliloti*, although it probably has some other function(s). Of interest, the *fixX* gene is reported to be essential for N_2 fixation (Dusha *et al.*, 1987; Earl *et al.*, 1987), although no biochemical studies have been performed on the product of either this or the other Fd-like genes.

Other studies have been less conclusive. Although *Bradyrhizobium japonicum* has a *nifB*-linked *frxA* gene encoding a homologue of *S. meliloti* 2[4Fe-4S]-type FdxN, its disruption has little effect on diazotrophic capacity (Ebeling *et al.*, 1988). This negative result might be explained by another Fd-encoding gene, designated *fdxN*, that is located downstream of *aldA*, which encodes alanine dehydrogenase (Gottfert *et al.*, 2001). The whole genome sequence of *Mesorhizobium loti* MAFF303099 has disclosed that this organism contains seven Fd-like genes, five of which seem to be orthologues of *S. meliloti* genes, but no Fld (Kaneko *et al.*, 2000). One gene, *msl8750*, which is linked to *nifB*, is an orthologue of *S. meliloti fdxN*, however, no genetic analysis has yet been performed on this gene.

2.2.4. The 2[4Fe-4S]-type FdxN Protein and NifF Protein are Primary and Secondary Electron Carriers in Rhodobacter capsulatus *and possibly also in other Purple Photosynthetic Bacteria*

The *R. capsulatus* B10 derivative SB1003 contains six Fds, Fd I-VI, and an Fld (Table 1). Three Fds, Fd I-III, and the Fld have been purified from wild-type *R. capsulatus* cells (Hallenbeck *et al.*, 1982; Jouanneau *et al.*, 1990; 1993; Saeki *et al.*, 1990b; Yakunin *et al.*, 1993; Yakunin and Gogotov, 1983), whereas Fd VI has been purified from cells of a mutant of *R. capsulatus*, lacking Fd I (Naud *et al.*, 1994).

Table 1. Flavodoxin and ferredoxin genes in representative diazotrophs. Gene names are shown with protein names if available. Genes are underlined if they or their products have been demonstrated to function in electron transport to nitrogenase. Presence of ferredoxin genes in K. pneumoniae *is obscure.*

Species	*nifF*	*Fld*	*fdxN*	*fdxA*	*fdxB*	*Unnamed*	*fdxC*	*fdxD*	*Fdx*	*fixX*	*fdxH*	*pe petF*
	FMN	FMN	2[4Fe-4S]	[3Fe-4S] [4Fe-4S]	2[4Fe-4S]	2[4Fe-4S]	[2Fe-2S]	[2Fe-2S]	[2Fe-2S]	[4Fe-4S]?	[2Fe-2S]	[2Fe-2S]
	flavodoxin under contol of *nif*-promoter	flavodoxin not controlled by *nif*-promoter			located near *nifQ*				member of *isc*			
Anabaena PCC7120	–	*fldA*	*fdxN*			*asl2914*					*fdxH*	*petF, all2919, alr0784*
Azotobacter vinelandii	*nifF*		*fdxN, fixFd, vnfFd*	*fdxA* (FdI)				*fesII*	*isc*	*fixX*		
Klebsiella pneumoniae	*nifF*	*fldA*	?	?	?	?	?	?	?	?	?	?
Rhodobacter capsulatus	*nifF*		*fdxN* (FdI)	*fdxA* (FdII)	*fdxB* (FdIII)		*fdxC* (FdIV)	*fdxD* (FdV)	*fdxE* (FdVI)			
Sinorhizobium Meliloti	–	–	*fdxN*	*Y03875*	*fdxB*		–	–	*Fdx*	*fixX*	–	–

Genes for all seven electron carriers have been cloned and annotated as *fdxN, fdxA, fdxB, fdxC, fdxD* and *fdxE* for Fd I to Fd VI, respectively (Armengaud *et al.*, 1997; Grabau *et al.*, 1991; Moreno-Vivian *et al.*, 1989; Saeki *et al.*, 1990a, 1991; Schatt *et al.*, 1989; Schmehl *et al.*, 1993; Willison *et al.*, 1993), and *nifF* for Fld (Gennaro *et al.*, 1996).

The *fdxC* and *fdxN* genes (encoding Fd IV and Fd I, respectively) are contiguous and preceded by both a characteristic *nif*-consensus promoter and a NifA-binding site (Saeki *et al.*, 1993; Schmehl *et al.*, 1993), and their expression is regulated by availability of fixed nitrogen. The genes *fdxB* and *fdxD* (encoding Fd III and Fd V, respectively) are located in the vicinity of *nifQ* and *nifH*, respectively, and their expression is regulated in the same way as *fdxC* and *fdxN* (Armengaud *et al.*, 1994; Jouanneau *et al.*, 1993; Moreno-Vivian *et al.*, 1989; Willison *et al.*, 1993). Although *nifF* is distant from other *nif*-genes, it is preceded by both a characteristic *nif*-consensus promoter and a NifA-binding site and is regulated by availability of iron as well as by availability of fixed nitrogen (Gennaro *et al.*, 1996). In contrast, *fdxA* and *fdxE* (encoding Fd II and Fd VI, respectively) are expressed constitutively (Armengaud *et al.*, 1997; Duport *et al.*, 1992; Suetsugu *et al.*, 1991). Fd IV and Fd V, which are encoded by *fdxC* and *fdxD* respectively, have been purified from *E. coli* cells harboring expression plasmids and physico-chemically characterized (Armengaud *et al.*, 1994; Grabau *et al.*, 1991). These FdxC (Fd IV) and FdxD (Fd V) proteins, together with FdxE (Fd VI), are all [2Fe-2S]-type Fds, whereas FdxN (Fd I) and FdxB (Fd III) are 2[4Fe-4S]-type Fds and FdxA (Fd II) is a [3Fe-4S]-[4Fe-4S] type. The cysteine-sequence motifs of these Fds are typical (Figure 1).

Either in a coupled system with plant chloroplasts or after chemical reduction, purified FdxN, FdxA and NifF, can each mediate electron transfer to *R. capsulatus* Fe protein (Hallenbeck and Gennaro, 1998; Hallenbeck *et al.*, 1982; Naud *et al.*, 1996; Yakunin *et al.*, 1993; Yakunin and Gogotov, 1983), unlike either FdxB or FdxD (Armengaud *et al.*, 1994; Jouanneau *et al.*, 1993). The mid-point potentials of functional redox centers may be –490 and –474±5 mV, respectively, for the [4Fe-4S] cluster in FdxN (Naud *et al.*, 1996; Saeki *et al.*, 1996) and the SQ/HQ pair in NifF (Hallenbeck and Gennaro, 1998). Stopped-flow kinetic analysis has shown that the Fe protein and NifF form a tight complex with a K_D of 0.44 μM and that the order of affinity for the Fe protein is NifF > FdxN > FdxA (Hallenbeck and Gennaro, 1998). FdxN and Fe protein can be cross-linked *in vitro* with 1-ethyl-3-(3-dimethylaminopropyl)carbodiimide and a single substitution of the residue Asp-36 with His impairs the formation of cross-linked products (Naud *et al.*, 1996).

The physiological importance of FdxN and NifF is supported by gene-disruption and complementation studies. Mutants disrupted in the *fdxN* gene display very poor diazotrophic growth on normal medium containing iron, and their whole-cell acetylene-reduction activity is less than 5% of that of wild-type cells (Jouanneau *et al.*, 1995; Saeki *et al.*, 1991; 1996; Schmehl *et al.*, 1993). This phenotype cannot be complemented by up to 5-fold overexpression of *R. capsulatus* FdxA, indicating the specialized function of FdxN in N_2 fixation (Jouanneau *et al.*, 1995; Saeki *et al.*, 1996). When grown under *nif*-derepressed conditions, cells deleted in *fdxN* accumulate 2-3 times as much [2Fe-2S]-type FdxD protein than wild-type cells (Jouanneau *et al.*, 1995). Nonetheless, purified FdxD cannot mediate electron transport to purified nitrogenase *in vitro*, probably because of its relatively high mid-point redox potential of –220 mV (Armengaud *et al.*, 1994).

With permealized *fdxN* mutant cells, *in vitro* acetylene-reduction activity with dithionite is approximately 20% of that in similarly treated wild-type cells (Jouanneau *et al.*, 1995; Schmehl *et al.*, 1993). Further, pulse-chase labeling experiments demonstrate that this decrease in activity correlates with the instability of dinitrogenase in an *fdxN*-deficient background (Jouanneau *et al.*, 1995). These findings suggest that FdxN functions to stabilize dinitrogenase as well as fulfilling its role in electron transfer. Disruption of *nifF* results in complete loss of the capacity to grow diazotrophically on nitrogen-free medium containing limited iron, but has little effect on diazotrophic growth in normal medium containing iron. (Gennaro *et al.*, 1996). In the presence of iron, the *nifF*-deficient mutant displayed an *in vivo* rate of acetylene-reduction activity that was about 55% of that by wild-type cells, although, unlike a *fdxN*-deficient mutant, it contained as much nitrogenase enzyme as wild-type cells (Gennaro *et al.*, 1996). Taking these observations together, it is conceivable that: (a) under iron-replete conditions, FdxN is the major electron carrier and that NifF is a secondary carrier, which is needed to fulfill maximum nitrogenase activity; and (b) under iron-limited conditions, when formation of FdxN holoprotein might be limited, NifF functions as the sole electron carrier. Construction and characterization of a *nifF* / *fdxN* double mutant should resolve this matter, but this task remains to be done.

In addition to the Mo-nitrogenase system, *R. capsulatus* possesses an alternative Fe-only nitrogenase system encoded by the *anf* genes (Masepohl and Klipp, 1996; Schneider *et al.*, 1991). An *fdxN* mutant grew very poorly on N_2 under both Mo-replete and Mo-depleted conditions (Schüddekopf *et al.*, 1993), whereas a *nifF* mutant grew diazotrophically under both conditions (Gennaro *et al.*, 1996). These observations suggest that the two nitrogenase systems in *R. capsulatus* share electron-carrier proteins.

Four Fds and one Fld have been purified and characterized from another purple non-sulfur bacterium, *R. rubrum* (Cusanovich and Edmondson, 1971; Ishikawa and Yoch, 1995; Yoch and Arnon, 1975; Yoch *et al.*, 1975). *R. rubrum* Fd I, encoded by *fdxN*, is homologous to *R. capsulatus* FdxN (von Sternberg and Yoch, 1993) and so would likely be a major electron donor to the Fe protein. However, expression of *fdxN* in *R. rubrum* is not completely repressed in ammonium-replete cultures (von Sternberg and Yoch, 1993). There are no genetic studies on these systems.

The purple sulfur bacterium, *Allochromatium vinosum* (formerly *Chromatium vinosum*), possesses a 2[4Fe-4S]-type Fd encoded by an *fdx* gene (Imhoff *et al.*, 1998). Although structural and physical studies have been performed on this protein, its participation in N_2 fixation remains to be confirmed (Kyritsis *et al.*, 1998; Moulis *et al.*, 1996b).

2.2.5. The [2Fe-2S]-type FdxH Protein is the Primary, but not Sole, Electron Carrier in Anabaena *sp. PCC 7120 and probably also in other Cyanobacteria*

The filamentous heterocyst-forming cyanobacterium, *Anabaena* sp. PCC 7120, contains as many as six Fds and a Fld (Table 1). Among these, three electron carriers have been purified and extensively characterized, including structure determination. These carriers are: a constitutive [2Fe-2S]-type Fd, PetF, from vegetative cells (Rypniewski *et al.*, 1991); a heterocyst-specific [2Fe-2S]-type Fd, FdxH, (Jacobson et al., 1992; 1993); and a Fld (Paulsen *et al.*, 1990; Rao *et al.*, 1992). Of these, only the [2Fe-2S]-type FdxH protein mediates efficient *in vitro* electron transfer to Fe protein. Both PetF and Fld mediate the reaction only poorly (Razquin *et al.*, 1994; Schmitz *et al.*, 1993).

The genes encoding these three proteins have been cloned (Alam *et al.*, 1986; Böhme and Haselkorn, 1988; Leonhardt and Straus, 1989). Using the cloned *fdxH* gene, protein engineering has revealed that positively charged residues, especially Lys-10 and Lys-11, are important for efficient reaction with the Fe protein (Schmitz *et al.*, 1993). *Anabaena* sp. PCC 7120 also contains a split Fd gene (*fdxN*), which becomes an entity only after an excision and rearrangement event that occurs coincident with the formation of heterocysts (Golden *et al.*, 1987; 1988; Mulligan *et al.*, 1988; Mulligan and Haselkorn, 1989). The rearranged *fdxN* gene encodes a product with Cys sequence motifs similar to those of FdxN proteins from *S. meliloti* and *R. capsulatus*. In addition, the genome sequence of *Anabaena sp.* PCC 7120 shows that this cyanobacterium contains three more putative Fd genes: one ORF, *asl2914*, encodes a 2[4Fe-4S]-type protein, whereas the other two ORFs, *all2919* and *alr0784*, encode [2Fe-2S]-type proteins (Kaneko *et al.*, 2001).

Expression of *fdxH* in heterocysts is strongly induced in response to nitrogen starvation (Böhme and Haselkorn, 1988). However, its expression is little affected by iron-limitation, unlike the expression of the vegetative-cell gene, *petF* (Razquin *et al.*, 1994). Disruption of *fdxH* markedly decreases the capacity of cells to fix N_2. Under iron-replete conditions, their doubling time increases about 3-fold and whole-cell acetylene-reduction activity decreases to about 20% of that in wild-type cells (Masepohl *et al.*, 1997b). The decrease in the capacity for acetylene reduction is even more severe under iron-limited conditions (Masepohl *et al.*, 1997b). Because *Anabaena sp.* PCC 7120 produces Fld in response to iron limitation but not to nitrogen starvation (Razquin *et al.*, 1994), the poor capacity for diazotrophic growth in the *fdxH* mutant indicates that Fld may be a poor electron donor to Fe protein *in vivo*. Taken together, these observations suggest that the [2Fe-2S]-type FdxH protein is the primary electron donor to nitrogenase, but that there might be another protein that supports the residual diazotrophic capacity of the *fdxH* mutant. A candidate for this role is the product of the rearranged *fdxN* gene, but this remains to be investigated. However, disruption of the orthologous *fdxN* gene has been reported to have very little effect on the capacity for N_2 fixation in another heterocystous cyanobacterium, *Anabaena variabilis* ATCC29413 (Masepohl *et al.*, 1997a). Studies on double mutants might be informative.

Several *fdxH* genes, distinct from paralogous *petF* genes, have been cloned from both heterocystous and non-heterocystous cyanobacteria (Schrautemeier and Böhme, 1992; Schrautemeier *et al.*, 1994; 1995). The products of these *fdxH* genes possess the conserved Lys-10 and Lys-11 residues that are important for reaction with the Fe protein. Interestingly, the heterocystous species, *A. variabilis* ATCC29413, possesses two sets of *nif*-gene clusters with distinct *fdxH* genes, *fdxH1* and *fdxH2*. The identity between the two products is 78.6% (Schrautemeier *et al.*,

1995). Although both *fdxH* genes are expressed in response to nitrogen starvation, *fdxH1* is expressed only in heterocysts, whereas *fdxH2* is expressed only in vegetative cells early on in starvation. Significantly, the non-heterocystous cyanobacterium, *Plectonema boryanum PCC73110*, possesses an *fdxH* gene that encodes a product 73.5% identical to the product of *fdxH2* in *A. variabilis* ATCC29413 (Schrautemeier *et al.*, 1994). Both *fdxH* from *P. boryanum* and *fdxH2* from *A. variabilis* ATCC29413 are located downstream from homologues of *R. capsulatus fdxB*. The *P. boryanum* FdxH protein, produced in *E. coli*, mediates effective electron transfer to the Fe protein from *A. variabilis* ATCC29413, however, the *P. boryanum* PetF protein, also produced in *E. coli*, does not. It is, therefore, plausible that [2Fe-2S]-type FdxH proteins are major electron donors to nitrogenase, even in non-heterocystous cyanobacteria (Schrautemeier *et al.*, 1994).

2.2.6. The 2[4Fe-4S]-type Fd and Fld in Clostridium pasteurianum

Numerous studies on iron-sulfur proteins and Fld in *C. pasteurianum* have been compiled (see Meyer, 2000) and include the structures of both the Fe protein and the 2[4Fe-4S]-type Fd (Bertini *et al.*, 1995; Schlessman *et al.*, 1998). It is generally accepted that the 2[4Fe-4S]-type Fd, but not a [2Fe-2S]-type Fd, mediates electron transfer to the Fe protein, although definitive genetic evidence for this is lacking (Mortenson *et al.*, 1963). Although the gene for the 2[4Fe-4S]-type Fd has been cloned and characterized (Graves *et al.*, 1985; Graves and Rabinowitz, 1986), disruption studies have not been performed. In addition, no genetic analysis of the Fld gene has yet been performed.

2.3. Identification of Primary Electron Mediators - A Challenging Task?

Diazotrophs clearly display preference for specific electron-carrier proteins. In most cases, preferred electron carriers are restricted to *nif*-specific Flds, which are encoded by *nifF*, the 2[4Fe-4S]-type Fds (often encoded by genes designated *fdxN*), and the [2Fe-2S]-type Fds, encoded by *fdxH*. However, many diazotrophs have alternative electron carrier(s) that may function to optimize the performance of, or back-up, the primary carrier. Consequently, diazotrophs often contain multiple Fds and Flds. This situation is evident in organisms whose whole genome sequences have been determined and is true even for *K. pneumoniae* with its second Fld gene, *fldA* (Achenbach and Genova, 1997). Thus, it is difficult to predict functions of Fd-encoding and Fld-encoding genes based simply on sequence information.

3. REDUCTION OF ELECTRON-DONOR PROTEINS FOR THE FE PROTEIN

3.1. Pyruvate-Fld (Fd) Oxidoreductase

Pyruvate-Fd oxidoreductase (EC 1.2.7.1), often abbreviated as PFOR, PFO or POR, catalyzes the following reaction with either Fd or Fld as electron acceptor. Those PFORs that have a homodimeric structure are found in many eubacteria and amitochondriate eukaryotes and show a mutual identity of 50% or more (Horner *et al.*, 1999; Hrdy and Müller, 1995; Wahl and Orme-Johnson, 1987). Those PFORs of hetero-tetrameric structure also occur in archaea and show a total identity of 20 to 30% with the homodimeric enzymes (Kerscher and Oesterhelt, 1982; Kletzin and Adams, 1996; Smith *et al.*, 1994). Not all PFORs are related to N_2 fixation.

$$\text{pyruvate} + \text{CoASH} \leftrightarrow \text{acetyl-SCoA} + CO_2 + 2e^- + H^+ \quad E_o' = -540 \text{ mV}$$

3.1.1. The NifJ Protein is required for N_2 Fixation in K. pneumoniae

Disruption of *nifJ* in *K. pneumoniae* prevents diazotrophic growth in the same way as the disruption of *nifF*. Extracts of cells carrying a mutation in *nifJ* fail to display acetylene-reduction activity with pyruvate. Nevertheless, they are capable of reducing acetylene after addition of an artificial reductant, such as dithionite (Hill and Kavanagh, 1980; Nieva-Gomez *et al.*, 1980; Roberts *et al.*, 1978). The *nifJ* gene is located within the major *nif*-gene cluster and its expression is upregulated by nitrogen starvation (Arnold *et al.*, 1988). The purified *nifJ*-gene product catalyzes reduction of Fld at the expense of pyruvate and is an essential component in assays of pyruvate-coupled acetylene reduction by *K. pneumoniae* (Shah *et al.*, 1983). In *K. pneumoniae*, therefore, the pyruvate Fld oxidoreductase is an exclusive enzyme in supplying reducing equivalents, via Fld, to nitrogenase. The purified *K. pneumoniae* NifJ protein shows substrate specificity for Fld as an electron acceptor. Fld from either *K. pneumoniae* or *A. vinelandii* was effective as electron donor in the coupled-nitrogenase reaction, but Fds from either *C. pasteurianum* or *R. rubrum*, as well as viologen dyes, were not (Shah *et al.*, 1983).

NifJ has a relative molecular mass (M_r) of 240kDa and consists of two identical subunits of 1171 residues with both iron-sulfur clusters and thiamine pyrophosphate (TPP) present. Although there has been a dispute concerning the numbers of iron and sulfur atoms, the NifJ protein probably has three [4Fe-4S]-type clusters and one TPP per subunit. This estimate is based on the crystal structure of a pyruvate-Fd oxidoreductase from *Desulfovibrio africanus*, which shows 55% identity (and complete conservation of the amino acids required to ligate the iron-sulfur clusters and TPP) with known homodimeric PFORs (Chabriere *et al.*, 1999; George *et al.*, 1985; Pieulle *et al.*, 1997). Two of the [4Fe-4S] clusters are ligated by a domain similar to a 2[4Fe-4S]-type Fd, whereas another [4Fe-4S] cluster is ligated by a set of four Cys residues that are uniquely conserved among PFORs. The TPP cofactor is non-covalently bound by a conserved Gly-Asp-Gly motif together with a conserved Phe residue that stacks with the thiazole ring of TPP (Chabriere *et al.*, 1999). Decarboxylation of pyruvate progresses by interaction with the thiazole ring moiety (Chabriere *et al.*, 2001).

3.1.2. Pyruvate-Fld (Fd) Oxidoreductase in Clostridium pasteurianum

A PFOR that is homologous to *K. pneumoniae* NifJ protein occurs in *C. pasteurianum* and is assumed to support Fd (or Fld) reduction (Mortenson *et al.*, 1963). In total, three distinct PFOR genes, GI 5911352 (Horner *et al.*, 1999), GI 3139139

(Moulis *et al.*, 1996a), and GI 5911352, have been registered for *C. pasteurianum* in the DDBJ/EMBL/GenBank database, although mutational analysis has not been carried out with any of them. In addition for *C. pasteurianum*, an alternative pathway for Fd reduction has been proposed that consists of an NAD-linked glyceraldehyde phosphate reductase and an NADH-Fd reductase (Jungermann *et al.*, 1973). However, no subsequent support for this pathway has been obtained.

3.1.3. Pyruvate-Fld (Fd) Oxidoreductases that are not Essential for N_2 Fixation

Nitrogenase activity in crude extracts of *R. rubrum* can be supported by pyruvate and stimulated by addition of CoA and Fd (Ludden and Burris, 1981). In corroboration, PFORs have been purified from two photosynthetic bacteria, *R. rubrum* (Brostedt and Nordlund, 1991) and *R. capsulatus* (Yakunin and Hallenbeck, 1998), and demonstrated to support the nitrogenase reaction *in vitro*. However, the *R. rubrum* enzyme is not essential for N_2 fixation, because disruption of the PFOR gene resulted in almost no decline in the capacity to fix N_2 (Lindblad *et al.*, 1996). Moreover, expression of the PFOR is observed both under diazotrophic conditions and, at a lower but significant level, under ammonium-replete conditions (Brostedt and Nordlund, 1991). PFOR in *R. capsulatus* might similarly not be essential for N_2 fixation, despite its *in vitro* reactivity with FdxN and NifF, because the expression of this enzyme seems to be constitutive and is upregulated by only 1.3-fold under diazotrophic conditions (Yakunin and Hallenbeck, 1998).

A gene homologous to *K. pneumoniae nifJ* has been identified and disrupted in *Anabaena* sp. PCC 7120 (Bauer *et al.*, 1993). Expression of this *nifJ* gene is regulated by availability of iron. The *Anabaena nifJ* mutant could grow diazotrophically under iron-replete conditions but not when iron was limiting (Bauer *et al.*, 1993). This result indicates that NifJ is not required for N_2 fixation unless iron is limited. On the other hand, determination of the whole genome sequence has revealed that *Anabaena* sp. PCC 7120 contains a second *nifJ* gene (Kaneko *et al.*, 2001). The originally identified gene, *nifJ1*, corresponds to ORF *alr2803*, whereas the second copy of *nifJ* corresponds to *alr1911*. Transcriptional analysis using RT-PCR revealed that the second *nifJ* gene is expressed in *Anabaena* sp. PCC 7120 cells grown on N_2 under iron-replete conditions (Schmitz *et al.*, 2001). It would, therefore, be worthwhile investigating the role of its product in N_2 fixation. It might also be noted that aerobic expression of homologues of *nifJ* is observed in the unicellular non-diazotrophic cyanobacteria, *Synechocystis* sp. PCC 6803 and *Synochococcus* sp. PCC 6301 (Schmitz *et al.*, 2001), and that cyanobacterial *nifJ* products are more homologous to *R. rubrum* NifJ than to NifJ in *K. pneumoniae* and *C. pasteurianum* (Horner *et al.*, 1999).

3.2. Fd-NADP$^+$ Oxidoreductase: FNR and FPR

Fd-NADP$^+$ oxidoreductase (EC 1.18.1.2) catalyzes the reversible reaction:

$$2\ \mathrm{Fd}_{reduced}^{-} + \mathrm{NADP}^{+} + 2\mathrm{H}^{+} \leftrightarrow 2\ \mathrm{Fd}_{oxidized} + \mathrm{NADPH} + \mathrm{H}^{+}$$

Because most Fds have an E_o' that is more negative than that of the NADPH / $NADP^+$ couple (–340 mV), the forward reaction is energetically favorable. The forward reaction occurs, for example, with plant and cyanobacterial Fd-$NADP^+$ oxidoreductases (often abbreviated FNR) that use Fd reduced by photosystem I (Knaff, 1996). Nevertheless, enzymes (often abbreviated FPR) that apparently favor the reverse reaction are present in both *A. vinelandii* and *E. coli* (Bianchi *et al.*, 1993; Isas and Burgess, 1994; Liochev *et al.*, 1994; Morimyo, 1988).

3.2.1. FNR in Heterocystous Cyanobacteria

Results of biochemical and molecular biological experiments on the heterocystous cyanobacteria, *Anabaena* spp. PCC 7119 and PCC 7120, suggest strongly that Fd-$NADP^+$ oxidoreductase (FNR), which is encoded by the *petH* gene, mediates reduction of FdxH by NADPH generated, in turn, by degradation of carbohydrate through the oxidative pentose-phosphate cycle (Bothe and Neuer, 1988; Haselkorn and Buikema, 1992; Houchins, 1985). Unfortunately, disruption studies of *petH* are impossible because it is the only gene encoding FNR in *Anabaena* sp. PCC 7120 (Kaneko *et al.*, 2001) and is an essential gene for photosynthesis (Knaff, 1996). However, evidence that PetH has a role in N_2 fixation comes from the observation that cell-free extracts of heterocysts reduce acetylene when supplemented with any one of glucose, fructose, fructose-6-phosphate, fructose-1,6-bisphosphate, glucose-6-phosphate, 6-phosphogluconate, ribose-5-phosphate or dihydroxyacetone phosphate, all of which contribute to the generation of NADPH (Böhme, 1987; Böhme and Schrautemeier, 1987). In addition, the enzymic activity of PetH in heterocysts is about 10-times greater than that in vegetative cells (Razquin *et al.*, 1994). An apparent lack of involvement of PFOR (the NifJ protein) in acetylene reduction is suggested by the observation that pyruvate, on its own, did not support acetylene-reduction activity by extracts of *A. variabilis*, in contrast to the situation in *K. pneumoniae*. However, in the presence of a stoichiometric amount of oxaloacetate, pyruvate could support acetylene reduction by these extracts (Böhme and Schrautemeier, 1987). The concentration of PetH protein increases in response to nitrogen stress, a consequence of the fact that the *petH* gene possesses two promoters, one of which is constitutively expressed and the other regulated by the transcriptional regulator, NtcA (Valladares *et al.*, 1999).

The *petH* gene products in *Anabaena* species have a M_r of 49 kDa and consist of about 440 amino-acid residues with a non-covalently bound flavin adenine dinucleotide (FAD) cofactor. Cyanobacterial enzymes have an amino-terminal extension and are, therefore, larger than FNRs, which have a M_r of about 33 kDa. Recently, the crystal structure of the electron-transfer complex between PetH and the vegetative cell Fd, PetF, (both from *Anabaena* sp. PCC 7119) has been determined (Morales *et al.*, 2000). The structure of the corresponding complex between maize proteins has also been solved (Kurisu *et al.*, 2001). In these structures, the molecular surface of PetF, which participates in intermolecular contact, is different from the surface that, in the heterocystous Fd, FdxH, includes the residues Lys-10 and Lys-11, which are important for reaction with the Fe protein. Although Lys-10 and Lys-11 are not conserved between PetF and FdxH, these findings nonetheless suggest that FdxH may use different surfaces for its interaction with FNR and the Fe protein.

3.2.2. FPR is Unrelated to N_2 Fixation in A. vinelandii

An NADPH-Fd oxidoreductase (FPR) was purified as a polypeptide that is overproduced in an *fdxA* mutant of *A. vinelandii* (Isas and Burgess, 1994). The protein has an M_r of 29 kDa and contains FAD. Its gene, *fpr*, has been cloned (Isas *et al.*, 1995) and the crystal structure of the protein has been solved (Prasad *et al.*, 1998). *A. vinelandii* FPR is a homologue of *E. coli* NADPH-Fd oxidoreductase but is less similar to PetH and other plant FNRs (Isas *et al.*, 1995). Although upregulation of FPR was observed under diazotrophic conditions, no activity was observed *in vitro* in a coupled nitrogen-fixation system incubated with NADPH, an ATP-regeneration system, purified FPR, and a [3Fe-4S][4Fe-4S]-type FdxA (Jung *et al.*, 1999). Therefore, it is evident that FPR and FdxA do not constitute a physiological electron-transport pathway to nitrogenase in *A. vinelandii.*

3.3. Rnf Complex: A Putative Ion-translocating NADH-Fd (Fld) Oxidoreductase

Unlike PFOR and FNR, the Rnf complex described below remains a putative Fd(Fld)-reducing system whose enzyme activity has not been directly measured. However, several lines of evidence have indicated that this membrane-bound protein complex is an energy-coupling oxidoreductase that can transfer electrons to nitrogenase in *R. capsulatus* and some other bacteria.

*3.3.1. Identification of Peculiar N_2 Fixation Genes (*rnf*) in* R. capsulatus

In the *Rhodospirillaceae*, the photosynthetic reaction center uses light energy to perform a charge separation and to drive the photosynthetic electron-transfer chain. However, the redox potential of the primary quinone acceptor molecule (Q_A) of the photosystem is about –150 mV. This value is not negative enough to achieve reduction of either Fd or Fld (Prince and Dutton, 1978). Even NAD^+ is probably not directly reduced by the photosystem but by a process that involves reversed electron transport driven by an energized membrane (Knaff, 1978). In addition, although *R. capsulatus* possesses PFOR, this enzyme is not directly involved in N_2 fixation (Yakunin and Hallenbeck, 1998). Consequently, electrons for nitrogenase must be generated by some other energetically feasible means, especially when this organism grows on its commonly used carbon source, malate.

A candidate for generating reductant was found by nucleotide sequencing and genetic analysis around the *fdxN* gene, which encodes the major electron donor to nitrogenase, Fd I. The *fdxN* gene is surrounded by several genes required for N_2 fixation but not homologous to other known *nif* genes (Saeki *et al.*, 1993; Schmehl *et al.*, 1993). They are shown in Figure 2. Schmehl and co-workers designated these unique genes as *rnf* (for ***Rhodobacter*** **n**itrogen **f**ixation). Among the ten *rnf* genes that have been identified, two genes, *rnfFX-rnfFY*, are located immediately downstream of *fdxN* and are cotranscribed with *fprA-fdxC-fdxN*. The other eight genes, *rnfA-rnfB-rnfC-rnfD-rnfG-rnfE-rnfH-rnfI*, probably form an operon with a single *nif*-consensus promoter. The properties of these genes and their products are summarized in Table 2.

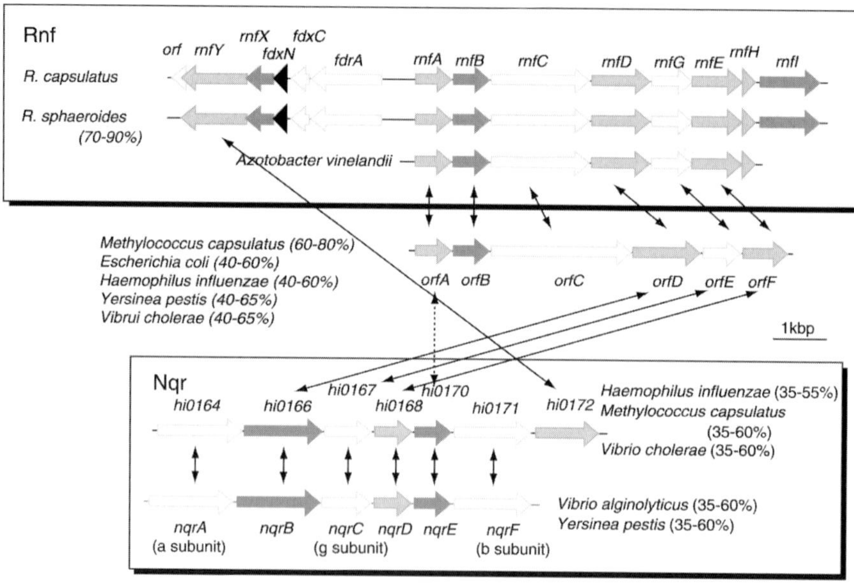

Figure 2. Organization of rnf *genes in* Rhodobacter capsulatus *and homologues in various bacteria. The upper panel displays orthologues of* rnf *and adjacent genes in two diazotrophs,* Rhodobacter sphaeriodes *and* Azotobacter vinelandii. *The lower panel displays representative* nqr *operons in marine bacteria and pathogenic bacteria. Four of the six* nqr *genes,* nqrB, nqrC, nqrD *and* nqrE, *have similarity with* rnf *genes,* rnfD, rnfG, rnfE *and* rnfA, *respectively. Paralogues of* rnfABCDGE *in non-diazotrophic bacteria are displayed in between the two panels. Values in parenthesis after species description indicate the range of identity with the* rnf *genes in* R. capsulatus. *The arrows link homologous genes with the same shading. Data have been taken from public databases.*

Disruption of any of the *rnf* genes, except *rnfH* for which mutants have not yet been obtained, results in an inability of *R. capsulatus* to grow by N_2 fixation under illuminated anaerobic conditions (Saeki *et al.*, 1993; Schmehl *et al.*, 1993). Notably, however, *rnf* mutants display a marked discrepancy between *in vivo* (measured with whole cells under anaerobically illuminated conditions) and *in vitro* (measured with permeabilized cells supplemented with dithionite and an ATP-regenerating system) acetylene-reduction activities. The *in vivo* activity of *rnf* mutants was less than 2% of that of wild-type cells, whereas the *in vitro* activity was about 20-25% (Schmehl *et al.*, 1993). The decline in *in vitro* activity is consistent with the decreased amount of cellular nitrogenase, which has been confirmed by immunological analysis (Schmehl *et al.*, 1993). Therefore, like the FdxN protein, *rnf* gene products might be involved in stabilization of nitrogenase components (Jouanneau *et al.*, 1995). However, the severe deterioration of *in vivo* activity in *rnf* mutants cannot be explained simply by the decrease in cellular nitrogenase and might indicate that *rnf*-gene products participate in electron transfer to nitrogenase.

Table 2. Properties of rnf *and adjacent genes, and their products.*

Gene	Effects of disruption[a-c,e]		Product properties			
	Nif	Anf	Residues	M_r	pI	Product property
rnfA	–	–	193	20,422	9.13	Transmembrane protein. Similar to NqrE.[b-d]
rnfB	–	+	187	19,089	4.72	Peripheral membrane protein. 2[4Fe-4S] and [2Fe-2S] clusters?[b-e]
rnfC	–	–	519	55,584	9.47	Peripheral membrane protein. 2x [4Fe-4S] clusters?[c-e]
rnfD	–	–	360	38,206	9.53	Transmembrane protein. Similar to NqrB.[c,g]
rnfG	–	–	217	22,846	5.93	Membrane anchored FMN-protein? Similar to NqrC.[e]
rnfE	–	–	243	25,865	8.33	Transmembrane protein. Similar to NqrD.[c,g]
rnfH	n.t.	n.t.	85	9,539	5.13	Unknown.[e]
rnfI	–	n.t.	340	34,979	11.30	Periplasmic? Similar to a flavoprotein subunit of flavocytochrome *c* sulfide dehydrogenase.[g]
fprA	+	+	435	47,910	5.66	FMN-protein. Interacts with FdxC.[b,c,f]
fdxC	+/–	n.t.	95	10,162	4.72	Ferredoxin, [2Fe-2S] cluster.
fdxN	+/–	+/–	64	6,732	3.89	Ferredoxin I, 2x [4Fe-4S] clusters.
rnfFX	–	n.t.	159	16,187	8.80	Fe-S cluster?[g,h]
rnfFY	–	–	366	38,307	4.84	Periplasmic? Secretory signal.[g,h]

[a]Schüddekopf *et al.*, 1993; [b]Saeki *et al.*, 1993; [c]Schmehl *et al.*, 1993; [d]Kumagai *et al.*, 1997; [e]Jouanneau *et al.*, 1997; [f]Jouanneau *et al.*, 2000; [g]Mito and Saeki, unpublished results; [h]Kumagai and Saeki, 2000; n.t., not tested.

Support for an electron-transfer role comes from the observation that, under growth conditions with L-serine as nitrogen source, which derepresses the expression of *nif*-related genes, *rnf* mutants were more resistant than wild-type cells to metronidazole (Schmehl *et al.*, 1993). This antimicrobial drug is believed to be converted to one or more lethal derivatives by accepting electrons from reduced Fd (Schmidt *et al.*, 1977). The probable participation of *rnf*-gene products in electron transfer to nitrogenase is further supported by overexpression of the *rnfABCDGEH* operon in wild-type cells of *R. capsulatus*. This manipulation elevated *in vivo* nitrogenase activity by 50-100% but did not affect either *in vitro* activity or intracellular concentrations of nitrogenase components (Jouanneau *et al.*, 1998).

In addition to its ability to fix N_2 photosynthetically under illuminated anaerobic conditions, *R. capsulatus* also fixes N_2 in the dark, either microaerobically (Meyer *et al.*, 1978) or anaerobically, with respiration supplemented by appropriate oxidants, such as dimethylsulfoxide (DMSO) (Madigan *et al.*, 1979). Functioning *rnf* genes are essential for N_2 fixation during anaerobic DMSO-dependent growth (Saeki and Kumagai, 1998). Whether it is also needed under microaerobic conditions remains to be demonstrated.

Because *R. capsulatus* has an alternative Fe-only nitrogenase system (*anf*) in addition to its Mo-nitrogenase system (*nif*), the involvement of *rnf*-gene function with the *anf* system was analyzed in strains deleted in *nifHDK*. Except for *rnfB*, all *rnf* genes that have been examined are essential for diazotrophic growth in the absence of molybdenum (Masepohl and Klipp, 1996; Schüddekopf *et al.*, 1993). Why *rnfB* is required for only the Mo-nitrogenase system is unclear.

3.3.2. Properties of rnf *Gene Products in* R. capsulatus

A remarkable feature of the *rnf*-gene products is that at least six, RnfA, RnfB, RnfC, RnfD, RnfG, and RnfE, are membrane bound and constitute a protein complex (Jouanneau *et al.*, 1998; Kumagai *et al.*, 1997; Mito and Saeki, unpublished results). The highly hydrophobic RnfA, RnfD, and RnfE have six, eight and at least four membrane-spanning regions, respectively, whereas RnfG, which has a N-terminal hydrophobic stretch, is probably anchored to the membrane. The remaining two more hydrophilic products, RnfB and RnfC, probably bind to membranes through protein-protein interactions with other Rnf proteins. The intracellular concentrations of RnfB and RnfC decrease significantly in all non-polar mutants of *rnfA, rnfB, rnfC, rnfD, rnfG* and *rnfE* (Jouanneau *et al.*, 1998; Kumagai *et al.*, 1997). In contrast, however, no significant decrease in either RnfB or RnfC has been observed in mutants of *rnfI, rnfFX* or *rnfFY*, suggesting that RnfI, RnfFX, and RnfFY do not belong to the putative membrane complex (Mito and Saeki, unpublished results).

The above features of *rnf* mutants and their derivatives are consistent with the assumption that some of the *rnf* products constitute an electron-transport pathway to nitrogenase. If this assumption is correct, at least one of the Rnf proteins must possess both a redox center(s) that interacts with FdxN and/or NifF and a site that interacts with some electron donor (or acceptor in the case of the reverse reaction). Furthermore, if Rnf products mediate reduction of FdxN (or NifF) at the expense of some electron donor that has a more positive redox potential, there must be a biochemical mechanism that sustains the energetically unfavorable reaction.

Both the *rnfB*- and *rnfC*-gene products possess iron-sulfur cluster(s), which were inferred by their amino-acid sequences and directly demonstrated in products expressed in *E. coli*. The *rnfB*-gene product contains twelve Cys residues of which eight constitute tandem Cys-X_2-Cys-X_2-Cys-X_3-Cys motifs usually found in 2[4Fe-4S]-type Fds (Figure 1), whereas the remaining four constitute another motif, Cys-X_2-Cys-X_4-Cys-$X_{(16-17)}$-Cys, which is also found in subunits of corrinoid/FeS proteins and carbon monoxide dehydrogenases (Eggen *et al.*, 1991; Lu *et al.*, 1993; Saeki *et al.*, 1993). Overexpression of *rnfB* in *E. coli* results in an accumulation of brown inclusion bodies formed by a polypeptide of the expected size. Absorption and EPR spectroscopic analysis of N-terminally $(His)_6$-tagged protein, after being solublized from such inclusion bodies, suggested that RnfB is highly unstable but possesses at least one [2Fe-2S]-type cluster with an EPR signal at $g_{x,y,z}$ of 1.85, 1.93, 2.065 (Jouanneau *et al.*, 1998). Likewise, the *rnfC*-gene product contains eight Cys residues that constitute a tandem Cys-X_2-Cys-X_2-Cys-X_3-Cys motif homologous to that in 2[4Fe-4S]-type Fds. Overexpression of *rnfC* in *E. coli* also results in production of a highly unstable brown protein of the expected size. Spectroscopic analysis of partially purified C-terminally $(His)_6$-tagged protein suggested that RnfC possesses at least one [4Fe-4S] cluster (Jouanneau *et al.*, 1998). In accordance with these analyses, an EPR signal, g=1.84, has been observed in membrane fractions anaerobically prepared from *nif*-derepressed wild-type cells but not in fractions from cells carrying a polar mutation in *rnfA* (Figure 3; T. Ohnishi and K. Saeki, unpublished results). It is, therefore, probable that RnfB and RnfC contain iron-sulfur center(s) *in vivo*. Of note, both RnfC and RnfG have a predicted FMN-binding site (as described below).

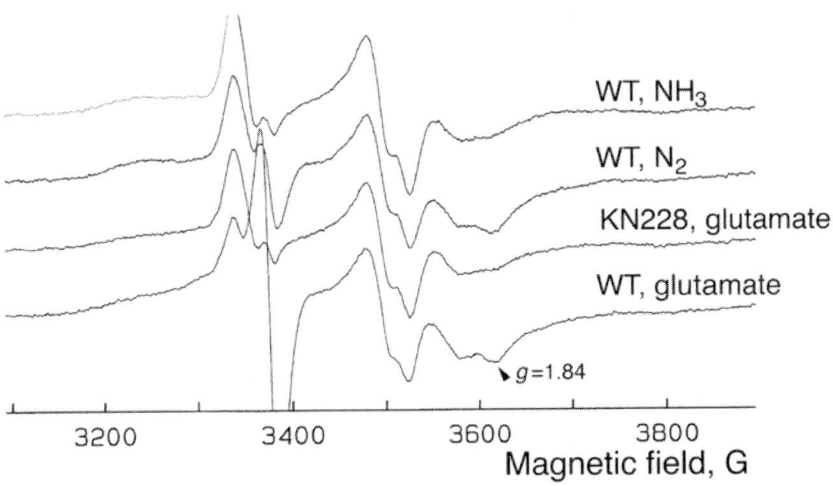

Figure 3. Electron paramagnetic resonance spectra of membrane fractions prepared from Rhodobacter capsulatus *wild-type (WT) and from a polar* rnfA-*mutant strain (KN228). Cells were grown anaerobically under illumination with either N_2 or glutamate and then disrupted in order to collect a chromatophore membrane fraction. The anaerobically prepared membranes were reduced with dithionite in the presence of both 0.1 mM methyl viologen and 0.1 mM benzyl viologen.*

3.3.3. Functional Role for rnf *Gene Products by Sequence Comparisons*

The presence of sites that interact with electron carriers has been inferred by comparing segments of the amino-acid sequences of *rnf* products with those of other redox enzymes. Kumagai *et al.* (1997) reported that RnfC protein possesses a potential NADH-binding motif that shows homology with the probable NADH-binding site found in H^+-translocating NADH-quinone oxidoreductases (H^+-NQRs) and in NAD^+-dependent hydrogenases (NAD^+-H_2ases). This motif, Gly-X_2-Gly-X-Gly-Gly-Ala-X-Phe-Pro-Ser/Thr-X_2-Lys-X_5-Lys, does not completely conform to the commonly accepted consensus NADH-binding motif, Gly-X-Gly-X_2-Gly-X_3-Gly (Scrutton *et al.*, 1990), but the underlined residues are almost completely conserved among *R. capsulatus* RnfC and its close homologues (Figure 2), the NuoF or Nqo1 subunits of bacterial H^+-NQR, the 51kDa flavoprotein subunit of mitochondrial complex I, and NAD^+-H_2ases (Patel *et al.*, 1991; Yagi, 1993). In addition, the predicted secondary structure of this region in RnfC forms a βαβ-fold. Interestingly, the proposed NADH-binding site in RnfC is followed by a sequence segment that is similar to the proposed FMN-binding site in H^+-NQRs and NAD^+-H_2ases (Patel *et al.*, 1991; Weiss *et al.*, 1991), although RnfC does not show overall similarity to these enzymes. Hence, NADH is a potential electron donor for FdxN (or NifF). However, it is not yet biochemically proved whether NADH binds RnfC and effectively donates electrons to FdxN and/or NifF.

Because the redox potentials of FdxN and NifF are, respectively, about –490 mV and –470 mV and much more negative than the redox potential of NADH (–320 mV), reduction of these electron-carrier proteins by NADH would require either a high ratio of NADH/NAD^+ or some form of energy consumption. Clues to a possible energy-transduction mechanism emerged from the similarity of hydrophobic Rnf proteins to the putative subunits of Na^+-translocating NADH-quinone oxidoreductases (Na^+-NQRs). These proteins function as an energy-conserving primary sodium pump that generates an electrochemical Na^+ gradient (Hayashi *et al.*, 2001a; Steuber, 2001). Na^+-NQR was first found in the marine bacterium, *Vibrio alginolyticus* (Tokuda and Unemoto, 1981), and later detected in another marine bacterium, *Vibrio harveyi* (Zhou *et al.*, 1999), and the pathogenic bacteria, *Haemophilus influenzae* (Hayashi *et al.*, 1996) and *Vibrio cholerae* (Barquera *et al.*, 2002). The best studied Na^+-NQR in *V. alginolyticus* is regarded as a membrane-protein complex, consisting of six subunits that are encoded by six genes in the *nqrABCDEF* operon (Nakayama *et al.*, 1998). Subunit sequences and the gene order are highly conserved in *V. harveyi, V. cholerae,* and *H. influenzae*, so the structure of Na^+-NQR is also believed to be conserved. None of the subunits of Na^+-NQRs shows any apparent similarity with the subunits of H^+-NQRs. Investigation of completely and partially sequenced microbial genomes has revealed that close homologues of the *nqrABCDEF* operon are present in a number of pathogenic bacteria, including *Neisseria gonorrhoeae, Neisseria meningitidis, Porphyromonas gingivalis, Pseudomonas aeruginosa,* and *Yersinia pestis* (Hase and Barquera, 2001; Zhou *et al.*, 1999).

Three of the gene products, NqrA, NqrC and NqrF, are relatively hydrophilic proteins that were identified in the initial three-subunit Na^+-NQR preparation that showed NADH-dehydrogenase and ubiquinone reductase activities (Hayashi and

Unemoto, 1984). The NqrA, NqrC and NqrF products correspond, respectively, to the α, γ, and β subunits of Na^+-NQR. NqrF (the β subunit) is the so-called NADH dehydrogenase fragment that is reported to contain a [2Fe-2S]-type cluster and FAD. The two other subunits, NqrC (the γ subunit) and the hydrophobic NqrB, both possess covalently bound FMN (Hayashi *et al.*, 2001b; Nakayama *et al.*, 2000). Although experimental data are lacking for the function of the subunits other than NqrF, it is hypothesized that the highly hydrophobic subunits, NqrB, NqrD and NqrE, span the cytoplasmic membrane multiple times and participate in ion-translocation (Rich *et al.*, 1995). These three hydrophobic *nqr* products from various species show about 25-40% sequence identity with the polytopic membrane proteins, RnfA, RnfD and RnfE, in *R. capsulatus* (Figure 2), suggesting functional and topological similarity. In addition, the relatively hydrophilic *nqrC* proteins are about 20% identical with RnfG in *R. capsulatus*, with significant conservation of the sequence around the Thr residue that binds FMN (Figure 4).

Combining all the above evidence, it has been proposed that Rnf proteins constitute an energy-coupling protein complex that reduces FdxN (or NifF) at the expense of NADH (Kumagai *et al.*, 1997). The proteins may be organized into two subcomplexes with distinct functions. One relatively hydrophilic subcomplex consists of RnfB and RnfC, which has some resemblance to the FP subcomplex of H^+-NQR and interacts with NADH and FdxN (or NifF). The other transmembrane subcomplex consists of RnfA, RnfD, RnfE and RnfG, resembles the hydrophobic portion of Na^+-NQR, and exploits a transmembrane ion-motive force. It should be stressed, however, that this hypothesis is as yet untested and the function of each subcomplex remains to be experimentally demonstrated.

Vibrio alginolyticus	NqrC	`220-QGSEHGVDGLSGATLTSNGVQHTF`
Rhodobacter capsulatus	RnfG	`171-KRDGGVFDQFSGATITPRAVVKTI`

Figure 4. Alignment of the amino-acid sequence around the FMN-linked Thr of NqrC and RnfG. Flavin-linked Thr is highlighted by inverse contrast. Residues conserved among various NqrC orthologues are doubly underlined.

3.3.4. Orthologues and Paralogues of rnf*-Gene Products in Various Organisms*

When the *rnf* genes were first genetically identified, no apparent homologues were known. With time, however, a few orthologues and several paralogues have been found in various microbial genomes. A complete set of the 10 *rnf* genes, together with *fprA, fdxC* and *fdxN*, is conserved in the genome of *Rhodobacter sphaeroides* 2.4.1, another purple non-sulfur photosynthetic bacterium, and closely related to *R. capsulatus* (JGI, DOE Joint Genome Institute; http:// www.jgi.doe.gov/index.html). Each orthologue in *R. sphaeroides* shows similarity of more than 70% with *R. capsulatus*, although no genetic analysis has been performed with this organism. In contrast to this complete conservation in the two *Rhodobacter* species, no *rnf* homologues could be identified in the genomes of two other purple non-sulfur photosynthetic bacteria, the completely sequenced *Rhodopseudomonas palustris*

CGA009 and the partially sequenced *R. rubrum* ATCC 11170 (both at the JGI website), suggesting that there is divergence of the electron-transport system to nitrogenase among members of the *Rhodospirillaceae*. The absence of *rnf* in *R. rubrum* S-1 is also supported by Southern hybridization analysis using fragments of *R. capsulatus rnf* genes as probes (K. Saeki, unpublished results).

Apparent orthologues of *rnf* have, however, been found in two non-photosynthetic diazotrophs, *A. vinelandii* OP (JGI website; Rubio *et al.*, 2002) and *Pseudomonas stutzeri* A15 (nucleotide sequence accession number GI:9843874). An orthologous *rnfABCDGEH* operon is present in *A. vinelandii* and the 3' portion of the operon, *rnfCDGEH*, is found in *P. stutzeri*. It remains to be seen whether the *rnf* orthologues have role(s) in nitrogen fixation in these non-phototrophic obligate aerobes. It is also noteworthy that both photosynthetic *R. palustris* and non-photosynthetic *A. vinelandii* have a *fixABCX* operon in addition to the *rnf* operon, although the physiological role of *fixABCX* is also not yet known.

Interestingly, but enigmatically, as genome sequencing of various organisms has proceeded, a number of non-diazotrophs have been found to possess six consecutive ORFs whose predicted products show 40-80% similarity with the six products of the *rnfABCDGE* operon (Figure 2). Although no physiological function has been assigned to these *rnf* paralogues, it is reported that the paralogous operon in *E. coli* is co-transcribed with a downstream *nth* gene that encodes endonuclease III, an oxidative base-excision repair DNA glycosylase (Gifford and Wallace, 2000), which contains a [4Fe-4S]-type cluster (Kuo *et al.*, 1992).

4. CONCLUSIONS

Accumulated mutational studies indicate that many diazotrophs have multiple pathways that mediate electron flow to nitrogenase (Figure 5). These pathways include multiple electron-transfer proteins (Fds and Flds) and multiple oxidoreductases (PFOR, FNR, the Rnf complex and other enzymes to be identified). Such redundancy, like the redundancy of Mo-nitrogenase and the alternative nitrogenase systems, might reflect the importance of N_2 fixation for survival of bacteria. Another explanation is that, by possessing subsystems (or subengines), diazotrophs either maximize or fine-tune their capacity for N_2 fixation to cope with various conditions.

ACKNOWLEDGEMENTS

The author thanks Dr. Y. Fujita for helpful discussion on cyanobacterial systems. The work of the author's group is supported in part by a Grant-in-Aid for Scientific Research from the Ministry of Education, Culture, Sports, Science and Technology of Japan (#14540594 and #15013232) and by a grant from the Program for Promotion of Basic Research Activities for Innovative Bioscience (PROBRAIN).

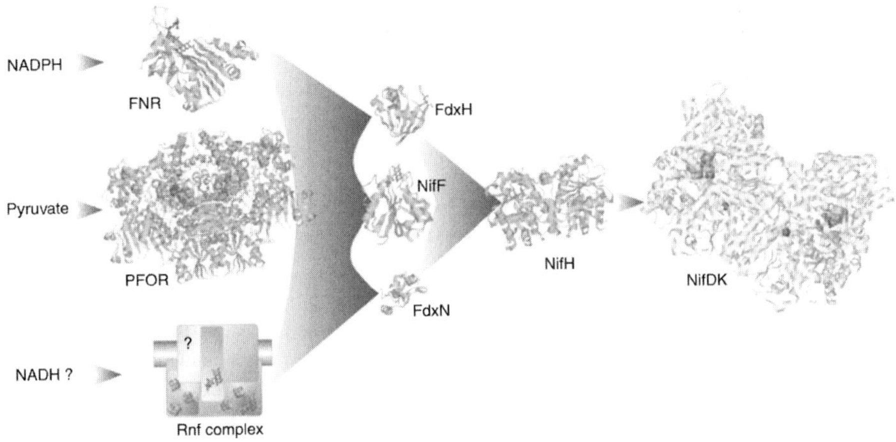

Figure 5. Schematic pathway of electron donation to nitrogenase. Structures of representative proteins are shown. FNR, ferredoxin-NADP$^+$ oxidoreductase (PDB coordinate used, 1QUE); PFOR, pyruvate ferredoxin oxidoreductase (1B0P); FdxH, heterocyst-specific [2Fe-2S]-type ferredoxin (1FRD); NifF, flavodoxin (1FLV); FdxN, 2[4Fe-4S]-type ferredoxin (1BLU); NifH, Fe protein (1G5P); NifDK, MoFe protein (1M1N). The putative structure of the Rnf complex is drawn from available information (Jouanneau et al., 1998; Kumagai et al., 1997; Mito and Saeki, unpublished results).

REFERENCES

Achenbach, L. A., and Genova, E. G. (1997). Transcriptional regulation of a second flavodoxin gene from *Klebsiella pneumoniae*. *Gene, 194*, 235-240.

Alam, J., Whitaker, R. A., Krogman, D. W., and Curtis, S. E. (1986). Isolation and sequence of the gene for ferredoxin I from the cyanobactium *Anabaena* sp. strain PCC 7120. *J. Bacteriol., 168*, 1265-1271.

Armengaud, J., Meyer, C., and Jouanneau, Y. (1994). Recombinant expression of the *fdxD* gene of *Rhodobacter capsulatus* and characterization of its product, a [2Fe-2S] ferredoxin. *Biochem. J., 300*, 413-418.

Armengaud, J., Meyer, C., and Jouanneau, Y. (1997). A [2Fe-2S] ferredoxin (FdVI) is essential for growth of the photosynthetic bacterium *Rhodobacter capsulatus*. *J. Bacteriol., 179*, 3304-3309.

Arnold, W., Rump, A., Klipp, W., Priefer, U. B., and Pühler, A. (1988). Nucleotide sequence of a 24,206-base-pair DNA fragment carrying the entire nitrogen fixation gene cluster of *Klebsiella pneumoniae*. *J. Mol. Biol., 203*, 715-738.

Barquera, B., Hellwig, P., Zhou, W. D., Morgan, J. E., Hase, C. C., Gosink, K. K., *et al.* (2002). Purification and characterization of the recombinant Na$^+$-translocating NADH:quinone oxidoreductase from *Vibrio cholerae*. *Biochemistry, 41*, 3781-3789.

Bauer, C., Scappino, L., and Haselkorn, R. (1993). Growth of the cyanobacterium *Anabaena* on molecular nitrogen: NifJ is required when iron is limited. *Proc. Natl. Acad. Sci. U.S.A., 90*, 8812-8816.

Beinert, H., and Kiley, P. J. (1999). Fe-S proteins in sensing and regulatory functions. *Curr. Opin. Chem. Biol., 3*, 152-157.

Benemann, J. R., Yoch, D. C., Valentine, R. C., and Arnon, D. I. (1969). The electron transport system in nitrogen fixation by *Azotobacter*. I. Azotoflavin as an electron carrier. *Proc. Natl. Acad. Sci. U.S.A., 64*, 1079-1086.

Benemann, J. R., Yoch, D. C., Valentine, R. C., and Arnon, D. I. (1971). The electron transport system in nitrogen fixation by *Azotobacter*. 3. Requirements for NADPH-supported nitrogenase activity. *Biochim. Biophys. Acta, 226*, 205-212.

Bennett, L. T., Jacobson, M. R., and Dean, D. R. (1988). Isolation, sequencing, and mutagenesis of the *nifF* gene encoding flavodoxin from *Azotobacter vinelandii*. *J. Biol. Chem., 263*, 1364-1369.

Bertini, I., Donaire, A., Feinberg, B. A., Luchinat, C., Piccioli, M., and Yuan, H. (1995). Solution structure of the oxidized 2[4Fe-4S] ferredoxin from *Clostridium pasteurianum*. *Eur. J. Biochem., 232*, 192-205.

Bianchi, V., Reichard, P., Eliasson, R., Pontis, E., Krook, M., Jornvall, H., *et al.* (1993). *Escherichia coli* ferredoxin NADP$^+$ reductase: activation of *E. coli* anaerobic ribonucleotide reduction, cloning of the gene (*fpr*), and overexpression of the protein. *J. Bacteriol., 175*, 1590-1595.

Böhme, H. (1987). Regulation of electron flow to nitrogenase in a cell free system from heterocysts of *Anabaena variabilis*. *Biochim. Biophys. Acta, 819*, 121-128.

Böhme, H., and Haselkorn, R. (1988). Molecular cloning and nucleotide sequence analysis of the gene coding for heterocyst ferredoxin from the cyanobacterium *Anabaena* sp. strain PCC 7120. *Mol. Gen. Genet., 214*, 278-285.

Böhme, H., and Schrautemeier, B. (1987). Electron donation to nitrogenase in a cell free system from heterocysts of *Anabaena variabilis*. *Biochim. Biophys. Acta, 819*, 115-120.

Bothe, H., and Neuer, G. (1988). Electron donation to nitrogenase in heterocysts. *Methods Enzymol., 167*, 496-501.

Brostedt, E., and Nordlund, S. (1991). Purification and partial characterization of a pyruvate oxido-reductase from the photosynthetic bacterium *Rhodospirillum rubrum* grown under nitrogen-fixing conditions. *Biochem. J., 279*, 155-158.

Bruschi, M., and Guerlesquin, F. (1988). Structure, function and evolution of bacterial ferredoxins. *FEMS Microbiol. Rev., 4*, 155-175.

Cammack, R. (1992). Iron-sulfur clusters in enzymes: themes and variations. *Adv. Inorganic Chem., 38*, 281-322.

Chabriere, E., Charon, M. H., Volbeda, A., Pieulle, L., Hatchikian, E. C., and Fontecilla-Camps, J. C. (1999). Crystal structures of the key anaerobic enzyme pyruvate:ferredoxin oxidoreductase, free and in complex with pyruvate. *Nature Struct. Biol., 6*, 182-190.

Chabriere, E., Vernede, X., Guigliarelli, B., Charon, M. H., Hatchikian, E. C., and Fontecilla-Camps, J. C. (2001). Crystal structure of the free radical intermediate of pyruvate:ferredoxin oxidoreductase. *Science, 294*, 2559-2563.

Cusanovich, M. A., and Edmondson, D. E. (1971). The isolation and characterization of *Rhodospirillum rubrum* flavodoxin. *Biochem. Biophys. Res. Commun., 45*, 327-336.

Duport, C., Jouanneau, Y., and Vignais, P. M. (1992). Transcriptional analysis and promoter mapping of the *fdxA* gene which encodes the 7Fe ferredoxin (FdII) of *Rhodobacter capsulatus*. *Mol. Gen. Genet., 231*, 323-328.

Dusha, I., Kovalenko, S., Banfalvi, Z., and Kondorosi, A. (1987). *Rhizobium meliloti* insertion element ISRm2 and its use for identification of the *fixX* gene. *J. Bacteriol., 169*, 1403-1409.

Duyvis, M. G., Wassink, H., and Haaker, H. (1998). Nitrogenase of *Azotobacter vinelandii*: Kinetic analysis of the Fe protein redox cycle. *Biochemistry, 37*, 17345-17354.

Earl, C. D., Ronson, C. W., and Ausubel, F. M. (1987). Genetic and structural analysis of the *Rhizobium meliloti fixA, fixB, fixC,* and *fixX* genes. *J. Bacteriol., 169*, 1127-1136.

Ebeling, S., Noti, J. D., and Hennecke, H. (1988). Identification of a new *Bradyrhizobium japonicum* gene (*frxA*) encoding a ferredoxinlike protein. *J. Bacteriol., 170*, 1999-2001.

Eggen, R. I., Geerling, A. C., Jetten, M. S., and de Vos, W. M. (1991). Cloning, expression, and sequence analysis of the genes for carbon monoxide dehydrogenase of *Methanothrix soehngenii*. *J. Biol. Chem., 266*, 6883-6887.

Freigang, J., Diederichs, K., Schafer, K. P., Welte, W., and Paul, R. (2002). Crystal structure of oxidized flavodoxin, an essential protein in *Helicobacter pylori*. *Protein Sci., 11*, 253-261.

Fukuyama, K. (2001). Ferredoxins containing one [4Fe-4S] center. In T. Poulos (Ed.), *Handbook of metalloproteins* (pp. 543-552). Chichester: John Wiley and Sons.

Gadkari, D., Morsdorf, G., and Meyer, O. (1992). Chemolithoautotrophic assimilation of dinitrogen by *Streptomyces thermoautotrophicus* UBT1: Identification of an unusual N_2-fixing system. *J. Bacteriol., 174*, 6840-6843.

Galibert, F., Finan, T. M., Long, S. R., Pühler, A., Abola, P., Ampe, F., *et al.* (2001). The composite genome of the legume symbiont *Sinorhizobium meliloti*. *Science, 293*, 668-672.

Gao-Sheridan, H. S., Pershad, H. R., Armstrong, F. A., and Burgess, B. K. (1998). Discovery of a novel ferredoxin from *Azotobacter vinelandii* containing two [4Fe-4S] clusters with widely differing and very negative reduction potentials. *J. Biol. Chem., 273*, 5514-5519.

Gennaro, G., Hübner, P., Sandmeier, U., Yakunin, A., and Hallenbeck, P. (1996). Cloning, characterization, and regulation of *nifF* from *Rhodobacter capsulatus*. *J. Bacteriol., 178*, 3949-3952.

George, D. G., Hunt, L. T., Yeh, L. S., and Barker, W. C. (1985). New perspectives on bacterial ferredoxin evolution. *J. Mol. Evol., 22*, 20-31.

Gifford, C. M., and Wallace, S. S. (2000). The genes encoding endonuclease VIII and endonuclease III in *Escherichia coli* are transcribed as the terminal genes in operons. *Nucleic Acids Res., 28*, 762-769.

Golden, J. W., Carrasco, C. D., Mulligan, M. E., Schneider, G. J., and Haselkorn, R. (1988). Deletion of a 55-kilobase-pair DNA element from the chromosome during heterocyst differentiation of *Anabaena* sp. strain PCC 7120. *J. Bacteriol., 170*, 5034-5041.

Golden, J. W., Mulligan, M. E., and Haselkorn, R. (1987). Different recombination site specificity of two developmentally regulated genome rearrangements. *Nature, 327*, 526-529.

Gottfert, M., Rothlisberger, S., Kundig, C., Beck, C., Marty, R., and Hennecke, H. (2001). Potential symbiosis-specific genes uncovered by sequencing a 410-kilobase DNA region of the *Bradyrhizobium japonicum* chromosome. *J. Bacteriol., 183*, 1405-1412.

Grabau, C., Schatt, E., Jouanneau, Y., and Vignais, P. M. (1991). A new [2Fe-2S] ferredoxin from *Rhodobacter capsulatus*. Coexpression with a 2[4Fe-4S] ferredoxin in *Escherichia coli*. *J. Biol. Chem., 266*, 3294-3299.

Graves, M. C., Mullenbach, G. T., and Rabinowitz, J. C. (1985). Cloning and nucleotide sequence determination of the *Clostridium pasteurianum* ferredoxin gene. *Proc. Natl. Acad. Sci. U.S.A., 82*, 1653-1657.

Graves, M. C., and Rabinowitz, J. C. (1986). *In vivo* and *in vitro* transcription of the *Clostridium pasteurianum* ferredoxin gene. Evidence for 'extended' promoter elements in Gram-positive organisms. *J. Biol. Chem., 261*, 11409-11415.

Haaker, H., and Klugkist, J. (1987). The bioenergetics of electron transport to nitrogenase. *FEMS Microbiol. Rev., 46*, 57-71.

Hageman, R. V., and Burris, R. H. (1978a). Kinetic studies on electron transfer and interaction between nitrogenase components from *Azotobacter vinelandii*. *Biochemistry, 17*, 4117-4124.

Hageman, R. V., and Burris, R. H. (1978b). Nitrogenase and nitrogenase reductase associate and dissociate with each catalytic cycle. *Proc. Natl. Acad. Sci. U.S.A., 75*, 2699-2702.

Hallenbeck, P. C., and Gennaro, G. (1998). Stopped-flow kinetic studies of low potential electron carriers of the photosynthetic bacterium, *Rhodobacter capsulatus*: Ferredoxin I and NifF. *Biochim. Biophys. Acta, 1365*, 435-442.

Hallenbeck, P. C., Jouanneau, Y., and Vignais, P. M. (1982). Purification and molecular properties of a soluble ferredoxin from *Rhodopseudomonas capsulata*. *Biochim. Biophys. Acta, 681*, 168-176.

Hase, C. C., and Barquera, B. (2001). Role of sodium bioenergetics in *Vibrio cholerae*. *Biochim. Biophys. Acta, 1505*, 169-178.

Haselkorn, R., and Buikema, W. J. (1992). Nitrogen fixation in cyanobacteria. In G. Stacey, R. H. Burris, and H. J. Evans (Eds.), *Biological nitrogen fixation* (pp. 166-190). New York: Chapman and Hall.

Hayashi, M., Nakayama, Y., and Unemoto, T. (2001a). Recent progress in the Na^+-translocating NADH-quinone reductase from the marine *Vibrio alginolyticus*. *Biochim. Biophys. Acta, 1505*, 37-44.

Hayashi, M., Nakayama, Y., and Uneomoto, T. (1996). Existence of Na^+-translocating NADH-quinone reductase in *Haemophilus influenzae*. *FEBS Lett., 381*, 174-176.

Hayashi, M., Nakayama, Y., Yasui, M., Maeda, M., Furuishi, K., and Unemoto, T. (2001b). FMN is covalently attached to a threonine residue in the NqrB and NqrC subunits of Na^+-translocating NADH-quinone reductase from *Vibrio alginolyticus*. *FEBS Lett., 488*, 5-8.

Hayashi, M., and Unemoto, T. (1984). Characterization of the Na^+-dependent respiratory chain NADH:quinone oxidoreductase of the marine bacterium, *Vibrio alginolyticus*, in relation to the primary Na^+ pump. *Biochim. Biophys. Acta, 767*, 470-478.

Hidalgo, E., and Demple, B. (1994). An iron-sulfur center essential for transcriptional activation by the redox-sensing SoxR protein. *EMBO J., 13*, 138-146.

Hill, S., and Kavanagh, E. P. (1980). Roles of *nifF* and *nifJ* gene products in electron transport to nitrogenase in *Klebsiella pneumoniae*. *J. Bacteriol., 141*, 470-475.

Hinkson, J. W., and Bulen, W. A. (1967). A free radical flavoprotein from *Azotobacter*. Isolation, crystallization, and properties. *J. Biol. Chem., 242*, 3345-3351.
Horner, D., Hirt, R., and Embley, T. (1999). A single eubacterial origin of eukaryotic pyruvate: ferredoxin oxidoreductase genes: Implications for the evolution of anaerobic eukaryotes. *Mol. Biol. Evol., 16*, 1280-1291.
Houchins, J. P. (1985). Electron transfer chains of cyanobacterial heterocysts. In P. W. Ludden and J. E. Burris (Eds.), *Nitrogen fixation and CO_2 metabolism* (pp. 261-268). New York: Elsevier Science.
Hrdy, I., and Müller, M. (1995). Primary structure and eubacterial relationships of the pyruvate:ferredoxin oxidoreductase of the amitochondriate eukaryote *Trichomonas vaginalis*. *J. Mol. Evol., 41*, 388-396.
Iismaa, S. E., and Watson, J. M. (1989). The *nifA* gene product from *Rhizobium leguminosarum* biovar *trifolii* lacks the N-terminal domain found in other NifA proteins. *Mol. Microbiol., 3*, 943-955.
Imhoff, J., Suling, J., and Petri, R. (1998). Phylogenetic relationships among the Chromatiaceae, their taxonomic reclassification and description of the new genera *Allochromatium, Halochromatium, Isochromatium, Marichromatium, Thiococcus, Thiohalocapsa* and *Thermochromatium*. *Int. J. Syst. Bacteriol., 48*, 1129-1143.
Isas, J. M., and Burgess, B. K. (1994). Purification and characterization of a $NADP^+$/NADPH-specific flavoprotein that is overexpressed in FdI^- strains of *Azotobacter vinelandii*. *J. Biol. Chem., 269*, 19404-19409.
Isas, J. M., Yannone, S. M., and Burgess, B. K. (1995). *Azotobacter vinelandii* NADPH:ferredoxin reductase cloning, sequencing, and overexpression. *J. Biol. Chem., 270*, 21258-21263.
Ishikawa, Y., and Yoch, D. C. (1995). Amino acid sequence of ferredoxin II from the phototroph *Rhodospirillum rubrum*: Characteristics of a 7Fe ferredoxin. *Photosyn. Res., 46*, 371-376.
Iwasaki, T., Suzuki, T., Kon, T., Imai, T., Urushiyama, A., Ohmori, D., et al. (1997). Novel zinc-containing ferredoxin family in thermoacidophilic archaea. *J. Biol. Chem., 272*, 3453-3458.
Jacobson, B. L., Chae, Y. K., Böhme, H., Markley, J. L., and Holden, H. M. (1992). Crystalization and preliminary analysis of oxidized recombinant, heterocyst [2Fe-2S] ferredoxin from *Anabaena* 7120. *Arch. Biochem. Biophys., 294*, 279-281.
Jacobson, B. L., Chae, Y. K., Markley, J. L., Raymont, I., and Holden, H. M. (1993). Molecular structure of the oxidized, recombinant, heterocyst [2Fe-2S] ferredoxin from *Anabaena* 7120 determined to 1.7-Å resolution. *Biochemistry, 32*, 6788-6793.
Jacobson, M. R., Brigle, K. E., Bennett, L. Y., Setterquist, R. A., Wilson, M. S., Cash, V. L., *et al.* (1989). Physical and genetic map of the major *nif* gene cluster from *Azotobacter vinelandii*. *J. Bacteriol., 171*, 1017-1027.
Joerger, R. D., and Bishop, P. E. (1988). Nucleotide sequence and genetic analysis of the *nifB-nifQ* region from *Azotobacter vinelandii*. *J. Bacteriol., 170*, 1475-1487.
Joerger, R. D., Loveless, T. M., Pau, R. N., Mitchenall, L. A., Simon, B. H., and Bishop, P. E. (1990). Nucleotide sequences and mutational analysis of the structural genes for nitrogenase 2 of *Azotobacter vinelandii*. *J. Bacteriol., 172*, 3400-3408.
Jouanneau, Y., Jeong, H. S., Hugo, N., Meyer, C., and Willison, J. C. (1998). Overexpression in *Escherichia coli* of the *rnf* genes from *Rhodobacter capsulatus* - characterization of two membrane-bound iron-sulfur proteins. *Eur. J. Biochem., 251*, 54-64.
Jouanneau, Y., Meyer, C., Gaillard, J., Forest, E., and Gagnon, J. (1993). Purification and characterization of a novel dimeric ferredoxin (FdIII) from *Rhodobacter capsulatus*. *J. Biol. Chem., 268*, 10636-10644.
Jouanneau, Y., Meyer, C., Gaillard, J., and Vignais, P. M. (1990). Purification and characterization of a 7Fe-ferredoxin from *Rhodobacter capsulatus*. *Biochem. Biophys. Res. Commun., 171*, 273-279.
Jouanneau, Y., Meyer, C., Naud, I., and Klipp, W. (1995). Characterization of an *fdxN* mutant of *Rhodobacter capsulatus* indicates that ferredoxin I serves as electron donor to nitrogenase. *Biochim. Biophys. Acta, 1232*, 33-42.
Jung, Y.-S., Roberts, V. A., Stout, C. D., and Burgess, B. K. (1999). Complex formation between *Azotobacter vinelandii* ferredoxin I and its physiological electron donor NADPH-ferredoxin reductase. *J. Biol. Chem., 274*, 2978-2987.
Jungermann, K., Thauer, R. K., Leimenstoll, G., and Decker, K. (1973). Function of reduced pyridine nucleotide-ferredoxin oxidoreductases in saccharolytic clostridia. *Biochim. Biophys. Acta, 305*, 268-280.

Kaneko, T., Nakamura, Y., Sato, S., Asamizu, E., Kato, T., Sasamoto, S., et al. (2000). Complete genome structure of the nitrogen-fixing symbiotic bacterium *Mesorhizobium loti*. *DNA Res.*, *7*, 331-338.

Kaneko, T., Nakamura, Y., Wolk, C. P., Kuritz, T., Sasamoto, S., Watanabe, A., et al. (2001). Complete genomic sequence of the filamentous nitrogen-fixing cyanobacterium *Anabaena* sp. strain PCC 7120. *DNA Res.*, *8*, 205-213.

Kerscher, L., and Oesterhelt, D. (1982). Pyruvate:ferredoxin oxidoreductase - new findings on ancient enzyme. *Trends Biochem. Sci.*, *7*, 371-374.

Kletzin, A., and Adams, M. (1996). Molecular and phylogenetic characterization of pyruvate and 2-ketoisovalerate ferredoxin oxidoreductases from *Pyrococcus furiosus* and pyruvate ferredoxin oxidoreductase from *Thermotoga maritima*. *J. Bacteriol.*, *178*, 248-257.

Klipp, W., Reiländer, H., Schlüter, A., Krey, R., and Pühler, A. (1989). The *Rhizobium meliloti fdxN* gene encoding a ferredoxin-like protein is necessary for nitrogen fixation and is cotranscribed with *nifA* and *nifB*. *Mol. Gen. Genet.*, *216*, 293-302.

Klugkist, J., Voorberg, J., Haaker, H., and Veeger, C. (1986). Characterization of three different flavodoxins from *Azotobacter vinelandii*. *Eur. J. Biochem.*, *155*, 33-40.

Knaff, D. (1978). Reducing potentials and the pathway of NAD reduction. In W. R. Sistrom (Ed.), *The photosynthetic bacteria* (pp. 629-640). New York: Plenum Press.

Knaff, D. (1996). Ferredoxin and ferredoxin-dependent enzymes. In C. F. Yocum (Ed.), *Oxygenic photosynthesis: The light reaction* (pp. 333-361). Dordrecht: Kluwer Academic Publishers.

Knight, Jr., E., and Hardy, R. W. (1966). Isolation and characteristics of flavodoxin from nitrogen-fixing *Clostridium pasteurianum*. *J. Biol. Chem.*, *241*, 2752-2756.

Knight, Jr., E., and Hardy, R. W. (1967). Flavodoxin. Chemical and biological properties. *J. Biol. Chem.*, *242*, 1370-1374.

Kumagai, H., Fujiwara, T., Matsubara, H., and Saeki, K. (1997). Membrane localization, topology, and mutual stabilization of the *rnfABC* gene products in *Rhodobacter capsulatus* and implications for a new family of energy-coupling NADH oxidoreductases. *Biochemistry*, *36*, 5509-5521.

Kuo, C. F., McRee, D. E., Fisher, C. L., O'Handley, S. F., Cunningham, R. P., and Tainer, J. A. (1992). Atomic structure of the DNA repair [4Fe-4S] enzyme endonuclease III. *Science*, *258*, 434-440.

Kurisu, G., Kusunoki, M., Katoh, E., Yamazaki, T., Teshima, K., Onda, Y., *et al.* (2001). Structure of the electron transfer complex between ferredoxin and ferredoxin-NADP$^+$ reductase. *Nature Struct. Biol.*, *8*, 117-121.

Kyritsis, P., Hatzfeld, O. M., Link, T. A., and Moulis, J.-M. (1998). The two [4Fe-4S] clusters in *Chromatium vinosum* ferredoxin have largely different reduction potentials. Structural origin and functional consequences. *J. Biol. Chem.*, *273*, 15404-15411.

Leonhardt, K. G., and Straus, N. A. (1989). Sequence of the flavodoxin gene from *Anabaena variabilis* 7120. *Nuc. Acids Res.*, *17*, 4384.

Lindblad, A., Jansson, J., Brostedt, E., Johansson, M., Hellman, U., and Nordlund, S. (1996). Identification and sequence of a *nifJ*-like gene in *Rhodospirillum rubrum*: Partial characterization of a mutant unaffected in nitrogen fixation. *Mol. Microbiol.*, *20*, 559-568.

Liochev, S., Hausladen, A., Beyer, Jr, W., and Fridovich, I. (1994). NADPH:ferredoxin oxidoreductase acts as a paraquat diaphoras and is a member of the *soxRS* regulon. *Proc. Natl. Acad. Sci. U.S.A.*, *91*, 1328-1331.

Lu, W.-P., Schiau, I., Cunningham, J., and Ragsdale, S. W. (1993). Sequence and expression of the gene encoding the corrinoid/iron-sulfur protein from *Clostridium thermoaceticum* and reconstitution of the recombinant protein to full activity. *J. Biol. Chem.*, *268*, 5605-5614.

Ludden, P. W., and Burris, R. H. (1981). *In vivo* and *in vitro* studies on ATP and electron donors to nitrogenase from *Rhodospirillum rubrum*. *Arch. Microbiol.*, *130*, 155-158.

Madigan, M. T., Wall, J. D., and Gest, H. (1979). Dark anaerobic dinitrogen fixation by a photosynthetic microorganism. *Science 204, 1429-1430.*

Martin, A. E., Burgess, B. K., Iismaa, S. E., Smartt, C. T., Jacobson, M. R., and Dean, D. R. (1989). Construction and characterization of an *Azotobacter vinelandii* strain with mutations in the genes encoding flavodoxin and ferredoxin I. *J. Bacteriol.*, *171*, 3162-3167.

Masepohl, B., Görlitz, K., Monnerjahn, U., Mosler, B., and Böhme, H. (1997a). The ferredoxin-encoding *fdxN* gene of the filamentous cyanobacterium *Anabaena variabilis* ATCC 29413 is not essential for nitrogen fixation. *New Phytol.*, *136*, 419-423.

Masepohl, B., and Klipp, W. (1996). Organization and regulation of genes encoding the molybdenum nitrogenase and the alternative nitrogenase in *Rhodobacter capsulatus*. *Arch. Microbiol.*, *165*, 80-90.

Masepohl, B., Kutsche, M., Riedel, K.-U., Schmehl, M., Klipp, W., and Pühler, A. (1992). Functional analysis of the cysteine motifs in the ferredoxin-like protein FdxN of *Rhizobium meliloti* involved in symbiotic nitrogen fixation. *Mol. Gen. Genet., 233*, 33-41.

Masepohl, B., Schölisch, K., Görlitz, K., Kutzki, C., and Böhme, H. (1997b). The heterocyst-specific *fdxH* gene product of the cyanobacterium *Anabaena* sp. PCC 7120 is important but not essential for nitrogen fixation. *Mol. Gen. Genet., 253*, 770-776.

Masepohl, B., Witty, J. F., Riedel, K.-U., Klipp, W., and Pühler, A. (1993). *Rhizobium meliloti* mutants defective in symbiotic nitrogen fixation affect the oxygen gradient in alfalfa (*Medicago sativa*) root nodules. *J. Exp. Bot., 44*, 419-426.

Matsubara, H., and Saeki, K. (1992). Structural and functional diversity of ferredoxins and related proteins. *Adv. Inorg. Chem., 38*, 223-280.

Mayhew, S. G., and Ludwig, M. I. (1975). *Flavodoxins and electron-tranferring flavoproteins*. New York: Academic Press.

Meyer, J. (2000). Clostridial iron-sulphur proteins. *J. Mol. Microbiol. Biotechnol., 2*, 9-14.

Meyer, J., Kelley, B. C., and Vignais, P. M. (1978). Aerobic nitrogen fixation by *Rhodopseudomonas capsulata*. *FEBS Lett., 85*, 224-228.

Morales, R., Charon, M.-H., Kachalova, G., Serre, L., Medina, M., Gomez-Moreno, C., *et al.* (2000). A redox-dependent interaction between two electron-transfer partners involved in photosynthesis. *EMBO Reports, 1*, 271-276.

Moreno-Vivian, C., Hennecke, S., Pühler, A., and Klipp, W. (1989). Open reading frame 5 (ORF5), encoding a ferrdoxinlike protein, and *nifQ* are cotranscribed with *nifE, nifN, nifX,* and ORF4 in *Rhodobacter capsulatus*. *J. Bacteriol., 171*, 2591-2598.

Morgan, T. V., Lundell, D. J., and Burgess, B. K. (1988). *Azotobacter vinelandii* ferredoxin I: cloning, sequencing, and mutant analysis. *J. Biol. Chem., 263*, 1370-1375.

Morimyo, M. (1988). Isolation and characterization of methyl viologen-sensitive mutants of *Escherichia coli* K-12. *J. Bacteriol., 170*, 2136-2142.

Mortenson, L. E., Valentine, R. C., and Carnahan, J. E. (1962). An electron transport factor from *Clostridium pasteurianum*. *Biochem. Biophys. Res. Commun., 7*, 448-452.

Mortenson, L. E., Valentine, R. C., and Carnahan, J. E. (1963). Ferredoxin in phosphoroclastic reaction of pyruvic acid and its relation to nitrogen fixation in *Clostridium pasteurianum*. *J. Biol. Chem., 238*, 794-800.

Moulis, J. M., Davasse, V., Meyer, J., and Gaillard, J. (1996a). Molecular mechanism of pyruvate-ferredoxin oxidoreductases based on data obtained with the *Clostridium pasteurianum* enzyme. *FEBS Lett., 380*, 287-290.

Moulis, J. M., Sieker, L. C., Wilson, K. S., and Dauter, Z. (1996b). Crystal structure of the 2[4Fe-4S] ferredoxin from *Chromatium vinosum*: evolutionary and mechanistic inferences for [3/4Fe-4S] ferredoxins. *Protein Sci., 5*, 1765-1775.

Mulligan, M. E., Buikema, W. J., and Haselkorn, R. (1988). Bacterial-type ferredoxin genes in the nitrogen fixation regions of the cyanobacterium *Anabaena* sp. strain PCC 7120 and *Rhizobium meliloti*. *J. Bacteriol., 170*, 4406-4410.

Mulligan, M. E., and Haselkorn, R. (1989). Nitrogen fixation (*nif*) genes of the cyanobacterium *Anabaena* species strain PCC 7120: the *nifB-fdxN-nifS-nifU* operon. *J. Biol. Chem., 264*, 19200-19207.

Nakamura, M., Saeki, K., and Takahashi, Y. (1999). Hyperproduction of recombinant ferredoxins in *Escherichia coli* by coexpression of the *ORF1-ORF2-iscS-iscU-iscA-hscB-hscA-fdx-ORF3* gene cluster. *J. Biochem., 126*, 10-18.

Nakayama, Y., Hayashi, M., and Unemoto, T. (1998). Identification of six subunits constituting Na^+-translocating NADH-quinone reductase from the marine *Vibrio alginolyticus*. *FEBS Lett., 422*, 240-242.

Nakayama, Y., Yasui, M., Sugahara, K., Hayashi, M., and Unemoto, T. (2000). Covalently bound flavin in the NqrB and NqrC subunits of Na^+-translocating NADH-quinone reductase from *Vibrio alginolyticus*. *FEBS Lett., 474*, 165-168.

Naud, I., Meyer, C., David, L., Breton, J., Gaillard, J., and Jouanneau, Y. (1996). Identification of residues of *Rhodobacter capsulatus* ferredoxin I important for its intraction with nitrogenase. *Eur. J. Biochem., 237*, 399-405.

Naud, I., Vinçon, M., Garin, J., Gaillard, J., Forest, E., and Jouanneau, Y. (1994). Purification of a sixth ferredoxin from *Rhodobacter capsulatus*: primariy structure and biochemical properties. *Eur. J. Biochem., 222*, 933-939.

Nieva-Gomez, D., Roberts, G. P., Klevickis, S., and Brill, W. J. (1980). Electron transport to nitrogenase in *Klebsiella pneumoniae. Proc. Natl. Acad. Sci. U.S.A., 77*, 2555-2558.

Otaka, E., and Ooi, T. (1989). Examination of protein sequence homologies: V. New perspectives on evolution between bacterial and chloroplast-type ferredoxins inferred from sequence evidence. *J. Mol. Evol., 29*, 246-254.

Patel, S. D., Aebersold, R., and Attardi, G. (1991). cDNA-derived amino acid sequence of the NADH-binding 51-kDa subunit of the bovine respiratory NADH dehydrogenase reveals striking similarities to a bacterial NAD^+-reducing hydrogenase. *Proc. Natl. Acad. Sci. U.S.A., 88*, 4225-4229.

Paulsen, K. E., Stankovich, M. T., Stockman, B. J., and Markley, J. L. (1990). Redox and spectral properties of flavodoxin from *Anabaena* 7120. *Arch. Biochem. Biophys., 280*, 68-73.

Pieulle, L., Magro, V., and Hatchikian, E. C. (1997). Isolation and analysis of the gene encoding the pyruvate-ferredoxin oxidoreductase of *Desulfovibrio africanus*, production of the recombinant enzyme in *Escherichia coli*, and effect of carboxy-terminal deletions on its stability. *J. Bacteriol., 179*, 5684-5692.

Prasad, G. S., Kresge, N., Muhlberg, A. B., Shaw, A., Jung, Y. S., Burgess, B. K., *et al.* (1998). The crystal structure of NADPH:ferredoxin reductase from *Azotobacter vinelandii. Protein Sci, 7*, 2541-2549.

Prince, R. C., and Dutton, P. L. (1978). Protonation and the reducing potential of the primary electron acceptor. In W. R. Sistrom (Ed.), *The photosynthetic bacteria* (pp. 439-453). New York: Plenum Press.

Raina, R., Bageshwar, U. K., and Das, H. K. (1993). The ORF encoding a putative ferredoxin-like protein downstream of the *vnfH* gene in *Azotobacter vinelandii* is involved in the vanadium-dependent alternative pathway of nitrogen fixation. *Mol. Gen. Genet., 236*, 459-462.

Raina, R., Reddy, M. A., Ghosal, D., and Das, H. K. (1988). Characterization of the gene for the Fe-protein of the vanadium dependent alternative nitrogenase of *Azotobacter vinelandii* and construction of a Tn*5* mutant. *Mol. Gen. Genet., 214*, 121-127.

Rao, S. T., Shaffie, F., Yu, C., Satyshur, K. A., Stockman, B. J., Markley, J. L., *et al.* (1992). Structure of the oxidized long-chain flavodoxin from *Anabaena* 7120 at 2 Å resolution. *Protein Sci., 1*, 1413-1427.

Razquin, P., Schmitz, S., Fillat, M., Peleato, M., and Böhme, H. (1994). Transcriptional and translational analysis of ferredoxin and flavodoxin under iron and nitrogen stress in *Anabaena* sp. strain PCC 7120. *J. Bacteriol., 176*, 7409-7411.

Reyntjens, B., Jollie, D. R., Stephens, P. J., Gao-Sheridan, H. S., and Burgess, B. K. (1997). Purification and characterization of a *fixABCX*-linked 2[4Fe-4S] ferredoxin from *Azotobacter vinelandii. J. Biol. Inorg. Chem., 2*, 595-602.

Ribbe, M., Gadkari, D., and Meyer, O. (1997). N_2 Fixation by *Streptomyces thermoautotrophicus* involves a molybdenum-dinitrogenase and a manganese-superoxide oxidoreductase that couple N_2 reduction to the oxidation of superoxide produced from O_2 by a molybdenum-CO dehydrogenase. *J. Biol. Chem., 272*, 26627-26633.

Rich, P. R., Meunier, B., and Ward, F. B. (1995). Predicted structure and possible ionmotive mechanism of the sodium-linked NADH-ubiquinone oxidoreductase of *Vibrio alginolyticus. FEBS Lett., 375*, 5-10.

Riedel, K.-U., Jouanneau, Y., Masepohl, B., Pühler, A., and Klipp, W. (1995). A *Rhizobium meliloti* ferredoxin (FdxN) purified from *Escherichia coli* donates electrons to *Rhodobacter capsulatus* nitrogenase. *Eur. J. Biochem., 231*, 742-746.

Roberts, G. P., MacNeil, T., MacNeil, D., and Brill, W. J. (1978). Regulation and characterization of protein products coded by the *nif* (nitrogen fixation) genes of *Klebsiella pneumoniae. J. Bacteriol., 136*, 267-279.

Rodriguez-Quinones, F., Bosch, R., and Imperial, J. (1993). Expression of the *nifBfdxNnifO*Q region of *Azotobacter vinelandii* and its role in nitrogenase activity. *J. Bacteriol., 175*, 2926-2935.

Rubio, L. M., Rangaraj, P., Homer, M. J., Roberts, G. P., and Ludden, P. W. (2002). Cloning and mutational analysis of the gamma gene from *Azotobacter vinelandii* defines a new family of proteins capable of metallocluster binding and protein stabilization. *J. Biol. Chem., 277*, 14299-14305.

Rypniewski, W. R., Breiter, D. R., Benning, M. M., Wesenberg, G., Oh, B. H., Markley, J. L., *et al.* (1991). Crystalization and structure determination to 2.5 Å resolution of the oxidized [2Fe-2S] ferredoxin isolated from *Anabaena* 7120. *Biochemistry, 30*, 4126-4131.

Saeki, K., and Kumagai, H. (1998). The *rnf* gene products in *Rhodobacter capsulatus* play an essential role in nitrogen fixation during anaerobic DMSO-dependent growth in the dark. *Arch. Microbiol., 169*, 464-467.

Saeki, K., Miyatake, Y., Young, D. A., Marrs, B. L., and Matsubara, H. (1990a). A plant ferredoxin-like gene is located upstream of ferredoxin I gene (*fdxN*) of *Rhodobacter capsulatus*. *Nucleic Acids Res., 18*, 1060.

Saeki, K., Suetsugu, Y., Tokuda, K., Miyataka, Y., Young, D. A., Marrs, B. L., *et al.* (1991). Genetic analysis of functional difference among distinct ferredoxins in *Rhodobacter capsulatus*. *J. Biol. Chem., 266*, 12889-12895.

Saeki, K., Suetsugu, Y., Yao, Y., Horio, T., Marrs, B. L., and Matsubara, H. (1990b). Two distinct ferredoxins from *Rhodobacter capsulatus*: Complete amino acid sequences and molecular evolution. *J. Biochem., 108*, 475-482.

Saeki, K., Tokuda, K., Fujiwara, T., and Matsubara, H. (1993). Nucleotide sequence and genetic analysis of the region essential for functional expression of the gene for ferredoxin I, *fdxN*, in *Rhodobacter capsulatus*: Sharing of one upstream activator sequence in opposite direction by two operons related to nitrogen fixation. *Plant Cell Physiol., 34*, 185-199.

Saeki, K., Tokuda, K., Fukuyama, K., Matsubara, H., Nadanami, K., Go, M., *et al.* (1996). Site-specific mutagenesis of *Rhodobacter capsulatus* ferredoxin I, FdxN, that functions in nitrogen fixation: Role of extra residues. *J. Biol. Chem., 271*, 31399-31406.

Schatt, E., Jouanneau, Y., and Vignais, P. M. (1989). Molecular cloning and sequence analysis of the structural gene of ferredoxin I from the photosynthetic bacterium *Rhodobacter capsulatus*. *J. Bacteriol., 171*, 6218-6226.

Schlessman, J. L., Woo, D., Joshua-Tor, L., Howard, J. B., and Rees, D. C. (1998). Conformational variability in structures of the nitrogenase iron proteins from *Azotobacter vinelandii* and *Clostridium pasteurianum*. *J. Mol. Biol., 280*, 669-685.

Schmehl, M., Jahn, A., Meyer zu Vilsendorf, A., Hennecke, S., Masepohl, B., Schuppler, M., *et al.* (1993). Identification of a new class of nitrogen fixation genes in *Rhodobacter capsulatus*: A putative membrane complex involved in electron transport to nitrogenase. *Mol. Gen. Genet., 241*, 602-615.

Schmidt, G. W., Matlin, K. S., and Chua, N. H. (1977). A rapid procedure for selective enrichment of photosynthetic electron transport mutants. *Proc. Natl. Acad. Sci. U.S.A., 74*, 610-614.

Schmitz, O., Gurke, J., and Bothe, H. (2001). Molecular evidence for the aerobic expression of *nifJ*, encoding pyruvate:ferredoxin oxidoreductase, in cyanobacteria. *FEMS Microbiol. Lett., 195*, 97-102.

Schmitz, S., Schrautemeier, B., and Böhme, H. (1993). Evidence from directed mutagenesis that positively charged amino acids are necessary for interaction of nitrogenase with the [2Fe-2S] heterocyst ferredoxin (FdxH) from the cyanobacterium *Anabaena* sp., PCC7120. *Mol. Gen. Genet., 240*, 455-460.

Schneider, K., Müller, A., Schramm, U., and Klipp, W. (1991). Demonstration of a molybdenum- and vanadium-independent nitrogenase in a *nifHDK*-deletion mutant of *Rhodobacter capsulatus*. *Eur. J. Biochem., 95*, 653-661.

Schrautemeier, B., and Böhme, H. (1992). Coding sequence of a heterocyst ferredoxin gene (*fdxH*) isolated from the nitrogen-fixing cyanobacterium *Calothrix* sp. PCC 7601. *Plant Mol. Biol., 18*, 1005-1006.

Schrautemeier, B., Cassing, A., and Böhme, H. (1994). Characterization of the genome region encoding an FdxH-Type ferredoxin and a new 2[4Fe-4S] ferredoxin from the nonheterocystous, nitrogen-fixing cyanobacterium *Plectonema boryanum* PCC 73110. *J. Bacteriol., 176*, 1037-1046.

Schrautemeier, B., Neveling, U., and Schmitz, S. (1995). Distinct and differently regulated Mo-dependent nitrogen-fixing systems evolved for heterocysts and vegetative cells of *Anabaena variabilis* ATCC29413: characterization of the *fdxH1/2* gene regions as part of the *nif1/2* gene clusters. *Mol. Microbiol., 18*, 357-369.

Schwartz, C. J., Giel, J. L., Patschkowski, T., Luther, C., Ruzicka, F. J., Beinert, H., *et al.* (2001). IscR, an Fe-S cluster-containing transcription factor, represses expression of *Escherichia coli* genes encoding Fe-S cluster assembly proteins. *Proc. Natl. Acad. Sci. U.S.A., 98*, 14895-14900.

Schüddekopf, K., Hennecke, S., Liese, U., Kutsche, M., and Klipp, W. (1993). Characterization of *anf* genes specific for the alternative nitrogenase and identification of *nif* genes required for both nitrogenases in *Rhodobacter capsulatus*. *Mol. Microbiol., 8*, 673-684.

Scrutton, N. S., Berry, A., and Perham, R. N. (1990). Redesign of the coenzyme specificity of a dehydrogenase by protein engineering. *Nature, 343*, 38-43.

Shah, V. K., Stacey, G., and Brill, W. J. (1983). Electron transport to nitrogenase. Purification and characterization of pyruvate:flavodoxin oxidoreductase. The *nifJ* gene product. *J. Biol. Chem., 258*, 12064-12068.

Shethna, Y. I., Wilson, P. W., and Beinert, H. (1966). Purification of a non-heme iron protein and other electron transport components from *Azotobacter* extracts. *Biochim. Biophys. Acta, 113*, 225-234.

Sieker, L. C., and Adman, E. T. (2001). The 2[4Fe-4S] Ferredoxins. In T. Poulos (Ed.), *Handbook of metalloproteins* (pp. 574-592). Chichester: John Wiley and Sons.

Simondsen, R. P., and Tollin, G. (1980). Structure-function relations in flavodoxins. *Mol. Cell. Biochem., 33*, 13-24.

Smith, E. T., Blamey, J. M., and Adams, M. W. (1994). Pyruvate ferredoxin oxidoreductases of the hyperthermophilic archaeon, *Pyrococcus furiosus*, and the hyperthermophilic bacterium, *Thermotoga maritima*, have different catalytic mechanisms. *Biochemistry, 33*, 1008-1016.

St. John, R. T., Johnston, H. M., Seidman, C., Garfinke, D., Gordon, J. K., Shah, V. K., *et al.* (1975). Biochemistry and genetics of *Klebsiella pneumoniae* mutant strains unable to fix N_2. *J. Bacteriol., 121*, 759-765.

Steigerwald, V., Pihl, T., and Reeve, J. (1992). Identification and isolation of the polyferredoxin from *Methanobacterium thermoautotrophicum* strain ΔH. *Proc. Natl. Acad. Sci. U.S.A., 89*, 6929-6933.

Steuber, J. (2001). Na^+ translocation by bacterial NADH:quinone oxidoreductases: An extension to the complex-I family of primary redox pumps. *Biochim. Biophys. Acta, 1505*, 45-56.

Stout, C. D. (2001). Ferredoxins containing two different Fe/S centers of the forms [4fe-4S] and [3Fe-4S]. In T. Poulos (Ed.), *Handbook of metalloproteins* (pp. 560-573). Chichester: John Wiley and Sons.

Suetsugu, Y., Saeki, K., and Matsubara, H. (1991). Transcriptional analysis of two *Rhodobacter capsulatus* ferredoxins by translational fusion to *Escherichia coli lacZ*. *FEBS Lett., 292*, 13-16.

Tagawa, K., and Arnon, D. I. (1962). Ferredoxins as electron carriers in photosynthesis and in the biological production and consumption of hydrogen gas. *Nature, 195*, 537-543.

Thorneley, R. N., and Deistung, J. (1988). Electron-transfer studies involving flavodoxin and a natural redox partner, the iron protein of nitrogenase. Conformational constraints on protein-protein interactions and the kinetics of electron transfer within the protein complex. *Biochem. J., 253*, 587-595.

Tokuda, H., and Unemoto, T. (1981). A respiration-dependent primary sodium extrusion system functioning at alkaline pH in the marine bacterium *Vibrio alginolyticus*. *Biochem. Biophys. Res. Commun., 102*, 265-271.

Valladares, A., Muro-Pastor, A. M., Fillat, M. F., Herrero, A., and Flores, E. (1999). Constitutive and nitrogen-regulated promoters of the *petH* gene encoding ferredoxin:$NADP^+$ reductase in the heterocyst-forming cyanobacterium *Anabaena* sp. *FEBS Lett., 449*, 159-164.

van Lin, B., and Bothe, H. (1972). Flavodoxin from *Azotobacter vinelandii*. *Arch. Microbiol., 82*, 155-172.

von Sternberg, R., and Yoch, D. C. (1993). Molecular cloning and sequencing of the ferredoxin I *fdxN* gene of the photosynthetic bacterium *Rhodospirillum rubrum*. *Biochim. Biophys. Acta, 1144*, 435-438.

Wahl, R. C., and Orme-Johnson, W. H. (1987). Clostridial pyruvate oxidoreductase and the pyruvate-oxidizing enzyme specific to nitrogen fixation in *Klebsiella pneumoniae* are similar enzymes. *J. Biol. Chem., 262*, 10489-10496.

Weiss, H., Friedrich, T., Hofhaus, G., and Preis, D. (1991). The respiratory-chain NADH dehydrogenase (complex I) of mitochondria. *Eur. J. Biochem., 197*, 563-576.

Wientjens, R. (1993). The involvement of the *fixABCX* genes and the respiratory chain in the electron transport to nitrogenase in *Azotobacter vinelandii.* Ph. D. Thesis, Agricultural Univeristy of Wageningen, Wageningen, The Netherlands.

Willison, J. C., Pierrard, J., and Hübner, P. (1993). Sequence and transcript analysis of the nitrogenase structural gene operon (*nifHDK*) of *Rhodobacter capsulatus*: Evidence for intramolecular processing of *nifHDK* mRNA. *Gene, 133*, 39-46.

Yagi, T. (1993). The bacterial energy-transducing NADH-quinone oxidoreductases. *Biochim. Biophys. Acta, 1141*, 1-17.

Yakunin, A. F., Gennaro, G., and Hallenbeck, P. C. (1993). Purification and properties of a *nif*-specific flavodoxin from the photosynthetic bacterium *Rhodobacter capsulatus*. *J. Bacteriol., 175*, 6775-6780.

Yakunin, A. F., and Gogotov, I. N. (1983). Properties and regulation of sysnthesis of two ferredoxins from *Rhodopseudomonas capsulata*. *Biochim. Biophys. Acta, 725*, 298-308.

Yakunin, A. F., and Hallenbeck, P. C. (1998). Purification and characterization of pyruvate oxidoreductase from the photosynthetic bacterium *Rhodobacter capsulatus*. *Biochim. Biophys. Acta,1409*, 39-49.

Yates, M. G. (1972). Electron transport to nitrogenase in *Azotobacter chroococcum*: *Azotobacter* flavodoxin hydroquinone as an electron donor. *FEBS Lett., 27*, 63-67.

Yoch, D. C., and Arnon, D. I. (1972). Two biologically active ferredoxins from the aerobic nitrogen-fixing bacterium *Azotobacter vinelandii*. *J. Biol. Chem., 247*, 4514-4520.

Yoch, D. C., and Arnon, D. I. (1975). Comparison of two ferredoxins from *Rhodospirillum rubrum* as electron carriers for the native nitrogenase. *J. Bacteriol., 121*, 743-745.

Yoch, D. C., Arnon, D. I., and Sweeney, W. V. (1975). Characterization of two soluble ferredoxins as distinct from bound iron-sulfur proteins in the photosynthetic bacterium *Rhodospirillum rubrum*. *J. Biol. Chem., 250*, 8330-8336.

Yoch, D. C., and Carithers, R. P. (1979). Bacterial iron-sulfur proteins. *Microbiol. Rev., 43*, 384-421.

Zanetti, G., Aliverti, A., and Binda, C. (2001). 2Fe-2S Ferredoxins. In T. Poulos (Ed.), *Handbook of metalloproteins*. Chichester: John Wiley and Sons.

Zheng, L., Cash, V. L., Flint, D. H., and Dean, D. R. (1998). Assembly of iron-sulfur clusters. Identification of an *iscSUA-hscBA-fdx* gene cluster from *Azotobacter vinelandii*. *J. Biol. Chem., 273*, 13264-13272.

Zhou, W., Bertsova, Y. V., Feng, B., Tsatsos, P., Verkhovskaya, M. L., Gennis, R. B., *et al.* (1999). Sequencing and preliminary characterization of the Na^+-translocating NADH:ubiquinone oxidoreductase from *Vibrio harveyi*. *Biochemistry, 38*, 16246-16252.

Chapter 12

PROSPECTS

J. R. GALLON[a] AND B. MASEPOHL[b]
[a]*Biochemistry Research Group, School of Biological Sciences, University of Wales Swansea, Swansea, SA2 8PP, U. K.*
[b]*Ruhr-Universität Bochum, Fakultät für Biologie, Lehrstuhl für Biologie der Mikroorganismen, D-44780 Bochum, Germany*

"Science should leave off making pronouncements: the river of knowledge has too often turned back on itself."
James Jeans (1930)

N_2 fixation is a process of fundamental importance to life on Earth. Although N_2 is a very simple molecule, the molecular machinery that converts N_2 to ammonium is complex and the study of biological nitrogen fixation has, for many years, occupied the attention of a large number of scientists from a wide range of disciplines. However, even though no scientist would question the value of this research for its own sake, it is a fact of life that the advancement of human knowledge on its own is rarely sufficient to justify funding from those who hold the purse strings. In consequence, scientists have been forced, over the years, to identify commercial applications of N_2-fixation research. An obvious application is the improvement of systems that benefit agriculture, especially in developing countries. Others include solar energy conversion (related to the H_2 evolution that accompanies the reduction of N_2 to ammonium), production of cheap biomass and high value biochemicals from N_2-fixing cells and, more recently, the control of nuisance diazotrophs. It is, however, no coincidence that the application in favour at any particular time reflects the political climate at that time. Thus, when fuel is expensive, solar energy conversion is touted as a rationale for research on N_2 fixation whereas, when the overuse of industrially fixed nitrogen is perceived as a problem, the development of novel N_2-fixing crop plants, which will free us from the need for nitrogenous fertilizer, is promoted. What is a fashionable application today may be out of favour tomorrow - although it may well re-emerge in the future! However, despite

W. Klipp, B. Masepohl, J. R. Gallon and W. E. Newton (eds.), Genetics and Regulation of Nitrogen Fixation in Free-Living Bacteria, 291-296.

changes in its commercial and political significance, the fundamental scientific importance of N_2 fixation remains constant and the continued need to understand this process fully, along with the organisms that catalyse it, will clearly spur research for many years to come.

The ability to fix N_2 is found among representatives of two of the primary kingdoms of living organisms, the prokaryotic Archaea and Eubacteria. No member of any of the eukaryote kingdoms has unequivocally been shown to fix N_2. Over the years, there have been a number of reports of N_2 fixation by individual eukaryote algae and fungi but, so far, none of these has survived detailed scrutiny. So, even though there is no theoretical barrier to N_2 fixation in eukaryote organisms, the ability to fix N_2 appears to be confined to prokaryotes.

Among the Archaea, known N_2-fixing species currently remain restricted to the methanogenic Euryarchaeota. In contrast, N_2 fixation is well-documented for species from several major phylogenetic groups among the Eubacteria, covering green sulphur bacteria, Gram-positive firmicutes, heliobacteria, cyanobacteria and all four subdivisions of proteobacteria. By far most work concerning genetics and regulation of N_2 fixation has been done using proteobacteria and, in consequence, there is currently a clear imbalance concerning the number of proteobacterial diazotrophs analyzed compared to the number of non-proteobacterial species. It is wise, therefore, to bear in mind that, outside the proteobacterial field, general statements and predictions concerning the genetics and regulation of N_2 fixation may be underpinned either by no more than a few examples or perhaps by a single observation. This situation emphasizes the need to shift the center of gravity of scientific interest more into the direction of non-proteobacterial species.

Until now, many species outside the proteobacteria have been remarkably resistant to genetic analysis. Noteworthy among these are non-heterocystous cyanobacteria, which are belatedly becoming recognised as major players in the global game of N_2 fixation. The reasons for the difficulty in applying the techniques of molecular genetics to these organisms are not clearly understood but both the extensive polysaccharide slime that surrounds many non-heterocystous cyanobacteria and their possession of very active restriction enzymes undoubtedly play a significant part. These barriers to DNA isolation, genetic transformation and analysis need to be investigated in detail. Moreover, rather than simply attempting to apply complex genetic techniques to non-heterocystous cyanobacteria (even if they work as, for example, with heterocystous cyanobacteria), we should perhaps start with very simple, basic genetic experiments. Whatever the approach, it is clear that there is an urgent need to study the genetics and regulation of N_2 fixation in these cyanobacteria and, indeed, in other non-proteobacterial diazotrophs.

The organization of nitrogen fixation (*nif*) genes, the structure of their products, and several aspects of control of synthesis and/or activity of nitrogenase by ammonium, suggest that N_2 fixation is based on similar modules in most representatives of both kingdoms analyzed so far. The only known exception is *Streptomyces thermoautotrophicus*, which contains an O_2-tolerant, superoxide-dependent nitrogenase that, uniquely, does not reduce acetylene. The high degree of conservation of all the other, O_2-sensitive, nitrogenases (including the Mo-independent, alternative nitrogenases) is explained by both an early development of biological N_2 fixation in phylogenetic terms and by horizontal gene transfer. However, it should be emphasized that proteobacteria and Archaea follow different strategies for ammonium regulation of *nif*-gene expression. In free-living diazotrophic proteobacteria, transcription of N_2-fixation genes is activated by a central transcriptional activator, NifA (or the NifA-like proteins, VnfA and AnfA), only in the absence of ammonium. In contrast, in Archaea, *nif*-gene expression is controlled by a novel repressor protein, NrpR, which prevents transcription of *nif* genes in the presence of ammonium. As in proteobacteria, N_2 fixation in cyanobacteria requires a transcriptional activator, in this case NtcA. However, NtcA and NifA belong to different families of regulatory proteins. Therefore, although ammonium regulation of nitrogenase gene expression is a common feature of all diazotrophs (except *S. thermoautotrophicus,* where nitrogenase gene expression is constitutive, but nitrogenase activity seems to be repressed by ammonium), the underlying mechanism seems to differ remarkably among different phylogenetic groups. The protobacterial 'classical' model cannot, therefore, simply be extrapolated to all diazotrophs. Hence, in this area too, there is a need for further work outside the proteobacteria.

The P_{II}-like signal-transduction proteins (GlnB, GlnK, GlnZ, GlnJ, GlnY, NifI1, NifI2) represent a highly conserved module, which is also present in Eukaryotes. Although many bacteria analyzed so far contain two P_{II}-like proteins, GlnB and GlnK, both the number of P_{II}-encoding genes and the function of their products may differ from species to species. For example, in some species, GlnB is essential for activation of NifA in the absence of ammonium, whereas, in other species, GlnB is involved in the down-regulation of NifA activity in the presence of ammonium. There are several examples in which P_{II}-like proteins appear to act directly on NifA; however, in a few species, P_{II}-mediated regulation of NifA activity involves another regulatory protein, NifL. Moreover, some P_{II}-like proteins are known to interact with AmtB-like ammonium transporters, which are considered to be involved in ammonium sensing. Even so, the roles of P_{II}-like proteins and AmtB-like ammonium transporters are far from completely understood.

In most cases, *glnB* is transcribed constitutively, whereas expression of *glnK* is activated only under nitrogen-limited conditions. There are, however, several lines of evidence that GlnB and GlnK can form heterotrimers and so it is possible that heterotrimers could facilitate fine-tuning of the signal-transduction cascade. In this context though, it is valid to ask whether data based on laboratory experiments reflect the situation in nature. For example, in the laboratory, most experiments are performed under conditions favouring maximal growth rates. Except during the specialized environmental conditions that lead to bloom formation, these laboratory conditions may be far from the natural situation, where most bacteria have to cope with one or more growth-limiting factors. Furthermore, it is pertinent to question the environmental significance of laboratory experiments involving either the effect of adding high concentrations of ammonium to N_2-fixing cultures or, in the case of photosynthetic diazotrophs, the relevance of instantaneous switching between high levels of illumination and complete darkness. Of course, such experiments are extremely valuable in probing the physiological, biochemical, and genetic events underpinning N_2 fixation in these organisms, but their findings are not necessarily of environmental relevance. There is, therefore, a need for scientists at the molecular end of the spectrum of research into N_2 fixation to design experiments that more accurately model the true environmental situation experienced by diazotrophs. The incorporation of physiological parameters into mathematical models that predict the response of these parameters to particular sets of natural conditions is a particularly exciting development in this area. For example, careful laboratory measurements of the responses of N_2 fixation to factors, such as temperature, availability of fixed nitrogen, other nutrients (carbon, phosphorus, iron, molybdenum *etc.*) and, in the case of phototrophic diazotrophs, illumination, can be used to construct models that can be compared to the behaviour of natural populations in a changing environment. As yet, such studies are in their infancy but we can confidently expect much more of this type of work during the next few years.

Nevertheless, for most physiological experiments, only one single condition is changed at a time. Consequently, virtually no information is available about how two or more different environmental signals might interact synergistically. Yet this is what happens in the real world; diazotrophic bacteria in their natural environment have to integrate and respond to a multiplicity of environmental signals. Their response would also involve interaction with other organisms in their immediate vicinity, a topic that is discussed later in this chapter. Therefore, a need exists for much more complex laboratory experiments to be conducted than have hitherto been performed. There is also a need to examine, in the laboratory, the effects on N_2 fixation of some hitherto neglected environmental conditions. For example, there have been few experiments that deal with the influence of temperature on N_2-fixation in free-living diazotrophs and virtually none that deal with pressure. *In vitro*, nitrogenase is relatively thermotolerant. Cyanobacterial nitrogenase has, for example, been shown to be catalytically active at 60°C and, because many N_2-fixing methanotrophs grow optimally above 80°C, their nitrogenase may be even more thermotolerant. In aerotolerant diazotrophs, it is usually the thermal inactivation of processes, which protect nitrogenase from inactivation by O_2, that sets the upper limit for N_2-fixation activity.

The overall decrease in volume that accompanies the conversion of N_2 to ammonium suggests that this process would be stimulated under elevated pressure. This situation is, for example, exploited in the industrial Haber-Bosch process, which operates at pressures as high as 100 MPa. Consistent with this fact, N_2 fixation by the unicellular cyanobacterium, *Gloeothece*, is stimulated by a factor of 1.64 at the much more modest pressure of 0.25 MPa. It should also be noted that the N_2-fixing marine cyanobacterium, *Trichodesmium*, is routinely found at depths greater than 50 m, where the hydrostatic pressure would be at least 0.5 MPa. However, it would be especially interesting to study N_2 fixation in organisms, such as *Methanopyrus kandleri*, which was isolated from thermal vents at a depth of 2 km (equivalent to about 20 MPa), and which grows only above 84°C.

Other environmental factors that have received a cursory study but no systematic evaluation of their effect on N_2 fixation include salinity (especially important for marine diazotrophs), desiccation, and the availability of elements, such as S, Fe, Mo, V, W, Mg and Ca.

There are many open reading frames (*orfs*) in *Azotobacter vinelandii* and other diazotrophs that are interspersed between known *nif* genes and forming part of *nif*-regulated operons. Genome projects have revealed that many of these *orfs* are highly conserved, but no function has yet been assigned to them and mutational analysis has not yet revealed a Nif phenotype under laboratory growth conditions. An unresolved question then is whether or not the products of these *orfs* have any function during diazotrophic growth under natural conditions.

In many, but not all, free-living diazotrophs, both the nitrogenase enzyme complex and the regulatory NifA protein seem to be fairly stable after addition of ammonium to N_2-fixing cultures. At least under laboratory conditions, stability may clearly exceed generation time with the result that, after several cell divisions, bacteria that have never themselves been exposed to conditions of N-limitation nevertheless contain nitrogenase. How does this situation arise? Do these bacteria possess some sort of long-term memory? In this context, it should not be forgotten that several cyanobacteria exhibit an entrained circadian pattern of N_2 fixation long after the initial signal (alternating light and darkness) has been removed. Their 'clock' proteins can, therefore, be thought of as providing just such a long-term memory.

In addition to regulatory aspects specific for N_2 fixation, detailed examination of more general regulatory networks, whihc are involved in the control of DNA structure by histone-like proteins, the influence of chaperones on nitrogenase synthesis, and the degradation of Nif proteins by proteases, has just started. In this arena, transcriptomics and proteomics will produce many new insights. Modern techniques in mass spectrometry, such as MALDI-TOF, ESI, and tandem mass spectrometry, can provide detailed structural information about proteins of interest. The application of these methods to nitrogenase and also to other proteins identified as having a role in N_2 fixation will undoubtedly be critical to the success of this approach.

A considerable body of knowledge has accumulated about the molecular interactions that occur between the symbiotic and associative bacteria and their respective host plants. However, much less is known about the micro-environments of free-living diazotrophs. In nature, no diazotroph exists in isolation and interactions with other organisms can be either advantageous (by their respiration, associated organisms may, for example, generate a micro-oxic environment that benefits N_2 fixation) or disadvantageous (for example, in competing for essential nutrients). For example, *Azotobacter* species have developed special adaptations to deal with molybdenum starvation; they possess a Mo-storage protein and the Mo-independent alternative nitrogenases seem to be advantageous for diazotrophic growth of *Azotobacter* in micro-colonies. Many phototrophs occur as components of complex microbial mats and the interactions between organisms in these systems is only beginning to be investigated.

In the laboratory, we study free-living diazotrophic bacteria as individual cells in either a batch or continuous culture. Indeed, great pride is taken in the purity of such cultures. However, under natural conditions, most bacteria live in so-called "biofilms", held together usually by extracellular polysaccharides. Communication within a biofilm involves quorum-sensing systems, which act as a measure of population density and allow communication not only between bacteria of the same species but also between organisms from different genera. Several diazotrophs have been shown to release quorum-sensing molecules but it is less certain whether (and how) they respond to their own quorum sensing molecules or to those released by other partners in the biofilm. For example, it remains speculative whether accumulation of a particular quorum-sensing molecule produced by the activity of non-diazotrophic bacteria may serve as a nutrient and/or nitrogen-starvation signal for diazotrophic bacteria in the biofilm, thereby triggering synthesis of nitrogenase.

So, research into N_2 fixation is by no means static and there are many fronts along which it can advance. Some will, undoubtedly, be more attractive than others to those who will fund this research, but all are of equal scientific merit. Predicting what will be commercially important in the future is a nightmare. For example, in contrast to predictions made in the 1970′s, energy costs have not escalated to the extent anticipated and industrial fertilizers continue to make a substantial contribution to world agriculture. Nevertheless, the first steps have recently been taken along the road that (despite the meanderings of the river of knowledge) should allow us to introduce *nif* genes into plants and these probes deserve to be encouraged. In this context, unicellular cyanobacteria remain an attractive model for understanding how nitrogenase may function if it is ever to be successfully transferred to the chloroplasts of higher plants. So, despite the problems of working with these organisms, they continue to be worthy of much more detailed study.

SUBJECT INDEX